D0204225

The First Year in the Life of Estuarine Fishes in the Middle Atlantic Bight

The First Year in the Life of Estuarine Fishes in the Middle Atlantic Bight

KENNETH W. ABLE AND MICHAEL P. FAHAY

RUTGERS UNIVERSITY PRESS
NEW BRUNSWICK, NEW JERSEY

Library of Congress Cataloging-in-Publication Data

Able, Kenneth W., 1945–
The first year in the life of estuarine fishes in the Middle Atlantic Bight / Kenneth W. Able, Michael P. Fahay.
p. cm.
Includes bibliographical references and index.
ISBN 0-8135-2500-4 (alk. paper)
1. Estuarine fishes—Mid-Atlantic Bight.
I. Fahay, Michael P. II. Title.
QL638.M524A35 1998
597.177′8634—dc21 97-30312
CIP

British Cataloging-in-Publication information available

Senior authorship was determined by a coin toss in the parking lot of the Vince Lombardi rest stop on the New Jersey Turnpike, in the midst of the Hackensack Meadowlands, surrounded by heavy urbanization and prime spawning habitat for *Fundulus heteroclitus.*

Manufactured in the United States of America

To Sue
and
To Cindy

Contents

Tables

Preface

An unprecedented research effort has been directed at the early life-history stages of fishes during the last three decades. This period has seen the first two Larval Fish Conferences (Blaxter 1974; Lasker and Sherman 1981), a workshop on eggs, larvae, and juveniles in Atlantic coast estuaries (Pacheco 1973), the CalCOFI ichthyoplankton program (1951–1994), the Ahlstrom Symposium on ontogeny and systematics of fishes (Moser et al. 1984), several regional ichthyoplankton guides (Fahay 1983; Okiyama 1988; Matarese et al. 1989; Moser 1996), and, in the Middle Atlantic Bight, the *R/V Dolphin* 1965–1966 ichthyoplankton surveys (Clark et al. 1969) followed by the MARMAP 1977–1987 surveys (Sherman 1980). These efforts have ranged from extensive surveys of oceanic waters to behavioral studies of juvenile fishes in estuaries. The incorporation of otolith studies to back-calculate aspects of the early life history (e.g., Secor et al. 1995) and laboratory studies of factors influencing survival of these early stages have also increased steadily. Few efforts, however, have bridged the morphological, physiological, ecological, and behavioral transition between pelagic larva and settled juvenile. We therefore decided to focus on this transition within the estuarine fish fauna of the Middle Atlantic Bight, hoping in the process that our summaries might have applications to young-of-the-year fishes in other ecosystems as well. We felt confident we could accomplish this focused approach because we each had had experience in various aspects of early life-history studies, either in cooperative efforts with each other or with many of our colleagues.

These colleagues helped us bring together these data in a variety of ways. Unpublished data were provided by: Pete Berrien (eggs), Ralph Bush (*Etropus*), Brian Campbell (*Syngnathus*), Eric Duval (*Gobiosoma*), Stacy Hagan (*Anchoa*), Dale Haroski (*Strongylura, Ammodytes*), Don McMillan (*Stenotomus*), Jill Morrison (*Hippocampus*), Kelly Smith (*Gasterosteus*) Stu Wilk (Newark Bay), Kim Wilson (Hudson River), and Dave Wyanski and Tim Targett (*Gobiosoma*). Public Service Electric & Gas provided data on Delaware Bay fishes. The collection of ichthyoplankton from continental shelf waters of the Middle Atlantic Bight was facilitated by a number of sea-going MARMAP investigators from the National Marine Fisheries Service laboratory at Sandy Hook, and we especially thank Doris Finan, John Sibunka, and Alyce Wells for their dedication. A small army of field workers at the Rutgers University Marine Field Station made extensive collections of young-of-the-year fishes in the Great Bay–Little Egg Harbor estuary, and their persistence and dedication is hereby gratefully acknowledged. Colin and Peter Able provided assistance with numerous field collections at Corson's Inlet. In the sincere hope that we have not overlooked anyone, we thank Carla Curran, Roger Hoden, John Manderson, Rich McBride, Matt Pearson, Susan Sogard, Kelly Smith, and Kim Wilson. Lab observations were provided by Matt Pearson, Lynn Wulff, and Jeanette Bowers-Altman. Stacy Hagan and Dave Witting assisted in all phases of data manipulation. Stacy Hagan prepared many of the figures. Nathan Able designed the book cover. We also thank Karl Nordstrom for fruitful discussions of estuarine geomorphology. The Nauset Team (Charlie Roman; Ken Heck; Mark Lazzari; Susan Kaiser, one of our very fine illustrators; and Capt H. Morgan) continue to support us and habitat-related estuarine research. Geoff Moser provided Nancy Arthur with lab space and a microscope. Anne Studholme and Jeff Cross provided M. P. F. not only time to work on this project, but also continued encouragement.

Thorough species account reviews were provided by Mark Chittenden (*Alosa* spp.,

Morone spp.), Bruce Collette (*Strongylura*), Bruce Comyns (*Urophycis*), Dave Conover (*Menidia*), Bob Cowen (*Peprilus*), Mike Fine (*Opsanus*), Jon Hare (*Brevoortia, Pomatomus*), Bill Hettler (*Brevoortia*), Ed Houde (*Anchoa*), Janet Duffy (*Cynoscion*), Eddie Matheson (*sciaenids*), Rich McBride (*Chaetodon, Prionotus, Caranx*), Tom Munroe (*Clupea*), Jack Musick (*gadids, phycids*), Melissa Neuman (*Scophthalmus*), Gary Shepherd (*Centropristis, Stenotomus, Paralichthys*), Mark Tupper (*Tautogolabrus*), Stan Warlen (*Brevoortia*), and Dave Witting (*Pseudopleuronectes*). Reviews of portions of the penultimate draft were graciously provided by Chris Chambers, Al Stoner, and Dave Witting. Their comments improved our organization considerably. We also gratefully acknowledge comments on early drafts of the entire manuscript by C. L. Smith. Smitty also provided abundant encouragement and support, perhaps even convincing us the project was worth completing.

Our two fish illustrators, Susan Kaiser and Nancy Arthur, understood that they were breaking some new ground in concentrating on a stage between larvae and adults. They (and we) were learning together, and perhaps their efforts will influence future juvenile fish illustration styles. We appreciate their labors very much. For rounding up and providing appropriate specimens, we thank Eugenie Böhlke and Bill Saul (Academy of Natural Sciences of Philadelphia), Cindy Fahay, Doris Finan, Karsten Hartel (Museum of Comparative Zoology, Harvard University), Mark Lazzari, and Linda Stehlik. Hampton E. Griffith illustrated the *Gabionellus boleosoma* for a manuscript in preparation by Dave Wyanski and Tim Targett.

We thank the following for very able editorial assistance. Pete Berrien developed the larval mapping programs, Jessie Fahay and Angie Podlinski helped with page collation, Karen Reeds and Doreen Valentine (Rutgers University Press) provided editorial guidance. Finally, many thanks to Barbara Zlotnik who thoroughly checked literature citations and brought together the final mass of chapters.

Funding sources included Hudson River Foundation, Rutgers University Marine Field Station, Institute of Marine and Coastal Sciences, Rutgers University, NOAA Sea Grant, NOAA National Undersea Research Program, Army Corps of Engineers, and Tropical Fish Hobbyist Publications, Inc.

Senior authorship was determined by a coin toss in the parking lot of the Vince Lombardi rest stop on the New Jersey Turnpike, in the midst of the Hackensack Meadowlands, surrounded by heavy urbanization and prime spawning habitat for *Fundulus heteroclitus*.

Abbreviations

We have limited the use of acronyms and cryptic abbreviations wherever possible, but a few should be explained.

A	anal fin
ANSP	Academy of Natural Sciences of Philadelphia
BL	body length
C	temperature in degrees centigrade
C&D Canal	Chesapeake and Delaware Canal
cm	centimeter
D	dorsal fin
FL	fork length
LEO-15	Long-Term Ecosystem Observatory in 15 meters
LSHCB	Little Sheepshead Creek Bridge
m	meter
MARMAP	Marine Resources Monitoring, Assessment, and Prediction
mm	millimeter
NL	notochord length
NMFS	National Marine Fisheries Service
NOAA	National Oceanic and Atmospheric Administration
Pect	pectoral fin
Plv	pelvic fin
ppm	parts per million
ppt	parts per thousand
PSE&G	Public Service Electric & Gas
RUMFS	Rutgers University Marine Field Station
SHML	Sandy Hook Marine Laboratory
SL	standard length
TL	total length
Vert	vertebrae

The First Year in the Life of Estuarine Fishes in the Middle Atlantic Bight

CHAPTER 1

Introduction

Even so apparently simple a matter as describing in detail the distribution of any species as a function of developmental stage and season has been done for only a handful of species.

—*Haedrich 1983*

Estuaries occur where oceanic and fresh waters blend, a situation in which extremes in environmental conditions are pronounced. In order to live in these conditions, fishes must be tolerant of abrupt changes in temperature, salinity, oxygen concentrations, and turbidity levels, as well as seasonal changes in these and other physical conditions. Partly because of these requirements, the number of permanent resident species in estuaries is low (Day 1981; Dando 1984). Conversely, productivity in estuaries is extremely high, with the result that a small number of species (most of which are small in body size) comprises a large percentage of the ichthyofauna, both in numbers and biomass (Haedrich 1983; Day et al. 1989). In Narragansett Bay, for example, 10 of 99 species recorded account for 90% of individuals collected (Oviatt and Nixon 1973). Six species comprise 91% of the total in Block Island Sound (Merriman and Warfel 1948), and 10 species comprise 90% of the total in Long Island Sound (Richards 1963). The small number of permanent species is augmented by transient species that temporarily inhabit estuaries (Gunter 1941, 1945; McHugh 1967). These include diadromous species—freshwater species that occasionally occur in estuaries—and primarily marine species that spawn at sea, but whose young use estuaries as nurseries. For many species of marine fishes, the juvenile stages are far more tolerant of environmental variability than are the adults (Holliday 1971), and as a result, adult stages of most transients are absent from estuaries. The estuarine fish fauna, therefore, includes both resident and transient components and a wide range of sizes and ages, thus creating a great diversity in life-history stages and a more limited diversity of species. Another way to illustrate this is that the number of species found in the estuary is many fewer than the number on the adjacent continental shelf. For example, in New England waters, the number of estuarine species is only 10% of the number of marine species (Haedrich and Hall 1976). This relatively low species total reflects the fact that only a small number of species per family has successfully adapted to estuarine conditions, at least compared to the adjacent ocean. Those that have successfully invaded estuaries, however, appear to occupy a small number of broad niches (Haedrich 1983). The sizes of estuaries may influence the number of these niches, as supported by the analysis of a number of California estuaries by Horn and Allen (1976), who found a positive relationship between size of embayment (more specifically the size of the mouth opening) and the number of species present. A recent study analyzed 28 U.S. West Coast estuaries and identified estuary mouth depth and area of the seawater zone as good predictors of species richness (Monaco et al. 1992). This relationship has not been formally assessed for estuaries of the U.S. East Coast, however. The number of resident and transient species we report here from our focused Great Bay–Little Egg Harbor study area includes 61 species (table 4.2) plus 50 rarely collected species (table 77.3). This total (111) is approximately the same as, or somewhat higher than, the totals reported from several other East Coast studies (table 1.1), but we caution that certain of these higher totals are products of increased sampling effort with a wide variety of

Table 1.1. Numbers of Fish Species Reported from Several Estuarine Systems in the Middle Atlantic Bight

Estuary	No. of species	Source
Nauset Marsh, Massachusetts	35	K. W. Able, M. P. Fahay, K. Heck and C. Roman, pers. observ.
Narragansett Bay, Rhode Island	99	Oviatt and Nixon 1973
Mystic River, Connecticut	51	Pearcy and Richards 1962
Connecticut River, Connecticut	44	Marcy 1976a
Great South Bay, New York	29	Monteleone 1992
Hudson/Raritan, New York/New Jersey	113	Texas Instruments 1973
Barnegat Bay, New Jersey	107	Tatham et al. 1984
Great Bay, New Jersey	107*	Present study
Delaware Bay, New Jersey/Delaware	98	Wang and Kernehan 1979
Assawoman Bay, Maryland	110	Schwartz 1964a
Chincoteague Bay, Virginia	99	Schwartz 1961
Eastern Shore, Virginia	70	Richards and Castagna 1970
Chesapeake Bay, Maryland/Virginia	285	Musick 1972
Albemarle Sound, North Carolina	80	Epperly 1984
Pamlico Sound, North Carolina	117	Epperly 1984

*Includes 61 regularly collected (table 4.2) plus 50 rarely collected (Table 77.3).

gears, as was the case in our Great Bay–Little Egg Harbor studies. Because these totals are results of uneven sampling effort between estuaries, we have not yet attempted to analyze the relationship between species richness and physical attributes of Middle Atlantic Bight estuaries (table 2.3).

According to several accounts, estuaries provide essential nursery habitat for approximately two-thirds of the economically important fish species along the East Coast of the United States, both for recreational and commercial fisheries (McHugh 1966; Clark 1967; Stroud 1971; Weinstein 1979; Bozeman and Dean 1980; Boesch and Turner 1984). A more recent estimate indicates some differences between coastal regions, with the Middle Atlantic Bight having relatively fewer species than other regions (Chambers 1992). However, it remains clear that many fisheries are concentrated on species that utilize estuaries as nurseries. Despite this economic importance, estuaries are often highly impacted by habitat degradation, declining water quality, and other anthropogenic effects (Kennish 1992). Because fisheries biologists have not quantified the linkage between habitat and fish production, and because they lack knowledge of the importance of specific habitats used by the early stages of estuarine fishes (Hoss and Thayer 1993), it is often difficult for resource managers to justify increased protection or enhancement of estuarine habitats.

Fisheries managers have ascribed "estuarine dependence" to species based on the occurrence (or collection) of early life-history stages in estuarine habitats, although the adult stages may occur either in estuaries or elsewhere (e.g., the continental shelf). The assumption accruing from these two collecting results is that estuaries must therefore provide essential nursery habitat. Fisheries biologists in South Africa have supplied a slightly different definition. There a species is defined as "estuarine dependent" if the local population would be adversely affected by the loss of estuarine habitats on the subcontinent (Whitfield 1994). In our research for the present study, we have found weaknesses in three areas that need to be addressed before a resolution of estuarine dependency can be found. First, well-defined habitats on the continental shelf (perhaps similar in structure and physical attributes to those in the estuary) need to be sampled with gear appropriate for the collection of small, often cryptic stages, in order to evaluate those habitats as equally essential nurseries. Second, the effects of habitat loss on local populations must be assessed, thus providing fisheries managers with a link between habitat quality (or lack thereof) and the production of fish year classes. Third, more detail needs to be provided on temporal and spatial use of habitat types where early stages are collected. At the most simplistic level, an appreciation of the value of estuaries as nurseries (and the resulting effect on survival of year classes) depends on an understanding of the life-history patterns of species comprising the fauna, yet our present knowledge of these patterns is incomplete for all but a very few (Haedrich 1983).

During the past decade, significant progress has been made in our understanding of fish recruitment. Despite this progress, many unknowns persist, and we still lack a unifying, coherent understanding of the factors controlling survival during the early life history of fishes. Much of the emphasis since the pioneering work of Hjort (1914, 1926; recently summarized in May 1974; Blaxter 1988; Pepin 1991; Miller 1994; Chambers and Trippel 1997) has focused on the search for a critical period of mortality during egg and larval development that might explain the extreme population fluctuations so frequently observed. To date, this search for a single critical period has been unsuccessful (Leggett 1986; Houde 1987; Blaxter 1988; but also see Li and Mathias 1987). Recent efforts have considered the possibility that more than one period of increased mortality may occur during the early life history of fishes (Graham et al. 1984; Walline 1985; Victor 1986; Veer and Bergman 1987; Campana 1996). Demersal species with pelagic larvae may face the added stress associated with settlement to the bottom (McGurk 1984; Houde 1987).

Much of the use of estuarine nurseries by fishes is begun during a transitional morphological period. During this transition, a fish undergoes anatomical changes associated with allometric growth and profound changes involving physiology and development. Shifts in habitats are often concurrent with these changes. The extent to which concurrence of developmental change and environmental shift affects mortality is largely unknown. The statement by Hempel (1965) that we know less about the larval-juvenile transition than any other part of the life history remains true 30 years later (Bailey and Houde 1989).

There is an increasing realization that processes occurring immediately before and after juveniles settle to the bottom (following a pelagic larval stage) may influence recruitment (Cushing 1996). This idea is best developed for tropical and temperate reef fishes (Jones 1987a,b, 1990; Shulman and Ogden 1987; Robertson 1988a,b; Forrester 1990; Carr 1991; Hunt von Herbing and Hunte 1991; Jones 1991; Levin 1991, 1993; Tupper and Hunte 1994; Williams et al. 1994; Booth 1995; Cowen and Sponaugle 1997) and in other temperate habitats as well (Breitburg 1991; Malloy and Targett 1991; Szedlmayer et al. 1992; Tupper and Boutilier 1995a,b; Campana 1996). In many cases, the search for the mortality that fuels year-class variation is expected in postsettlement young-of-the-year juveniles (Sissenwine 1984; Smith 1985; Houde 1987; Elliot 1989; Doherty 1991; Beverton and Iles 1992) and extends to the first winter (Post and Evans 1989; Conover and Present 1990; Hales and Able in press).

The lack of adequate research on mortality during metamorphosis and settlement of benthic fishes may be due to the difficulty in sampling these small individuals quantitatively (de Lafontaine et al. 1992). Given the shortcomings in our understanding of the first year in the life of fishes—especially Middle Atlantic Bight estuarine species that provide the basis for many of our fisheries—it is our purpose in this book to marshal the published evidence, reexamine a wealth of unpublished data, and augment these with results of our own recent studies that have focused on collections during the first year of life, and synthesize patterns in morphology of early stages and their use of estuarine habitats. Within the estuary-ocean continuum of our study area, we have

Table 1.2. Checklist of Middle Atlantic Bight Estuarine Fishes Treated in This Study

Scientific name (author)	Common name
Carcharhinidae	
Mustelus canis (Mitchill)*	smooth dogfish
Rajidae	
Raja eglanteria Bosc	clearnose skate
Elopidae	
Elops saurus Linnaeus	ladyfish
Albulidae	
Albula vulpes (Linnaeus)	bonefish
Anguillidae	
Anguilla rostrata (Lesueur)*	American eel
Muraenidae	
Gymnothorax sp. Jordan and Davis	honeycomb moray
Ophichthidae	
Myrophis punctatus Lütken	speckled worm eel
Ophichthus gomesi (Castelnau)	shrimp eel
Congridae	
Conger oceanicus (Mitchill)*	conger eel
Clupeidae	
Alosa aestivalis (Mitchill)*	blueback herring
A. mediocris (Mitchill)*	hickory shad
A. pseudoharengus (Wilson)*	alewife
A. sapidissima (Wilson)*	American shad
Brevoortia tyrannus (Latrobe)*	Atlantic menhaden
Clupea harengus Linnaeus*	Atlantic herring
Sardinella aurita Valenciennes	Spanish sardine
Engraulidae	
Anchoa hepsetus (Linnaeus)*	striped anchovy
A. mitchilli (Valenciennes)*	bay anchovy
Engraulis eurystole (Swain and Meek)	silver anchovy
Osmeridae	
Osmerus mordax (Mitchill)*	rainbow smelt
Synodontidae	
Synodus foetens (Linnaeus)*	inshore lizardfish
Gadidae	
Gadus morhua Linnaeus	Atlantic cod
Microgadus tomcod (Walbaum)*	Atlantic tomcod
Pollachius virens (Linnaeus)*	pollock
Phycidae	
Urophycis chuss (Walbaum)*	red hake
U. regia (Walbaum)*	spotted hake
U. tenuis (Mitchill)*	white hake
Enchelyopus cimbrius (Linnaeus)	fourbeard rockling
Merlucciidae	
Merluccius bilinearis (Mitchill)	silver hake
Ophidiidae	
Ophidion marginatum (DeKay)*	striped cusk-eel
O. welshi (Nichols and Breder)	crested cusk-eel
Batrachoididae	
Opsanus tau (Linnaeus)*	oyster toadfish
Hemiramphidae	
Hyporhamphus meeki Banford and Collette	silverstripe halfbeak
Belonidae	
Strongylura marina (Walbaum)*	Atlantic needlefish
Cyprinodontidae	
Cyprinodon variegatus Lacepède*	sheepshead minnow
Fundulidae	
Fundulus heteroclitus (Linnaeus)*	mummichog
F. luciae (Baird)*	spotfin killifish
F. majalis (Walbaum)*	striped killifish
Lucania parva (Baird and Girard)*	rainwater killifish
Poeciliidae	
Gambusia holbrooki Girard*	eastern mosquitofish
Atherinidae	
Membras martinica (Valenciennes)*	rough silverside
Menidia beryllina (Cope)*	inland silverside
M. menidia (Linnaeus)*	Atlantic silverside
Gasterosteidae	
Apeltes quadracus (Mitchill)*	fourspine stickleback
Gasterosteus aculeatus Linnaeus*	threespine stickleback
G. wheatlandi Putnam	blackspotted stickleback
Pungitius pungitius (Linnaeus)	ninespine stickleback
Fistulariidae	
Fistularia tabacaria Linnaeus	bluespotted cornetfish

Table 1.2. *Continued*

Scientific name (author)	Common name
Syngnathidae	
Hippocampus erectus Perry*	lined seahorse
Syngnathus fuscus Storer*	northern pipefish
Triglidae	
Prionotus carolinus (Linnaeus)*	northern searobin
P. evolans (Linnaeus)*	striped searobin
Cottidae	
Myoxocephalus aenaeus (Mitchill)*	grubby
Percichthyidae	
Morone americana (Gmelin)*	white perch
M. saxatilis (Walbaum)*	striped bass
Serranidae	
Centropristis striata (Linnaeus)*	black sea bass
Epinephelus striatus (Bloch)	Nassau grouper
E. niveatus (Valenciennes)	snowy grouper
Mycteroperca microlepis (Goode and Bean)	gag
Apogonidae	
Phaeoptyx pigmentaria (Poey)	dusky cardinalfish
Pomatomidae	
Pomatomus saltatrix (Linnaeus)*	bluefish
Carangidae	
Caranx crysos (Mitchill)	blue runner
C. hippos (Linnaeus)*	crevalle jack
Decapterus macarellus (Cuvier)	mackerel scad
Selene vomer (Linnaeus)	lookdown
Seriola zonata (Mitchill)	banded rudderfish
Trachinotus carolinus (Linnaeus)	Florida pompano
T. falcatus (Linnaeus)	permit
Lutjanidae	
Lutjanus griseus (Linnaeus)*	gray snapper
Gerreidae	
Unidentified species	mojarras
Haemulidae	
Orthopristis chrysoptera (Linnaeus)	pigfish
Sparidae	
Lagodon rhomboides (Linnaeus)	pinfish
Stenotomus chrysops (Linnaeus)*	scup
Sciaenidae	
Bairdiella chrysoura (Lacepède)*	silver perch
Cynoscion regalis (Bloch and Schneider)*	weakfish
Leiostomus xanthurus Lacepède*	spot
Menticirrhus saxatilis (Bloch and Schneider)*	northern kingfish
Micropogonias undulatus (Linnaeus)*	Atlantic croaker
Pogonias cromis (Linnaeus)*	black drum
Chaetodontidae	
Chaetodon capistratus Linnaeus	foureye butterflyfish
C. ocellatus Bloch*	spotfin butterflyfish
Mugilidae	
Mugil cephalus Linnaeus*	striped mullet
M. curema Valenciennes*	white mullet
Sphyraenidae	
Sphyraena barracuda (Walbaum)	great barracuda
S. borealis DeKay*	northern sennet
Labridae	
Tautoga onitis (Linnaeus)*	tautog
Tautogolabrus adspersus (Walbaum)*	cunner
Stichaeidae	
Lumpenus lumpretaeformis (Walbaum)	snakeblenny
L. maculatus (Fries)	daubed shanny
Ulvaria subbifurcata (Storer)	radiated shanny
Pholidae	
Pholis gunnellus (Linnaeus)*	rock gunnel
Uranoscopidae	
Astroscopus guttatus Abbott*	northern stargazer
Blenniidae	
Chasmodes bosquianus (Lacepède)	striped blenny
Hypleurochilus geminatus (Wood)	crested blenny
Hypsoblennius hentz (Lesueur)*	feather blenny

Continued

Table 1.2. *Continued*

Scientific name (author)	Common name	Scientific name (author)	Common name
Ammodytidae		*E. crossotus* Jordan and Gilbert	fringed flounder
Ammodytes americanus DeKay*	American sand lance	*Paralichthys dentatus* (Linnaeus)*	summer flounder
Gobiidae		Pleuronectidae	
Gobionellus boleosoma (Jordan and Gilbert)*	darter goby	*Pseudopleuronectes americanus* Walbaum*	winter flounder
Gobiosoma bosc (Lacepède)*	naked goby	Soleidae	
G. ginsburgi Hildebrand and Schroeder*	seaboard goby	*Trinectes maculatus* (Bloch and Schneider)*	hogchoker
Scombridae		Cynoglossidae	
Scomber japonicus Houttuyn	chub mackerel	*Symphurus plagiusa* (Linnaeus)	blackcheek tonguefish
S. scombrus Linnaeus	Atlantic mackerel	Monacanthidae	
Scomberomorus maculatus (Mitchill)	Spanish mackerel	*Aluterus schoepfi* (Walbaum)	orange filefish
Stromateidae		*A. scriptus* (Osbeck)	scrawled filefish
Peprilus triacanthus (Peck)*	butterfish	*Monacanthus hispidus* (Linnaeus)	planehead filefish
P. alepidotus (Linnaeus)	harvestfish	Ostraciidae	
Scophthalmidae		*Lactophrys* sp.	unidentified boxfish
Scophthalmus aquosus (Mitchill)*	windowpane	Tetraodontidae	
Paralichthyidae		*Chilomycterus schoepfi* (Walbaum)	striped burrfish
Citharichthys spilopterus Günther	bay whiff	*Sphoeroides maculatus* (Bloch and Schneider)*	northern puffer
Etropus microstomus (Gill)*	smallmouth flounder		

*Species treated in individual chapters ($N = 70$). Details of the remaining, rarely collected, species ($N = 50$) are in table 77.3.

identified 120 species of estuarine fishes that fit one of several criteria for inclusion. Table 1.2 is a checklist of 70 species that we treat in detail—based on their patterns of use, relative abundance, or availability of data—as well as 50 rarely collected species. Our syntheses of early life-history patterns is presented in the final chapter, along with suggestions for future research on the importance of estuarine habitats.

CHAPTER 2

Study Area

LIMITS AND PHYSICAL CHARACTERISTICS

We have defined three study areas to relate life histories of estuarine fishes during their first year with physical characteristics of their environment. The largest of these encompasses continental shelf waters of the Middle Atlantic Bight. Within this area we are concerned with patterns of reproduction, largely based on gonadosomatic studies or the temporal and spatial distributions of eggs and larvae. A more restricted study area focuses on the central portion of this bight and estuarine-riverine systems between the Hudson River Estuary and Delaware Bay. Here we emphasize the temporal and spatial distribution of late larvae and juvenile fishes up to age 1. Our most detailed focus is on an estuarine-inlet system in southern New Jersey where we have the most comprehensive and long-term experience and data on ingress, habitat use, growth rates, and egress patterns of estuarine fishes. In the following descriptions of these study areas, we address aspects of the physical environment that might affect the behavior or occurrences of early life-history stages of fishes during their first year. To this end, we emphasize circulation, temperatures, salinities, and short-term (but regularly occurring) oceanographic phenomena.

The Middle Atlantic Bight is a North Atlantic Ocean neritic area limited by Cape Cod, Massachusetts and the adjacent Georges Bank to the north and Cape Hatteras, North Carolina to the south (fig. 2.1). Our coverage includes the estuarine borders of this bight and its continental shelf to a depth of 200 m. The width of this shelf ranges from 240 km off southern New England to 50 km at its southern extent off North Carolina. Several large estuarine embayments are located along the bight's coastline, including Narragansett Bay, Long Island Sound, Raritan–Sandy Hook Bays, Delaware Bay and Chesapeake Bay, as well as several smaller systems. The most important of these are dominated by single (or several) rivers, and the estuary represents the upper portion of a former river that once extended out onto the continental shelf (Patrick 1994). The South Atlantic Bight is located to the south of Cape Hatteras. North and east of Georges Bank lie the Gulf of Maine and the Scotian Shelf. East of the 200-m depth contour lies the Slope Sea (*sensu* Csanady and Hamilton 1988) and, farther east, the Gulf Stream, which originates in the South Atlantic Bight and flows closely past Cape Hatteras before veering toward the northeast. The Sargasso Sea, a tropical region of the North Atlantic Ocean, is located beyond the Gulf Stream and extends south to the Bahamas and beyond. All of these neighboring oceanic, coastal, and neritic regions influence in some way the character of our study area and its fauna.

We define the central part of the Middle Atlantic Bight as that portion of the bight bordered by western Long Island and the coastlines of New York, New Jersey, and Delaware (fig. 2.2). Within this region, we include the New York Bight (*sensu* Cowen et al. 1993) situated in the apex of the Middle Atlantic Bight. Submarine sand ridges are common features on the inner continental shelf in this region (McBride and Moslow 1991), and this feature is included in the LEO-15 study site (see below). Two major riverine estuaries, associated with the Hudson and Delaware Rivers, are major components of this region. Other important estuaries include Great South Bay and Raritan Bay, New York; Sandy Hook Bay, Barnegat Bay, Little Egg Harbor, Great Bay, and numerous sounds and bays in southern New Jersey; and Indian River Bay, Delaware.

Our most focused study area comprises a contiguous ecosystem including the Mullica River, Great Bay–Little Egg Harbor estuaries,

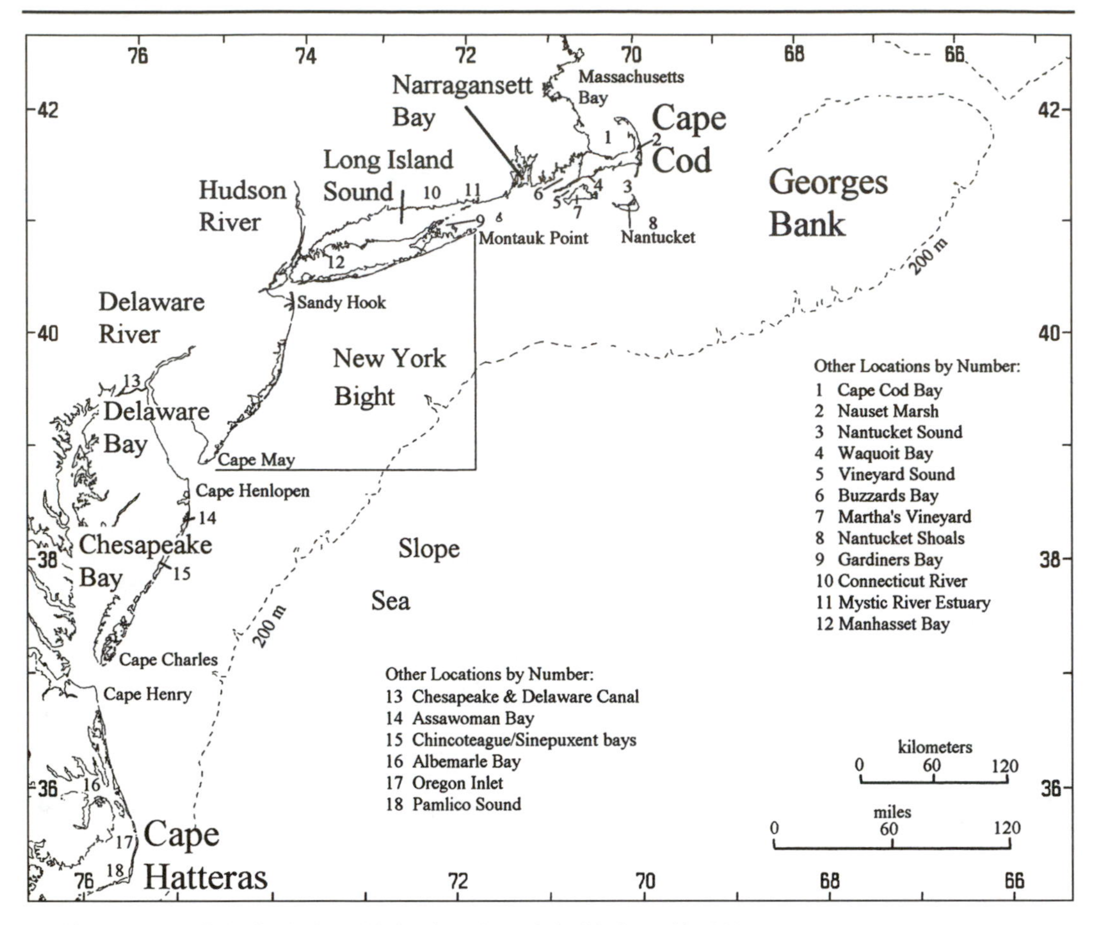

2.1 Middle Atlantic Bight Study Area. Limits of continental shelf indicated by 200-m contour.

Little Egg Inlet, and the Beach Haven Ridge (fig. 2.3). The Mullica River and its tributaries drain an area of about 400,000 ha of relatively undeveloped pinelands in southern New Jersey. This area has been designated the Pinelands National Reserve and is relatively pristine (Good and Good 1984). Waters from this extensive drainage enter into Great Bay, a drowned river valley, and the adjacent Little Egg Harbor, a barrier beach estuary. Together, these comprise the Mullica River–Great Bay National Estuarine Research Reserve (Psuty et al. 1993; Able et al. 1996b). The shorelines of these relatively shallow (1.7 m average depth at mean low water) polyhaline marsh systems total 283 km and consist of extensive stands of saltmarsh cordgrass (*Spartina alterniflora*). They share qualities with many other estuaries in the Middle Atlantic Bight, including a moderate tidal range from <0.7 m in Little Egg Harbor to 1.1 m near the mouth of Great Bay (Chizmadia et al. 1984; Durand 1984; Able et al. 1992; Psuty et al. 1993; Able et al. 1996b). Stations with varying physical characteristics (table 2.1; fig. 2.3) were regularly sampled with otter and beam trawls between June 1988 and October 1989. The results of these collections contribute to our discussions of habitat relationships in young-of-the-year of several species. The Beach Haven Ridge (Stahl et al. 1974) is located on the inner continental shelf just outside Little Egg Inlet. It is also the LEO-15 study site (von Alt and Grassle 1992), and collections here are the source of much of our data concerning seasonality of reproduction and settlement of fishes on the inner continental shelf.

CIRCULATION

Oceanic circulation within the Middle Atlantic Bight (fig. 2.4) is slow moving. The general transport is from northeast to southwest on the

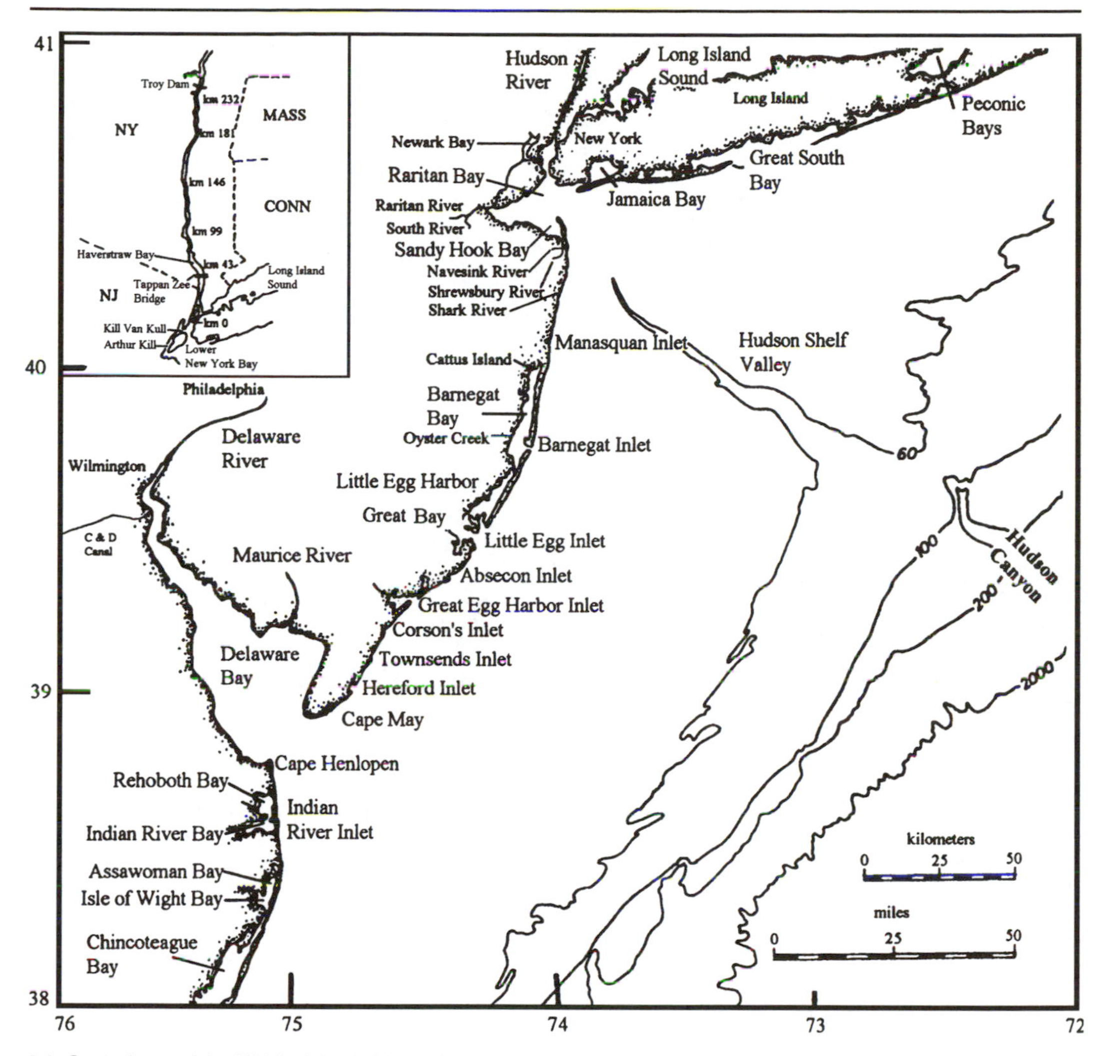

2.2 Central part of the Middle Atlantic Bight with major locations mentioned in text. Southern New Jersey Inland Bays (see table 4.2) are estuarine systems behind Absecon through Hereford inlets. Delaware Inland Bays include Rehoboth and Indian River bays. Continental shelf depth contours indicated in meters.

shelf and in adjacent Slope Sea waters during most of the year (Ingham 1982). Average monthly wind conditions and their effect on surface water circulation (Ekman Transport) change seasonally in the bight (table 2.2). Superimposed on the general drift are rotary tidal currents and short-term events influenced by weather conditions (Cook 1988). Extreme conditions can affect both the circulation and hydrography of the study area as well as its fauna. These can occur as major storms, severe droughts, oxygen depletions, major plankton blooms, or ephemeral phenomena such as warm-core rings, current reversals, or coastal upwelling events.

Middle Atlantic Bight shelf water is divided into three major bands: (1) a low-salinity, nearshore band affected primarily by runoff from major estuaries; (2) a midshelf band (between the 20- and 100-m contours); and (3) an outer shelf band near the shelf edge, which receives intrusions of warmer, more saline water from the Slope Sea (Ingham 1982). Most of the freshwater runoff into the Middle Atlantic Bight occurs during the spring and originates from four major plumes associated with the Hudson River, the Connecticut River, the Delaware Bay, and the Chesapeake Bay (Bigelow and Sears 1935; Ketchum and Corwin 1964; Charnell and Hansen 1974; Pape 1981; Durski

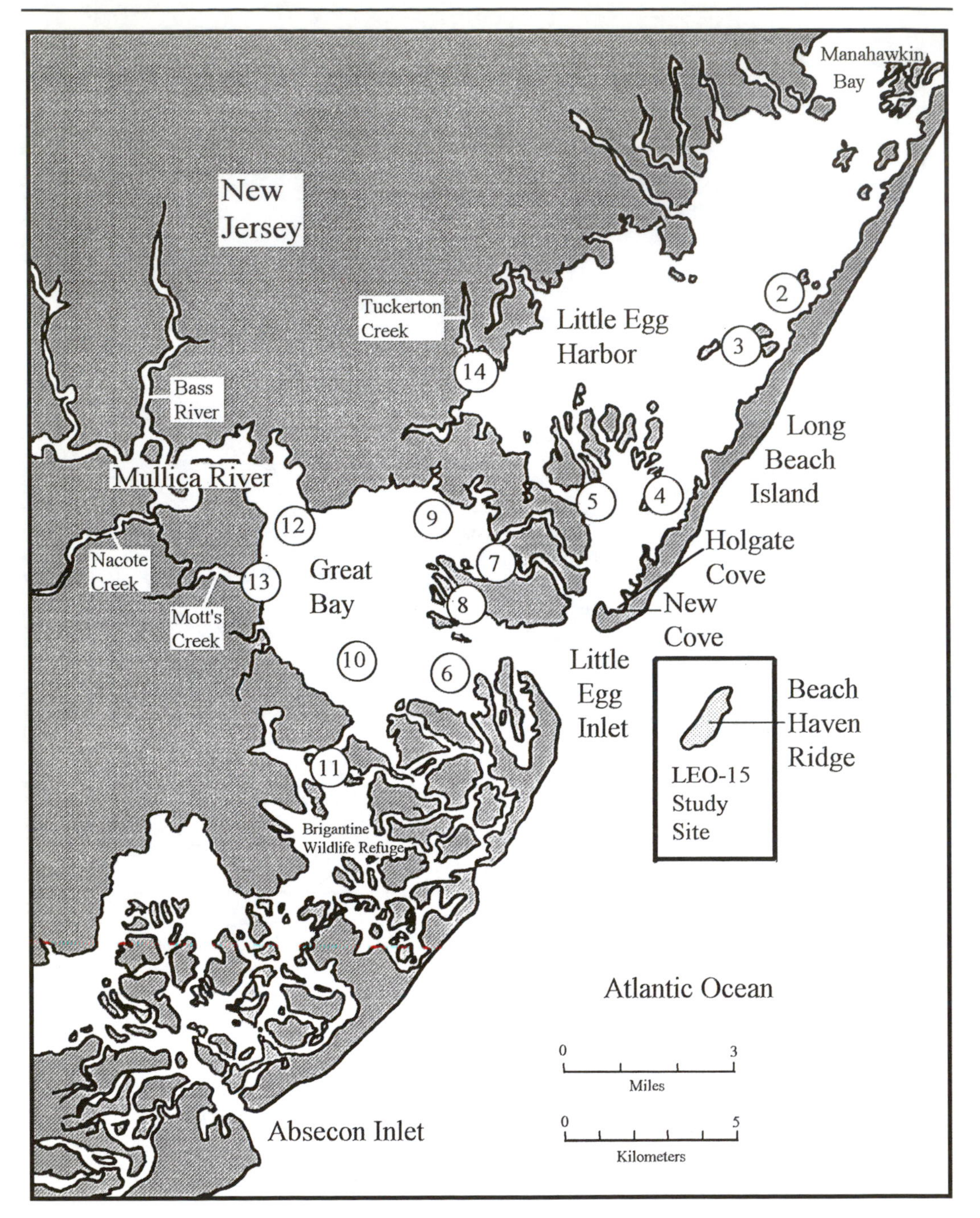

2.3 Mullica River, Great Bay–Little Egg Harbor, Little Egg Inlet, Beach Haven Ridge study area. The rectangle indicates the Long-Term Ecosystem Observatory at 15 meters (LEO-15) study area. Circled numbers indicate stations where otter trawl and beam trawl tows were regularly made (Station No. 1 infrequently sampled and not included). See Table 2.1 for physical characteristics of these stations.

Table 2.1. Characteristics of Great Bay–Little Egg Harbor Stations Regularly Sampled with 4.2-m Otter Trawl and 1-m Beam Trawl

Station	Location	Relative structure*	Substrate % silt	Depth range (m)	Major habitat type
2	Ham Island	6	16.1	0.6–1.2	Eelgrass/sponge
3	Marshelder Island	6	13.6	0.6–0.9	Eelgrass
4	Intracoastal Waterway	1	0.9	4.0–7.0	Sand/shell
5	Marshelder Channel	3	3.8	2.7–3.7	Shell/sand
6	Grassy Channel	4	19.2	1.2–4.3	Sand/hydroid
7	Little Sheepshead Creek	5	34.0	3.7–6.1	Sponge/shell/sand/silt/peat
8	Newman's Thorofare	1	0.8	3.0–7.9	Sand/rubble
9	Cape Horn	1	4.7	0.6–2.0	Sand/amphipod tubes
10	Intracoastal Waterway	1	54.4	1.7–3.7	Silt/sand
11	Little Bay	4	16.5	0.9–1.5	Sand/sea lettuce
12	Graveling Point	3	84.1	2.1–2.5	Shell/silt/clam beds
13	Mott's Creek	4	90.5	1.5–2.5	Silt/peat/hydroids
14	Tuckerton Creek	2	98.1	1.5–3.0	Silt/detritus

SOURCE: After Szedlmayer and Able 1996.
NOTE: See figure 2.3 for locations. Station 1 was infrequently sampled and deleted from the analysis.
*A qualitative estimate of physical heterogeneity ranging from 1 (flat, sand habitat with little or no other structure) to 6 (dense eelgrass with red sponge).

1996). Most of these plumes flow into shelf water and then south along the coastline of the bight.

Bottom currents in the bight vary in direction from region to region, but the results relative to inshore-offshore movements are similar. From southern New England to the offings of Chesapeake Bay, these bottom currents move toward shore (Bumpus 1973; Pape 1981). Strong near-bottom currents flowing toward New York Harbor and Long Island Sound have also been described (Charnell and Hansen 1974; Hardy et al. 1976). There is an apparent divergence in bottom drift near the Hudson Shelf Valley (fig. 2.2), where northeast of the valley, net drift is to the north; southwest of the valley, it is westward (Charnell and Hansen 1974; Hardy et al. 1976). Within the Hudson Shelf Valley, very strong currents flow mostly landward during winter and seaward during summer (Keller et al. 1973; Nelson et al. 1978; Ingham 1982). All of these patterns potentially affect the distribution and drift of fish larvae (e.g., Malchoff 1993).

HYDROGRAPHY

Meteorological and hydrographic conditions in the Middle Atlantic Bight are among the most variable in the world. Ocean temperatures fluctuate by more than 20–25 C between winter and summer (Parr 1933; Grosslein and Azarovitz 1982; Cook 1988), and range seasonally from near freezing to well over tropical limits. Sea surface temperatures vary seasonally, ranging from about 2 C along the coast during February and March (in the New York Bight) to highs near 30 C (but usually 25–26 C) during late summer (off Cape Hatteras, N.C.). Winter water-column temperatures off New York range from 2–3 C near the coast to 7–11 C at the 200-m contour offshore. Comparable values for the offings of Cape May are 3–4 C to 9–10 C, and, for the offings of Chesapeake Bay, 5–5.5 C to 9–10 C (Ingham 1982). In typical winters, bottom temperatures between the 150- and 250-m contours are about 9–11 C throughout the bight (Cook 1988). Changing seasonal patterns of cooling and warming on the bottom result in a temperature of about 10 C being

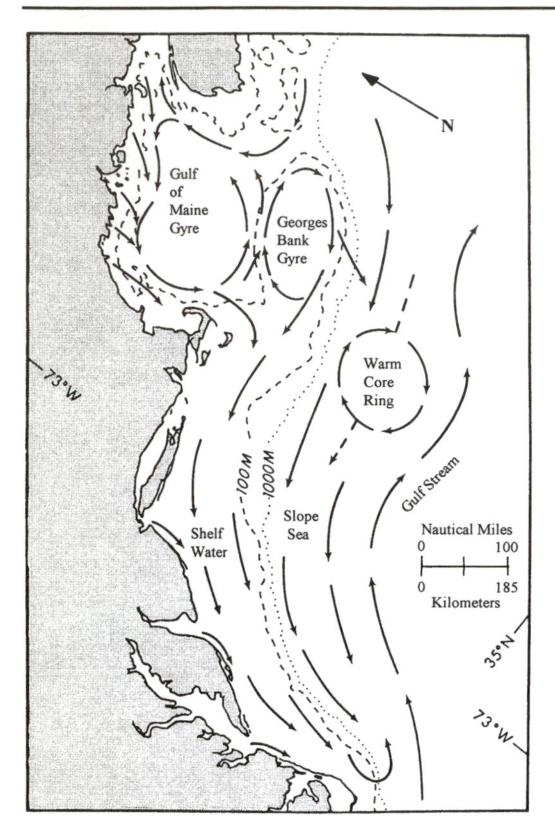

2.4 General surface circulation in the Middle Atlantic Bight and adjoining areas (after Ingham 1982).

always available somewhere within the Middle Atlantic Bight, and this is important to the seasonal distributions of fishes (Parr 1933).

Warming begins in April, and shelf water in the bight becomes highly stratified during the summer and fall when a strong thermocline separates the warm surface mixed layer from a cold cell. This dome-shaped cold cell forms on the shelf bottom between Georges Bank and Cape Hatteras during late-winter months and persists through the summer, gradually warming in late summer and fall (Bumpus 1973). Its usual position is between the 40- and 100-m isobaths, and its average thickness is about 35 m (or from the sea bottom to the underside of the thermocline). Its total volume represents about 30% of the total water mass in the Middle Atlantic Bight during summer (Ingham 1982). In summer months, surface layers (where most developing fish larvae occur) overlie this cell and warm considerably, resulting in a strongly stratified water column. The thermocline between these warm- and cold-water masses occurs at depths between 10 and 25 m, depending on distance from shore, and in nearshore areas it intersects the bottom at depths between 10 and 15 m. The offshore edge of this thermocline intersects the bottom at depths between 80 and 100 m (Houghton et al. 1988). Vertical mixing accompanies the breakdown of this stratification in the fall, and bottom temperatures reach their maximum 1 or 2 months after surface waters reach theirs (Ketchum and Corwin 1964). After erosion of the thermocline during the fall, conditions are vertically isothermal during the winter and early spring when conditions are generally coldest along the coast and become gradually warmer offshore (Ingham 1982).

Salinities in the Middle Atlantic Bight are a product of freshwater runoff entering the bight at the surface near the coast, combining with high-salinity slope water entering over the bottom from offshore (Bigelow and Sears 1935).

Table 2.2. Monthly Prevailing Winds and Resulting Ekman Transport at Two Locations in the Middle Atlantic Bight

	Jan	Feb	Mar	Apr	May	Jun	Jul	Aug	Sep	Oct	Nov	Dec
Central Middle Atlantic Bight (39N × 72W)												
Wnd	NNW	NNW	NW	W	SW	SW	SW	WSW	N	N	NW	NW
Ekm	SW	SW	SW	S	SE	SE	SE	SE	W	W	SW	SW
Southern Middle Atlantic Bight (36N × 75W)												
Wnd	NW	NW	WNW	WSW	SW	SW	SW	SW	NE	NE	NW	NW
Ekm	SW	SW	SSW	SSE	SE	SSE	SSE	SE	NW	WNW	WSW	WSW

SOURCE: Adapted from Cook 1988.
NOTE: Wnd = direction from which winds blow. Ekm = direction of resulting Ekman Transport.

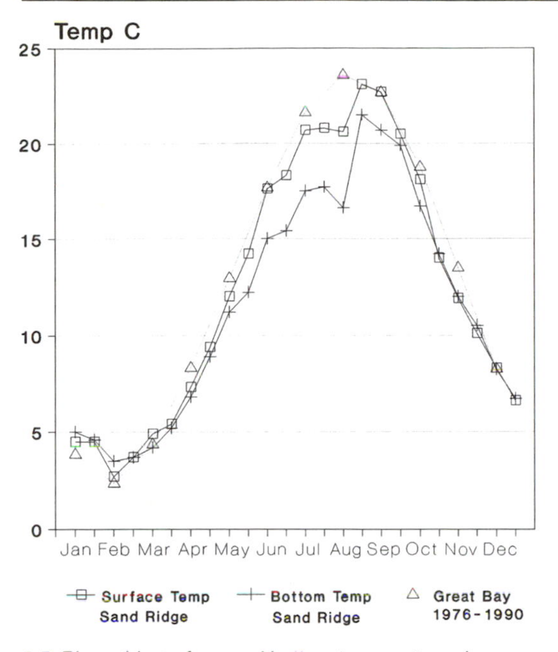

2.5 Bimonthly surface and bottom temperature observations from Beach Haven Ridge (1972–1974) and mean surface temperatures from the adjacent Great Bay (1976–1990). Data from Milstein et al. (1977), Able et al. (1992).

Values range from 30 ppt near shore to 35 ppt on the bottom near the shelf edge. The average throughout the bight ranges from 32.5 ppt at the surface to 35 ppt along the bottom. Salinities are lowest during spring runoff of freshwater, increase through remaining seasons, and reach the highest levels during winter (Cook 1988). The effect of freshwater runoff on ocean salinities varies through the bight. The Hudson River plume lowers salinities as far as 145 to 160 km offshore, whereas the effect of the Chesapeake Bay plume rarely reaches more than 80 km (Cook 1988).

Hydrographic conditions in our most focused study area (Great Bay, N.J.) vary considerably through the year. Mean monthly temperatures range from 2.1 to 23.9 C (fig. 2.5), but extremes can extend from −0.8 to 25.2 C. Salinities are usually between 28.0 and 29.8 ppt, but short-term events cause overall extremes of 23.6 to 34.5 ppt (Able et al. 1992). At the LEO-15 study site, located on the Beach Haven Ridge immediately outside Little Egg Inlet, mean monthly sea surface temperatures range from 2.5 to 21.5 C, and salinities from 28.8 to 30.6 ppt (Thomas and Milstein 1973).

SHORT-TERM PHENOMENA

Short-term oceanographic phenomena such as Gulf Stream eddies and rings, current reversals, and coastal upwelling may influence the transport and subsequent distribution and abundance of young-of-the-year fishes in the Middle Atlantic Bight. The offshore boundary of Middle Atlantic Bight shelf water is called the Shelf/Slope Front and is usually located at the approximate location of the 200-m isobath (fig. 2.1). This front extends from Georges Bank to Cape Hatteras and is present year-round (Beardsley and Flagg 1976). Offshore of this front, the slope water is warmer and saltier, and consists of a mixture of continental shelf and Gulf Stream water. Warm-core rings occasionally spin off the western edge of the Gulf Stream, and these slowly drift in a southwesterly direction within the slope water mass off the bight. These rings enclose highly saline, warm waters from the Gulf Stream, may measure 150 to 230 km across, and reach depths of 2,000 m (Cheney 1978). From 6 to 11 such rings are produced annually (Fitzgerald and Chamberlin 1981). They are important because they may remove "streamers" of shelf water by "advection," may inject slope water and modified Gulf Stream water onto the shelf, and may modify circulation on the shelf. These features occur most frequently in the vicinity of Georges Bank, where they are younger, larger, and stronger in rotational flow than when they enter waters off the central part of the Middle Atlantic Bight (Cook 1988). Rings (and other features in the Slope Sea) also contain and transport fish larvae from southern waters and deliver them to shelf waters of the Middle Atlantic Bight from whence some travel to estuarine nurseries (Cowen et al. 1993; Hare and Cowen 1993; J. A. Hare, M. P. Fahay, and R. K. Cowen, in prep.).

Current reversals have been cited as mechanisms allowing for the retention of blue crab larvae in the neighborhood of major Middle Atlantic Bight estuaries (Epifanio et al. 1989; Epifanio 1995), and they undoubtedly play a role in the distribution of fish larvae as well. They typically occur when prevailing winds during summer in the central Middle Atlantic Bight are from the west and southwest (table 2.2). When these conditions continue over a

Table 2.3. Physical Characteristics of Estuarine Systems Considered in This Study

Estuary	Length (km)	Average width (km)	Average depth (m)	Mouth dimensions Depth (m)	Mouth dimensions Width (km)	Stratification* High	Stratification* Low	Estuarine drainage (km²)	Area (km²) by salinity zone† Tidal fresh	Mixing	Seawater
Nauset Marsh, MA	$\sim$7	$\sim$2	$\sim$2	3.5	0.25	VH	VH	<1	0	0	10
Buzzards Bay, MA	48	11	10.6	23	10.64	VH	VH	1,491	0	5	585
Narragansett Bay, RI	47	15	10.0	30	21.79	VH	VH	3,444	8	52	368
Connecticut River, CT	88‡	0.5	3.8	7.6	1.39	HS	MS	2,823	47	5	0
Long Island Sound, NY	320	20	19.5	91	34.08§	VH	VH	18,725	75	427	2,815
Gardiners Bay, NY	50	14	6.1	30	20.26	VH	VH	1,036	0	5	505
Great South Bay, NY	116	3.5	2.2	8.5	0.76‖	VH	VH	2,188	0	202	189
Hudson/Raritan, NY, NJ	257	2.7	4.6	15	9.67	MS	VH	21,929	96	435	241
Barnegat Bay, NJ	64	4.5	1.4	6.1	.41	VH	VH	3,496	0	78	186
Great Bay, NJ#	25	4	2.5	7.6	3.57	VH	VH	1,476	32	98	47
Delaware Bay, NJ, DE	222	24.1	7.4	37	17.91	VH	VH	12,302	59.6	518	1,412
Chincoteague Bay, VA	51	7.2	1.3	10	1.45	VH	VH	777	0	0	355
Chesapeake Bay, VA**	319	24.8	8.5	24	17.40	MS	MS	56,863	396	9,226	298
Albemarle Sound, NC	172	16.9	5.8	4.6	9.57††	VH	VH	15,032	1,880	508	0
Pamlico Sound, NC	193	25.3	5.5	7.6	1.94‡‡	VH	VH	14,996	249	4,898	104

SOURCE: NOAA 1985; Roman and Able 1989; Durand and Nadeau 1972; Able et al. 1996a.

NOTE: Southern New Jersey Inland Bays and Delaware Inland Bays (referred to in Table 4.2) consist of multiple estuarine systems and are not included in this table.

*Indicates degree of stratification in mixing zone, measured during 3-month period of high freshwater runoff (High) and 3-month period of low runoff (Low). HS = highly stratified, presence of salt wedge, little mixing between surface and bottom; MS = moderately stratified, surface layers less saline than bottom, moderate amount of mixing mostly because of tidal turbulence; VH = Vertically homogeneous, no stratification and turbulent tidal mixing.

†Indicates average amount of estuarine area covered by differing salinities: Tidal fresh = 0.0–0.5 ppt; Mixing = 0.5–25.0 ppt; Seawater = >25.0 ppt.

‡Within estuarine drainage portion of river only.

§Measured at eastern end of sound.

‖Average width of three inlets.

#Includes Mullica River and drainage basin. Does not include Little Egg Harbor.

**Mainstem bay only; not including tributaries.

††No connection to ocean; measured at interface with Pamlico Sound.

‡‡Average width of five inlets.

period of days, surface currents over the midshelf off Chesapeake and Delaware Bays reverse their direction and flow to the north (Boicourt 1982).

Another result of prevailing west and southwest summer winds is upwelling of cold bottom water. In the central part of the Middle Atlantic Bight, temperatures on the shelf bottom within the cold cell reach a minimum (3.8–4.7 C) during early June. Maximums occur after fall turnover when they may reach 12 to 13 C in late November (Ingham 1982). Nearer the coast, the front may approach (and reach) the New Jersey, New York, and Delaware shores, or even enter estuaries during upwelling events precipitated by strong west or southwest winds, which are usually strongest during summer (table 2.1) (Scott and Csanady 1976; Hicks and Miller 1980; Ingham 1982; Neuman 1996). In addition to lowering temperatures, this inshore incursion of cold cell waters may result in anoxic conditions causing mass mortalities of fishes and other marine organisms (Steimle and Sindermann 1978; Swanson and Parker 1988; Glenn et al. 1996). Under the opposite wind conditions, the cold cell may also migrate offshore, forcing the Shelf/Slope Front to move seaward and exposing the offshore shelf bottom to colder temperatures (Csanady 1978).

REPRESENTATIVE ESTUARINE SYSTEMS IN THE MIDDLE ATLANTIC BIGHT

A need exists for fisheries biologists and habitat managers to characterize individual estuaries with emphases on factors that affect fishes (Miller 1984). To begin this process, and to better understand latitudinal and other sources of variation in estuarine young-of-the-year fishes in the Middle Atlantic Bight (see chap. 4, table 4.2), we have examined the physical characteristics of several estuaries throughout the bight (table 2.3). These characteristics vary considerably. Size (surface area) ranges from Nauset Marsh (about 14 km^2) to Chesapeake Bay (about 8,000 km^2). Average depth extremes are 2 m or less in Nauset Marsh and Barnegat Bay to 19.5 m in Long Island Sound. Salinities also vary in these systems, and these are expressed as the area covered by tidal freshwaters (0–0.5 ppt), mixing (0.5–25.0 ppt), and seawater (>25.0 ppt). Some systems (e.g., Nauset Marsh, Gardiners Bay, and Chincoteague Bay) lack appreciable freshwater input, while at the other extreme, Albemarle Sound is dominated by freshwaters and those with intermediate salinities. Nearly all of these systems are vertically homogeneous in the mixing zone, during periods of both high and low runoff. Our focused study area, the Great Bay–Little Egg Harbor estuarine system, was described above. Below we briefly characterize other systems throughout the bight, from north to south (figs. 2.1, 2.2).

Nauset Marsh is a shallow, 950-ha barrier island estuary on the outer shore of Cape Cod with direct exchange to the Atlantic Ocean through a single inlet (table 2.3; fig. 2.1). Depths range to 5 m (mostly < 2 m), and the tidal amplitude is 1.5 m inside the inlet. Annual salinities range from 24 to 34 ppt and temperatures from −2.0 to 27 C. Freshwater input is minimal and is derived primarily from groundwater seepage. Important habitats include *Spartina alterniflora* marsh, tidal kettle ponds, shallow tidal channels, intertidal mud- and sandflats, eelgrass beds, and peat reefs (Able et al. 1989; Able and Fahay pers. observ.)

Buzzards Bay ranges from 5 m in depth at the northern end to about 30 m depth at the mouth (table 2.3, fig. 2.1). Bottom sediments are mostly sand and silt, but there are also areas of rocky bottom and rock ledges (Lux and Wheeler 1992). Temperatures in the Slocum River (an estuary leading into the bay) range from about 2.0 to 25.0 C, and salinities here can range from near 0 ppt in the upper reaches to about 35 ppt near the open bay (Hoff and Ibara 1977). Hydrographic and physical conditions are similar in nearby Waquoit Bay, which faces on Vineyard Sound, and in the vicinity of Woods Hole, and we combine fish observations from these localities (see table 4.2).

Narragansett Bay is representative of the southern New England coastline, comprising an area about 47 km long by 15 km wide (table 2.3; fig. 2.1). The bottom is rocky, and depths average 10 m in this estuary. Most of the freshwater input is constrained by impoundments, such that an oligohaline zone is absent and high salinity values (25 to 30 ppt) predominate (Nixon 1985). Temperatures range from 0 to

20 C through the year. Important fish habitats include eelgrass beds, rocky substrates, pilings, stony beaches, tidal flats (sand or mud), and salt marsh meadows and creeks (Patrick 1994).

The Connecticut River is the longest river in New England, flowing from the Canada–New Hampshire border to Saybrook on Long Island Sound (table 2.3; fig. 2.1). Within the Middle Atlantic Bight, it is second only to the Susquehanna River in volume of water carried. Its length (660 km) exceeds that of the Hudson River by 160 km, and its monthly average discharge volume (333 m^3 per sec) also exceeds that of the Hudson River (273 m^3 per sec) (Merriman and Thorpe 1976). The lower section of this river is tidally influenced, and saline waters reach upstream a maximum distance of about 23 km, except during spring when runoff is at a maximum (Boyd 1976). In contrast to other estuaries considered here, the Connecticut River system lacks a zone with salinities greater than 25.0 ppt (NOAA 1985). The bottom in the lower section is largely silt and sand, and extensive marsh vegetation or seagrass beds are lacking.

Long Island Sound is about 320 km long and averages 20 km wide (table 2.3; fig. 2.1). It is a system with two openings: a western one that is connected to the ocean via the East River and Lower New York Harbor, and an eastern one mingling with open ocean water in Block Island Sound. Strong tidal currents occur at both ends. Most of the freshwater input is from the Connecticut River, located near the eastern end, or mouth, of the system. The sound is situated between a mixture of sandy beaches and rocky outcrops on the Connecticut shore and primarily sandy beaches along the Long Island shore. Salt marshes along its margin are relatively few. Bottom sediments range from mud, silt, and clay in the central and western basins to coarse sand in the eastern basin. There is a gradient of environmental quality within the sound, ranging from relatively pristine in the eastern end to heavily degraded in the western (Schubel 1986). Surface temperatures in the central part of the sound range from about 2 to 23 C, and bottom temperatures are about the same. Salinities range from about 24 to 28 ppt but are generally much lower in the vicinity of the Connecticut River mouth and near the western end (Riley 1956).

Gardiners Bay (table 2.3; fig. 2.1) is located at the extreme eastern end of Long Island and differs in physical characteristics from Long Island Sound because it is directly influenced by offshore waters of the Atlantic Ocean and Block Island Sound. Temperatures here are generally colder than in Long Island Sound (Hickey et al. 1975). It is situated at the "clean" end of the environmental quality gradient found in the latter sound, but, in contrast to most other estuarine systems considered here, it lacks a tidal freshwater zone where salinities are < 0.5 ppt (NOAA 1985).

Great South Bay is a shallow embayment on the south shore of Long Island (table 2.3; fig. 2.2). The average depth is 2.2 m, and maximum depths of 7.7 m are found in navigation channels (Monteleone 1992). The bay opens directly through an inlet to the Atlantic Ocean and it is another example of a system with no tidal freshwater zone (salinities < 0.5 ppt) (NOAA 1985). Natural substrates include sand, mud, and detritus, but these are dominated by clay and silt. Dense beds of eelgrass (*Zostera marina*) occur, as do sporadic patches of certain algae, including *Ulva lactuca* (Briggs and O'Connor 1971). In a 2-year study, temperatures ranged from below 0 C during January and February, when the bay was ice-covered, to 27 C in July and August. Salinities during the same period averaged 27.1 ppt, and ranged from a springtime low of 20.0 ppt to above 29.0 ppt near the inlet (Monteleone 1992).

Both the Hudson and Delaware River Estuaries are large, drowned river valleys with large population and industrial centers in their drainage basins. As a result, these estuaries are heavily impacted, and habitat degradation is commonplace (MacKenzie 1992; Burger 1994; Dove and Nyman 1995; Stanne et al. 1996). Conditions in these highly impacted areas appear to be improving as a result of enhanced sewage treatment and reduced industrial inputs (e.g., Weisberg and Burton 1993; Weisberg et al. 1996).

The Hudson River (fig. 2.2) is tidal from The Battery on Manhattan Island (km 0) to Troy

Dam (km 243). The salt front (salinity ~1 ppt) during spring runoff in February–May oscillates between km 18 and km 80 (Dew and Hecht 1994b). At the mouth of the Hudson is a large embayment divided into the Raritan and Sandy Hook Bays, and several back bays located behind large islands, such as Newark Bay. The shorelines in this area are heavily populated and, combined with dense industrialization, contribute to severe chemical and mechanical pollution. In short, this is one of the most severely impacted estuarine systems in the world.

The system composed of Raritan and Sandy Hook Bays is fed by runoff from the Hudson and Raritan Rivers as well as several smaller rivers including the Navesink and Shrewsbury (table 2.3; fig. 2.2). These bays are relatively shallow but have an irregular topography including banks, dredged ship channels, shoals, holes, and borrow pits (the site of mined sand and gravel). The bottom is primarily fine sand with patches of shell debris in the northeastern part, but it is soft and muddy in most of the remaining areas (Woodhead et al. 1987). Thick concentrations of a macroalga, *Ulva lactuca*, are seasonally abundant features of this region, as are restricted areas of *Spartina* salt marsh and small, remnant patches of eelgrass (*Zostera marina*) (MacKenzie and Stehlik 1988). Temperatures range from about 0 C in winter to about 25 C in late summer. Salinities in the bays range from 25 to 32 ppt, with the lowest values during the spring (Woodhead et al. 1987). Lower salinities are found in the tributaries.

Barnegat Bay is a shallow, lagoon-type estuary typical of the back bay system of a barrier island coastline (table 2.3; fig. 2.2). It is about 64 km long by 4.5 km wide. Its depth ranges from 1 to 6 m (Kennish and Olsson 1975), and 73% is less than 2 m deep at mean low water (Barnes 1980). Several small creeks feed into the bay, but freshwater inflow is very low. The average salinity is 25 ppt (Chizmadia et al. 1984). Sand and sandy-mud bottoms predominate, and the vegetation includes beds of eelgrass as well as summer occurrences of *Ulva lactuca* and other macroalgae (Moeller 1964; Chizmadia et al. 1984). Temperatures range from as low as −1.4 C in winter to a high of 28 C in summer, and because of its shallowness, daily variations occur quickly, in response to changes in air temperatures (Chizmadia et al. 1984).

The shallow, coastal embayments that we refer to as "Southern New Jersey Inland Bays" are typified by the embayments behind Hereford Inlet (fig. 2.2). Freshwater input is negligible into these embayments, and the salinity is generally between 28 and 32 ppt through the year. Temperatures range from about 0 C during winter to 26 C during summer, and, on the marsh surface, can reach 40 C during midsummer days (Allen et al. 1978). Habitats near the inlet are typically sandy and have strong currents. Elsewhere, the upper embayment consists of sand-mud or soft mud bottoms, and shallow creeks and sounds drain extensive *Spartina alterniflora* marshes.

Much of the lower Delaware River is tidally influenced and flows into Delaware Bay, a relatively shallow, turbid estuary (table 2.3; fig. 2.2). Other constituents of the system include the Chesapeake and Delaware (C & D) Canal, which connects the upper bay with Chesapeake Bay, and several creeks and small rivers (Wang and Kernehan 1979). Most of these tributaries are impounded a few kilometers upstream from the river or bay, where the incursion of tidal and brackish water stops. Brackish water extends up the Delaware River as far as Philadelphia and tidal freshwaters as far as Trenton. Extremely poor environmental conditions in past decades have made the area near Philadelphia one of the most severely polluted estuaries in the United States. Conditions have improved in recent years, however, and this has resulted in enhanced production of local fish resources (Weisberg et al. 1996).

The Indian River drainage basin is composed of Indian River Bay (a drowned river valley estuary) and Rehoboth Bay (a bar-built estuary) (fig. 2.2). The basin is connected to the Atlantic Ocean by an inlet 61 m wide by 4.6 m (average) deep. Each bay is surrounded by extensive marsh surface and is fed by numerous small tributaries (Derickson and Price 1973).

The largest bay along the eastern shore of Virginia and Maryland is Chincoteague Bay (table 2.3; fig. 2.2) but, despite its size, it is typical of many systems along this portion of the coast. It is separated from the Atlantic Ocean by the narrow Assateague Island with inlets at the northern and southern ends. The area is shallow and very turbid, and there are abundant beds of *Ulva lactuca,* other algae, and extensive salt marshes, but no extensive eelgrass beds. Sandy shoals typify the eastern side of the bay; clay and mud bottoms predominate in the deeper western part. During summers with little rainfall, the system tends to become a reverse estuary, where salinities are higher than neighboring coastal waters (Richards and Castagna 1970). Extreme temperatures of −1.7 to 31.5 C have been recorded (Schwartz 1961; Richards and Castagna 1970).

The major estuary in the southern part of the bight is Chesapeake Bay (table 2.3; fig. 2.1). The shoreline of this system totals about 7,400 km and is fed by 19 principal rivers and 400 lesser creeks. Salinity in this system ranges from zero, at the mouths of its several rivers, through about 15 ppt at the midpoint, to 30 ppt, or higher, at the bay mouth. Although there are deep holes, the average depth throughout the system is only about 8.5 m. Vast expanses of flats slope gently from both shores toward the deeper channel. Temperatures measured in submerged aquatic vegetation beds in the lower part of the bay range from near 0 C in January to about 28 C in July (Orth and Heck 1980; Olney and Boehlert 1988). Important fish habitats include sand beaches, intertidal mud or sand flats, piers, rocks, and jetties, seagrass meadows composed of eelgrass and other submerged aquatic vegetation, oyster bars, and salt marshes (Lippson and Lippson 1984).

The southernmost estuaries include Albemarle and Pamlico (along with Currituck, Croatan, Roanoke, and Core) Sounds, which together comprise an estuary with an area of about 7,101 km^2 (table 2.3; fig. 2.1). This complex is a coastal lagoon separated from the Atlantic Ocean by a barrier beach perforated by Oregon, Hatteras, Ocracoke, Drum, and Barden inlets. Salinities range from oligohaline (0.5–5.0 ppt) in Albemarle Sound, to polyhaline (18–30 ppt) in eastern Pamlico Sound (Epperly and Ross 1986). Temperatures in the Neuse River, as it enters Pamlico Sound, range from 1.8 to 32.0 C (Tagatz and Dudley 1961). Extensive salt marshes, eelgrass, and other submerged aquatic vegetation beds account for most of the fish habitat in this complex of sounds, although the tributary creeks also contain considerable amounts of detritus (Epperly and Ross 1986).

CHAPTER 3

Methodology

The data contributing to the species accounts in this book are derived from three major sources: (1) published literature; (2) archived specimens, informally published technical reports, and raw data; and (3) original information assembled by our associates and us over the last two decades, with the collecting emphasis placed on the continental shelf of the Middle Atlantic Bight and the Great Bay–Little Egg Harbor estuary in the central part of this bight. Almost all of this sampling effort was directed at early life-history stages of fishes.

LITERATURE

Several published sources have contributed to our understanding of the life histories of Middle Atlantic Bight fishes, although the focus of these is not necessarily on estuarine species, nor are events during the first year in their life completely addressed. Some of the faunal works are extralimital (but still valuable), such as the treatment of the fishes of the Gulf of Maine (Bigelow and Schroeder 1953), a work currently in revision (B. B. Collette and G. K. McPhee in prep.). Within the Middle Atlantic Bight, several regional efforts provide a wealth of summary information on the fishes of Chesapeake Bay (Hildebrand and Schroeder 1928), Delaware Bay (Wang and Kernehan 1979), and Barnegat Bay (Kennish and Lutz 1984). In addition to these faunal works, we have also drawn on numerous estuarine-specific and species-specific studies for our synthesis. For example, the compilations of information on early development of southern Middle Atlantic Bight fishes (Fritzsche 1978; Hardy 1978a, b; Johnson 1978; Jones et al. 1978; Martin and Drewry 1978) have proven especially useful. Several other studies addressing specific systems within the central part of the Middle Atlantic Bight are invaluable sources for describing spatial and temporal occurrences of young-of-the-year stages, as well as for providing evidence that many aspects of fishes' early life histories are not well known or described (table 3.1). Within the central part of the Middle Atlantic Bight, recent studies have identified the fauna (Able 1992), summarized occurrences and relevant sources for the identification of egg and larval stages (Fahay 1993), and provided a listing of fish-related studies from some of the estuaries in the region (Able and Kaiser 1994).

ARCHIVES, INFORMAL PUBLICATIONS, AND RAW DATA

Historical unpublished data sets have been consulted for many species (table 3.2). Many of these originated from environmental impact studies associated with proposed or existing power plants. Notable among these are data reports and summaries from the Mullica River–Great Bay–Little Egg Harbor–Little Egg Inlet–Beach Haven Ridge corridor (fig. 2.3), which includes habitats from a relatively pristine river, an estuary, and a study site on the inner continental shelf. These studies have particular value because many voucher specimens are deposited in the Academy of Natural Sciences of Philadelphia (ANSP) and are available for reexamination, length measurements, and verification of occurrences (Able 1992). Other extensive material available at ANSP includes unpublished collections of juvenile fishes taken with small mesh trawls on the continental shelf and in Delaware Bay by A. E. Parr and associates in the 1930s (Able 1992). A similarly comprehensive source includes studies throughout Delaware Bay that were associated with the Public Service Electric and Gas nuclear power facility in Salem, New Jersey (table 3.2). All of these data are used in a variety of ways but most frequently are incorporated into the length frequency distributions for individual species.

Table 3.1. Important Studies and Sources of Information Concerning the Occurrence and Distribution of Early Stages of Fishes from the Central Part of the Middle Atlantic Bight

Source	Study areas*	Collecting gear
Nichols and Breder 1927	No. New Jersey–Massachusetts	†
Perlmutter 1939	Long Island, sound, ocean shores	Plankton nets, various trawls
Wheatland 1956	Long Island Sound	Clarke-Bumpus plankton net
Schwartz 1961	Chincoteague, Sinepuxent bays	Otter trawl
Richards 1963	Long Island Sound	Shrimp trawl
Schwartz 1964a	Isle of Wight, Assawoman bays	Otter trawl
Croker 1965	Sandy Hook Bay	Plankton nets
Perlmutter et al. 1967	Hudson River	Seine
Derickson 1970	Rehoboth & Indian River bays	Seine, otter trawl
Thomas 1971	Lower Delaware River	Otter trawl, seine
Briggs and O'Connor 1971	Great South Bay	Seine
Allen et al. 1978	Hereford Inlet	Otter trawl, seine, traps, fyke net, gill net, plankton nets
Dovel 1981	Lower Hudson River	Plankton nets
Milstein 1981	Southern New Jersey, coastal	Otter trawl
Talbot and Able 1984	High salt marshes, New Jersey	Pit traps, dip nets
Tatham et al. 1984	Barnegat Bay	Trawls, seines, plankton nets, gill nets ‡
Morse et al. 1987	Continental shelf	Bongo plankton nets
Rountree and Able 1992a	So. New Jersey marsh creeks	Weir, seine
Comyns and Grant 1993	Continental shelf	Plankton nets, neuston nets

* For study areas, see figure 2.2.
† Based on published reports of other studies and unpublished data from a wide variety of sources.
‡ Also intake screens and discharge areas of power-plant cooling system

ORIGINAL INFORMATION

Interpretations of life-history patterns depend strongly on the degree of development in relation to size in early stages and the subsequent effect of gear avoidance and selectivity when targeted at those stages. We have been involved with an extensive array of projects and sampling gears in the last two decades to collect eggs, larvae, and juvenile fishes over the Middle Atlantic Bight continental shelf and along the Mullica River–Great Bay–Little Egg Harbor–Little Egg Inlet–Beach Haven Ridge corridor. A primary consideration for this multiple sampling gear approach was to avoid size-related or temporal gaps in collections that might influence our interpretations. As a result of these efforts, we have amassed original information on early stages of fishes based on more than 28,000 samples and an uncounted number of specimens (table 3.3). These sampling series have provided extensive information on the occurrences of young-of-the-year, including inferences on habitat changes associated with ontogeny, and have also been the primary source for our interpretations of growth rates, although the latter are often augmented by collections from the sources cited above. Details of sampling methodology and techniques have been described for several of these efforts (table 3.3) and we will only briefly describe each of them here.

Larvae were collected in the ocean during the NMFS MARMAP surveys, which provided excellent spatial and temporal coverage of continental shelf waters between Nova Scotia and Cape Hatteras, North Carolina from 1977 through 1987 (Sherman 1980, 1988). This sampling was focused on the entire water column

Table 3.2. Informally Published Technical Reports Consulted in the Preparation of Summaries of Life Histories of Estuarine Fishes in the Central Part of the Middle Atlantic Bight

Source	Area studied
de Sylva et al. 1962	Delaware River Estuary
Daiber and Smith 1970	Delaware Bay
Schuler 1971	Delaware River and Bay
Hamer 1972	Mullica River–Great Bay Estuary
Thomas et al. 1972	Beach Haven Ridge
Thomas and Milstein 1973	Little Egg Inlet and adjacent ocean
Schuler 1974	Delaware River and Bay
Tatham et al. 1974	Little Egg Inlet and adjacent ocean
Thomas and Milstein 1974	Southern New Jersey coast
Thomas et al. 1974	Little Egg Inlet and adjacent ocean
Ashton et al. 1975	Delaware River Basin
Thomas et al. 1975	Little Egg Inlet and adjacent ocean
Milstein et al. 1977	Little Egg Inlet and adjacent ocean
Swiecicki and Tatham 1977	Little Egg Inlet and adjacent ocean
Tatham et al. 1977	Oyster Creek (Barnegat Bay)
Himchak 1981, 1982a, b, 1983, 1984	Several New Jersey rivers
Pacheco 1983, 1984	Sandy Hook Bay
Public Service Electric & Gas Co. 1984	Delaware River and Estuary
Himchak and Allen 1985	Navesink River
Woodhead et al. 1987	Lower Hudson–Raritan Estuary
Kahnle and Hattala 1988	Hudson River Estuary
O'Herron et al. 1994	Delaware Estuary

and was accomplished with double-oblique tows from surface to bottom or 200 m depth, whichever was less, using 60-cm bongo plankton samplers fitted with 0.505-mm mesh nets. Data derived from these surveys provide information on the timing of reproduction, distribution, and abundance of eggs and larvae and insight into maximum sizes attained in the pelagic larval stage.

We augmented these larval collections with plankton sampling at the LEO-15 study site on the Beach Haven Ridge (fig. 2.3), where the temporal effort was year-round, but concentrated on the summer and fall months. At this location, samples were collected with an opening-and-closing Tucker trawl (1 m^2, 0.505-mm mesh) with multiple codends that were fished in a step-oblique manner from the surface to the bottom (K. W. Able, M. P. Fahay, D. A. Witting, R. S. McBride, and L. S. Hales, in prep.). The first two oceanic collection efforts help to separate larvae that are spawned in the ocean but do not come into the estuary—at least as larvae—when compared to results of our third larval sampling program. In this 6-year study, larvae were sampled weekly with fixed plankton nets suspended in Little Sheepshead Creek during night flood tides, and these samplers were directly targeted at larvae ingressing from adjacent oceanic waters as well as larvae resulting from local, estuarine spawning (D. A. Witting, K. W. Able, and M. P. Fahay in prep.). These samplers had 1-m diameter mouths, and were fitted with 1.0-mm mesh netting. At least three replicate half-hour sets were made on each sampling date, and a total

Table 3.3. Sampling Effort for Young-of-the-Year Fishes over the Middle Atlantic Bight Continental Shelf, Inner Continental Shelf (Beach Haven Ridge), and the Great Bay–Little Egg Harbor Estuary

Gear	Location	Water depth (m)	No. of stations	Duration/ frequency	Focus of sampling (life-history stage)	No. of samples	Source*
Bongo plankton net, 0.505-mm mesh	Middle Atlantic Bight Continental Shelf	11–1400	~180	1977–1987 6–8/year	Eggs & larvae	11,438	Sibunka & Silverman 1984, 1989; Morse et al. 1987; Berrien & Sibunka in press
Tucker trawl 0.505-mm mesh	Inner Continental Shelf (Beach Haven Ridge)	2–19	2	Jul 1991–Nov 1992/ monthly	Larvae	138	
1-m plankton net 1.0-mm mesh	Little Sheepshead Creek Bridge	4	1	Feb 1989–Nov 1994/ weekly	Ingressing larvae	1,351	Witting 1995; D. A. Witting, K. W. Able, and M. P. Fahay in prep.
Dip-net (night-light)	RUMFS boat basin, Great Bay	2–3	1	Apr 1986–Aug 1992/ aperiodic	Pelagic late larvae	153	Able et al. 1997
2-m beam trawl 3.0-mm mesh	Beach Haven Ridge & Great Bay	2–19	30	Jul 1991–Oct 1994 monthly	Early settled Juveniles to adults	363	
Experimental trap	RUMFS boat basin	2–3	7	Jul 1992–Dec 1994 daily	Juveniles	3,400	
Throw trap (1 m^2)	Great Bay–Little Egg Harbor	0.2–0.4	6	May–Sep 1988– May–Sep 1989/ biweekly	Juveniles	436	Sogard & Able 1991

Killitrap	RUMFS boat basin	2–3	1	Nov 1990–Dec 1994/ daily	Juveniles	6,221	
5.3-m otter trawl 6.0-mm mesh	Great Bay–Little Egg Harbor	0.6–5.0	21	Jun 1988–Nov 1990 monthly	Juveniles/adults	1,465	Szedlmayer & Able 1996
1-m beam trawl 3.0-mm mesh	Great Bay–Little Egg Harbor	0.6–5.0	36	May 1992–Oct 1995 monthly	Juveniles/adults	1,675	
Gear comparison (seine, beam-, otter trawls)	Great Bay–Little Egg Harbor	0.1–5.0	26	May 1991–Oct 1991 monthly	Juveniles/adults	507	
Weir, seine	Tidal creek, Great Bay	0.5–2.0	3	Apr 1987–Apr 1991	Juveniles/adults	192	Rountree et al. 1992
6.1-m seine 4-mm mesh	Great Bay/Little Egg Harbor	0.1–1.0	10 & 6	May 1990–Apr 1991 monthly Jun 1994–Oct 1995 monthly	Juveniles/adults	408 195	
Throw trap	Marsh surface, Great Bay	<1.0	30	May 1990–Jul 1991 (summer only)/ biweekly	Juveniles	840	K. J. Smith & K. W. Able in prep.
Pop net	RUMFS boat basin, Great Bay	2–3	1	Aug 1995–May 1996/weekly	Juveniles/adults	224	S. M. Hagan & K. W. Able in prep.

* Sources of sampling methodology and gear descriptions are included if they are available.

of 1,351 such sets were made between 1989 and 1994.

Many larvae pass through a pelagic-juvenile stage before descending to the bottom. Fish in these stages are tied strongly to surface layers, have fully formed fin rays, are capable swimmers, and are especially difficult to sample adequately, especially with gear focused on the water column. We have attempted to address this possible void of information by sampling with night lights. This year-round sampling effort (table 3.3) was undertaken from the RUMFS boat basin dock, in Great Bay, on an opportunistic basis, and resulted in the collection of late-stage larvae and pelagic juveniles, many of which have escaped collection with other sampling gears.

Few gears adequately sample recently settled fishes. Our efforts to sample these stages included beam trawls of two sizes and a variety of traps (table 3.3). We used a 2-m beam trawl at the LEO-15 study site on Beach Haven Ridge, and both 2-m and 1-m beam trawls at several estuarine sites. Both gears were deployed monthly over several years. A large number of samples of this difficult life-history stage were also obtained from the daily deployment of modified killitraps and experimental traps at fixed locations in Great Bay.

Juvenile fishes were sampled with a wide array of gears, in a variety of habitats, and over a long time period (table 3.3). In an attempt to focus on specific habitats within the estuary, and young-of-the-year use of these habitats, we used otter trawls and beam trawls on 13 repeatedly sampled stations described in chapter 2. These stations were chosen based on their depths, substrate type, amount of structured habitat, and proximity to the estuary mouth. Collections at these various stations are expressed on a catch-per-unit-of-effort basis and described in a histogram in pertinent species chapters.

Several sampling problems have compromised the ability of fisheries scientists to fully understand the first year in fishes' life histories. These include collecting-gear inefficiency, limited temporal or spatial sampling scale, and the difficulty in sampling recently settled, or just transforming, juveniles. In addressing these problems, our goal was to sample thoroughly during all months of the year with a variety of gears capable of producing a complete record of all size intervals during fishes' first year. Gear avoidance by larvae, especially larger individuals with well-developed sensory systems and swimming abilities, is especially problematic. This is evident in the disparity in sizes and catch rates between day and night collections with bongo nets (Morse 1989). By combining the results from the three ichthyoplankton samplers described above, and augmenting these with night-lighting collections, we were able to encompass larval stages at mean sizes ranging from 0.7 cm SL to 4.8 cm TL (fig. 3.1). The combination of beam trawls and traps yielded early settled individuals in the next increments, at mean sizes ranging from 4.8 to 5.9 cm TL. Finally, our juvenile fish samplers (seines and otter trawls) collected a wide array of sizes, with means increasing from 6.4 to 10.7 cm TL (fig. 3.1).

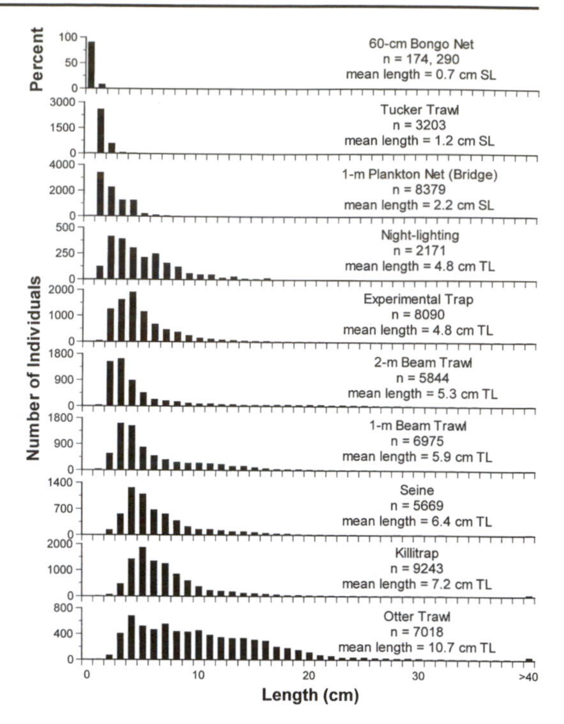

3.1 Composite length frequencies of young-of-the-year fishes collected by a variety of gears over the Middle Atlantic Bight continental shelf, on the inner continental shelf in the vicinity of Beach Haven Ridge, and in the Great Bay–Little Egg Harbor estuary. Number of specimens measured indicated by *n*. Y-axis indicates percent in the first panel, number of observations in the remaining panels.

The spatial and temporal sampling scales with the above-described gears were designed to sample all developmental stages across all available habitats during all time periods when they might have been present. Because spawning occurs during all months of the year (see chap. 4), and because we were attempting to establish when important ontogenetic events took place in the estuary, we endeavored to sample during as much of the year as possible. Our temporal coverage was most complete for ichthyoplankton over the shelf and in the estuary and for demersal juveniles in the estuary (fig. 3.2). Sampling for planktonic larvae and early-settlement-stage juveniles on the inner continental shelf also occurred year-round, but was more focused on summer and fall months when we detected most activity. Our spatial scale of sampling covered the entire continental shelf (for larvae) and as many discrete estuarine habitats as were available (for larvae and juveniles). If there remain discrete undersampled zones in the Middle Atlantic Bight where larvae or early juveniles might regularly occur, we suggest that these zones are associated with the immediate surf zone along the exposed coast, especially in those parts of this zone located between rock jetties extending offshore from the beach or mudflats at high tide, with a covering of inches of water. These zones are not accessible with boats, and the use of conventional sampling gear is not feasible.

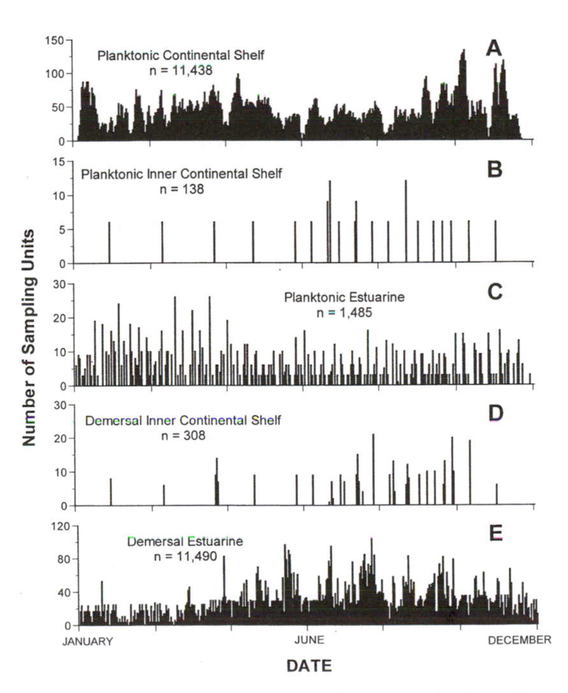

3.2 Temporal distribution of sampling events by date, life-history stage, and general location, for original data analyzed for the present study.

OUTLINE OF A TYPICAL SPECIES ACCOUNT

Selection of a species for a detailed account (chap. 5–76) is largely based on the amount and completeness of the information available for the first year of life. Generally, the more abundant species, which have been more frequently collected, are the species included. Economically important species have been included because they have been the focus of earlier research. As a result, most of the common species in the Middle Atlantic Bight have been incorporated into this synthesis. Species only rarely collected from estuaries in the central portion of the bight are listed, with details, in a separate table (see chap. 77). Our determination of young-of-the-year stages is based on sizes determined from the progression of length modes in monthly length-frequency histograms, a method that compares well with otolith analysis of age (Gauldie 1996).

Each species account begins with the scientific name, author of that name, and currently used common name. Scientific nomenclature, common names, and the sequence in which families are listed generally follow Robins et al. (1991). Departures from the latter classification will be found in the gadiforms, hemiramphids, fundulids, pleuronectiforms, and tetraodontiforms, where phylogenetic studies (based largely on ontogenetic evidence) support separating certain larger families into smaller, more distinct ones with common, derived characters (e.g., Parenti 1981; Markle 1982; Ahlstrom et al. 1984; Collette et al. 1984b; Dunn and Matarese 1984; Fahay and Markle 1984; Leis 1984; Chapleau and Keast 1988; Eschmeyer 1990) or reallocating certain species among genera (Cooper 1996). For each species account, we attempted to provide information in the following categories.

Distribution This section includes the known distribution for the species, in estuaries and elsewhere, based largely on the available literature. It also includes general habitat information, when it is available. For many species, seasonal movement and migrations may be included. Where discrete stocks or subpopulations are recognized, this is also indicated. Details of the distribution of early life-history stages in Middle Atlantic Bight estuaries are provided in tabular form (see table 4.2). Details of the distribution by estuary for some species were based on those included in the recent compilation by Stone et al. (1994). For species that we added, this information was based on individual publications for each estuary. Some groups that were lumped together in the Stone et al. (1994) treatment (Cyprinodontidae, Atherinidae, Mugilidae, Gobiidae) have been considered by individual species in our analysis. Where necessary and appropriate, we discuss the distributions of early life-history stages in more detail, and indicate where generalizations may lead to false conclusions regarding their occurrences.

Reproduction Included in the section on reproduction is the available information from the literature on the time and location of spawning, typically based on egg or larval distributions from plankton collections, either on the continental shelf or in estuaries. In some instances, we provide original (or previously published) data on the timing of gonad maturation and spawning.

Description In this section we provide a concise description of static values (e.g., meristic characters) for eggs and larvae with emphasis on diagnostic characters and reference, where available, to a source of more detailed information. Special attention is paid to transforming or recently settled juveniles because they are often poorly known. This includes, for most species, an original illustration with details of collection. All of these illustrated specimens have been catalogued in the Academy of Natural Sciences of Philadelphia (ANSP), and the account includes the museum accession number.

The First Year Here we focus on aspects of the early life history, that is, the first 12 months of life. In some instances we present data on length of older individuals because it helps to define the size limits of the first year. Much of the interpretation in this section is based on original data sources or reanalyzed historical data sets and the published literature. The first year includes treatment of egg and larval distribution and abundance, and, in a few instances, survival, metamorphosis and settlement, growth, habitat use, emigration, and overwinter mortality.

Egg and larval distributions include original data for the entire Middle Atlantic Bight continental shelf, the Great Bay–Little Egg Harbor estuary, and the inner continental shelf in the vicinity of Beach Haven Ridge. Larval distribution relative to size, season, and location is reported whenever possible. Larval duration is based on available information from otolith studies. Aspects of movements, particularly estuarine ingress and egress, are treated relative to size and season. The dynamic aspects of morphological metamorphosis (e.g., onset of scale formation, eye migration in flatfishes, etc.) and settlement are reported, largely relative to size and location. Evaluations of daily growth, when presented for the first time, are calculated from monthly changes in modal length frequencies and should be recognized as relatively crude estimates. Size at end of the first summer, end of the winter, and at 12 months of age are also reported, based on the same evidence. Length-frequency histograms are separated into "ocean" and "estuary," pending availability of data from both areas. When original collections were made at the Beach Haven Ridge (LEO-15 study site) they were included in the ocean data, although this site is an area consistently influenced or affected by waters from the estuary and is thus within the estuarine zone (*sensu* Smith 1966). Estuarine habitat information is typically descriptive, but occasionally habitat-specific growth, as a measure of habitat quality, is presented. Seasonal movements and other aspects of behavior are included wherever possible. If they are available, aspects of winter biology—that is, distribution, growth, and temperature-induced mortality—conclude the chapter.

CHAPTER 4

Characteristics of the Middle Atlantic Bight Ichthyofauna

The ichthyofauna of the Middle Atlantic Bight is temperate, but influenced by tropical and boreal species that make seasonal migrations into the bight. Many species are transients, and only a very few are restricted to the Middle Atlantic Bight (Parr 1933, Grosslein and Azarovitz 1982). Most of these endemics occur near the coastal zone. More typical of the bight is a group of about 60 migratory species, many of which are important both to commercial and recreational interests. This group is dominated by species that tend to occur near the coast and in northern parts of their range during summer and offshore and in the southern part of their range during winter. Some spend the colder months south of the Middle Atlantic Bight. The estuarine fauna has a large migratory component as well. Some of these migrations are between estuaries and the continental shelf, some occur seasonally across the shelf from estuaries to the continental shelf edge, and some involve migrations out of the Middle Atlantic Bight, either to the south or to the north and east. The life histories of fishes from the Middle Atlantic Bight are complex, in large part because of these migrations (Grosslein and Azarovitz 1982). Often, different migration patterns are exhibited by different-aged fishes, such that wintering areas for young-of-the-year will differ from those of older, adult fishes. The mobility of certain species is further complicated by a net movement with age within the range, with the result that older and younger fish will dominate in different parts of the range.

The ichthyofauna of the central part of the Middle Atlantic Bight comprises 336 marine and estuarine species (including 19 elasmobranchs) (Able 1992; Fahay 1993). Included in the estuaries of this region are species-rich families such as Clupeidae (herrings), Fundulidae and Cyprinodontidae (killifishes), Sciaenidae (drums), and Scophthalmidae and Paralichthyidae (left-eyed flatfishes). The ontogenetic composition of this ichthyofauna is also well described (Fahay 1993), such that we can begin to measure the importance of various habitats to early life-history stages.

Patterns of reproduction within the Middle Atlantic Bight are exemplified by the character of the ichthyoplankton (Morse et al. 1987; Smith and Morse 1988). Spawning over the continental shelf is at its lowest during winter, when few species spawn, although an explosion of the *Ammodytes dubius* population in the mid-1970s resulted in increased ichthyoplankton abundance during winter throughout the Middle Atlantic Bight (Sherman et al. 1984). Both egg abundance and diversity increase during the early spring, when species such as *Scomber scombrus, Limanda ferruginea,* and *Cynoscion regalis* begin to spawn, the former two in the northern part of the Middle Atlantic Bight, the latter in the southern part of the bight. Throughout the bight, abundance and diversity reach a peak during mid- to late-summer when many species reproduce. Included in this group are many covered in the present treatment of estuarine species (e.g., *Scophthalmus aquosus, Tautoga onitis, Prionotus carolinus,* and *Pomatomus saltatrix*). Spawning is curtailed during the fall, but *Paralichthys dentatus* eggs reach peak abundance in southern New England shelf waters, and *Brevoortia tyrannus* and *Prionotus* spp. egg concentrations occur in the southern part of the Middle Atlantic Bight (Smith and Morse 1988).

In continental shelf waters in the central part of the Middle Atlantic Bight, the taxonomic composition of dominant fish larvae changes seasonally. Larvae that transform into juveniles that use estuarine nurseries in this part of the bight are produced throughout the year, and many of these are among the most abundant taxa during certain months (table 4.1). Winter

Table 4.1. Species Composition of Most Abundant Larval Fishes Collected in Continental Shelf Waters in the Central Part of the Middle Atlantic Bight during NMFS-MARMAP Surveys, 1977–1987

Taxon	Jan	Feb	Mar	Apr	May	Jun	Jul	Aug	Sep	Oct	Nov	Dec
Ammodytes spp.	1	1	1	1	2	—	—	—	—	—	—	4
Gadus morhua	2	3	2	3	7	—	—	—	—	—	—	7
*Paralichthys dentatus**	3	2	9	—	—	—	—	—	—	5	2	1
*Brevoortia tyrannus**	4	—	—	—	—	—	—	—	—	—	5	5
Merluccius bilinearis	5	9	—	—	—	7	9	9	—	6	4	2
Maurolicus muelleri	6	—	—	—	—	—	—	—	—	—	—	8
*Leiostomus xanthurus**	7	—	—	—	—	—	—	—	—	—	—	—
*Pollachius virens**	8	4	7	—	—	—	—	—	—	—	—	—
Gobiidae	9	—	—	—	—	—	—	—	—	—	8	—
*Clupea harengus**	10	—	—	—	—	—	—	—	—	—	—	—
*Micropogonias undulatus**	11	—	—	—	—	—	—	—	—	—	—	—
*Pholis gunnellus**	—	5	4	11	—	—	—	—	—	—	—	—
Myoxocephalus octodecemspinosus	—	6	3	10	—	—	—	—	—	—	—	—
Paralepididae	—	7	—	—	—	—	—	—	—	—	—	—
*Anguilla rostrata**	—	8	—	—	—	—	—	—	—	—	—	—
Notolepis rissoi	—	10	10	—	—	—	—	—	—	—	—	—
*Pseudopleuronectes americanus**	—	—	5	5	—	—	—	—	—	—	—	—
*Myoxocephalus aenaeus**	—	—	6	9	—	—	—	—	—	—	—	—
Cottidae	—	—	8	7	—	—	—	—	—	—	—	—
Benthosema glaciale	—	—	—	2	9	—	—	—	—	—	—	—
Limanda ferruginea	—	—	—	4	1	1	10	—	—	—	—	—
Liparis spp.	—	—	—	6	5	—	—	—	—	—	—	—
Melanogrammus aeglefinus	—	—	—	8	10	—	—	—	—	—	—	—
Scomber scombrus	—	—	—	—	3	3	—	—	—	—	—	—
Enchelyopus cimbrius	—	—	—	—	4	2	—	—	—	—	10	—
*Scophthalmus aquosus**	—	—	—	—	6	6	11	—	—	3	3	3
Glyptocephalus cynoglossus	—	—	—	—	8	5	—	—	—	—	—	—
Lophius americanus	—	—	—	—	—	4	7	—	—	—	—	—
*Tautogolabrus adspersus**	—	—	—	—	—	8	3	7	—	—	—	—
Hippoglossina oblonga	—	—	—	—	—	9	1	2	4	8	—	—
*Urophycis chuss**	—	—	—	—	—	10	4	—	—	—	—	—
*Peprilus triacanthus**	—	—	—	—	—	—	2	1	8	—	—	—
*Pomatomus saltatrix**	—	—	—	—	—	—	5	6	—	—	—	—
Engraulidae	—	—	—	—	—	—	6	8	10	—	—	—
Citharichthys arctifrons	—	—	—	—	—	—	8	4	2	2	6	—
Urophycis spp.	—	—	—	—	—	—	—	3	1	—	—	—
*Etropus microstomus**	—	—	—	—	—	—	—	5	3	7	—	—
*Prionotus carolinus**	—	—	—	—	—	—	—	10	5	9	—	—

Table 4.1. *Continued*

Taxon	Jan	Feb	Mar	Apr	May	Jun	Jul	Aug	Sep	Oct	Nov	Dec
*Ophidion marginatum**	—	—	—	—	—	—	—	—	6	—	—	—
Lepophidium profundorum	—	—	—	—	—	—	—	—	7	—	9	—
*Centropristis striata**	—	—	—	—	—	—	—	—	9	—	—	—
Ophidiidae	—	—	—	—	—	—	—	—	—	4	—	—
Bothus spp.	—	—	—	—	—	—	—	—	—	10	11	—
*Urophycis regia**	—	—	—	—	—	—	—	—	—	1	1	6
Ceratoscopelus maderensis	—	—	—	—	—	—	—	—	—	—	7	—
Diaphus spp.	—	—	—	—	—	—	—	—	—	—	—	9

NOTE: Larvae are ranked (top 10–11 per month) according to numbers collected per 10 m^2 of sea surface. Dashes indicate few or no collections of that species in that month.
*Taxa pertinent to species treated herein.

larvae are dominated by *Ammodytes* sp., *Gadus morhua, Pholis gunnellus,* and *Paralichthys dentatus.* During the spring, *Ammodytes* sp. larvae remain abundant but are joined by larvae of *Limanda ferruginea* and *Scomber scombrus.* During the summer months, several taxa become abundant, including *Enchelyopus cimbrius, Hippoglossina oblonga, Peprilus triacanthus, Tautogolabrus adspersus,* and a mixture of *Urophycis* species. *Urophycis regia* larvae are abundant during the fall, along with larvae of *Citharichthys arctifrons, Etropus microstomus,* and *Scophthalmus aquosus.* Many juveniles occurring in Middle Atlantic Bight estuaries are the result of spawning that does not occur in the bight. One well-studied example of this is the spring cohort of bluefish (*Pomatomus saltatrix*) that is spawned south of Cape Hatteras but uses Middle Atlantic Bight estuaries as nurseries (Kendall and Walford 1979; McBride and Conover 1991; Hare and Cowen 1993). The inclusion of Georges Bank in our area of interest allows comparisons between species that use estuaries in the Middle Atlantic Bight and those same species that reproduce on Georges Bank but do not have access to estuaries during their life cycle (e.g., *Pseudopleuronectes americanus, Scophthalmus aquosus*). Two species of eels (*Anguilla rostrata* and *Conger oceanicus*) spawn near the Bahamas in the Sargasso Sea, a tropical region of the North Atlantic Ocean, separated from the Middle Atlantic Bight by the Gulf Stream and Slope Sea, yet their late larval and juvenile stages are common constituents of estuaries in this bight.

Patterns of estuarine use by early life-history stages of fishes (based on published records) vary spatially in the Middle Atlantic Bight. We have plotted these reported occurrences within 16 estuaries to demonstrate some of this variation (table 4.2). We caution, however, that these occurrences are not quantitative values and, in some cases, a positive occurrence may be based on the collection of a single specimen of a certain life-history stage. We have endeavored to include evaluations of these occurrences in the appropriate species accounts to correct misleading generalizations. Some obvious trends in these distributions are due solely to latitudinal differences in the fauna. Some species reach the southern limit of their range in the Middle Atlantic Bight, extending as far as the vicinity of Chesapeake Bay (*Clupea harengus, Pollachius virens, Stenotomus chrysops,* and *Tautogolabrus adspersus*). Early stages of other species with boreal affinities are found only in the northernmost estuaries (*Osmerus mordax, Microgadus tomcod, Pholis gunnellus,* and *Myoxocephalus aenaeus*). The former two can be found as far south as the Hudson–Raritan estuary, whereas *P. gunnellus* and *M. aenaeus* extend their ranges to Great Bay and Delaware inland bays, respectively. Conversely, those with more southern affinities reach the northern limit of their range in estuaries near Cape Cod

Table 4.2. Distribution of Early Life-History Stages of Fishes by Representative Estuaries in the Middle Atlantic Bight

Species	Pam Alb NC	Ches Bay MD	East Shore VA/MD	Inlnd Bays DE	Del Bay DE/NJ	South Inlnd NJ	Great Bay NJ	Barn Bay NJ	Hudsn Raritn NJ	Great South NY	Gard Bay NY	L.I. Sound NY	Conn River CT	Narr Bay RI	Buzz Bay MA	Nauset Marsh MA
Mustelus canis			J				J		J		J				J	
Anguilla rostrata	J	J	J	J	J	J	J	J	J	J	J	J	J	J	J	J
Conger oceanicus	J		J	J	J	J	J		J	J		J				J
Alosa aestivalis	J	ELJ	J	ELJ	ELJ	J		ELJ	ELJ	J	J	LJ	ELJ	ELJ	ELJ	
A. mediocris		ELJ	J		J				J						J	
A. pseudoharengus	J	ELJ	J	ELJ	ELJ	ELJ		ELJ	ELJ	J	J	ELJ	ELJ	ELJ	ELJ	J
A. sapidissima	J	ELJ	J	ELJ	ELJ	J		J	ELJ	J	J	ELJ	ELJ	J	J	
Brevoortia tyrannus	LJ	ELJ	LJ	LJ	ELJ	ELJ	ELJ	ELJ	ELJ	ELJ	ELJ	ELJ	LJ	ELJ	ELJ	LJ
Clupea harengus		J	J	LJ	LJ	LJ	LJ	LJ	LJ	J	J	LJ	J	LJ	LJ	LJ
Anchoa hepsetus	ELJ	LJ	J	EJ	J	ELJ	ELJ		E	EJ				J		
A. mitchilli	LJ	ELJ	ELJ	ELJ	ELJ	ELJ	ELJ	ELJ	ELJ	ELJ	ELJ	ELJ	ELJ	ELJ	ELJ	J
Osmerus mordax									ELJ		J	ELJ	ELJ	ELJ	ELJ	
Synodus foetens	J	J	J	J	J	J	LJ	J	LJ			LJ				
Microgadus tomcod			J						ELJ	ELJ	ELJ	ELJ	ELJ	ELJ	ELJ	J
Pollachius virens		J	J	J	J	J	J	J	J	J	LJ	LJ	LJ	LJ	ELJ	J
Urophycis chuss		J	J	J	J	J	EJ	J	LJ	EJ	LJ	J	J	ELJ	ELJ	J
U. regia	J	J	J	J	LJ	J	J	J	J	J		J			J	
U. tenuis						J	J			J					J	J
Ophidion marginatum		J	LJ	LJ	LJ	LJ	J		L							
Opsanus tau	LJ	ELJ	ELJ	ELJ	ELJ	ELJ	ELJ	ELJ	ELJ	ELJ	ELJ	ELJ	ELJ	ELJ	ELJ	

Strongylura marina	J	LJ	J	J	J	J	LJ	J	J	LJ						
Cyprinodon variegatus	J	ELJ	ELJ	ELJ	ELJ	ELJ	ELJ	ELJ	ELJ	ELJ	ELJ	ELJ	ELJ	ELJ	ELJ	
Fundulus heteroclitus	J	ELJ	ELJ	ELJ	ELJ	ELJ	ELJ	ELJ	ELJ	ELJ	ELJ	ELJ	J	ELJ	ELJ	ELJ
F. luciae	ELJ	ELJ	ELJ	ELJ	ELJ	ELJ	ELJ	ELJ								
F. majalis	LJ	ELJ	ELJ	ELJ	ELJ	ELJ	ELJ	ELJ		J	ELJ		J		ELJ	ELJ
Lucania parva	J	ELJ	ELJ	ELJ	ELJ	ELJ	ELJ	ELJ	ELJ	J					ELJ	
Membras martinica	J	ELJ	LJ	ELJ	ELJ	ELJ	ELJ	J	ELJ							
Menidia beryllina	J	LJ	ELJ	ELJ	ELJ	ELJ	ELJ	ELJ	ELJ	J		J			ELJ	J
M. menidia	LJ	LJ	ELJ	ELJ	ELJ	ELJ	ELJ	ELJ	ELJ	ELJ	ELJ	ELJ	LJ	L	ELJ	LJ
Apeltes quadracus		ELJ	J	ELJ	ELJ	ELJ	ELJ	ELJ	ELJ	LJ	ELJ	ELJ	J		ELJ	ELJ
Gasterosteus aculeatus		ELJ		ELJ	ELJ	ELJ	ELJ	ELJ		LJ		LJ		L	ELJ	LJ
Hippocampus erectus	J	J	J	LJ		J	LJ	LJ		J		J				
Syngnathus fuscus	J	LJ	LJ	LJ	LJ	LJ	LJ	LJ	LJ	LJ	LJ	LJ	LJ	LJ	LJ	LJ
Prionotus evolans	J	J	J	J	J	ELJ	LJ	J	J			EL		E	J	
P. carolinus	J	J	ELJ	ELJ	ELJ	ELJ	LJ	J	ELJ	ELJ	ELJ	ELJ	ELJ	ELJ	ELJ	
Myoxocephalus aenaeus				J	ELJ	J	LJ	ELJ	ELJ	L	ELJ	ELJ			ELJ	ELJ
Morone americana	J	ELJ	ELJ	J	ELJ	ELJ	L	ELJ	ELJ	J	ELJ	ELJ	ELJ	ELJ	ELJ	
M. saxatilis		ELJ	J	J	ELJ	J	J	J	ELJ	J	J	J	J	J	J	
Centropristis striata	LJ	J	J	J	J	LJ	LJ	LJ	J		LJ	LJ		LJ	ELJ	
Pomatomus saltatrix	J	J	LJ	J	LJ	J	LJ	LJ	ELJ	J	ELJ	J	J	LJ	LJ	
Caranx hippos	J	J		J	J	J	J	J	J		J	J			J	
Lutjanus griseus	J			J	J	J	J		J	J					J	
Stenotomus chrysops		J	J	J	J	J	LJ	J	ELJ	J	ELJ	ELJ	ELJ	ELJ	ELJ	

Continued

Table 4.2. *Continued*

Species	Pam Alb NC	Ches Bay MD	East Shore VA/MD	Inlnd Bays DE	Del Bay DE/NJ	South Inlnd NJ	Great Bay NJ	Barn Bay NJ	Hudsn Raritn NJ	Great South NY	Gard Bay NY	L.I. Sound NY	Conn River CT	Narr Bay RI	Buzz Bay MA	Nauset Marsh MA
Bairdiella chrysoura	LJ	ELJ	EJ	LJ	J	ELJ	LJ	J	LJ	LJ		LJ				
Cynoscion regalis	LJ	ELJ	ELJ	ELJ	ELJ	ELJ	LJ	ELJ	ELJ	ELJ	ELJ	ELJ	J	ELJ	ELJ	
Leiostomus xanthurus	LJ	LJ	LJ	ELJ	LJ	J	LJ	LJ	LJ	J	J	J	J	ELJ	J	
Menticirrhus saxatilis	LJ	J	J	LJ	ELJ	LJ	J	ELJ	ELJ	ELJ	ELJ		ELJ	ELJ		
Micropogonias undulatus	LJ	LJ	LJ	ELJ	LJ	LJ	LJ	LJ	LJ		J					
Pogonias cromis		ELJ	EJ	LJ	ELJ	J		J	J					J		
Chaetodon ocellatus		J	J	J		J	J		J						J	
Mugil cephalus	J	J	J	J	J	J	J	J	J	J				J	J	
M. curema	J	J	J	J	J	J	J	J	J	J	J	J			J	J
Sphyraena borealis		J	J	J	J	J	J		J	J	J	J			J	
Tautoga onitis		ELJ	J	ELJ	ELJ	ELJ	ELJ	ELJ	ELJ	ELJ	ELJ	ELJ	ELJ	ELJ	ELJ	J
Tautogolabrus adspersus			J		ELJ	ELJ	ELJ	ELJ	ELJ	ELJ	ELJ	ELJ	ELJ	ELJ	ELJ	LJ
Pholis gunnellus							LJ				E	EJ		L	J	LJ
Astroscopus guttatus		LJ		J	J	J	LJ	J	J							
Hypsoblennius hentz	LJ	LJ	LJ	LJ	L		LJ	ELJ								
Ammodytes americanus		L	LJ	ELJ	ELJ	ELJ	ELJ	ELJ	ELJ	ELJ	ELJ	ELJ	ELJ	ELJ	ELJ	LJ
Gobionellus boleosoma	LJ			L			LJ									
Gobiosoma bosc	LJ	ELJ	ELJ	ELJ	ELJ	ELJ	ELJ	ELJ	J			J			J	
G. ginsburgi	L	LJ	LJ	LJ	LJ	ELJ	LJ		J		L	J		L		
Peprilus triacanthus	LJ	ELJ	J	J	ELJ	LJ	LJ	LJ	ELJ	ELJ	ELJ	ELJ	J	ELJ	ELJ	J

Scophthalmus aquosus	LJ	ELJ	LJ	ELJ	J	ELJ	ELJ	ELJ	ELJ	ELJ	ELJ	ELJ	ELJ	ELJ	ELJ	J
Etropus microstomus	L		J	J	LJ	ELJ	LJ	J	J	E		J				J
Paralichthys dentatus	LJ	LJ	J	LJ	LJ	LJ	LJ	LJ	ELJ	J	J	LJ	J	ELJ	ELJ	J
Pseudopleuronectes americanus		ELJ	LJ	ELJ	ELJ	ELJ	ELJ	ELJ	ELJ	ELJ	ELJ	ELJ	ELJ	ELJ	ELJ	LJ
Trinectes maculatus	LJ	ELJ	ELJ	ELJ	ELJ	ELJ	ELJ	ELJ	ELJ	ELJ	ELJ	ELJ	ELJ	ELJ	ELJ	
Sphoeroides maculatus	LJ	LJ	J	LJ	J	LJ	LJ	ELJ	J	LJ	J	LJ		L	LJ	

SOURCES: Pamlico-Albemarle Sounds, North Carolina—45–47; Chesapeake Bay and tributaries, Maryland—1, 20, 23, 50–59; Eastern shore, Virginia and Maryland—1, 12, 18, 20, 34, 36, 53–59; Delaware inland bays, Delaware—1, 8, 20, 21, 22, 34, 38; Delaware Bay, Delaware and New Jersey—1, 8, 10, 15, 20, 21; Southern inland bays, New Jersey—1, 4, 7, 10, 20, 33; Great Bay, New Jersey—20, 26, 27, 30, 35, 48; Barnegat Bay, New Jersey—1, 3; Hudson–Raritan–Sandy Hook Bays, New Jersey—1, 6, 14, 16, 17, 19, 20, 24, 60; Great South Bay, New York—1, 20, 21, 28, 39; Gardiners Bay, New York—1, 9, 20, 21, 24; Long Island Sound, New York—1, 11, 13, 21, 25, 32; Connecticut River, Connecticut—1, 42, 43; Narragansett Bay, Rhode Island—1, 11, 20, 29, 31, 40; Buzzards Bay, Massachusetts—1, 5, 20, 24, 37, 41; Nauset Marsh, Massachusetts—44.

Table sources listed by number: 1. Stone et al. 1994; 2. Musick 1972; 3. Tatham et al. 1984; 4. Allen et al. 1978; 5. Lux and Nichy 1971; 6. Smith 1985; 7. Bean 1887; 8. Wang and Kernehan 1979 (includes de Sylva et al. 1962); 9. Hickey et al. 1975; 10. Himchak 1982b; 11. Roberts 1978; 12. Schwartz 1961; 13. Richards 1959; 14. Kahnle and Hattala 1988; 15. Weisberg and Burton 1993; 16. Wilk and Silverman 1976b; 17. Berg and Levinton 1985; 18. Cowan and Birdsong 1985; 19. Dovel 1981; 20. Pacheco 1973; 21. Perlmutter 1939; 22. Pacheco and Grant 1973; 23. Olney and Boehlert 1988; 24. Nichols and Breder 1927; 25. Wheatland 1956; 26. Rountree and Able 1992a; 27. Szedlmayer and Able 1996; 28. Monteleone 1992; 29. Herman 1963; 30. Witting 1995; D. A. Witting, K. W. Able, and M. P. Fahay in prep.; 31. Bourne and Govoni 1988; 32. Pearcy and Richards 1962; 33. McDermott 1971; 34. Schwartz 1964a; 35. Swiecicki and Tatham 1977; 36. Richards and Castagna 1970; 37. Sumner et al. 1913; 38. Derickson and Price 1973; 39. Briggs and O'Connor 1971; 40. Oviatt and Nixon 1973; 41. Lux and Wheeler 1992; 42. Marcy 1976a; 43. Marcy 1976b; 44. Able and Fahay unpubl. observ.; 45. Tagatz and Dudley 1961; 46. Ross and Epperly 1985; 47. Hettler and Barker 1993; 48. Fahay and Able 1989; 49. Barans 1972; 50. Virginia Institute of Marine Sciences 1962; 51. Orth and Heck 1980; 52. Hildebrand and Schroeder 1928; 53. Jones et al. 1978; 54. Hardy 1978a; 55. Hardy 1978b; 56. Johnson 1978; 57. Fritzsche 1978; 58. Martin and Drewry 1978; 59. Joseph et al. 1964; 60. Croker 1965.

NOTE: See figures 2.1, 2.2 and 2.3 for location of estuaries and table 2.3 for physical characteristics of these estuaries. Occurrence indications are based on published or unpublished sources and are not quantitative; thus the collection of a single specimen results in a positive indication. See individual species chapters for evaluations of certain of these generalizations. *Gambusia holbrooki* not included because of frequent introduction of congeners.

*E = eggs; L = larvae; J = juveniles.

(*Urophycis regia, Opsanus tau, Lucania parva, Cynoscion regalis, Leiostomus xanthurus, Menticirrhus saxatilis, Chaetodon ocellatus, Mugil cephalus, Sphyraena borealis, Trinectes maculatus,* and *Sphoeroides maculatus*). Other species are found only in estuaries in the southern part of the bight (*Fundulus luciae, Membras martinica, Micropogonias undulatus, Astroscopus guttatus, Hippocampus erectus,* and *Hypsoblennius hentz*). In certain species, eggs, larvae, and juveniles are found in northern estuaries, whereas only juveniles are found in southern estuaries (*Pollachius virens, Urophycis chuss,* and *Stenotomus chrysops*). The opposite trend is found in *Bairdiella chrysoura, Pogonias cromis,* and, perhaps, *Gobiosoma bosc.*

Several species occur in Middle Atlantic Bight estuaries as juveniles only (*Anguilla rostrata, Conger oceanicus, Caranx hippos, Lutjanus griseus, Chaetodon ocellatus, Mugil cephalus,* and *M. curema*), and these are all products of spawning in the South Atlantic Bight or oceanic waters east of the Bahamas (Sargasso Sea). Other species spawn in Slope Sea waters beyond the continental shelf, then appear in estuaries as juveniles (*Urophycis regia* and *U. tenuis*). More subtle distributional trends include those of *Peprilus triacanthus,* in which all early stages are found in more northern estuaries, but are restricted to the larger and deeper Delaware and Chesapeake Bays in the south. The eggs of *Paralichthys dentatus* have only been reported from a few deep, northern estuaries, although larvae and juveniles occur (uncommonly) throughout the bight. In some species (e.g., *Ophidion marginatum*), larvae and juveniles are simply found in estuaries in the central part of the bight adjacent to continental shelf areas where they spawn. For other species, there are curious omissions, such as the lack of reports of *Sphoeroides maculatus* eggs, although their larvae and juveniles occur throughout the bight. We conclude that our understanding of some of this information is influenced somewhat by sampling deficiencies (true also of other species such as *Mustelus canis* and *Gobionellus boleosoma*). Other apparent differences in distribution may be due to differences in the characteristics of individual estuaries (table 2.3). For example, eggs and larvae of *Alosa* spp. are absent in systems lacking a freshwater tidal component (e.g., eastern shore of Virginia, Gardiners Bay, Great South Bay). However, despite different estuarine characteristics, all three early stages of several species are ubiquitous in their occurrences, perhaps attesting to their adapability to a wide range of conditions (*Brevoortia tyrannus, Anchoa mitchilli, Opsanus tau, Fundulus heteroclitus, Menidia menidia, Pseudopleuronectes americanus,* and *Trinectes maculatus*).

The Great Bay–Little Egg Harbor estuary has been the site of extensive sampling for early stages of fishes during the last decade, and this has prompted several analyses including the present one. Several descriptions of specific aspects of the fish fauna within this system have been published, most focusing on young-of-the-year. These include fishes in vegetated habitats (Sogard and Able 1991, 1994), marsh creeks (Rountree and Able 1992a, 1993), marsh pools (Smith 1995), inlet beaches (Haroski and Able unpubl. observ.), polyhaline shores (Able et al. 1996), seasonal and habitat patterns of use (Szedlmayer and Able 1996); ichthyoplankton and pelagic juveniles collected near Little Egg Inlet (Witting 1995; D. A. Witting, K. W. Able, and M. P. Fahay, in prep.); and larval and juvenile fishes of the adjacent LEO-15 study site on the Beach Haven Ridge (K. W. Able and M. P. Fahay unpubl. observ.). The numerically dominant species are members of the families Engraulidae, Clupeidae, Fundulidae, Gobiidae, Syngnathidae, Atherinidae, and Sciaenidae. Seasonal and annual variation in abundance of many species is pronounced. An analysis of a long-term data set suggests a stable ichthyoplankton assemblage with marked seasonal components that occur in a regular progression (Witting 1995; D. A. Witting, K. W. Able, and M. P. Fahay, in prep.). Important annual variations in abundance, measured in regular and repeated sampling programs, can be influenced by long-term trends in abundance, as for recovering stocks of *Clupea harengus.* Variations in delivery systems to this estuary may be influenced by large-scale hydrographic perturbations, such as locally produced upwelling along the coast or features associated with the Slope Sea or Gulf Stream near the edge of the continental shelf.

Several rare species occur irregularly in the Great Bay–Little Egg Harbor system in young-of-the-year stages, and some of these may be more common in estuaries of the southern or northern parts of the Middle Atlantic Bight. Their morphological and ecological characteristics are important considerations for our synthesis of estuarine use by fishes, but because data on them are sparse, we have simply summarized details of their collection in chapter 77 (table 77.3).

CHAPTER 5

Mustelus canis (Mitchill) Smooth Dogfish

DISTRIBUTION

The widely distributed shark, *Mustelus canis,* has been reported from the Bay of Fundy, Canada, to Uruguay (Castro 1983). It is one of the most abundant inshore sharks in the Middle Atlantic Bight (Bigelow and Schroeder 1953; Castro 1983), where it has been reported from estuaries, bays, and inner continental shelf waters to waters as deep as the outer continental shelf. Records of early life-history stages for the Middle Atlantic Bight are scarce (table 4.2).

REPRODUCTION

Adults mature at about 85 cm, and gestation lasts about 10 months (Castro 1983). The adults migrate into the central Middle Atlantic Bight from more southern areas in the spring to bear young with litters of 18 to 20 pups. The presence of small pups with placental scars indicates that birth occurs in shallow inshore waters in New Jersey from May through early July (Rountree and Able 1996).

DESCRIPTION

Development is viviparous, and the young—which closely resemble adults—are born live at approximately 28 to 39 cm TL (Rountree and Able 1996). This species can be distinguished from other sharks in the Middle Atlantic Bight by its slender shape and the two large dorsal fins without spines and low, flat, pavementlike teeth.

THE FIRST YEAR

The newly born pups are most abundant in New Jersey estuaries from mid-May through early July, and the young-of-the-year occur there through October (Rountree and Able 1996). Based on modal length-frequency progression, they grow an average of 1.9 mm per day (range, 1.5–2.1 mm) and reach 55–70 cm by the end of October (fig. 5.1). Similar growth rates were estimated from tooth width and tooth replacement data (Moss 1972).

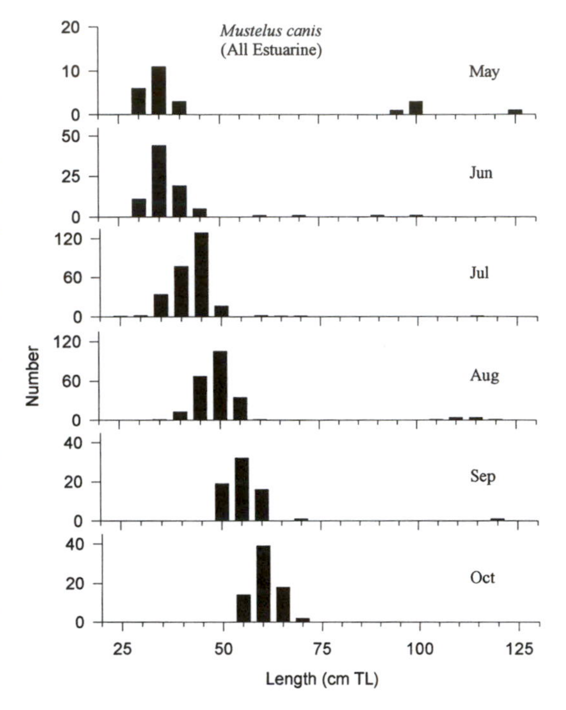

5.1 Monthly length frequencies of young-of-the-year *Mustelus canis* in the Great Bay–Little Egg Harbor estuary. Source: Rountree and Able (1996).

The young-of-the-year have been collected in a variety of habitats (Rountree and Able 1992a, 1996). The smallest young-of-the-year are found in subtidal, polyhaline marsh creeks, especially those with shoals at the mouth, in the spring and early summer. The larger young-of-the-year are in deeper waters of bays later in the summer. They are most frequently collected in the creeks on flood tides, especially at night (Rountree and Able 1993). Increased activity at night has also been observed in the laboratory (Casterlin and Reynolds 1979). The young-of-the-year do not aggregate by sex or exhibit different emigration patterns between sexes. Once they leave the estuaries and bays in the fall, they move south for the winter (Able and Rountree unpubl. data). In two instances, two young-of-the-year tagged in New Jersey were collected in the winter off North Carolina (Rountree and Able 1996).

CHAPTER 6

Anguilla rostrata (Lesueur) American Eel

6.1 *Anguilla rostrata* elver, 57.5 mm TL. Collected April 25, 1991, marsh pool, dip net, RUMFS, Great Bay, New Jersey. ANSP 175207. Illustrated by Susan Kaiser.

DISTRIBUTION

The range of *Anguilla rostrata* is the western North Atlantic Ocean between 5 N and 62 N (Bertin 1956). Its distribution includes coastal areas from Greenland to the northern coast of South America, rarely to Brazil. It is also found in the West Indies and Bermuda. *Anguilla rostrata* is abundant and common along the Atlantic coast of the United States, where it is found in estuaries, coastal streams, rivers, and landlocked lakes. It is a commercially important species (as adults, large juveniles, and elvers) in the Delaware Estuary and Raritan Bay (MacKenzie 1992), accounting for a valuable portion of the fisheries harvest. It is also extremely abundant in the lower Hudson River (Smith 1985). Adult females are found principally in freshwater, while males are found almost exclusively in salty or brackish waters. There is a trend for occupation of deeper waters during colder months in both sexes. Juveniles (elvers) have been reported from every estuary we have surveyed in the Middle Atlantic Bight (table 4.2).

REPRODUCTION

After a "yellow eel" stage lasting 9 to 19 years for females and 7 to 12 years for males, eels become mature, and both sexes undergo a fall migration downstream and eventually enter the sea (see table 6.1 for definitions of the several life-history stages of *A. rostrata*). This spawning migration is accompanied by a metamorphosis to a "silver eel" stage during which the eye enlarges, the internal organs atrophy, and the pectoral fin changes its shape. Eels have been recorded leaving Chesapeake Bay in November (Wenner and Musick 1974), but observations of adults at sea have been extremely rare (e.g., Wenner 1973; Robins et al. 1979). It is estimated that migration to the spawning area takes about 2 to 3 months (Eales 1968). Spawning occurs in the western tropical Atlantic between Bermuda and the Bahama Islands (the Sargasso Sea) during spring. Evidence of spawning is based primarily on the presence of

Table 6.1. Unique Terminology Pertaining to Life-History Stages of *Anguilla rostrata*

Term	Definition
Silver eel	Oceanic, spawning adult migrating to spawning area (also found in estuaries immediately before egress)
Leptocephalus	Oceanic, larval stage with leaflike body, small head
Glass eel	Oceanic/estuarine, transforming, unpigmented early juvenile
Elver	Estuarine, transformed early juvenile with pigmented body
Whip	Estuarine/freshwater, juvenile up to age 2
Yellow eel	Freshwater, female stage lasting 9–19 years; male 7–12 years
Green eel	Freshwater, summer stage prior to migration to ocean
Black eel	Freshwater, fall stage migrating to ocean

very small (e.g., <10 mm) larvae in the area. Smith (1968) presents more detailed life-history evidence.

DESCRIPTION

Ovarian eggs are slightly elliptical, and range from 0.59 to 1.25 mm in diameter (Edel 1975). Ripe or fertilized eggs are undescribed. Leptocephalus larvae have a laterally flattened, leaf-like body with a small head and prominent teeth. Pigment is lacking, unlike all other leptocephali likely to be collected in the Middle Atlantic Bight. Individual chevron-shaped myomeres are clearly visible and the gut extends about 75% of the body length. Further details of early development are summarized by Smith (1989b). Newly transformed young-of-the-year (or elvers) of *Anguilla rostrata* (Fig. 6.1) are easily recognized by their elongate shape, lack of pelvic fins, and long-based dorsal and anal fins merging with the caudal fin. Scales are absent until juveniles reach a length of about 16 cm. They differ from recently transformed *Conger oceanicus* in that the lower jaw extends farther anteriorly than the snout and the position of dorsal fin origin is different (see Appendix A). In both species, the unpigmented, oceanic stage acquires dark brown coloration after entering estuarine habitats. Vert: 102–111; D: mean 231; A: mean 199; Pect: 14–20; Plv: none.

THE FIRST YEAR

After hatching, developing larvae (leptocephali) drift at sea for up to a year during which they are transported away from the spawning area and north by the Gulf Stream (Power and McCleave 1983). The leptocephalus is characterized by specialized teeth, poorly developed internal organs, and a laterally compressed transparent body composed of a gelatinous central core (Smith 1984). Leptocephali transform into glass eels as they cross the Gulf Stream edge and approach the North American coast (Smith 1968). During this process, the specialized larval teeth are lost, and the gelatinous central matrix breaks down, resulting in a shorter body length and a rounded cross-section (Pfeiler 1986; for feeding, metabolism, and physiology of leptocephali, also see Pfeiler and Govoni 1993). During the 11-year

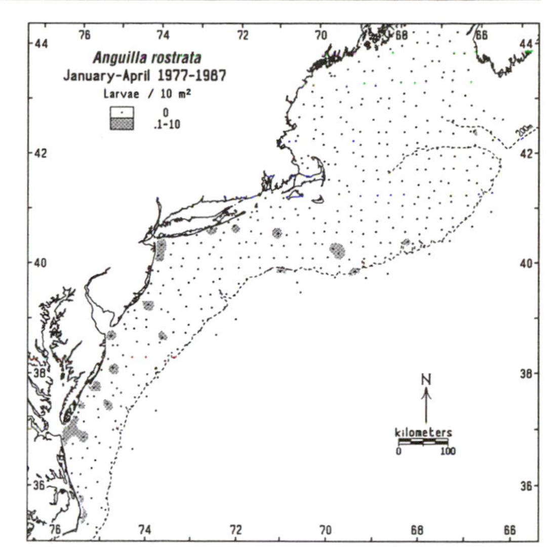

6.2 Locations of *Anguilla rostrata* glass eel collections between 1977 and 1987 during MARMAP study. Despite density label of .1–10 in legend, the average density of glass eels during the study was two per 10 m² of sea surface in shaded areas.

MARMAP study, only 34 glass eels were collected throughout the Middle Atlantic Bight (fig. 6.2). Most were collected near the coast. These specimens were most commonly found in February and March (fig. 6.3), but a few were also collected in the summer and fall months. Glass eels begin to enter Delaware Bay and other New Jersey tidal marshes, bays and coastal streams as early as November (Vladykov 1966), but in the Great Bay–Little Egg Harbor area, this ingress closely coincided with the largest collections on the continental shelf (fig. 6.3). During 6 years of ichthyoplankton sampling in this study area, this ingress peaked in February in 5 years and was delayed until March in 1 year (fig. 6.4). Abundance was similar across most years. As glass eels enter these brackish water habitats, they undergo another transformation in which they acquire dark brown, cutaneous pigmentation and begin the elver stage (fig. 6.1). Elvers have been collected from temperatures as low as −0.8 C (Jeffries 1960.) There is evidence in Japanese elvers of reverse thermotaxis, whereby they are attracted to lower temperatures when ambient temperatures are low and to higher temperatures when ambient temperatures are high (Hiyama 1953.) If this is also true for *A. rostrata* elvers, it may

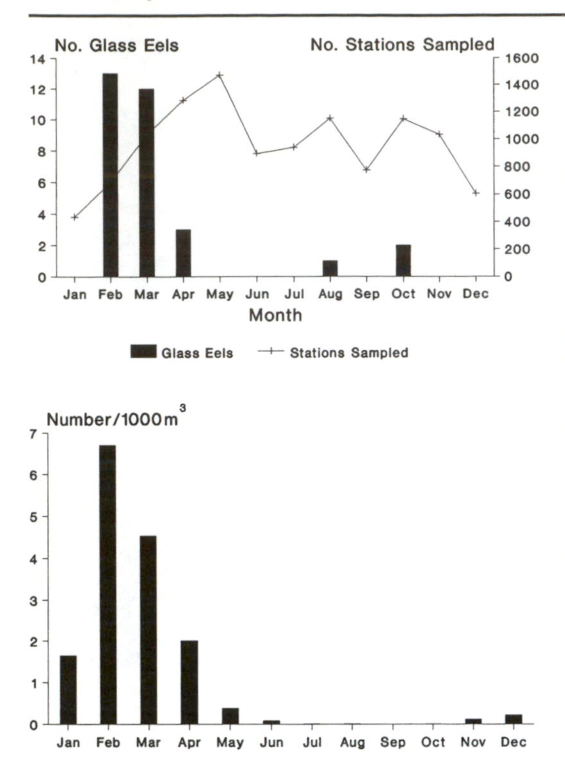

6.3 Monthly occurrences of *Anguilla rostrata* glass eels (upper) during MARMAP study over the Middle Atlantic Bight continental shelf, 1977–1987, and (lower) collected in 1-m, 1-mm plankton net in Little Sheepshead Creek Bridge study, Great Bay, New Jersey, 1989–1994 (n = 2,238).

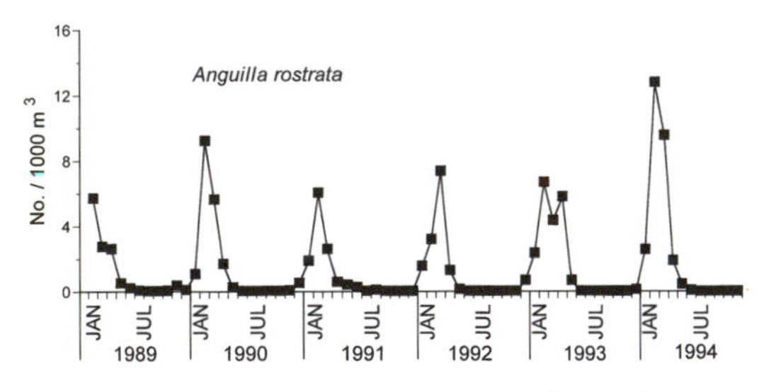

6.4 Annual pattern of *Anguilla rostrata* glass eels entering estuarine portion of Great Bay study area during Little Sheepshead Creek Bridge 1-m plankton net collections, 1989–1994 (n = 2,238).

explain the occurrence of winter forerunners, that is, those that enter estuarine waters colder than the ocean.

Growth rates have not been studied in populations from the central part of the Middle Atlantic Bight. Juvenile eels in both the glass eel and elver stages are remarkably consistent in size. Glass eels collected in continental shelf waters during the MARMAP study and several more recent survey cruises ranged from 51 to 64 mm SL (fig. 6.5). Elvers collected from several combined estuarine sources are slightly larger (fig. 6.6). The smallest of these correspond in length to glass eels collected offshore. Because these small elvers are the dominant size class from November through July, it is difficult to follow the progression of year classes, but individuals up to 25 cm may be young-of-the-year. In several Canadian studies, however, workers found lengths of young-of-the-year to range as large as 31 cm; age-1 individuals were found to range from 21 to 38 cm (Smith and Saunders 1955).

Habitat requirements include the open ocean for developing leptocephalus stages and continental shelf waters for glass eels, although it is not known whether they are strongly neustonic or might be associated with the bottom. Elvers (after acquisition of dark cutaneous pigment) occupy a wide range of coastal marine habitats, often in association with eelgrass (Eales 1968.) These varied habitats include tidal flats and marshes, harbors, barrier beach ponds, coastal rivers, creeks, and streams. Of the several col-

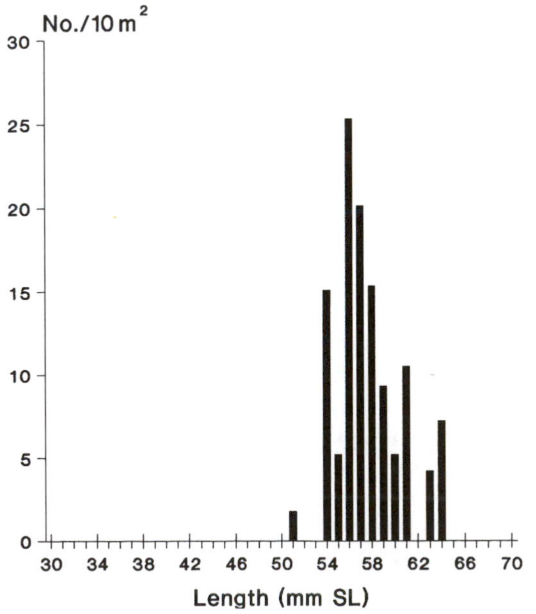

6.5 Lengths of all *Anguilla rostrata* glass eels collected during MARMAP study over the Middle Atlantic Bight continental shelf, 1977–1990 (n = 42).

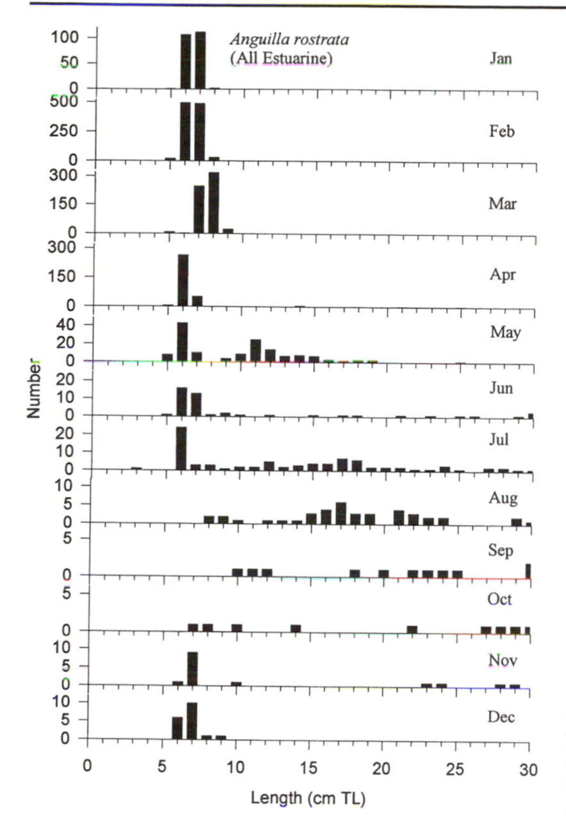

6.6 Monthly length frequencies of *Anguilla rostrata* young-of-the-year and older stages collected in the Great Bay study area, 1986–1994. Sources: de Sylva et al. 1962; RUMFS: otter trawl survey (n = 10); 1-m beam trawl (n = 29); 1-m plankton net (n = 1,636); experimental trap (n = 24); killitrap (n = 73); tidal creek weir (n = 192) (Rountree et al. 1992); night-light (n = 77); seine (n = 11); gear comparison (n = 7).

lection methods described in our study area (chap. 3; table 3.3), young-of-the-year have most often been collected in water-column ichthyoplankton sampling during night flood tides, although dip-netting under lights is also an effective means of sampling. Larger eels are cryptic and are most often collected as they enter traps, although many have been collected in a weir deployed in an intertidal creek.

With further development, young females then ascend into freshwaters while males remain in estuarine conditions. There is a delay in upstream migration from tidal to nontidal habitats, and it has been suggested that this delay is related to a period of physiological adjustment to freshwater (Haro and Krueger 1988.) Whips (juveniles larger than elvers) are able to withstand abrupt salinity changes and occur in euryhaline conditions (Cox 1916). In the Delaware Estuary, juveniles and adults have been collected from turbid shoreline sites with a substrate comprising finely ground plant-stem detritus (de Sylva et al. 1962.) They also occur abundantly in tidal creeks (Rountree et al. 1992) and tidal impoundments (Clark 1994).

Eels of all sizes occur in waters of all depths (Tesch 1977), but negative phototaxis becomes more pronounced with age. Therefore, older and larger eels avoid light more so than do young stages. This response translates into nocturnal activity and results in larger individuals being found at greater depths than are smaller individuals. They are also abundant in dark habitats under and around piers in the Lower Hudson River (Able et al. 1995b). They are, however, sensitive to harsh winter conditions. During fall cooling, eels move away from coastal waters into brackish and freshwaters. If freshwaters are not available, wintering eels seek out deeper parts of rivers and bays in which to spend the coldest parts of the winter (Smith and Saunders 1955). In these areas, they are known to burrow into mud (Cox 1916; Thomas 1968; MacKenzie 1992) or to hibernate in burrows equipped with ventilation holes (Eales 1968). During spring, they return to shallower coastal areas. Except during seasonal movements, they are strongly sedentary, and there is a tendency to remain in a home territory. Experimentally displaced eels have returned to a home territory from as far away as 80 km (Vladykov 1971). Adult eels remain in freshwater and estuarine habitats until they reach maturity and begin the spawning migration into the open ocean.

CHAPTER 7

Conger oceanicus (Mitchill) Conger Eel

7.1 *Conger oceanicus* elver, 62.0 mm TL. Collected May 14, 1990, dip net (night-light), RUMFS boat basin, Great Bay, New Jersey. ANSP 175208. Illustrated by Susan Kaiser.

DISTRIBUTION

Conger oceanicus (fig. 7.1), a bottom-dwelling species, occurs in coastal and continental shelf waters from Cape Cod to northern Florida (Bigelow and Schroeder 1953), rarely to the northern Gulf of Mexico and eastern Atlantic (Hood et al. 1988). It is found from the coast to 475 m depth and is usually associated with structured habitat such as piers, wrecks, jetties, reefs, or burrows shared with tilefish (Able et al. 1982; Grimes et al. 1986). Although it has not usually been considered an estuarine species, rare, single occurrences of young stages in estuaries have been reported (e.g., Pearson 1941; Hauser 1975; Moring and Moring 1986). These isolated reports have originated from estuaries throughout the Middle Atlantic Bight (table 4.2).

REPRODUCTION

Previous studies have found no ripe individuals among material collected from the coast to the continental shelf/slope break in the Middle Atlantic Bight (Hood et al. 1988; Eklund and Targett 1990), but early signs of maturation in females have been found in late spring and early summer (Hood et al. 1988). These maturing females have not been observed later in the year, suggesting that they migrate out of the area for reproductive purposes. Spawning occurs in the Sargasso Sea, east and northeast of the Bahama Islands (Schmidt 1931; Miller and McCleave 1994; M. J. Miller 1995). The spawning season is apparently long, perhaps extending from late summer through the winter, as evidenced by occurrences of small larvae (McCleave and Miller 1994). Direct observations of the oceanic spawning migration are meager, and there is no evidence of spent adults returning to coastal or shelf areas after spawning.

DESCRIPTION

Eggs are undescribed. During its early life history, the conger eel goes through four morphological stages (fig. 7.2). While in its oceanic engyodontic and euryodontic (based on different dentition) leptocephalus stages, it undergoes positive growth, reaching lengths of about 100 mm (Smith 1989a, b, c). The intestine occupies more than 90% of the length of the laterally flattened body during this early, oceanic stage. Pigment is limited to fine melanophores along the ventral edge of the body and gut, although the largest individuals may also display a few melanophores along the lateral midline. Most other leptocephali in the family Congridae have a prominent series of melanophores in this area (Smith 1989a, b, c). The leptocephalus then undergoes a metamorphosis that includes shrinkage in length along with modifications of many body proportions. Among these are a shrinking of the gut along with an anterior movement of the anus, an increase in head length, an anterior migration of the dorsal fin origin (fig. 7.2), and a reduction in body depth associated with a gradual change from a laterally flattened body to a cylindrical body shape.

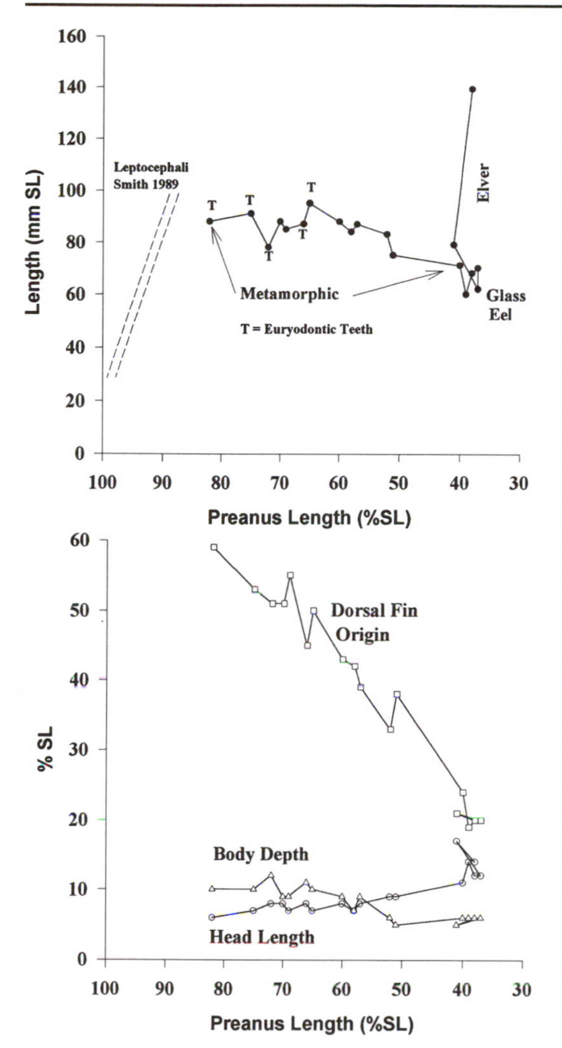

7.2 Changes in size and body proportions associated with early development in *Conger oceanicus.*

Larval teeth are also lost, and a labial flange develops along the upper lip. Pigment during this metamorphosis gradually increases from a simple line of melanophores along the lateral midline to a subtle dark shading of the dorsal surfaces. Two characters remain constant through metamorphosis: the total number of myomeres (approximately equal to the number of vertebrae) does not vary from 140 to 148, and the last vertical blood vessel does not change from its position at the 51st to 55th myomere. Metabolic rates remain low during the leptocephalus stage of a related congrid, then increase at metamorphosis (Pfeiler and Govoni 1993), suggesting that nutrient requirements are higher after this transformation. These modifications conclude at the glass eel stage, the smallest size (60–70 mm TL) of the animal while occupying estuarine waters. Positive growth then resumes, body proportions are stabilized, and darker pigmentation is acquired dorsally. The animal has now begun its elver stage, where it is recognizable as a conger eel (see Identification Key, Appendix A). Juveniles of *Conger oceanicus* (fig. 7.1) have eellike body shapes and lack pelvic fins. In the earliest juvenile (elver) stages they are drab brown in color and lack pattern. The snout tip protrudes beyond the lower jaw tip. Color and shape are similar in *Anguilla rostrata* of the same size, but the lower jaw protrudes farther than the snout tip. Vert: 143–147; D: mean 273; A: mean 187; Pect: 16–18; Plv: none.

THE FIRST YEAR

Early-stage leptocephali have been collected in the Sargasso Sea region, north and east of the Bahamas (Schmidt 1931; Castonguay and McCleave 1987; M. J. Miller 1995); and larger leptocephali (up to 85 mm) have been collected from the Florida Current and Gulf Stream south of Cape Hatteras, North Carolina (McCleave and Miller 1994). The mechanisms used by these leptocephali to exit the Gulf Stream and cross the continental shelf are undescribed. There are no records of leptocephali occurring in continental shelf waters of the Middle Atlantic Bight despite intense sampling effort (table 3.3). This omission is either due to their ability to avoid collecting gear (large leptocephali are adept and capable swimmers; M. P. Fahay pers. observ.) or to the fact that older larvae are strongly bottom-oriented and thus beyond the range of plankton nets. Beam trawls sampling the bottom outside Little Egg Inlet between May 20 and June 5, 1996 collected numerous late leptocephali and transforming specimens (85–117 mm TL, n = 142), suggesting that they are bottom-oriented on the inner continental shelf as they approach estuaries (M. Neuman pers. comm.). As they enter inlets, leptocephali have been collected with a variety of collecting devices including stationary plankton nets, dip nets under night-lights, and small-mesh traps. Despite the long spawning season, this ingress regularly occurs in Great

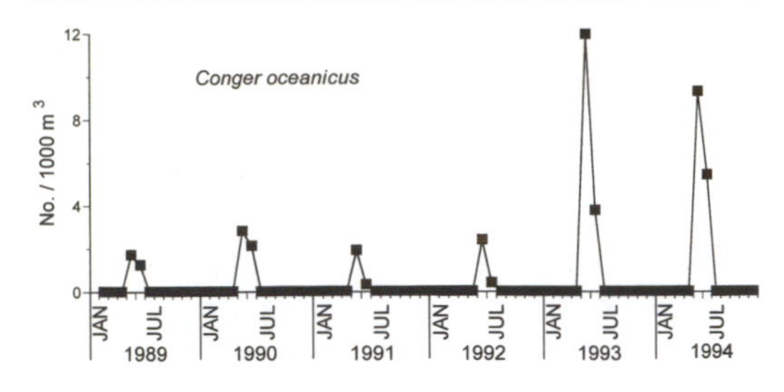

7.3 Annual occurrences of *Conger oceanicus* in Little Sheepshead Creek Bridge 1-m plankton net collections, 1989–1994 (n = 283).

Bay–Little Egg Harbor in May, June and July (fig. 7.3) and co-occurs with a morphological metamorphosis (fig. 7.2). The abundance of ingressing larvae is much greater in some years than in others. Whether a comparable and equally variable ingress occurs into other Middle Atlantic Bight bays and estuaries remains undescribed.

Describing a growth rate during the first growing season is difficult based on available evidence. Length frequencies derived from various trap and trawl collections indicate the ingress of metamorphosing larvae in May at a mode of about 100 mm TL (fig. 7.4). During June, glass eels with expected lengths of 60 to 70 mm are a difficult stage to collect and are not as common as larger leptocephali (about 100 mm) that continue to ingress. There are no well-defined length modes during late summer and fall, and those represented in the histogram are similar to the back-calculated size at age 1 from another study (Hood et al. 1988) and thus may all represent young-of-the-year. *Conger oceanicus* is rarely collected in any stage in estuarine waters during the winter or early spring. Without age-at-length data, it would be difficult to conclude that the coincidence of length modes in November and April (at about 250–350 mm TL) suggests a cessation in growth during the intervening winter.

Various habitats are used by various life-history stages. Metamorphosing leptocephali are pelagic (or bentho-pelagic) as they enter estuaries from May through July. Morphological metamorphosis and settlement to bottom habitats may not be immediate. Pelagic stages (70–104 mm; possibly including new immigrants from the ocean) continue to be sampled by dip nets under night-lights into July (Able et al. 1997). We have made very few collections of the glass eels and conclude either that this stage is highly cryptic as observed in the laboratory (G. Bell, D. A. Witting, and K. W. Able, pers. observ.), or the stage is very short-lived and thus underrepresented compared to stages preceding and following. For anguilliforms in general, the leptocephalus stage can last months to years, but transformation to the juvenile stage is usually completed in 2 to 3 weeks (Pfeiler 1986). Elvers, too, are cryptic and not frequently collected by any method.

We have few records of conger eels collected in the estuary by any method from December through March, yet we also lack evidence that they leave estuaries during this period. Although congers may burrow in the estuarine substrate and remain inactive through

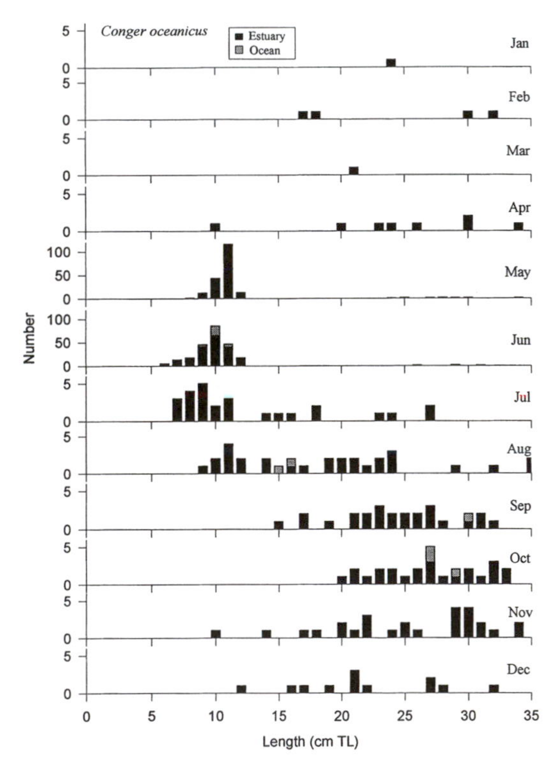

7.4 Monthly length frequencies of *Conger oceanicus* late larval, elver, and early adult stages collected in the Great Bay study area. Sources: ANSP (primarily Ichthyological Associates collections); Levy et al. 1988; RUMFS: otter trawl (n = 22); 1-m beam trawl (n = 9); 1-m plankton net (n = 208); experimental trap (n = 138); LEO-15 2-m beam trawl (n = 1); gear comparison (n = 7); killitrap (n = 78); seine (n = 14); night-light (n = 28); weir (n = 5).

the winter, most evidence suggests an offshore migration during this period (as suggested by Bigelow and Schroeder 1953). By-catch collections of conger eels in the black sea bass trap fishery on the adjacent continental shelf, for example, increase during November (Eklund and Targett 1991), which is consistent with an offshore movement by the population. The conger eel by-catch in the tilefish longline fishery near the edge of the continental shelf also is at a maximum during winter and spring (Hood et al. 1988) and at its lowest during summer, again suggesting a seasonal inshore-offshore movement by the population.

CHAPTER 8

Alosa aestivalis (Mitchill) Blueback Herring

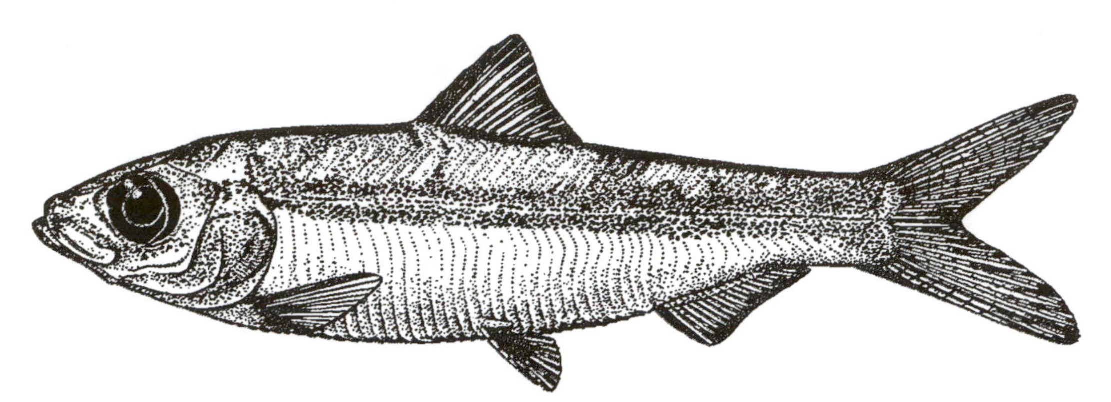

8.1 *Alosa aestivalis* juvenile, 45 mm TL. After Mansueti and Hardy 1967.

DISTRIBUTION

A schooling, anadromous species, *Alosa aestivalis* (fig. 8.1), is distributed along the East Coast of North America between Nova Scotia and Florida. It is most abundant in the Middle Atlantic and South Atlantic Bights (Neves 1981), where it occurs in coastal ocean waters, rivers, and the freshwater portions of estuaries. Juveniles occur in most estuaries throughout the Middle Atlantic Bight, but eggs and larvae are not found in those systems lacking significant freshwater input, such as Nauset Marsh (table 4.2).

REPRODUCTION

In our study area, males spawn at ages 3 to 4; females at ages 4 to 5 (Marcy 1969; Scherer 1972). Spawning occurs during spring or early summer (late March through July) in fresh or brackish waters. Reported spawning runs in the central part of the Middle Atlantic Bight include those in the Hudson and Delaware Rivers (Dovel 1981; Mayo 1982a) as well as a few smaller runs in rivers along the New Jersey coast including the Raritan, South, Navesink, Swimming, and Great Egg Harbor Rivers (Zich 1977; Himchak 1983; Himchak and Allen 1985). No evidence of spawning was found within a Mullica River-Great Bay study area (Milstein et al. 1977). In another survey near the mouth of Delaware Bay, adults were abundant in gill net collections in May and June, and the same study found ripe females as early as April (de Sylva et al. 1962). Spawning occurs mostly at night (Graham 1956) in fast currents over a hard substrate (Loesch and Lund 1977). Spawning can occur in temperatures as low as 14 C but optimum temperatures are 21 to 24 C (Mansueti and Hardy 1967). *Alosa aestivalis* does not run as far upstream to spawn as does *A. pseudoharengus,* and it generally spawns 3 to 4 weeks later than its congener. After spawning, adults retreat downstream to the ocean.

DESCRIPTION

Eggs are moderately adhesive, yellowish, and semitransparent and range from 0.87 to 1.11 mm in diameter (Kuntz and Radcliffe 1917; Mansueti 1962). The perivitelline space occupies approximately 25% of the egg's radius, and the oil globules are small, unequal in size and scattered (Kuntz and Radcliffe 1917). Larvae hatch at 3.1 to 5.0 mm TL, are elongate, and have pigment on the upper and lower gut surfaces. There are 11 to 13 myomeres between the dorsal and anal fins (M. P. Fahay pers. observ.). See Jones et al. (1978) for a more detailed synopsis of early developmental characters. Juveniles (fig. 8.1) have sharp scutes along the belly, nine or more rays in the pelvic fin, and 44 to 50 gill rakers on the lower limb of the first gill arch. The interior lining of the gut cavity is black or dark gray, and larger juveniles have a single dark spot posterior to the gill cover. The bony plate on the side of the head under the eye is wider than deep, and the eye diameter is less than the snout length. Vert: 47–53; D: 15–20; A: 15–21; Pect: 14–18; Plv: 9–11.

THE FIRST YEAR

The eggs are semipelagic (float near the bottom) except in water with no current where they are demersal (Lippson and Moran 1974). Incubation occupies 50 hours at 22 C and 80 to 94 hours at 20 to 21 C (Jones et al. 1978). Larvae are 3.1 to 5.0 mm TL at hatching. Transformation from larval to juvenile morphology occurs at about 20 mm TL and an age of 25 to 35 days (Watson 1968). Fin ray development is complete and the body has begun to deepen at this time. Scales first appear at about 25 to 29 mm TL and are fully developed at 45 mm TL (Hildebrand 1963). After transformation, juveniles usually move upstream in tidal rivers, where they remain until late summer or early fall (Scherer 1972). In Delaware Bay, however, larvae and/or juveniles have been observed distributed throughout the estuarine shore zone in March, moving toward the sea with the onset of warm weather in July, moving farther downstream by August, and leaving the estuary entirely by late fall (de Sylva et al. 1962). Most *A. aestivalis* individuals collected by Milstein et al. (1977) throughout their Great Bay and nearby coastal study area were young fishes, apparently results of spawning in estuaries either north or south of Great Bay. Most of these collections were in winter (26.9%) and spring (62.7%) when they were taken in ocean (18.4%) or estuarine (81.5%) habitats.

Growth is slow in young-of-the-year, based on the small sizes attained by fall. Perlmutter et al. (1967) collected small larvae from the upper Hudson River in June (15–44 mm SL) and August (14–49 mm SL). Most fish collected in late summer and early fall in our study area are 50 to 100 mm TL (fig. 8.2). This range closely approximates the sizes (36–71 mm TL) reported for young-of-the-year emigrating to the ocean in the fall in the Chesapeake Bay region (Hildebrand and Schroeder 1928). Most overwintering (December through March) young-of-the-year collected by Milstein (1981) were 60 to 110 mm FL. There is little, if any, growth over the first winter. Our own observations of lengths indicate that juveniles from the previous year enter estuaries during March at lengths of about 70 to 80 mm TL, at the same time as young-of-the-year from local spawning are appearing at smaller lengths (e.g., < 20 mm TL). Spring weir collections in a subtidal marsh creek in our study area ranged from 80 to 101 mm SL (Rountree and Able 1992a) and represent fishes reaching their first birthday. This is in general agreement with Chesapeake Bay observations made by Hildebrand and Schroeder (1928), who reported age 1 during spring ranging from 64 to 119 mm TL. Through spring and summer, therefore, young-of-the-year and age 1 are only separated by about 5 to 10 cm (fig. 8.2). These size ranges are also evident in collections from Delaware Bay

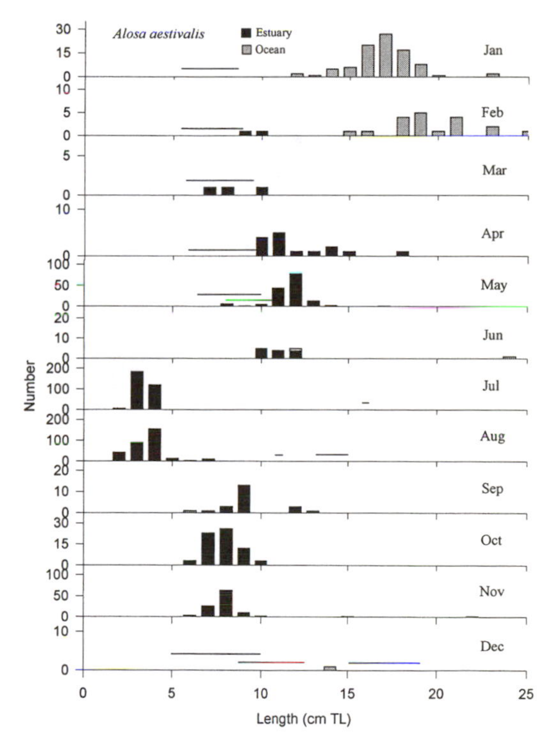

8.2 Monthly length frequencies of *Alosa aestivalis* collected in New Jersey/New York estuarine and adjacent coastal ocean waters. Sources: Perlmutter et al. 1967; Allen et al. 1978; Kahnle 1987; Kahnle and Hattala 1988; Wilk et al. 1992; RUMFS: otter trawl (n = 17); LEO-15 2-m beam trawl (n = 1); seine (n = 2); weir (n = 162). Horizontal bars indicate length ranges where frequencies are unknown. Sources: Clark et al. 1969; Pacheco and Grant 1973; Wilk and Silverman 1976b.

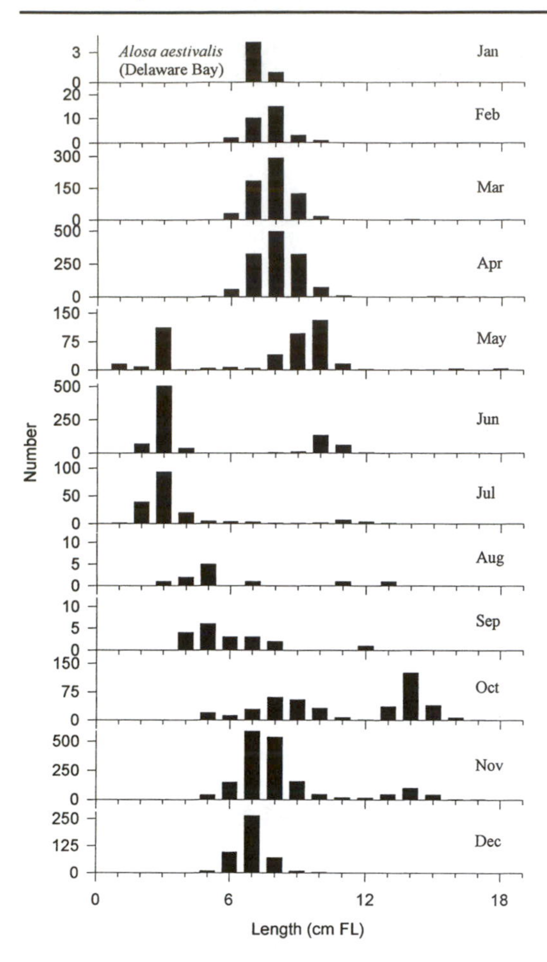

8.3 Monthly length frequencies of *Alosa aestivalis* collected from Delaware Bay, 1970–1978. Data courtesy of Public Service Electric & Gas and Environmental Consulting Services, Inc.

(fig. 8.3). Newly hatched larvae are first observed in May, growth tapers off in the fall when a mode of about 7 cm FL is reached, and there is little evidence of size increase until the following May, when young-of-the-year reach age 1, and growth resumes. Year classes here are also separated by about 7 cm. Numbers of fish in Delaware Bay are lowest during January and February, when most young-of-the-year are found in oceanic waters (see below).

Juveniles, while in freshwater, remain high in the water column and avoid near-bottom depths (Warinner et al. 1969). They can tolerate abrupt rises in salinity (Chittenden 1972b) but several studies have shown a lack of tolerance for sudden increases in temperature (Marcy 1973; Schubel et al. 1977). Emigration from freshwater habitats occurs in response to heavy rainfall, high water, and a decline in temperature (Pardue 1983). After leaving estuaries in the fall, young-of-the-year presumably overwinter in areas near their estuarine nurseries (Milstein 1981) which are segregated from overwintering areas described for older year classes (Neves 1981). *Alosa aestivalis* (all young-of-the-year) ranked third of 53 species in winter trawl hauls on the inner continental shelf off New Jersey reported by Milstein (1981). These were collected within 7.4 km of shore in an area consistently influenced or affected by waters from the estuary. Winter bottom temperatures in this nearshore area were generally warmer and more saline than those in the adjacent estuary and ranged from a high of 10.0 C in December to a low of 2.0 C in February. As temperatures warmed in March (4.4–6.5 C), the inner continental shelf occurrences decreased, presumably coincident with a spring migration back into estuaries. Winter collections at Indian River Inlet, Delaware continued to yield *A. aestivalis* young-of-the-year between December and May at lengths between 46 and 96 mm FL (Pacheco and Grant 1973), further attesting to their proclivity to remain near the nursery area through the winter. A similar phenomenon has been reported from the Chesapeake Bay, where some young-of-the-year remain in areas near their nurseries in deeper, more saline parts of the bay through the winter (Hildebrand and Schroeder 1928).

CHAPTER 9

Alosa mediocris (Mitchill) Hickory Shad

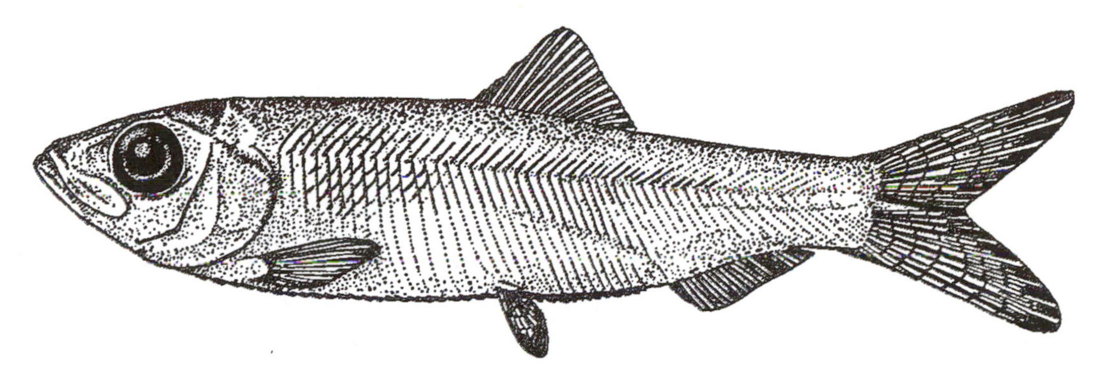

9.1 *Alosa mediocris* juvenile, 35.2 mm TL. After Mansueti 1962.

DISTRIBUTION

Alosa mediocris (fig. 9.1) is found along the coast of the United States between the Bay of Fundy and northern Florida (Mansueti 1962), but it is rare north of Cape Cod (e.g., Collette and Hartel 1988). Adults occur in ocean waters, apparently never far from land (Hildebrand and Schroeder 1928), but in these coastal habitats they are rare (e.g., Wilk et al. 1992). They are also rare in the Hudson River Estuary (Kahnle and Hattala 1988) and were absent in a study of the Lower Hudson-Raritan Estuary (Woodhead et al. 1987). No adults (and only one young-of-the-year) were collected in extensive seining collections throughout the nontidal Delaware River from 1960 to 1966 (M. Chittenden pers. comm.). They were only rarely collected during 3 years' sampling in lower New York Harbor and Sandy Hook Bays (Pacheco 1983, 1984; Wilk and Silverman 1976b). Smaller adults (average 250 mm SL) have been reported from Sandy Hook Bay during August (Nichols and Breder 1927). In the surf off Long Island, larger individuals (140–430 mm FL) were the sixth, fifth, and third most abundant species taken during seine surveys in May through November of 1961, 1962, and 1963, respectively (Schaefer 1967). Lengths of 58 individuals gill-netted in the ocean off Little Egg Inlet ranged from 245 to 375 mm FL during August (Thomas et al. 1974). Recent gill-net sampling in a Little Egg Harbor study area at night (Rountree and Able 1997) demonstrated that *A. mediocris* was the sixth most abundant species collected in tidal creeks and adjacent shallow bay habitats and were similar in size to the ocean collections. The single size cohort, ranging from 241 to 350 mm SL, was collected primarily in late July and early November and showed no particular preference between creek and bay habitats or between ebb and flood-tide cycles. Presumably, all the foregoing records pertain to fishes age 1 and older. Within the Middle Atlantic Bight, eggs and larvae have only been reported from the Chesapeake Bay, and juveniles are rarely reported from estuaries (table 4.2).

REPRODUCTION

This is an anadromous species that enters bays and freshwaters during spring to spawn. In certain Chesapeake Bay tributaries, spawning occurs in tidal freshwater, apparently between dusk and midnight in April through June (Mansueti 1962; Mansueti and Hardy 1967). No evidence has been found of spawning in Delaware Bay waters although ripe adults have been collected in the C&D Canal (Wang and Kernehan 1979). There is some evidence of spawning in the Delaware River based on the presence of larvae and "young" (Himchak 1981). Milstein et al. (1977) found no evidence of spawning in Great Bay and the adjacent ocean area near Little Egg Inlet. These authors considered the few specimens they collected as "migrant adults." After spawning, adults leave estuaries and return to the ocean. The collections of adults made in July in Little Egg Harbor (Rountree and Able 1997) may represent spent fishes returning to the ocean from freshwater spawning areas.

DESCRIPTION

The eggs are slightly adhesive and transparent and range from 0.96 to 1.65 mm in diameter. The perivitelline space occupies approximately 50% of the egg's radius, and there are a few small oil globules (Mansueti 1962). Larvae hatch between 5.2 and 6.5 mm. Larvae are elongate and have elongate melanophores along the venter of the gut. There are 7 to 8 myomeres between the dorsal and anal fins (M. P. Fahay pers. observ.). See Jones et al. (1978) for a more detailed synopsis of early developmental characters. Juveniles have sharp scutes along the belly (fig. 9.1). They also have nine rays in the pelvic fin and 18 to 23 gill rakers on the lower limb of the first gill arch. The interior lining of the gut cavity is black or dark gray, and there may be a dark spot behind the edge of the gill cover followed by a row of smaller spots. The bony plate on the side of the head, under the eye, is deeper than wide, and the lower jaw tip extends farther anteriorly than the snout tip. The body is slimmer than similar-sized *A. sapidissima.* Vert: 53–55; D: 15–20; A: 19–23; Pect: 15–16; Plv: 9.

THE FIRST YEAR

Incubation occupies 48 to 70 hours at 16 to 31 C (Jones et al. 1978). Very little is known about the early life history of this species. Young-of-the-year apparently leave nursery areas in tidal freshwaters and migrate to the ocean early in the summer, although some evidence from the Chesapeake Bay suggests age 1 individuals occur in bays irregularly throughout the year (Mansueti 1962). Little is known about the growth rate. Limited data suggests they attain lengths of 150 to 200 mm when they reach age 1 (Mansueti 1962).

CHAPTER 10

Alosa pseudoharengus (Wilson)
Alewife

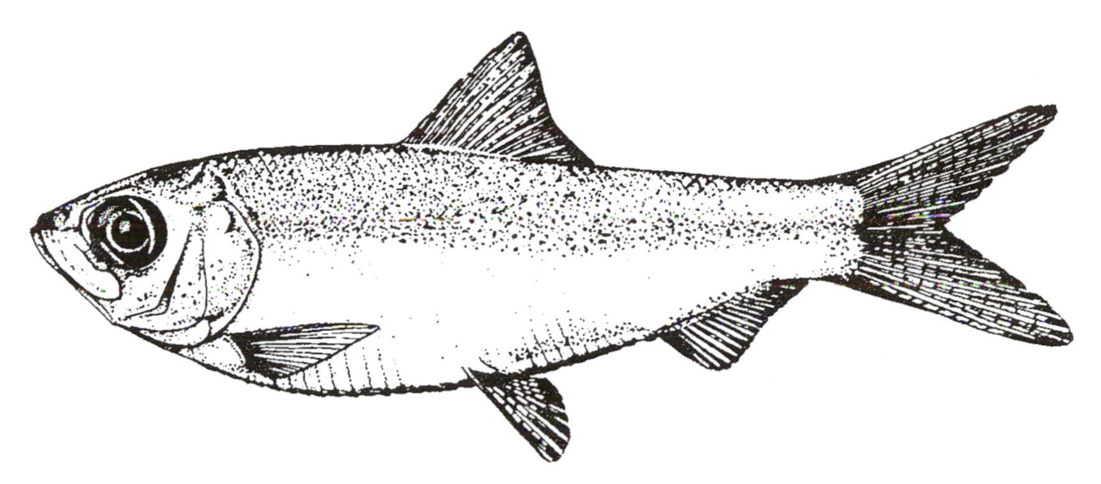

10.1 *Alosa pseudoharengus* juvenile, 42.0 mm TL. After Mansueti and Hardy 1967.

DISTRIBUTION

A schooling, anadromous species found along the coast of eastern North America from the Gulf of St. Lawrence to South Carolina, *Alosa pseudoharengus* (fig. 10.1) is most abundant between the Gulf of Maine and Chesapeake Bay (Berry 1964; Winters et al. 1973). It is also found in the Great Lakes, and there are landlocked populations in New York (Smith 1985). A nonmigratory dwarf population has been reported to occur in the mouth of the Susquehanna River (Foerster and Goodbred 1978) but has not been subsequently found. During autumn, most of the population winters near the bottom, at depths between 56 and 110 m, in continental-shelf waters north of our study area, off southern New England, Georges Bank, and Gulf of Maine (Neves 1981). During spring they are distributed over the entire shelf and concentrated between Rhode Island and Cape May (Mayo 1982b). While inhabiting shelf habitats, adults are most common in temperatures of 4 to 7 C. Early life-history stages occur in estuaries throughout the Middle Atlantic Bight, except in those with limited freshwater input, such as Nauset Marsh (table 4.2).

REPRODUCTION

Adults migrate into coastal rivers to spawn during the spring. Spawning runs in the central part of the Middle Atlantic Bight occur in the Hudson, Raritan, South, Navesink, Swimming, Great Egg Harbor, and Delaware Rivers (Perlmutter et al. 1967; Zich 1977; Dovel 1981; Himchak 1982a, 1983; Himchak and Allen 1985). Evidence of spawning has also been found in the Batsto River (U. Kils and K. W. Able pers. observ.), Mullica River, Nacote Creek, and in the Brigantine Wildlife Refuge (Milstein et al. 1977) and several small tributaries of the Delaware River estuary (Wang and Kernehan 1979). During the spawning migration, adults are highly tolerant of salinity changes (Chittenden 1972b). In studies in Connecticut and Chesapeake Bay, spawners were dominated by fishes age 3 to 8, and about two-thirds of these had spawned in more than 1 year (Joseph and Davis 1965; Marcy 1969). There is some evidence that adults return to their natal rivers to spawn (Thunberg 1971; Havey 1973). Males apparently mature at an earlier age (3–4) than females (4–5), but do not live as long (Joseph and Davis 1965). Mean size at maturity ranges from 265 to 278 mm for males and 284 to 308 mm for females (Mayo 1974). *Alosa pseudoharengus* runs farther upstream to spawn than does *A. aestivalis,* and it generally spawns 3 to 4 weeks earlier than its congener (Jones et al. 1978). Spawning sites are typically shallow, with sluggish flow, ranging from oxbows in large rivers to small streams, and there are also records of spawning in ponds located on barrier beaches (Bigelow and Welsh 1925; Kissil 1974; Wang and Kernehan 1979; Mullen et al. 1986). The spawning migration commences when river temperatures reach 10.5 C (Cianci 1969; Kissil 1969). Spawning begins at temperatures between 13 and 15 C and ends when waters are warmer than 27 C (Loesch 1969). An upper lethal temperature of 29.7 C has been reported for eggs in the Hudson River (Kellogg 1982). After spawning, an adult downstream movement is apparently triggered by an increase in water flow (Huber 1978).

DESCRIPTION

Eggs are moderately adhesive and range from 0.80 to 1.27 mm in diameter (Mansueti 1962). Adhesive properties are lost a few hours after spawning. The oil globules are very small and numerous (Norden 1967). Larvae hatch between 2.5 and 5.0 mm TL, are elongate, and have pigment on the upper anterior and lower posterior gut surfaces. There are 7 to 9 myomeres between the dorsal and anal fins (M. P. Fahay pers. observ.). See Jones et al. (1978) for a more detailed synopsis of early developmental characters. Juveniles have sharp scutes along the belly, 9 or more rays in the pelvic fin, and 39 to 41 gill rakers on the lower limb of the first gill arch (fig. 10.1). The interior lining of the gut cavity is clear or pale with dusky spots, and larger juveniles have a single dark spot posterior to the gill cover. The bony plate on the side of the head, under the eye, is wider than deep. The eye diameter is greater than the snout length. Vert: 46–50; D: 12–19; A: 15–21; Pect: 13–16; Plv: 10.

THE FIRST YEAR

The eggs are semipelagic (float near the bottom) except in water with no current, where they are demersal (Lippson and Moran 1974). When spawning occurs in sluggish habitats such as oxbows or quiet ponds, high levels of suspended sediments have been shown to cause high egg mortalities (Schubel and Wang 1973). Incubation time varies with temperature: 2.1 days at 28.9 C to 15 days at 7.2 C (Jones et al. 1978). Transformation from larval to juvenile morphology occurs at about 20 mm TL when fin ray development is complete and the body has begun to deepen. Scales first appear at about 25 to 29 mm TL and are fully developed at 45 mm TL (Hildebrand 1963). After this transformation juveniles in tidal water tend to move farther upstream from their spawning site (Warinner et al. 1969).

Our estimates of growth are based on collections made in the central part of the Middle

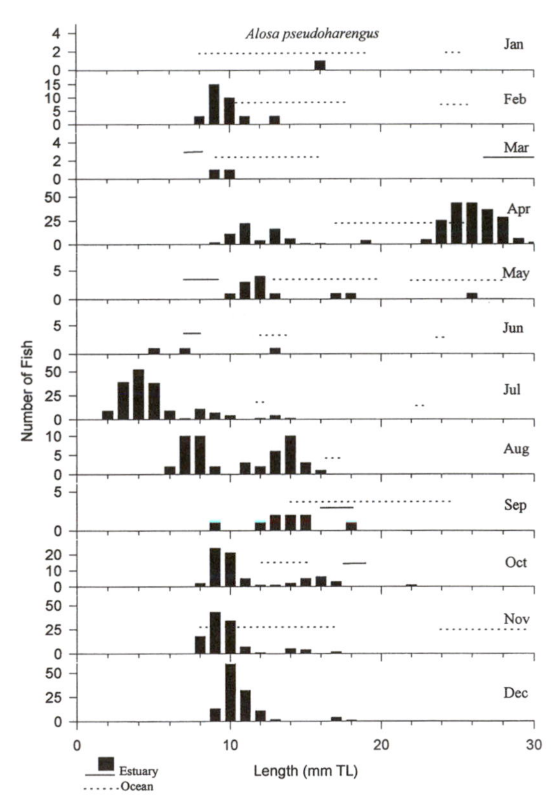

10.2 Monthly length frequencies of *Alosa pseudoharengus* juveniles and adults collected in New Jersey/New York estuarine and adjacent coastal ocean waters. Sources: Perlmutter et al. 1967; Pacheco and Grant 1973; Rohde and Schuler 1974; Pacheco 1983; Kahnle 1987; Kahnle and Hattala 1988; RUMFS weir collections (Rountree et al. 1992). Horizontal dotted lines refer to length ranges where frequencies are unknown. Sources: Schaefer 1967; Clark et al. 1969; Wilk and Siverman 1976a,b; Allen et al. 1978; Wilk et al. 1992.

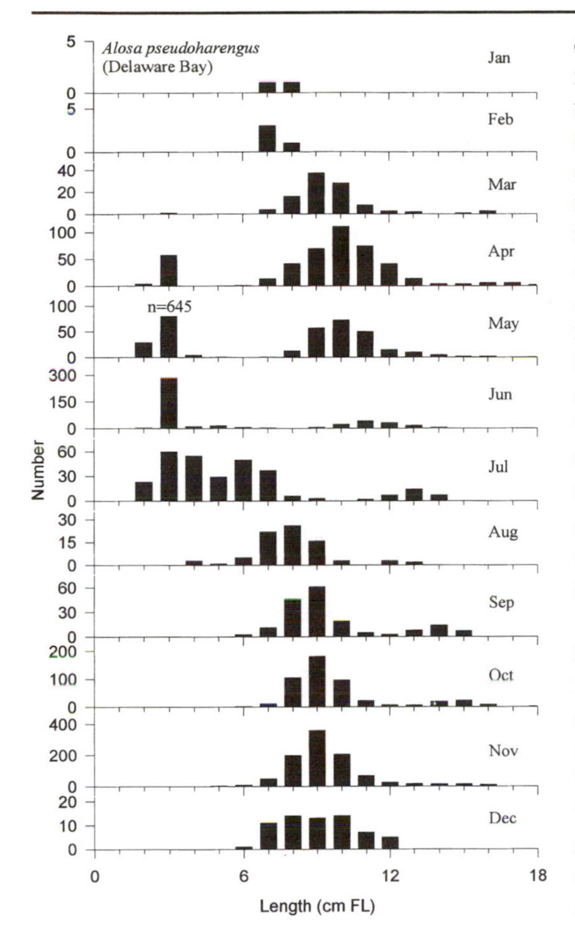

10.3 Monthly length frequencies of *Alosa pseudoharengus* collected from Delaware Bay, 1970–1978. Note 3-cm bar is truncated in May graph (n = 645). Data courtesy of Public Service Electric & Gas and Environmental Consulting Services, Inc.

Atlantic Bight (fig. 10.2) or are restricted to Delaware Bay (fig. 10.3). In the bight, the young-of-the-year exhibit a mode at about 4 cm during July and reach about 10 cm by the end of summer. Growth appears to be arrested during winter and resumes the following spring. At age 1, the average size is between 10 and 12 cm. These rates are in general agreement with studies made elsewhere where: (1) small larvae from the upper Hudson River were 10 to 59 and 45 to 74 mm SL in June and August, respectively (Perlmutter et al. 1967); (2) emigrants moving downstream during fall from a reservoir in Rhode Island ranged from 30 to 105 mm TL (Richkus 1975); (3) wintering young-of-the-year ranged from 60 to 170 mm FL on the inner continental shelf (Milstein 1981); (4) age 1+ males at the end of their second summer in the Connecticut River averaged 147 mm TL (Marcy 1969); (5) fish of age group II were about 180 mm TL by the end of the third summer (Netzel and Stanek 1966); and (6) spawning fish migrating up the Delaware River were 220 to 320 mm TL in late March and early April (Rohde and Schuler 1974). The growth rates of juveniles in two spawning ponds in Massachusetts ranged from 0.2 to 0.5 mm per day (Cole et al. 1980). Collections from Delaware Bay show similar patterns, with spring through summer growth, followed by a cessation of size increase during winter and resumption of growth as young-of-the-year reach age 1 in the spring. Age 0 and 1 year classes are separated by about 7 cm.

This species uses a full range of habitats from the edge of the continental shelf into freshwater during the first year. Eggs and larvae occur exclusively in freshwater. Juveniles occur in upper levels of the water column until October, when they spend more time near the bottom (Mullen et al. 1986). Juveniles have been shown to prefer temperatures of 20 to 22 C and salinities of 4 to 6 ppt in laboratory tests (Meldrim and Gift 1971), although they are collected from a much wider range in natural habitats.

Juveniles emigrate from fresh and brackish waters during late summer and fall. Waves of this downstream movement are apparently triggered by heavy rainfall (Cooper 1961), high water levels, and dropping temperatures (Richkus 1975) but may be inhibited somewhat on bright sunny days (Richkus 1974). After this fall emigration, young-of-the-year presumably overwinter in areas near their estuarine nurseries (Milstein 1981), segregated from major wintering areas described for older year classes (Neves 1981). *Alosa pseudoharengus* (all young-of-the-year) ranked 7th of 53 species in a winter trawl survey on the inner continental shelf off New Jersey (Milstein 1981). The study area where these young-of-the-year were collected was within 7.4 km of shore, ranged from 2.4 to 19.2 m in depth, and is an area consistently influenced or affected by waters from the estuary. Winter bottom temperatures in this nearshore area were generally warmer and more saline than those in the adjacent estuary and ranged from a high of 10.0 C in December to a low of 2.0 C in February. As temperatures warmed in March (4.4–6.5 C), the continental shelf occurrences of young-of-the-year decreased, presumably coinciding with a spring reentry into fresh and brackish waters at age 1. Young-of-the-year have been collected in the Indian River Inlet during February and May at lengths between 66 and 94 mm FL (Pacheco and Grant 1973) and have also overwintered in the upper part of the Delaware River Estuary (Wang and Kernehan 1979), further attesting to their proclivity to remain near the nursery area through the winter.

CHAPTER 11

Alosa sapidissima (Wilson) American Shad

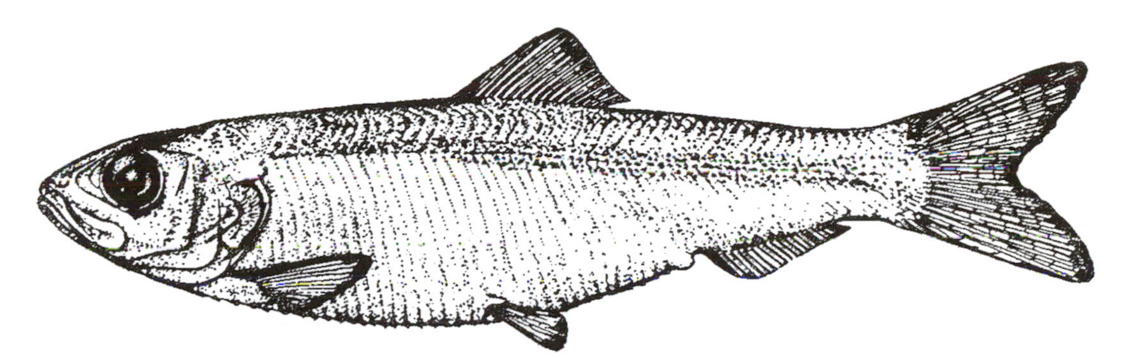

11.1 *Alosa sapidissima* juvenile, 39.5 mm TL. After Mansueti and Hardy 1967.

DISTRIBUTION

Alosa sapidissima (fig. 11.1) is an anadromous species found along the coast of North America from the Gulf of St. Lawrence to Florida (Robins and Ray 1986). The center of abundance is between Connecticut and North Carolina. Adults congregate in the Gulf of Maine during summer and fall (Walburg 1960) and then move south to overwinter in deeper waters off the Middle and South Atlantic Bights during winter (Talbot and Sykes 1958). Early life-history stages are found in several estuaries in the central part of the Middle Atlantic Bight, but eggs and larvae are absent in systems with limited freshwater input, such as Nauset Marsh, Gardiners Bay, and Great South Bay (table 4.2).

REPRODUCTION

Alosa sapidissima is an anadromous species that spends several years in the ocean before maturing and returning to natal rivers for spawning during the spring. Males mature at 3 to 5 years old; females at 4 to 6 years old (Leim 1924; Leggett 1976). The timing of spawning migrations, as well as spawning itself, is regulated by water temperatures (Chittenden 1969). Spawning begins when temperatures reach 12 C and ends when they exceed 20 C. Spawning occurs at night, generally in shallow waters with moderate currents (Marcy 1972). New Jersey is bracketed by important spawning runs in the Hudson and Delaware Rivers as well as smaller runs in tributaries to these and other rivers (Chittenden 1969; Smith 1985; Miller 1995). In the Hudson River, perhaps half of the spawning adults survive the upstream/downstream spawning migration and spawn again the following year. In the Delaware River, fewer than 5% do so (Talbot and Sykes 1958; Chittenden 1975; Leggett and Carscadden 1978), although this value may increase in response to improved dissolved oxygen conditions in recent years (Weisberg and Burton 1993). Productivity of *A. sapidissima* in the Delaware River declined sharply at the beginning of this century, and this decline has been attributed to reductions in suitable spawning and nursery habitat caused by pollution (Chittenden 1976). Recent improvements in water quality in the lower Delaware River have resulted in expansions in this habitat, and increases in the size of the spawning population have also followed (Weisberg and Burton 1993; Weisberg et al. 1996).

DESCRIPTION

The eggs are transparent, pale amber, or pink and are weakly demersal. Their diameter ranges from 2.5 to 3.8 mm, they lack oil globules; and there is a very wide perivitelline space occupying more than 50% of the radius (Bigelow and

Welsh 1925; Mansueti 1955; Marcy and Jacobson 1976). The elongate larvae hatch at 5.7 to 10.0 mm and have 55 to 57 myomeres, with about 7 between the dorsal and anal fin bases (M. P. Fahay pers. observ.). Pigment is light and variable, but usually occurs as two rows of melanophores along the venter or along the dorsal edge of the posterior portion of the gut. See Jones et al. (1978) for a more detailed synopsis of early developmental characters. Juveniles (fig. 11.1) have sharp scutes along the belly, 8 to 10 rays in the pelvic fin, and 59 to 76 gill rakers on the lower limb of the first gill arch. The interior lining of the gut cavity is pale or silvery, and there may be a dark spot behind the edge of the gill cover followed by a cluster of smaller spots. The bony plate on the side of the head, under the eye, is deeper than wide, and the snout tip extends farther anteriorly than the lower jaw tip. The body is deeper than similar-sized *A. mediocris.* Vert: 51–60; D: 14–21 (usually 18–19); A: 18–25 (usually 21–22); Pect: 13–18; Plv: 8–10.

THE FIRST YEAR

The embryos incubate for 2 days (at 27 C) to 17 days (at 12 C) (Ryder 1887; Marcy and Jacobson 1976). After hatching at 5 to 10 mm TL, larvae are pelagic for 2 to 3 weeks, and reach a length of 25 to 28 mm before transforming to the juvenile stage (Jones et al. 1978). The relative body depth increases through the larval and juvenile stages. Scales are first evident above the midlateral area at about 34.5 mm and are complete at 52 mm (Jones et al. 1978). Juveniles remain in Delaware River habitats through the first summer and begin to gradually disperse downstream by midsummer (Chittenden 1969). After reaching lengths of 75 to 125 mm in the fall, young-of-the-year pass through estuarine habitats and move out to the ocean. This migration by young-of-the-year is apparently triggered by decreasing water temperatures and occurs earlier in upstream locations. A peak in this downstream migration occurs in late October in rivers draining into Delaware Bay (Chittenden 1972b; Leggett 1977).

There is much within-river variation in size, with the smallest young-of-the-year always being found farther upstream, as has been demonstrated for the Delaware River (Chittenden 1969). A recent study in the Hudson River suggests the possibility that individuals are capable of enhanced growth under certain conditions, and these individuals may migrate to marine waters as early as late June and at ages of only 6 to 9 weeks (Limburg 1995). Young-of-the-year reach approximately 120 mm TL after a full year, and 240 mm TL after 2 years, based on data derived primarily from the Hudson River population (fig. 11.2) and from the Bay of Fundy (Leim 1924).

The range of acceptable environmental conditions is fairly well known for young-of-the-year. Prolonged exposure to temperatures below 4 to 6 C produces sublethal effects or death, especially during the fall and winter before young-of-the-year have acclimated (Chittenden 1972a). Young-of-the-year tolerate salinity in-

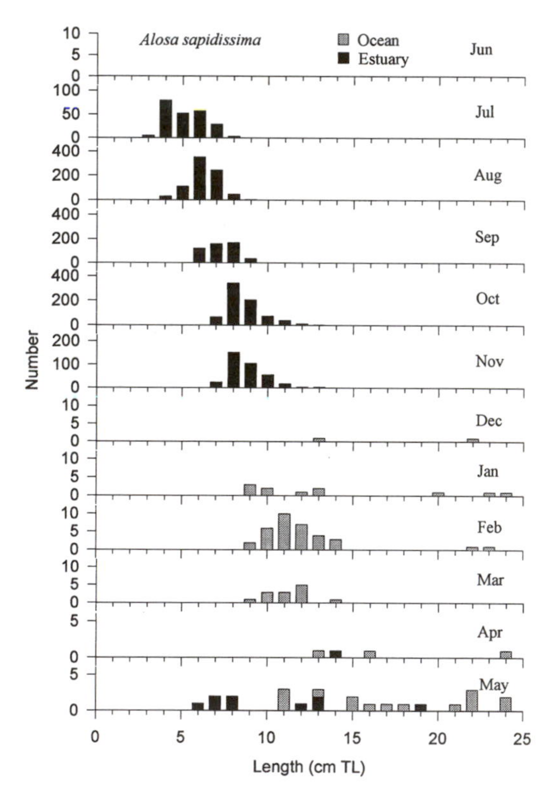

11.2 Monthly length frequencies of *Alosa sapidissima* juveniles and adults collected in New Jersey–New York estuarine and adjacent coastal ocean waters. Sources: Perlmutter et al. 1967; Thomas et al. 1975; Kahnle and Hattala 1988; Wilk et al. 1992; RUMFS: otter trawl ($n = 9$); weir ($n = 4$).

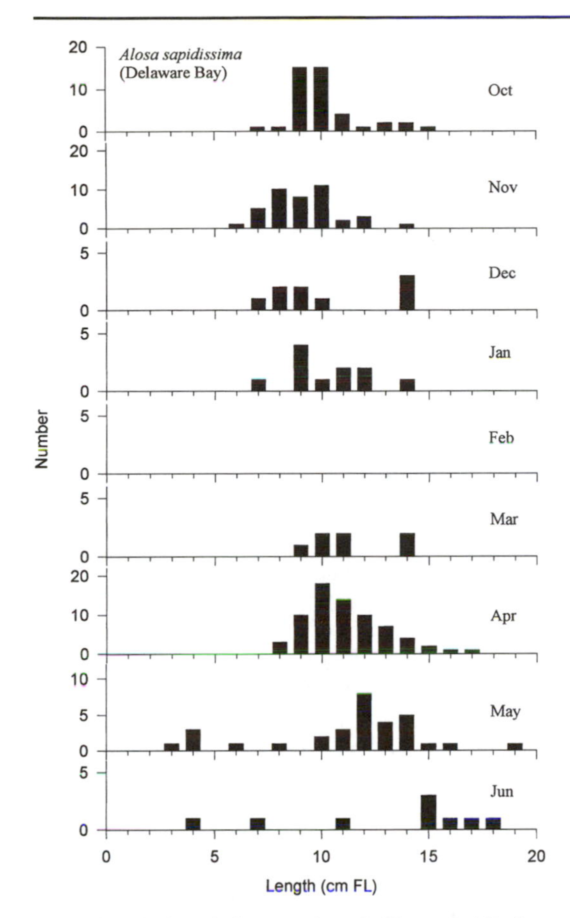

11.3 Monthly length frequencies of *Alosa sapidissima* collected between October and June in Delaware Bay, 1970–1980. Data courtesy of Public Service Electric & Gas and Environmental Consulting Services, Inc.

creases (5–30 ppt) much better than salinity decreases (30–0 ppt), where mortality results if the decrease is abrupt (Chittenden 1973b). In a laboratory study, young-of-the-year began to die when dissolved oxygen levels were below 1.92 mg per liter, and total mortality occurred at levels below 0.64 mg/liter (Chittenden 1973a). This study also found that minimum daily levels of 2.5 to 3.0 mg/liter were sufficient to permit migration through polluted areas.

Relatively little research has been done on the oceanic stages in the early life history. After leaving their natal rivers, juveniles remain at sea until reaching maturity. Some juveniles have been found to spend their first winter on the inner continental shelf off New Jersey (Milstein 1981) in association with two congeners, *A. pseudoharengus* and *A. aestivalis*. These fish ranged in length from less than 100 mm to about 150 mm through the winter, consistent with our current findings for lengths during those months (fig. 11.2). Some young-of-the-year remain in Delaware Bay waters through their first winter (fig. 11.3). The length ranges of these fishes were similar to those described above and they exhibited very little (or no) growth through this season.

CHAPTER 12

Brevoortia tyrannus (Latrobe) Atlantic Menhaden

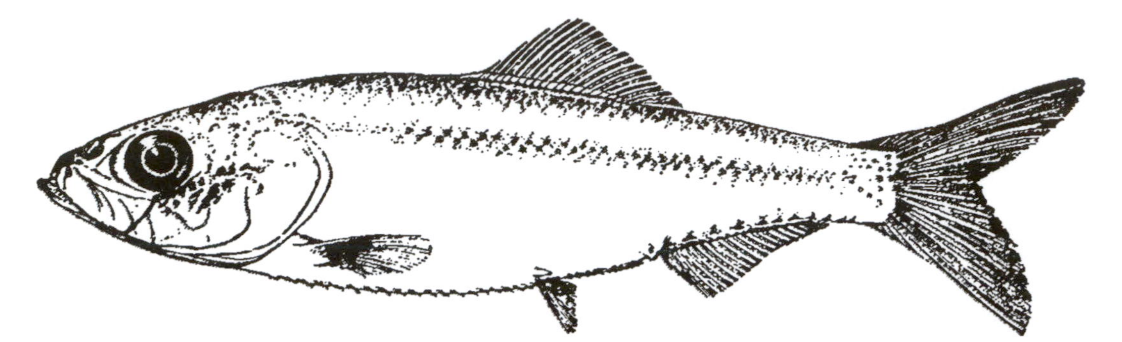

12.1 *Brevoortia tyrannus* juvenile, 41.0 mm TL. After Kuntz and Radcliffe 1917.

DISTRIBUTION

Brevoortia tyrannus (fig. 12.1) occurs commonly along the Atlantic coast of North America from the northern Gulf of Maine to waters off Indian River, Florida (Dahlberg 1970). Adults occur most abundantly in large schools in coastal areas, and these schools are especially abundant in and adjacent to major estuaries and bays. Seasonal migrations during spring and fall reportedly coincide with shifts in position of the 10 C isotherm (Reintjes 1969). They migrate north through the central part of the Middle Atlantic Bight during spring and return south during a fall migration (Nicholson 1971). These coastal migrations account for the occurrence of early life-history stages in all estuaries in the Middle Atlantic Bight (table 4.2). Most overwinter in waters south of Cape Hatteras, although some may also occur in Chesapeake Bay tributaries (June and Chamberlain 1959).

REPRODUCTION

Extensive egg collections in the Middle Atlantic Bight show that most spawning occurs over the inner continental shelf (Berrien and Sibunka in press), but some activity may extend into the lower regions of major bays and estuaries (e.g., Pearson 1941; Dovel 1971). Spawning occurs during nearly every month in some part of the range (McHugh et al. 1959). Based on extensive analyses in past decades, it has been determined that there is limited spawning activity during the northward spring migration, limited summer spawning as far north as Cape Cod (and into the Gulf of Maine), then increased spawning activity during the southward fall migration. This pattern is followed by intense spawning in the South Atlantic Bight during winter (Higham and Nicholson 1964). Collections of larvae in the Middle Atlantic Bight during the mid 1960s also substantiate this pattern (Kendall and Reintjes 1975). Spawning peaks in mid-May to early June and in September and October. This pattern has also been confirmed for Peconic Bays, New York (Ferraro 1980). In that study, spawning was most intense at temperatures between 15 and 18 C. Our analysis has confirmed that larvae in the Middle Atlantic Bight are abundant over the inshore part of the continental shelf in the fall (September–November) and are common during the winter (December–March) (fig. 12.2).

An extensive analysis of the age, size, and hatching dates for larvae collected in the Great Bay–Little Egg Harbor estuary indicates that these larvae are from different spawning areas (Warlen et al. in press). Larvae collected in the fall are likely from local spawning off New Jersey during the adult fall migration to the south. Larvae collected during the winter

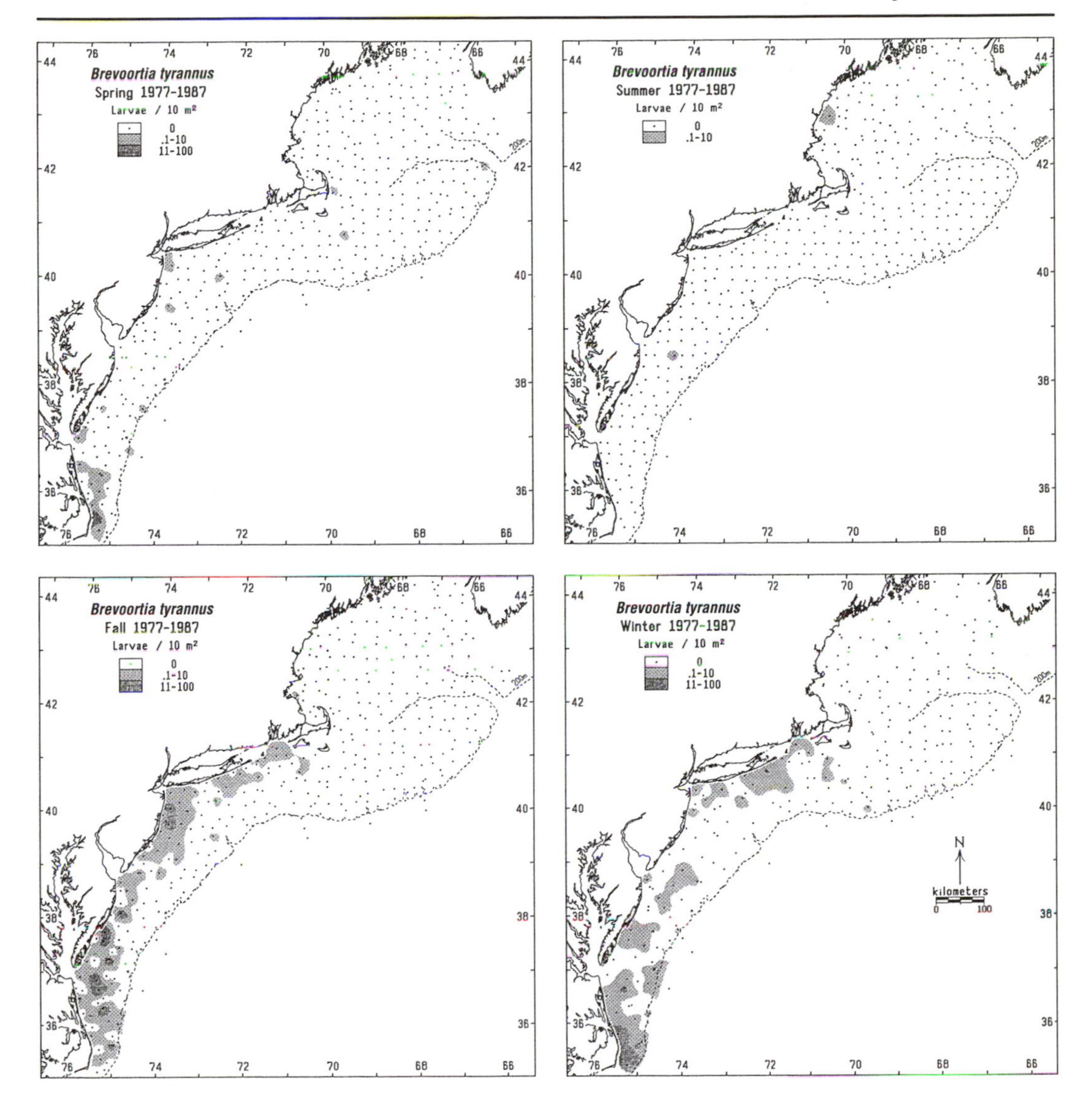

12.2 Seasonal distributions of larval *Brevoortia tyrannus* collected during MARMAP surveys, 1977–1987.

were probably spawned in the winter south of Cape Hatteras. The contribution of larvae from south of Cape Hatteras varied widely for the 3 years analyzed (1989–1990, 1990–1991, 1992–1993) and ranged from 10% to 87% (Warlen et al. in press) with the remainder contributed by local spawning off New Jersey.

DESCRIPTION

Eggs are pelagic and spherical and have a smooth chorion. Their diameter ranges from 1.3 to 1.9 mm. The yolk is segmented, and the perivitelline space is wide. The single oil globule ranges from 0.1 to 0.2 mm. Larvae hatch at 2.4 to 4.5 mm, with unpigmented eyes and an elongate body. The snout-anus length is 75% to 85% of standard length, and the anus is always posterior to the dorsal fin (Hettler 1984). Typical pigment includes a series of spots along the dorsal surface of the entire gut and along the ventral surface of the posterior gut. Juveniles (fig. 12.1) are recognized as herrings by the laterally compressed, relatively deep, and usually silvery body; the single, short-based dorsal fin; lack of adipose fin, and scales (scutes) of the belly midline sharp to the touch. The eye is smaller than in juveniles of *Alosa* species. There may be a single dark spot behind the edge of the gill cover. There are also seven rays in the pelvic fin (nine in *Alosa* species). Vert: 47–50; D: 18–24; A: 18–24; Pect: 13–19; Plv: 7.

THE FIRST YEAR

Larvae enter estuaries where they transform into juveniles (Pacheco and Grant 1965; Reintjes 1969). In the Great Bay–Little Egg Harbor, they are primarily postflexion stage (D. A. Witting, K. W. Able, and M. P. Fahay in prep.). In Delaware estuaries, larvae are typically 10 to 20 mm TL when they ingress (Wang and Kernehan 1979), and the height of this activity is December through May. In the Indian River, young occurred from September to June and were most abundant from December through May (Reintjes and Pacheco 1966). In New Jersey, the locally spawned larvae enter Great Bay–Little Egg Harbor in the fall, and this peak is often followed by a period of few or no collections (fig. 12.3); this has also been reported elsewhere (Lewis 1965). These periods of low catches are often associated with low water temperatures (1–5 C), followed by generally warming temperatures and larger numbers of larvae from south of Cape Hatteras, although there is some annual variation in this pattern (Warlen et al. in press). Thus, the larval contributions to this estuary, and presumably others in the region, are from different spawning areas at different times, but the relative contribution of these to the juvenile population is not known.

Transformation from the larval to the juvenile stage occurs at a length of 30 to 38 mm (Lewis et al. 1972). During this time, slender, scaleless, and largely unpigmented larvae become deep-bodied, large-headed juveniles and develop pigmentation and full complements of scales and ventral scutes. As the juveniles grow, they move upstream into lower-salinity waters (Lewis et al. 1972) and areas of maximum phytoplankton production (Friedland et al. 1996). Growth of juveniles has been estimated at 1 mm per day in some areas based on progression of length modes (Reintjes 1969). Growth is similar for juveniles in our study area through the spring and summer, and these young-of-the-year are 8 to 17 cm TL by September (fig. 12.4).

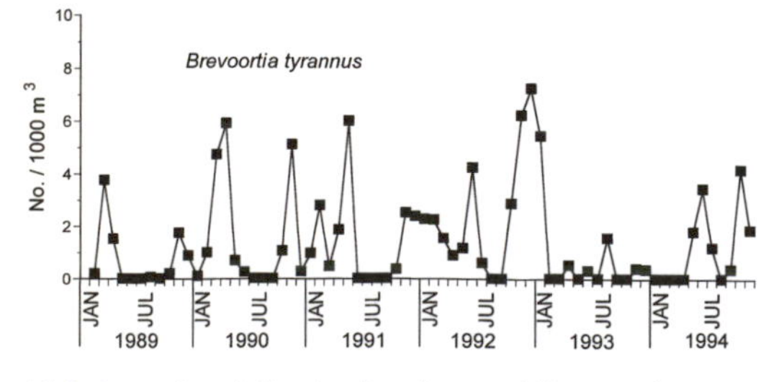

12.3 Annual variation in abundance of *Brevoortia tyrannus* larvae in Little Sheepshead Creek bridge 1-m plankton net collections ($n = 870$).

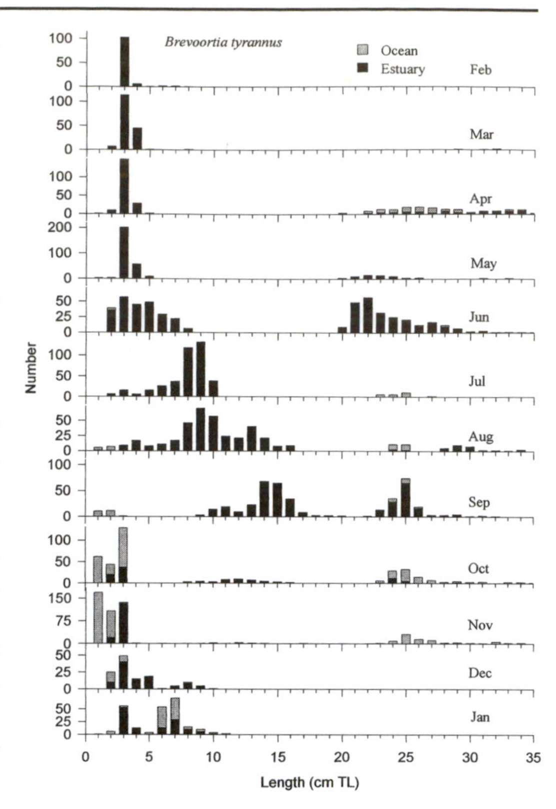

12.4 Monthly length frequencies of *Brevoortia tyrannus* young-of-the-year and small adults collected in the Great Bay study area. Sources: RUMFS: otter trawl ($n = 107$); 1-m plankton net ($n = 833$); LEO-15 Tucker trawl ($n = 146$); gear comparison ($n = 1$); seine ($n = 54$); night-light ($n = 14$); weir ($n = 252$).

The habitat of eggs and larvae is in the water column in coastal areas and estuaries. After transformation, young-of-the-year reside in estuaries through the summer (Ahrenholz et al. 1989) and have been reported to occur in salinities of less than 1 to 36 ppt. Important nursery areas in the central part of the Middle Atlantic

Bight include the C&D Canal and tidal tributaries of the lower Delaware River (Smith 1971; Wang and Kernehan 1979).

Most juveniles remain in estuarine habitats until September or October (Reintjes 1969) when declining temperatures initiate emigration to the ocean (Friedland and Haas 1988). Mortality has been reported when temperatures drop below 3 C for several days or when temperatures decrease rapidly to 4.5 C (Lewis 1965; Reintjes and Pacheco 1966). Therefore, as previously mentioned, larvae ingressing during the fall may be particularly susceptible to cold-induced mortality, while those ingressing during early spring are not. Most young-of-the-year migrate south and overwinter in offshore waters. Emigration to the ocean begins earliest in the north and progressively later to the south. Overwintering by juveniles in estuarine areas has been recorded between Chesapeake Bay and Florida (Reintjes 1969), but some have managed to overwinter in elevated temperatures of power-plant discharge plumes as far north as Delaware (Wang and Kernehan 1979).

CHAPTER 13

Clupea harengus Linnaeus
Atlantic Herring

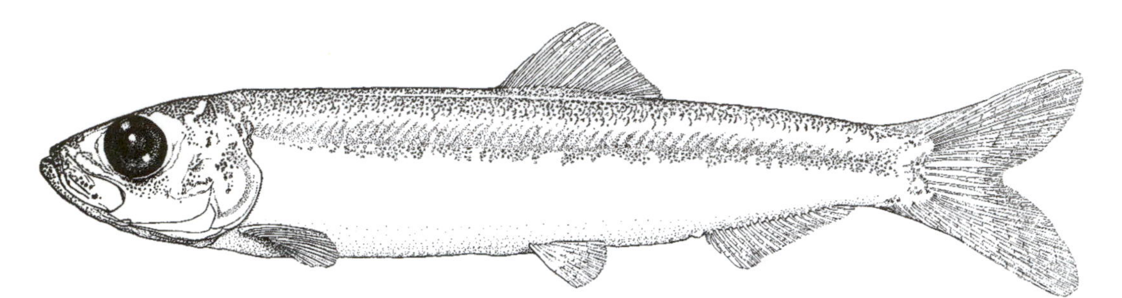

13.1 *Clupea harengus* juvenile, 46.0 mm SL. Collected May 6, 1993, dip net (night-light), RUMFS boat basin, Great Bay, New Jersey. ANSP 175209. Illustrated by Susan Kaiser.

DISTRIBUTION

A North Atlantic species, *Clupea harengus* (fig. 13.1), is found between western Greenland and Cape Hatteras (Scott and Scott 1988). The species is characterized by distinct populations, each with preferred spawning, feeding, and wintering areas (Iles and Sinclair 1982). These populations also have characteristic migration patterns between these areas, sometimes resulting in mingling of separate groups. Juveniles are also known to form separate concentrations in certain favored areas. The commercially exploitable stock nearest our study area occurs on Georges Bank. This stock has gone through several wide fluctuations in abundance, having been decimated by overfishing in the 1960s and 1970s but experiencing a resurgence in the 1980s. Evidence for this resurgence is provided by increased spawning activity, especially in the Nantucket Shoals/western Georges Bank region (Smith and Morse 1990). The Georges Bank stock uses the New York Bight as a wintering area between December and April, then returns to the bank in time to feed on a zooplankton bloom in the spring (Anthony 1982). Larvae and/or juveniles have been recorded from most estuaries north of Chesapeake Bay (table 4.2). There are also a few records of juveniles as far as 68 km up the Hudson River (Smith 1985).

REPRODUCTION

Clupea harengus spawns somewhere within its range throughout the year, with peaks of activity in spring and fall. Fishes in the Georges Bank stock spawn in September and October (Anthony 1982). Spawning fish sometimes form huge schools before depositing demersal eggs in shallow waters during spring and in deeper waters during fall. Based on collections of larvae, spawning occurred during every month in the MARMAP study area between the northern part of the Middle Atlantic Bight and Gulf of Maine but peaked in the fall (fig. 13.2). Spawning probably does not occur in continental shelf waters in the central part of the Middle Atlantic Bight because larvae were almost totally nonexistent there between 1977 and 1987, although a few were collected in the fall and winter (Morse et al. 1987). More recent spring and fall surveys have also not found many larvae in this study area. The origin of larvae that later appear in New Jersey inshore waters is unknown, but they may result from late-season, residual spawning by the overwintering group from Georges Bank, although these fish are not reported to spawn until their return to Georges Bank in the spring (Anthony 1982). The temporal occurrences of these young-of-the-year coincide with the common winter occurrence of this group in the New York Bight (Anthony 1982).

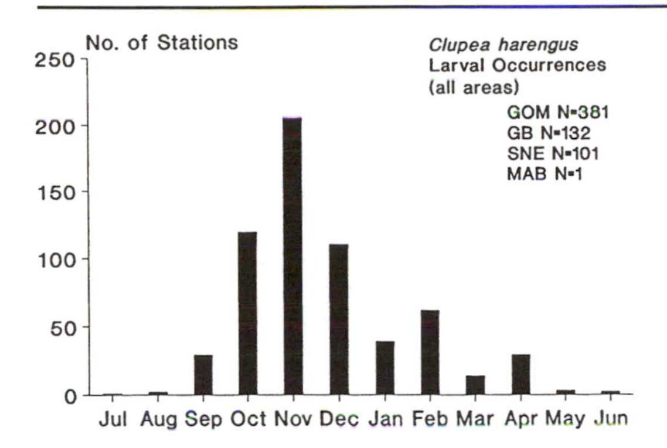

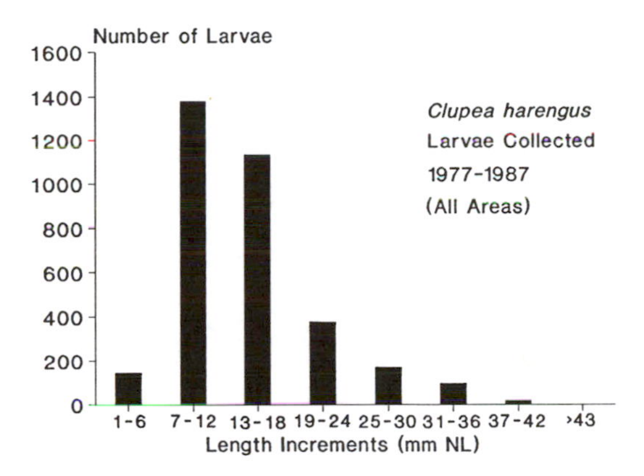

13.2 Lengths of larval *Clupea harengus* and months collected throughout MARMAP study area (Gulf of Maine through northern Middle Atlantic Bight), 1977–1987.

DESCRIPTION

Eggs are demersal and adhesive, and range from 1.0 to 1.4 mm in diameter. The perivitelline space is wide and there are no oil globules. Larvae hatch at 4 to 10 mm TL (usually >6.0 mm TL) with pigmented eyes but unformed mouth parts. The snout-anus length is about 80% of the elongate body. See Fahay (1983) for further details of larval development. Juveniles are recognized as clupeids by the laterally compressed, usually silvery body; single, short-based dorsal fin; lack of adipose fin; and scales (scutes) of belly midline sharp to the touch (fig. 13.1). They differ from other clupeid juveniles in the position of the dorsal fin origin near midbody (rather than closer to the head) and lack of a dark spot behind the gill cover. Vert: 55–57; D: 16–22 (usually 17–19); A: 15–21 (usually 17–19); Pect: 13–21; Plv: 6–10.

THE FIRST YEAR

Eggs remain attached to the bottom through the incubation period, which lasts 10 to 30 days depending on temperatures. Larvae hatch at 4 to 10 mm TL and those collected in continental shelf waters range from less than 6 to 42 mm NL (fig. 13.2). Development to transformation requires another 5 to 7 months, depending on temperatures. Larvae attain lengths of about 30 mm before beginning to transform into juveniles (Fahay 1983) but may attain a maximum larval size of 42.5 mm before transformation (Saila and Lough 1981). During this process, the gut length decreases relative to body length, the dorsal fin migrates forward on the body, the body deepens, the air bladder becomes prominent and all fin rays are ossified. European laboratory studies found this process begins 152 days after hatching at sea temperatures (7–12 C), 120 days in warmer lab temperatures (Doyle 1977). In the Georges Bank/Nantucket Shoals area, estimated duration of larval life is 210 days (Saila and Lough 1981).

Larvae enter the Great Bay–Little Egg Harbor estuary during the winter and spring. On the inner continental shelf near Little Egg Inlet, two larvae (37–39 mm) were collected during January, and ingressing larvae (19–63 mm) were taken in plankton net sampling in nearby Little Sheepshead Creek (fig. 13.3) between January and May (D. A. Witting, K. W. Able, and M. P. Fahay in prep.). Of 30 entering larvae examined, 29 were in Transformation Stage 4B, and 1 was in Stage 4C (*sensu* Doyle 1977). Larvae (23–45 mm) have also been recorded entering Indian River Inlet from January through April (Pacheco and Grant 1973). The source of

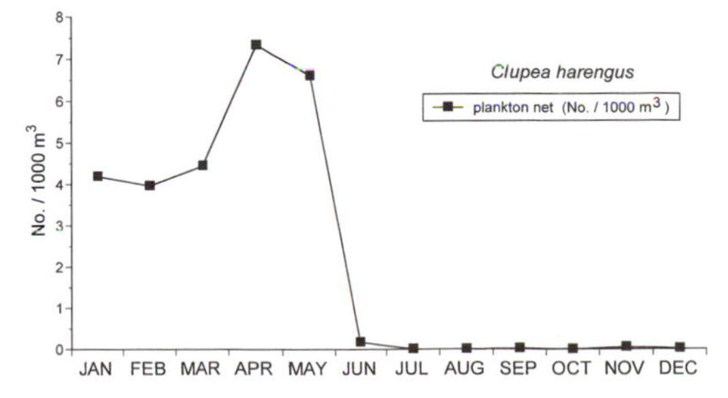

13.3 Monthly occurrences of *Clupea harengus* larvae in the Great Bay study area based on Little Sheepshead Creek Bridge 1-m plankton net collections 1989–1994 (n = 1,427).

transforming larvae entering New Jersey inlets remains enigmatic. We have seen that egg incubation requires up to a month, and development to transformation another 5 to 7 months. Most ingress of late transformation larvae occurs between January and May. Therefore, spawning producing these progeny must occur between July and September, at a time when adults are not recorded from the central part of the Middle Atlantic Bight. It is possible that the resurgence of spawning by this species over Georges Bank and Nantucket Shoals (Smith and Morse 1990) has also included increased reproductive activity in the northern part of the Middle Atlantic Bight, but we presently lack the sampling effort to demonstrate this. Larval occurrences in the Great Bay–Little Egg Harbor study area have been increasing in recent years (fig. 13.4), including in pop-net sampling from 1995 to 1996 (S. Hagan unpubl. observ.) and may be another signal that spawning activity in the central part of the Middle Atlantic Bight is increasing along with the resurgence. Establishing the birthdates of larvae ingressing New Jersey estuaries would be a logical first step in unraveling this mystery.

Growth rates during the initial juvenile stages vary between nursery areas and between spring- and fall-spawned cohorts. It is difficult to describe a growth rate in our Great Bay–Little

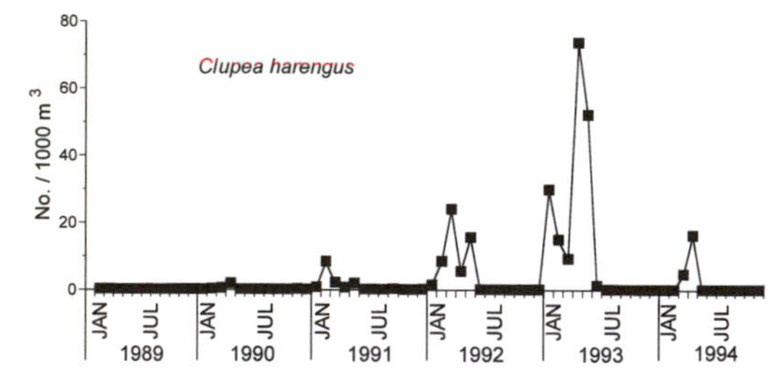

13.4 Annual variation in abundance of *Clupea harengus* larvae in Little Sheepshead Creek Bridge 1-m plankton net collections 1989–1994 (n = 1,427).

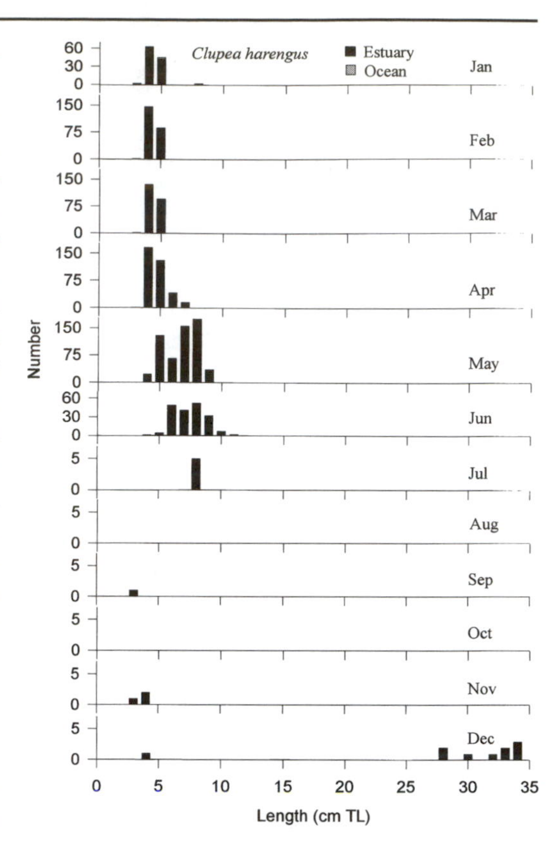

13.5 Monthly length frequencies of *Clupea harengus* collected in estuarine and adjacent coastal ocean waters of the central part of the Middle Atlantic Bight. Sources: de Sylva et al. 1962; Schaefer 1967; Pacheco and Grant 1973; Wilk and Silverman 1976a; Allen et al. 1978; RUMFS: otter trawl (n = 74); 1-m plankton net (n = 1,056); LEO-15, Tucker trawl (n = 2); LEO-15 2-m beam trawl (n = 2); killitrap (n = 2); seine (n = 10); night-light (n = 255); weir (n = 329).

Egg Harbor study area, since almost all available data on young-of-the-year end in early summer when they apparently leave the system at about 10 cm and are no longer available to collecting gear (fig. 13.5). Details of this emigration, and subsequent distribution of young-of-the-year in the central part of the Middle Atlantic Bight remain undescribed.

CHAPTER 14

Anchoa hepsetus (Linnaeus) Striped Anchovy

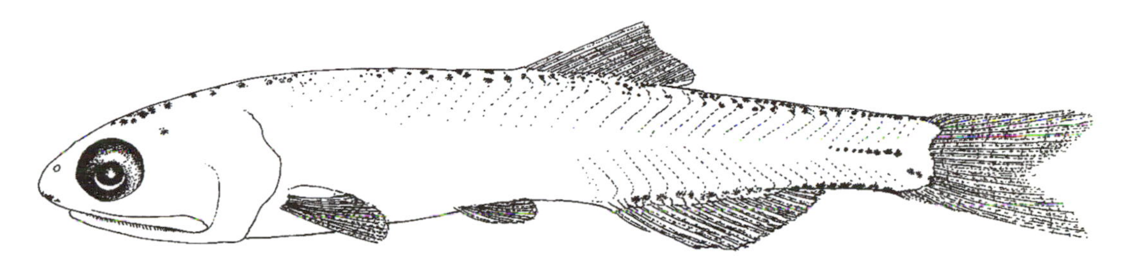

14.1 *Anchoa hepsetus* juvenile, 34.0 mm TL. Collected August 2, 1995, pop net, RUMFS boat basin, Great Bay, New Jersey. ANSP 175210. Illustrated by Nancy Arthur.

DISTRIBUTION

A pelagic species found from Nova Scotia to Uruguay, *Anchoa hepsetus* (fig. 14.1) is most abundant from Chesapeake Bay to the West Indies (Hildebrand 1963). Even in these areas, it is annually variable. In the Middle Atlantic Bight, the early life-history stages have been reported from most estuaries up to Great South Bay and Narragansett Bay (table 4.2). In many of these, it has been collected primarily from the polyhaline portions, as in Delaware Bay (de Sylva et al. 1962) and Great Bay–Little Egg Harbor (Able et al. 1996b; S. M. Hagan and K. W. Able unpubl. data).

REPRODUCTION

Spawning in the ocean occurs from April through September, with the greatest abundance occurring from southern New Jersey southward, based on the distribution and abundance of eggs (Berrien and Sibunka in press). Spawning begins as early as April near Cape Hatteras, then northward to Delaware Bay by May. In lower Delaware Bay, it occurs from May through August (Stevenson 1958). Eggs have been collected in May in Great South Bay, Long Island (Monteleone 1992); however, the smallest larvae have not been found before September in Great Bay–Little Egg Harbor (fig. 14.3). Spawning in the southern part of the Middle Atlantic Bight reportedly occurs within harbors, estuaries, and sounds (Hildebrand and Cable 1930).

DESCRIPTION

The eggs are strongly elliptical. The long axis is 1.2 to 1.6 mm, and the short axis is 0.7 to 0.9 mm. The perivitelline space is narrow, oil globules are absent, and the yolk is segmented. At hatching, larvae are 3.6 to 4.0 mm and taper from a bulbous head and yolk sac to a thin tail; eyes and mouth parts are nonfunctional. The anus is located about two-thirds the distance from the snout. Larger larvae have muscle-band striations in the posterior half of the intestine. Further details of egg and larval development are in Hildebrand and Cable (1930) and Fahay (1983). Juveniles are characterized by a laterally compressed, silvery body (fig. 14.1). The midventral line of the belly is smooth, unlike

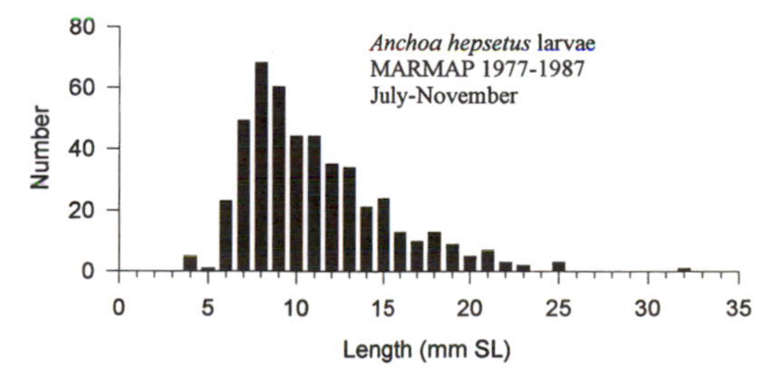

14.2 Composite length frequencies of *Anchoa hepsetus* larvae in the Middle Atlantic Bight.

most herrings, which have a ridge of sharp scutes. A single, short dorsal fin is located slightly behind the midpoint of the body, and the anal fin origin is under the rear of the dorsal fin base. The caudal fin is deeply forked. The mouth is positioned well under the rounded, overhanging snout, and the maxillary bone extends well past the rear margin of the eye. There are no strong spines, ridges, or bumps on the head or body. See Appendix A for separation from *A. mitchilli* and *Engraulis eurystole.* Vert: 40–44; D: 13–17; A: 18–23; Pect: 13–17; Plv: 7.

THE FIRST YEAR

Many aspects of the early life history, including events during the first year, are poorly known. The larvae (up to 25 mm) have been frequently collected in the ocean (fig. 14.2). In Great Bay–Little Egg Harbor young-of-the-year (1–10 cm TL) have been collected from July through November (fig. 14.3). In Hereford Inlet in southern New Jersey, estuary collections were dominated by individuals smaller than 60 mm (Allen et al. 1978). Some young-of-the-year have also been collected from Indian River Inlet during August (Wang and Kernehan 1979). In the Great Bay–Little Egg Harbor Estuary, presumed young-of-the-year (6–8 cm) have been collected from subtidal creeks in late summer (Rountree et al. 1992; Able et al. 1996b). In Chesapeake Bay, juveniles and adults migrate to deeper waters in fall and winter (Musick 1972) and leave the bay (Hildebrand and Schroeder 1928).

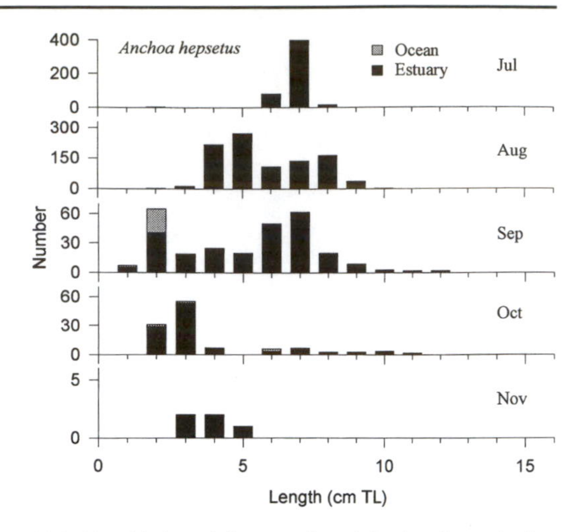

14.3 Monthly length frequencies of *Anchoa hepsetus* larvae and young-of-the-year collected in the Great Bay–Little Egg Harbor study area. Sources: RUMFS: otter trawl (n = 49); 1-m plankton net (n = 189); LEO-15 Tucker trawl (n = 38); LEO-15 2-m beam trawl (n = 4); gear comparison (n = 4); weir (n = 58); pop net (n = 1,519).

CHAPTER 15

Anchoa mitchilli (Valenciennes)
Bay Anchovy

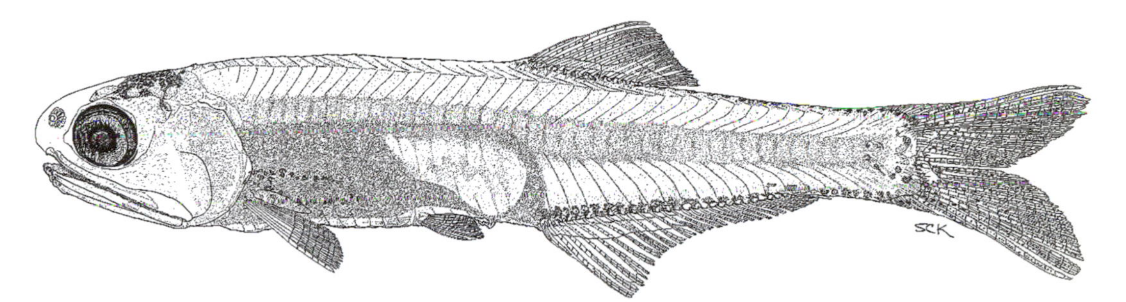

15.1 *Anchoa mitchilli* juvenile, 33.0 mm SL. Collected September 27, 1990, 1-m plankton net, Little Sheepshead Creek Bridge, Great Bay, New Jersey. ANSP 175211. Illustrated by Susan Kaiser.

DISTRIBUTION

Anchoa mitchilli (fig. 15.1) is a coastal species found from Maine to the coastal Gulf of Mexico (Hildebrand 1943; Bigelow and Schroeder 1953) and may be the most abundant fish species in the western north Atlantic (McHugh 1967; Morton 1989). It is found in a wide variety of habitats over a broad range of salinities, from the ocean to tidal freshwaters. In the Middle Atlantic Bight, it undergoes a seasonal migration along inner continental shelf waters to the southern portion of the bight and beyond during the fall and a return migration in the spring (Vouglitois et al. 1987). Young stages occur in every estuary in the Middle Atlantic Bight (table 4.2).

REPRODUCTION

Males and females may mature at 40 to 45 mm FL at approximately 10 months of age (Zastrow et al. 1991), but they are also reported to mature as small as 31 mm FL and at 2.5 to 3 months (Luo and Musick 1991). Individuals in Chesapeake Bay were reported to spawn every 1.3 to 4 days with up to 54 total spawnings per female (Luo and Musick 1991). Most spawning occurs in Middle Atlantic Bight estuaries between April and November as indicated by gonad development and the occurrence of eggs and larvae (Stevenson 1958; Croker 1965; Dovel 1981; Vouglitois et al. 1987; Castro and Cowen 1991; Luo and Musick 1991; Zastrow et al. 1991; Monteleone 1992; Wang and Houde 1995). This same time frame is evident on the inner continental shelf west of Nantucket Shoals (Berrien and Sibunka in press). In intensive sampling in Chesapeake Bay, a correlation analysis suggested that spawning was most intense in areas of high zooplankton abundance and where the ctenophore *Mnemiopsis leidyi,* a potential predator on eggs and larvae, was least abundant (Dorsey et al. 1996). Most spawning occurs in the evening (Ferraro 1980; Luo and Musick 1991; Zastrow et al. 1991). Egg densities on the inner continental shelf (Berrien and Sibunka in press) may be of the same order of magnitude as in the estuary (Vouglitois et al. 1987). In the latter study, egg abundance was high as far as 6 km offshore but declined at locations 15 to 18 km offshore.

DESCRIPTION

The eggs are slightly oval to nearly spherical (0.8 to 1.3 mm in the longest dimension), but size and shape vary with salinity, and this may account for the large size variation reported (Dovel 1971). The eggs have a narrow perivitelline space and a segmented yolk. Hatching occurs at 2.0 mm TL, and eye and mouth parts are nonfunctional. Larvae are bulbous anteriorly, and the anus is located approximately two-thirds of the distance from the snout. Larger larvae have muscle-band striations in the posterior part of the intestine. Further details of development are described by Kuntz (1914a) and Wang and Kernehan (1979). Juveniles are characterized by a laterally compressed, silvery body (fig. 15.1). The belly is smooth, unlike most clupeids, which have a ridge of sharp scutes. A single, short dorsal fin is located near midbody, and the anal fin origin is under the middle of the dorsal fin base. The mouth is well under the rounded, overhanging snout, and the maxillary bone extends well past the rear margin of the eye. There are no strong spines, ridges, or bumps on the head or body. See Appendix A for separation from *Anchoa hepsetus* and *Engraulis eurystole*. Vert: 38–44; D: 13–17; A: 24–30; Pect: 10–13.

THE FIRST YEAR

Hatching occurs in approximately 24 hours at summer temperatures (Mansueti and Hardy 1967; Monteleone 1992). Mortality during embryonic and larval development is high. In a Chesapeake Bay field study, 73% of the eggs died before hatching. For recently hatched larvae (24 hours after hatching), the mortality was 64% (Dorsey et al. 1996). Mesocosm experiments with eggs and yolk sac larvae in Chesapeake Bay indicated that 95% of a cohort died within 2 days of hatching (Houde et al. 1994). In Great South Bay, the average seasonal mortality rates for eggs varied annually but ranged from 69% to 98.2% per day, while larval (3–15-day-old) rates were 34.6% to 37.0% per day (Castro and Cowen 1991). These high mortality rates point out the probable importance of survival during the egg and larval stages in recruitment processes.

In Great Bay–Little Egg Harbor, there is considerable annual variation in abundance and size of larvae. Years of peak abundance frequently may be followed by years with very low levels (fig. 15.2). Similar extreme variation has been suggested for the Hudson River (Dovel 1981), Barnegat Bay (Vouglitois et al. 1987), Delaware Bay (Derickson and Price 1973), and Chesapeake Bay (Newberger and Houde 1995). These fluctuations are likely to result in recruitment variation because the populations are dominated by young-of-the-year and age 1

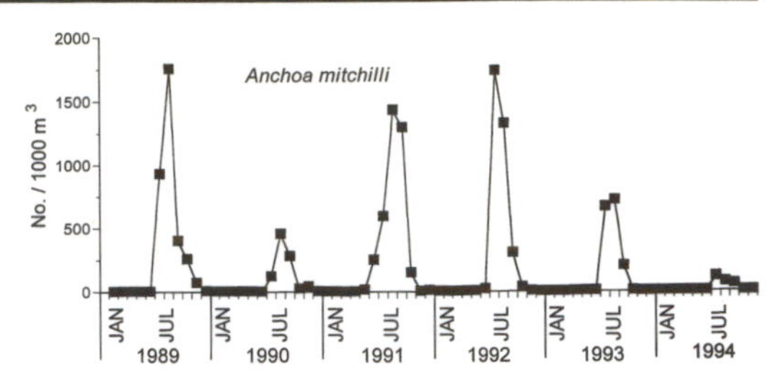

15.2 Annual variation in abundance of *Anchoa mitchilli* larvae in Little Sheepshead Creek Bridge 1-m plankton net collections 1989–1994 (n = 105,117).

individuals (Vouglitois et al. 1987; Newberger and Houde 1995; Wang and Houde 1995). In upper and mid-Chesapeake Bay, annual production of young-of-the-year was 856 g per 100 m^3, 87.9% of which was produced in the first 3 months of life (Wang and Houde 1995). Production by young-of-the-year (92.6%) accounted for nearly all of the annual production for this species. Annual mortality rates in Chesapeake Bay ranged from 89% to 95% per year, and growth averaged 0.47 mm per day for young-of-the-year (Newberger and Houde 1995). Production of *A. mitchilli* young-of-the-year is of such large magnitude that it could influence the total fish production of many estuaries.

Several estimates of growth are available for young-of-the-year. In Great South Bay, larval growth rates averaged 0.53 to 0.55 mm per day (Castro and Cowen 1991). In Chesapeake Bay, these varied from 0.32 to 0.47 mm per day (Zastrow et al. 1991). In Great Bay–Little Egg Harbor, the young-of-the-year are approximately 2 to 6 cm TL by October (fig. 15.3) and the same pattern is evident for Delaware Bay (fig. 15.4). In Barnegat Bay, the relative growth and survival varied between years with the number of larvae and juveniles smaller than 10 mm FL ranging from 31% to 70% to 10% in three successive years.

The young-of-the-year tolerate a wide range of estuarine habitats. Eggs have been collected over a wide salinity gradient in both Delaware Bay (Wang and Kernehan 1979) and Chesapeake Bay (Dovel 1971), but egg viability was highest when salinities were greater than 8.0 ppt (Wang and Kernehan 1979). In Great South Bay, egg density was greatest near an inlet, but

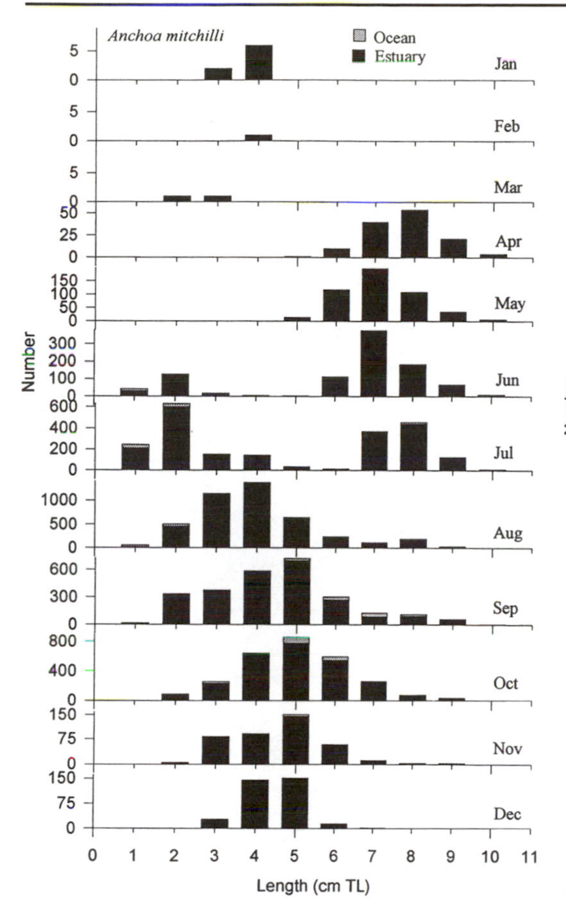

15.3 Monthly length frequencies of *Anchoa mitchilli* larvae and young-of-the-year collected in the Great Bay–Little Egg Harbor study area. Sources: RUMFS: otter trawl (*n* = 5,419); 1-m beam trawl (*n* = 107); 1-m plankton net (*n* = 4,567); experimental trap (*n* = 3); LEO-15 Tucker trawl (*n* = 125); LEO-15 2-m beam trawl (*n* = 135); gear comparison (*n* = 2,129); killitrap (*n* = 2); seine (*n* = 296); night-light (*n* = 108); weir (*n* = 1,475).

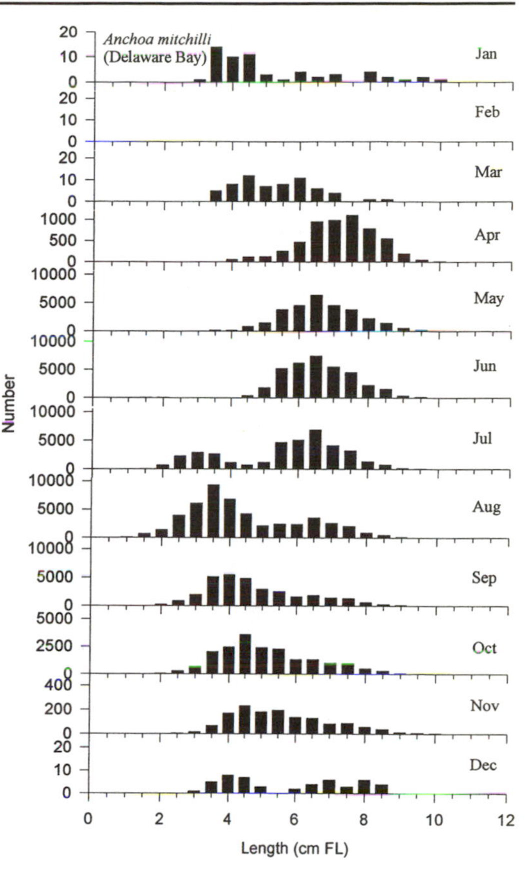

15.4 Monthly length frequencies (0.5-cm intervals) of *Anchoa mitchilli* collected from Delaware Bay, 1979–1982. Data courtesy of Public Service Electric & Gas and Environmental Consulting Services, Inc.

larvae were homogeneous throughout the bay (Monteleone 1992). The young-of-the-year are ubiquitous from the deeper portions of estuaries (Szedlmayer and Able 1996) to the shorelines (Able et al. 1996b).

The movements of young-of-the-year appear to have a regular pattern. In the Hudson River, many newly hatched larvae moved upstream from high-salinity areas (>10 ppt) to lower-salinity areas (<10 ppt). In early fall, larvae and juveniles began to move downstream into higher salinity areas (Dovel 1981; Loos and Perry 1991; Wang and Kernehan 1979). In Great Bay–Little Egg Harbor, the young-of-the-year are abundant from June to January although most leave by October (fig. 15.5).

At the end of the first summer, most individuals migrate out of Middle Atlantic Bight

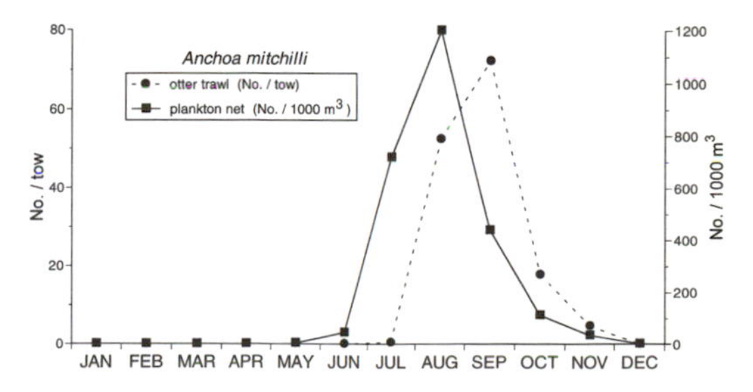

15.5 Monthly occurrences of *Anchoa mitchilli* larvae in the Great Bay study area based on otter trawl collections 1988–1989 (*n* = 19,070) and Little Sheepshead Creek Bridge 1-m plankton net collections 1989–1994 (*n* = 105,117).

estuaries and are very abundant on the inner continental shelf in the fall (Vouglitois et al. 1987). For Great Bay–Little Egg Harbor, estimates of the size composition of young-of-the-year in October or later may be confounded by the likelihood that larger individuals leave the estuary first. This would explain the decrease and eventual disappearance of larger individuals (6–7 cm TL) during November and December (fig. 15.3). A similar pattern has been suggested for Chesapeake Bay (Wang and Houde 1995).

Events in the ocean during the winter may significantly affect the number of young-of-the-year that return to Barnegat Bay the following spring because abundant year classes are not always evident the following year (Vouglitois et al. 1987).

CHAPTER 16

Osmerus mordax (Mitchill)
Rainbow Smelt

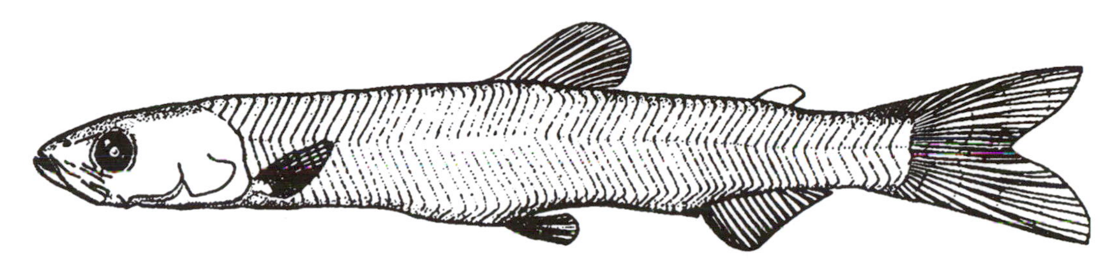

16.1 *Osmerus mordax* early juvenile, 36.0 mm TL. After Cooper 1978.

DISTRIBUTION

Osmerus mordax (fig. 16.1) occurs along the coast between Labrador and New Jersey (Scott and Crossman 1973). It is an anadromous species, ascending freshwater streams in the spring for spawning (Scott and Scott 1988). They avoid warm water and move offshore to deeper, cooler waters during summer (Bigelow and Schroeder 1953). They reach their southern range limit in northern New Jersey, although they once occurred in Delaware Bay (Robins and Ray 1986). They were once considered common spawning residents in the New York area (Nichols and Breder 1927). Recent records include anadromous populations from the Hudson River and numerous streams on Long Island, as well as several landlocked populations in New York (Smith 1985). Early life-history stages have been reported from Middle Atlantic Bight estuaries from the Hudson River and north (table 4.2).

REPRODUCTION

In the central part of the Middle Atlantic Bight, most information on spawning concerns landlocked populations, and there is little on those that utilize estuarine waters. Spawning presumably occurs in the Hudson River and in certain, small Long Island streams in February or March when water temperatures reach about 9 C (Smith 1985). Mature adults have been collected recently in Newark Bay from January through April and young-of-the-year have also been collected there during April (S. J. Wilk pers. comm.).

DESCRIPTION

Eggs are demersal, adhesive and about 1.0 mm in diameter. They have a granular, yellow yolk, and numerous oil globules. The adhesive outer covering becomes loosened by agitation, turns inside out, and becomes a pedestal-shaped attachment device that holds the developing egg above the substrate (Cooper 1978). Larvae hatch at 5.5 to 6.0 mm. Fine pigment spots are located along the ventral edge of the gut and tail. The snout-anus length is 65 to 75% TL, which is shorter than in similar appearing *Alosa* spp. larvae (Cooper 1978). In juveniles, a single, short-based dorsal fin, followed by a rayless adipose fin, is completely formed by 29 mm (fig. 16.1). Fin rays are complete at about 36 mm. The pelvic fin moves from a position anterior, to a position posterior, of the dorsal fin origin at sizes 22 to 36 mm (Cooper 1978). The maxilla extends slightly beyond the rear edge of the eye. There may be a silvery stripe along the side of the body. Vert: 60–64; D: 10–11; A: 15–17; Pect: 10; Plv: 8.

THE FIRST YEAR

Incubation times for eggs vary with temperature. In the Miramichi River, New Brunswick, eggs incubate for as little as 8 days (at 20 C) to 63 days (at about 4 C) (McKenzie 1964). After hatching at 5.5 to 6.0 mm, larvae drift downstream to brackish water. At 36 mm, scales cover the posterior half of the area between the vent and caudal fin. Pigment is sparse at this size, with some specimens showing scattered pigment on the caudal fin and perhaps a single spot on the posterior edge of the opercle. Specimens 40 mm and larger have increased pigment on the caudal fin, giving it a dusky appearance (Cooper 1978).

Larvae may grow to 20 to 40 mm in a few months and may reach 50 mm by August (Scott and Crossman 1973). In the Woods Hole, Massachusetts region, they may reach 45 to 75 mm by the end of the first summer (Bigelow and Schroeder 1953).

There is no information on habitat use by young-of-the-year in the central part of the Middle Atlantic Bight. A single cohort with a mode at about 30 mm was recently (1994) collected in Newark Bay during April (S. J. Wilk pers. comm.). In the St. Lawrence River, developing larvae are retained in turbid estuarine waters by using "selective tidal stream transport," by which they migrate to near-surface layers during flood tides and descend to the bottom during the ebb (Laprise and Dodson 1989), thereby minimizing net downstream displacement. Young-of-the-year in our area most likely migrate downstream and into the ocean by early summer as they do in southern New England waters (Bigelow and Schroeder 1953).

CHAPTER 17

Synodus foetens (Linnaeus) Inshore Lizardfish

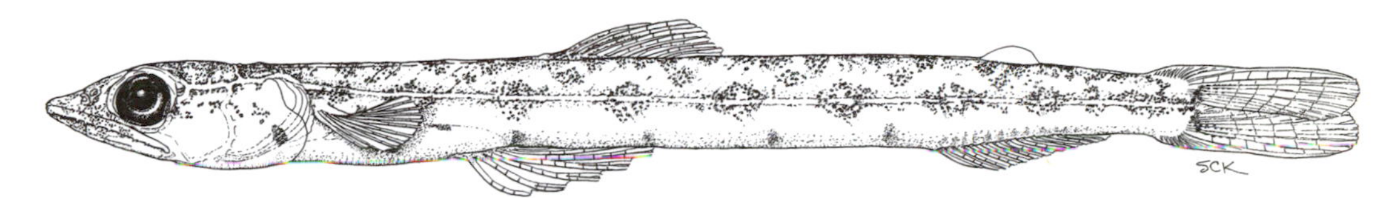

17.1 *Synodus foetens* early juvenile, 35.4 mm SL. Collected July 23, 1990, dip net (night-light), RUMFS boat basin, Great Bay, New Jersey. ANSP 175212. Illustrated by Susan Kaiser.

DISTRIBUTION

Synodus foetens (fig. 17.1) is found from Massachusetts to Brazil, including the Gulf of Mexico (Robins and Ray 1986). Several studies have remarked on interannual variation in abundance of this seasonal visitor in New Jersey (e.g., Allen et al. 1978). It is most common south of South Carolina (Breder 1948). They occur in bays, coastal areas, and the continental shelf to a depth of 200 m, but their favored habitat is shallow waters with sandy bottoms where they burrow into the substrate (Robins and Ray 1986). Juveniles occur regularly in estuaries in the southern part of the Middle Atlantic Bight (table 4.2), but larvae are only rarely reported.

REPRODUCTION

Aspects of reproduction are poorly known. Distribution of synodontid larvae near the edge of the continental shelf (fig. 17.2) indicates that they are probably transported into the Middle Atlantic Bight via the Gulf Stream from south of Cape Hatteras, where spawning presumably begins in spring and continues through summer and fall. There is no evidence to suggest that adult occurrences during some years are associated with a spawning migration. In the Tampa Bay, Florida, area the smallest larvae occurred during November and December (Springer and Woodburn 1960) although larvae less than 40 mm have been collected during the entire year (Gibbs 1959). In Middle Atlantic Bight continental shelf waters, small larvae have been collected continuously from March through October (fig. 17.2), suggesting a prolonged spawning period or prolonged ingress.

DESCRIPTION

Eggs are undescribed. Larvae hatch at about 2.5 mm and have a blunt snout and a prominent series of six melanophores located along the margin of the gut and lower body. The skin is inflated in larvae smaller than 10 mm, and fin rays are complete in some larvae at this size (Hoese 1965). The presence of an adipose fin between the dorsal and caudal fins separates juveniles of this species from all others considered here except *Osmerus mordax,* which is more laterally compressed. The flat head and large mouth resemble those of a lizard (fig. 17.1). Juveniles are elongate, and the anal fin base is longer than the dorsal fin base. Characteristic pigment includes eight, roughly diamond-shaped spots spaced along the midline of the body, which is a different pattern from the plain silvery coloration in smelts. Vert: 56–61; D: 10–13; A: 9–12; Pect: 10–15; Plv: 8.

THE FIRST YEAR

As larvae transform into juveniles, the snout elongates, the prominent melanophores along the gut internalize, and, with further growth, become overlain with dermal pigment and disappear. Larvae and transforming juveniles begin to settle to the bottom at sizes of about 30 to 40 mm (Hoese 1965). Recently settled individuals (30–35 mm) have been collected in

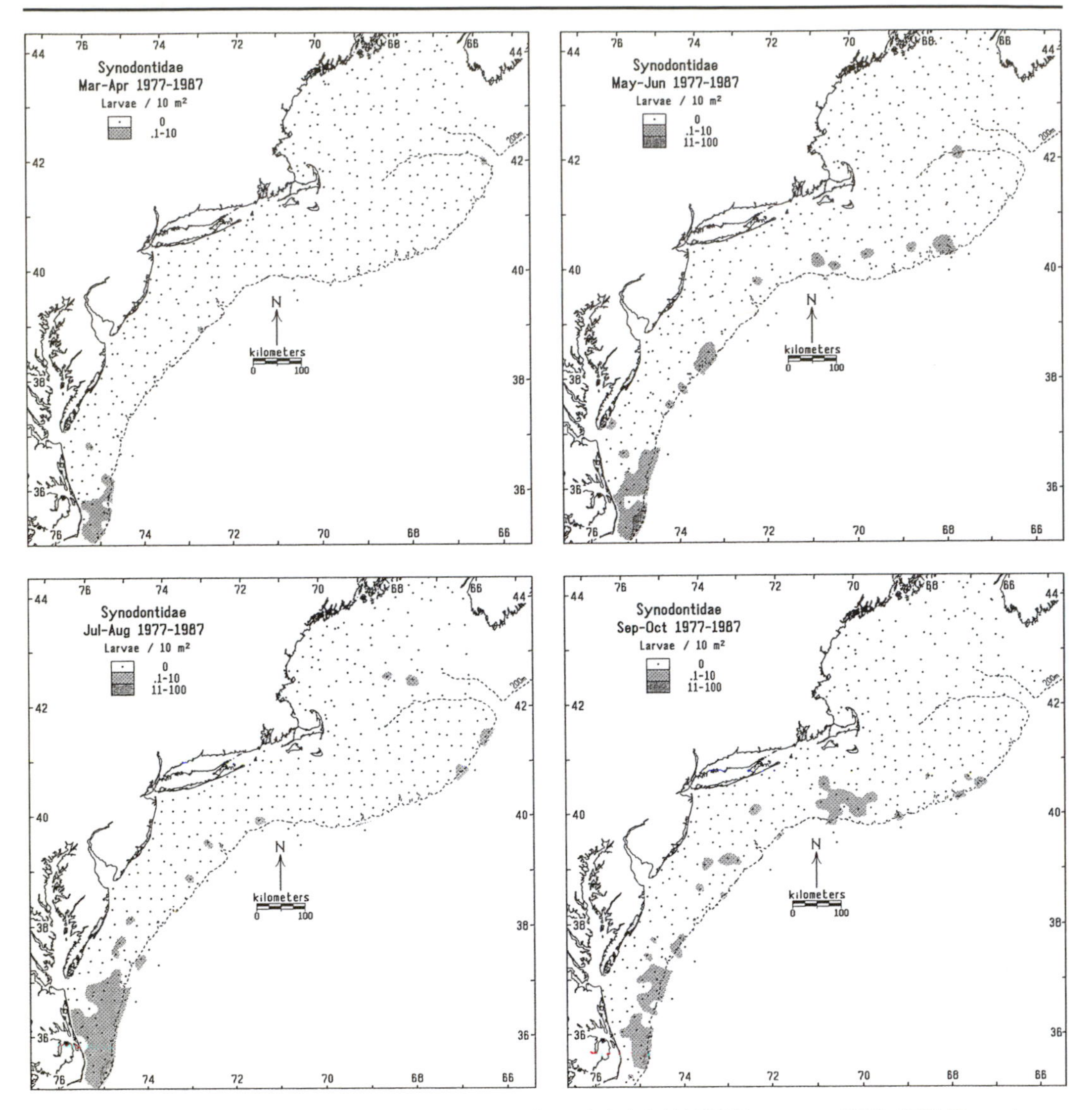

17.2 Bimonthly distributions of Synodontidae larvae collected during MARMAP surveys 1977–1987. These larvae were not identified to species level.

epibenthic sled hauls on the sea bottom near Hereford Inlet (Allen et al. 1978). The same authors report seine collections of pretransformation larvae (about 35 mm).

It is unknown whether settlement occurs in one habitat more so than in others, although there are reports of early settlers burrowing in muddy substrates (Breder 1962; Hoese 1965). It is also unknown whether all young-of-the-year enter estuaries, either before or after settlement. Collections of small juveniles in the Little Egg Inlet–Great Bay area peak in June and young-of-the-year continue to occur there through the summer and fall (fig. 17.3). In

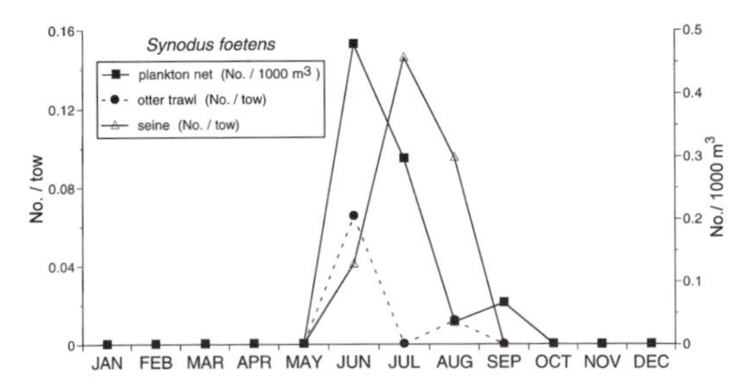

17.3 Monthly occurrences of *Synodus foetens* larvae and young-of-the-year in the Great Bay–Little Egg Harbor study area based on otter trawl collections 1988–1989 ($n = 8$), seine 1990–1991 ($n = 13$) and Little Sheepshead Creek Bridge 1-m plankton net collections 1989–1994 ($n = 28$).

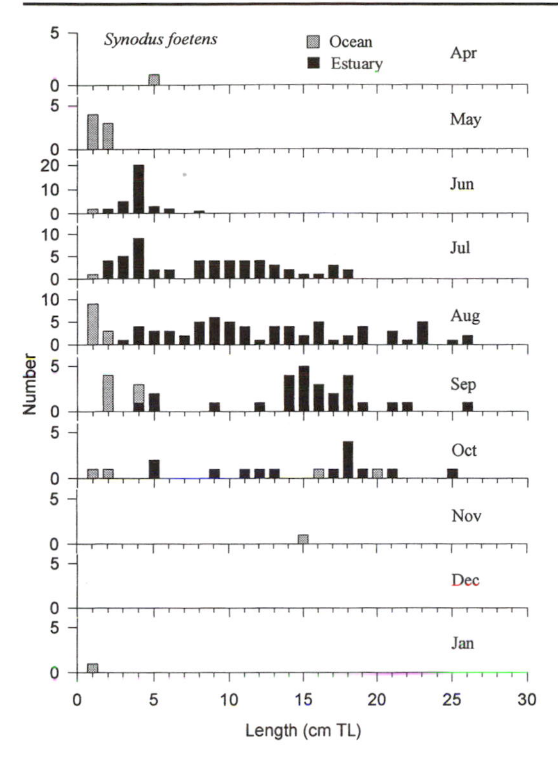

17.4 Monthly length frequencies of *Synodus foetens* young-of-the-year and adults collected in estuarine and adjacent coastal ocean waters of the central part of the Middle Atlantic Bight. Sources: Allen et al. 1978; MARMAP samples 1977–1987; RUMFS: otter trawl (n = 89); 1-m beam trawl (n = 3); 1-m plankton net (n = 26); LEO-15 2-m beam trawl (n = 4); gear comparison (n = 22); killitrap (n = 1); seine (n = 24); night-light (n = 18); weir (n = 1).

North Carolina, Hettler and Barker (1993) collected juveniles entering estuaries with mean lengths of 26.8 mm SL at Oregon Inlet and 33.5 mm SL at Okracoke Inlet during June, July, and August.

Determining growth rates from available length data (fig. 17.4) is equivocal without attendant age data, because it is not possible to determine the progression of modes of discrete length groups. Confounding this situation is the fact that adults, as well as young-of-the-year, apparently migrate into the Middle Atlantic Bight for the summer. Lacking evidence to the contrary, we suggest the individuals up to 25 cm collected in late summer and fall are all young-of-the-year.

Synodus foetens are known to bury in sandy substrates and from there to dart out after prey (Allen et al. 1978). Sampling directed at different habitat types in the Great Bay–Little Egg Harbor estuary indicates the species is most abundant in sea lettuce (*Ulva lactuca*) beds over sandy substrates, although it was also frequently collected over silty bottoms with structure ranging from none to shelly or peaty.

This species is not found north of Cape Hatteras during the winter months (January through May) (Anderson et al. 1966). Presumably, young-of-the-year as well as older year classes that have spent the summer in the Middle Atlantic Bight migrate south of the bight before the onset of winter, but there are no descriptions of this emigration.

CHAPTER 18

Microgadus tomcod (Walbaum) Atlantic Tomcod

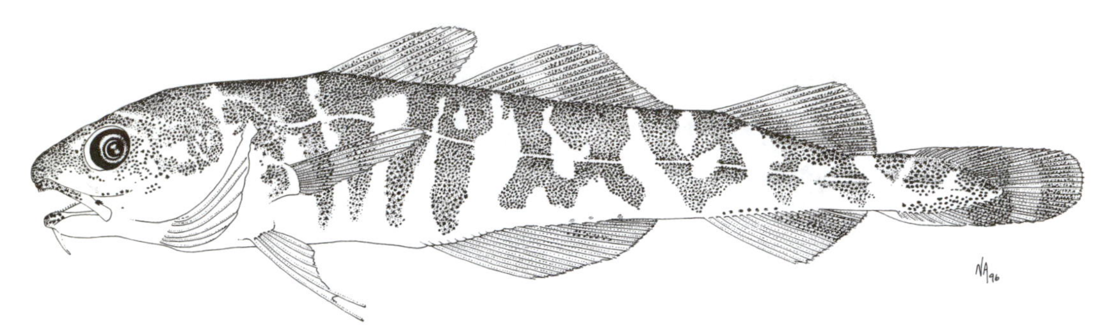

18.1 *Microgadus tomcod* juvenile, 53.3 mm TL. Collected June 22, 1993, pyramid trap, Pier 40, Hudson River, New York. ANSP 175213. Illustrated by Nancy Arthur.

DISTRIBUTION

Microgadus tomcod (fig. 18.1) is reported to occur from southern Labrador to Virginia (Robins and Ray 1986) or as far south as North Carolina (F. J. Schwartz pers. comm. in Scott and Scott 1988). In the last 30 years, the southernmost stock (occurring in our study area) has undergone a decrease in life span, abundance, and geographical range (Dew 1991). It now occurs only as far south as the northern part of New Jersey (Dew and Hecht 1994a). It was once considered abundant, especially from fall through early spring, in the New York–Sandy Hook Bay area (Nichols and Breder 1927) and through winter in the Sea Isle City and Asbury Park areas (Fowler 1906). In a comparison of relative abundances of species collected in our study area during 1929–1933 and 1972–1973 (Thomas and Milstein 1974), *M. tomcod* was ranked the fifth most abundant species trawled in the ocean and bays off southern New Jersey in the earlier period; in the more recent time period, it was not collected at all. It was once especially common during spring in the Barnegat Bay and Barnegat Inlet region but has not been collected there recently (Tatham et al. 1984). Its present status in the central part of the Middle Atlantic Bight is uncertain (Heintzelman 1971; Miller 1972). It is a coastal fish, ascending rivers into habitats with very low salinities, or strictly riverine fish in some systems including the Hudson River (Dew and Hecht 1994a). Recently, in our study area, it has been collected (but only rarely) in Sandy Hook Bay (Pacheco 1983, 1984) and more frequently in the Hudson River Estuary (Dew and Hecht 1976; Able et al. 1995b) and Newark Bay (S. J. Wilk pers. comm.). Within the Middle Atlantic Bight, early life-history stages have been reported from estuaries east and north of the Hudson River (table 4.2), but are not found in systems such as Nauset Marsh, where tidal freshwater zones are absent (table 2.3).

REPRODUCTION

The Hudson River marks the current southern limit to spawning (Grabe 1978). Spawning throughout its range is accomplished during the winter both by young-of-the-year (just less than age 1) and older year classes and involves an elaborate courtship behavior involving small groups (Howe 1971) or pairs (Klauda et al. 1988). Most spawning occurs in freshwater, near the upstream extent of saltwater intrusion. In the Hudson River, almost all of the spawners (93%–99%) are age-0 resident fishes approaching their first birthday (McLaren et al. 1988), and most of these do not survive beyond the first year.

DESCRIPTION

The demersal eggs are spherical or slightly oval, and range from 1.39 to 1.70 mm in diameter. They are apparently only weakly adhesive, the earliest stages adhering to each other more so than to substrates. Oil globules are small and not always evident, but, when present, number 3 to 12 (Hardy and Hudson 1975). Larvae hatch at 5 to 6 mm TL (Bigelow and Schroeder 1953; Booth 1967) or an average of 7.0 mm with a full yolk sac (Dew and Hecht 1994b). Larvae resemble those of other gadids in having a snout-anus length less than 50% TL with several distinct pigmented areas on top of the head, dorsal edge of gut, over the air bladder, on the venter anterior to the anus, rows of melanophores along both the dorsal and ventral edges of tail, and with a row developing along the midlateral line (Booth 1967). Juveniles have three dorsal and two anal fins (fig. 18.1). They also have a long barbel at the lower jaw tip, and a rounded caudal fin. The snout tip extends farther forward than the lower jaw. One of the rays in the pelvic fin is elongate. Juveniles of *Pollachius virens* have a similar fin structure but lack the long jaw barbel and elongate pelvic fin ray and have a slightly forked caudal fin (see fig. 19.1). The color pattern is likely to be highly blotchy (not uniformly dusky as in *P. virens*). Vert: 53–57; D_1: 11–15; D_2: 15–19; D_3: 16–21; A_1: 12–21; A_2: 16–20; Plv: 6.

THE FIRST YEAR

Most eggs are deposited in December and January (Booth 1967), and incubation occupies 24 to 60 days (Scott and Crossman 1973) or 61 to 70 days (Dew and Hecht 1994a,b), depending on temperature. Most hatching in the Hudson River occurs between mid-February and mid-March (Dew and Hecht 1994a). Hatching success is greatest in freshwater and declines with increasing salinity. Larval size at hatching is related to the duration of the incubation period, which in turn is related to temperature (Dew and Hecht 1994b; see Pepin 1991). Aside from the larvae of a few marine species, collected near the mouth of the river, *M. tomcod* is the only species represented in the winter ichthyoplankton in the Hudson River Estuary (Dew and Hecht 1994a). Larvae are capable swimmers and are positively phototactic immediately after hatching. They swim to the surface in order to fill their air bladders (Peterson et al. 1980), although studies in the Hudson River found that they remain near bottom and are only moved passively upward into the water column (Dew and Hecht 1994b). Larvae remain semipelagic (Dew and Hecht 1994b) until reaching lengths of about 12 mm TL when they settle and begin a demersal habit (Booth 1967). Larvae transform gradually into juveniles when full complements of fin rays are acquired at about 23 mm (Hardy 1978a).

First-year growth is characterized by an initial fast rate in spring and early summer, followed by a slow rate in midsummer and resumed faster rates during fall (Howe 1971). Early growth is enhanced by higher spring temperatures, but is then impeded in late May by temperatures above 13 C (Dew and Hecht 1994b). Modes of 40 to 49 and 60 to 69 mm TL have been reported for May and June, respectively (Young et al. 1991). A second fast growth rate resumes in the fall (Grabe 1978). They reach 9 cm TL by September (Howe 1971) and upper size limits of 15 to 16 cm for young-of-the-year during November and December have been reported (Grabe 1978). This growth-rate pattern (fast in the spring, slow during summer, fast in the fall) has also been observed in Newark Bay from 1993 to 1994 (S. J. Wilk pers, comm.), where young-of-the-year reach modes of 10 and 18 cm during September and November, respectively (fig. 18.2). Multiple year classes are also evident in collections from Newark Bay, suggesting that they live beyond the first year, in contrast to the situation described for the Hudson River.

Habitat where spawning and egg deposition occurs is typically freshwater near the upriver extent of saltwater intrusion during high tides. In the Hudson River, this intrusion can occur as far upriver as km 243, near Albany (Dew and Hecht 1994b), but during the reproductive season it is typically located between kms 18 (George Washington Bridge) and 80 (Con Hook). Bottom temperatures are likely to be below 3 C in spawning areas (Grabe 1978). Some studies have found that larvae migrate into saline water immediately after hatching (Peterson et al. 1980). In the Hudson River Estuary, larvae are gradually displaced downstream as they develop (Dew 1995), but the bulk of post-yolk-sac larvae remain concentrated just downstream from the salt front, the position of which is determined by the volume of spring freshwater flows. Therefore, as this flow varies, the highest concentrations of larvae are likely to be shifted both up- and downstream as development proceeds. Larvae and juveniles, combined, are most abundant in salinities of 4.5 to 8.7 ppt (Dew and Hecht 1976a).

A combination of river flow and temperatures may determine to a large extent the distribution of young-of-the-year within estuarine habitats of the Hudson River (Klauda et al. 1988). Juveniles are first observed in the Hudson River Estuary in March, are concentrated in the downstream portions during April and May, and begin to move back upstream between late April (Dew 1995) and July, possibly associated with the upstream encroachment of the salt front that occurs during summer when freshwater flow is lowest. Consistent with this pattern, young-of-the-year reach a peak in June under piers and among pilings in the lower Hudson River Estuary (Able et al. 1995b). Here the sizes are somewhat smaller than the Newark Bay cohort (fig. 18.3). Collections decrease after June, perhaps owing to increased gear avoidance associated with their development (Klauda et al. 1988) or with an upstream migration as described above, or movement into deeper water in the Hudson River Estuary (unpubl. observ.).

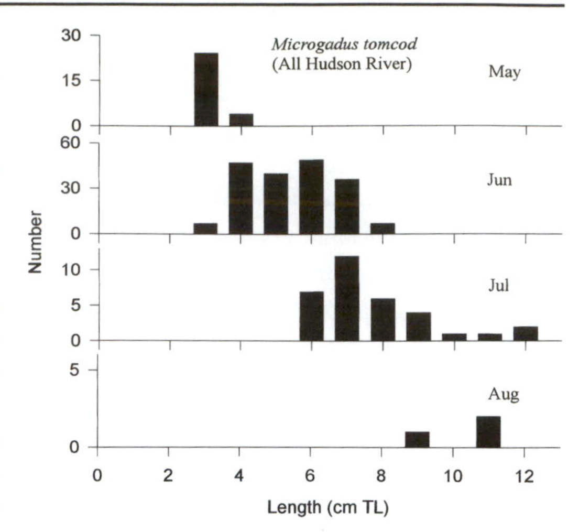

18.3 Monthly length frequencies of *Microgadus tomcod* collected in traps from the lower Hudson River Estuary, New York, 1993–1994. Source: Able et al. 1995b.

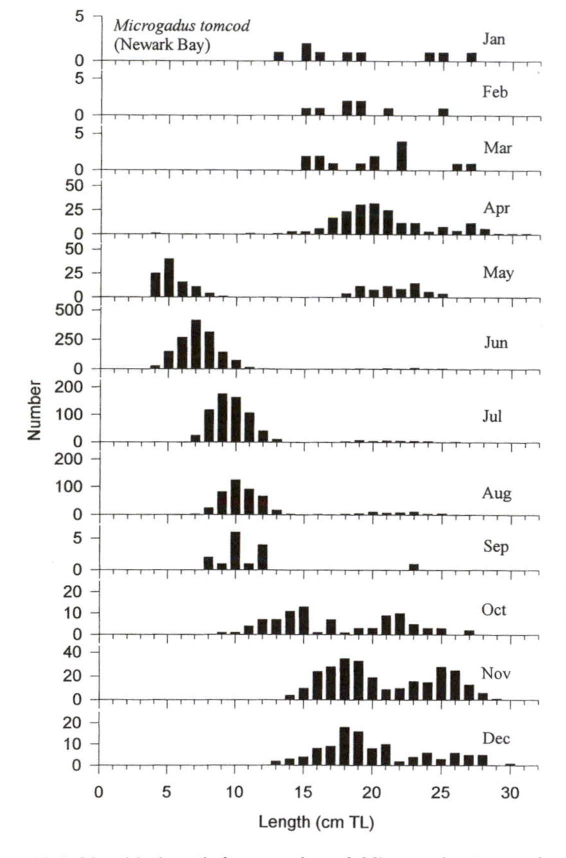

18.2 Monthly length frequencies of *Microgadus tomcod* collected from Newark Bay, May 1993–April 1994, with 8.5-m otter trawl. Data courtesy S. J. Wilk, NOAA, NMFS, Sandy Hook Marine Laboratory.

The evidence suggests that *M. tomcod* spends its entire life history in the Hudson River Estuary, rather than migrating seasonally to sea as it does in some other parts of its range (Lawler et al. 1975 cited in Grabe 1978). Thus, population fluctuations in this estuary are due to mortality and not to immigrations or emigrations (Dew and Hecht 1994b).

CHAPTER 19

Pollachius virens (Linnaeus)
Pollock

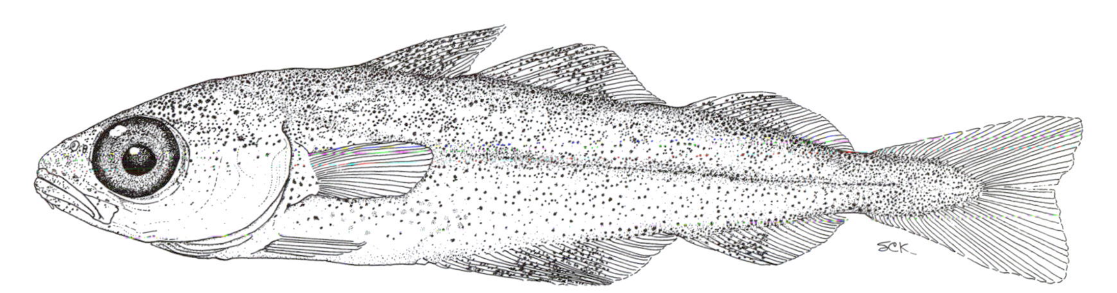

19.1 *Pollachius virens* juvenile, 39.5 mm SL. Collected April 15, 1988, seine, cove near RUMFS, Great Bay, New Jersey. ANSP 175214. Illustrated by Susan Kaiser.

DISTRIBUTION

Pollachius virens (fig. 19.1) is primarily a boreal species that occurs on both sides of the North Atlantic. In the western North Atlantic its distribution extends from western Greenland and Labrador to Cape Hatteras, North Carolina (Scott and Scott 1988), but it is uncommon south of New Jersey. In contrast to its near relatives (*Gadus morhua* and *Melanogrammus aeglefinus*), *P. virens* is strongly pelagic. On the Scotian Shelf, studies indicate it prefers a depth range between 110 and 181 m (Scott 1982), although this range can vary with season and food supply. Adult fishes occur in waters as cold as 0 C and apparently do not tolerate temperatures greater than 11 C (Bigelow and Schroeder 1953). Young stages (known in some areas as "harbor pollock") occur in bays and estuaries throughout the species' range, although many reports are largely anecdotal and lack firm, descriptive details (Nichols and Breder 1927; Bigelow and Schroeder 1953). In some parts of its range, this coastal and estuarine occurrence has been largely overlooked, in part because of difficulties in sampling this fast-swimming fish. Historical (as well as contemporary) collections reported on here indicate that the range of juveniles extends at least to New Jersey and less commonly to Chesapeake Bay (table 4.2).

REPRODUCTION

Both sexes reach maturity during their third year, at lengths of 50.5 cm in males and 47.9 cm in females (Mayo et al. 1989). Most reported landings in the central part of the Middle Atlantic Bight occur between March and June and comprise these larger individuals. Some of these large adults form huge spawning aggregations during the winter and spring. One such aggregation was described as a giant school more than 6.4 km in diameter south of Block Island, Rhode Island during May (Wilk et al. 1979). Occurrences of eggs (Berrien and Sibunka in press) and larvae (fig. 19.2) indicate that spawning begins in the fall in the northern part of the Middle Atlantic Bight when temperatures are below 9.4 C. During the winter, spawning is most intense around the periphery of the Gulf of Maine and on Georges Bank. The occurrence of eggs and very small larvae indicates reproduction also occurs, although less intensely, during spring months in the central part of the Middle Atlantic Bight. Eggs have been found as far south as the offing of Delaware Bay (Berrien and Sibunka in press), and larvae have been collected over the continental shelf as far south as the Virginia Capes area (fig. 19.2). The relatively small sizes and lack of fin ray development in larvae collected in the central part of the Middle Atlantic Bight

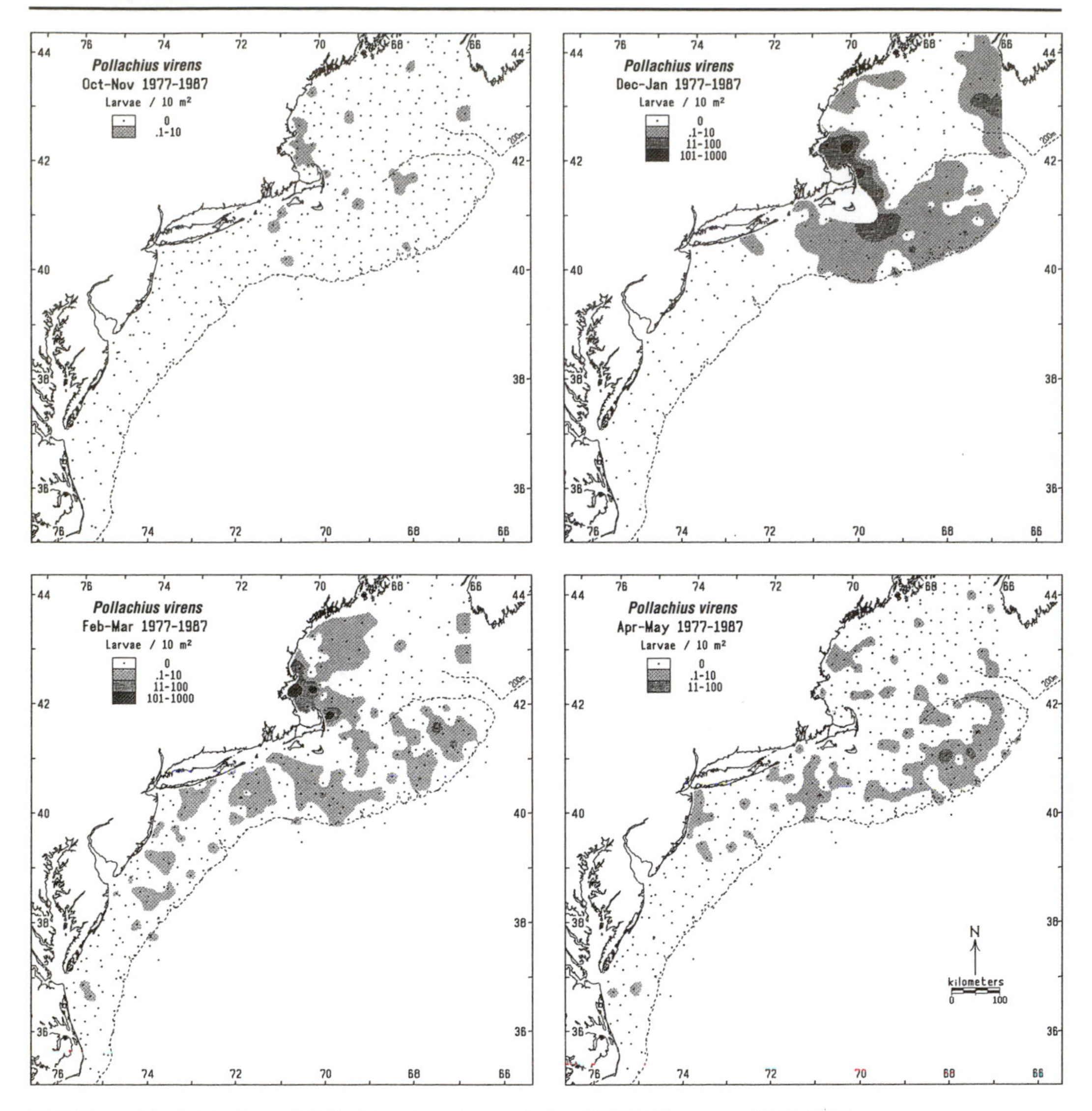

19.2 Bimonthly distributions of *Pollachius virens* larvae during MARMAP surveys, 1977–1987.

(fig. 19.3) indicate that they were spawned locally.

DESCRIPTION

The pelagic eggs are spherical and transparent. The diameter ranges from 1.0 to 1.2 mm, oil globules are lacking, and the perivitelline space is narrow. The larvae are 3.0 to 4.0 mm at hatching, have unpigmented eyes, unformed mouth parts, and, in the earliest stages, the anus opens laterally on the finfold (not at the edge). Characteristic larval pigment includes two dorsal and two ventral groups of melanophores on the body between the anus and caudal tip. Caudal, dorsal, and anal fin rays begin forming at about 9 to 11 mm, and transformation to the juvenile stage occurs at sizes larger than 25 mm. More details of early development are in Schmidt (1905, 1906) and Fahay (1983). Juveniles (fig. 19.1) have three dorsal fins that are separate at the bases of their rays and two anal fins that touch at their bases. If a barbel is present at the tip of the lower jaw, it is tiny. The barbel is lacking in older stages. The color pattern is dull olive green and lacks pattern. There are no greatly elongated rays in the pel-

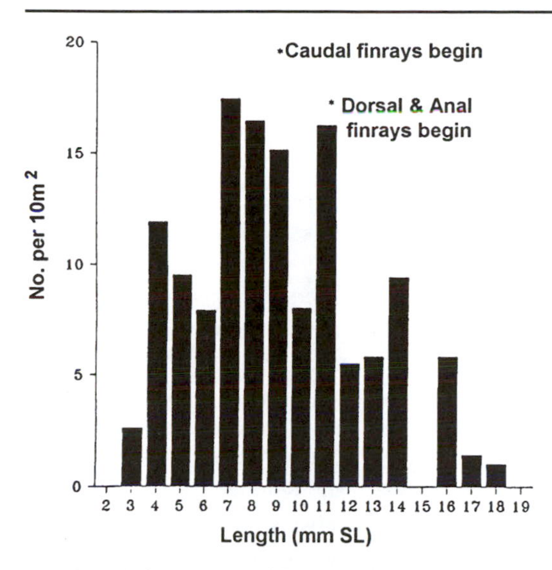

19.3 Length frequencies of *Pollachius virens* pelagic larvae collected over the central Middle Atlantic Bight continental shelf during MARMAP surveys, 1977–1987. Abundances of each size interval are indicated as number per 10 m^2 of sea surface, and the sizes at which caudal, dorsal, and anal fin rays begin to form are indicated with asterisks.

vic fin. Vert: 53–57; D_1: 13–14; D_2: 21–22; D_3: 24–28; A_1: 24–28; A_2: 20–21; Pect: 19–22; Plv: 6.

THE FIRST YEAR

In the central part of the Middle Atlantic Bight, larvae are distributed over the entire continental shelf from February to May (fig. 19.2). These larvae range in size up to 18 mm SL. Transformation from the larval to juvenile stage is gradual, with no sudden or dramatic morphological changes. There is no obvious settlement to the bottom and this species is not strongly bottom-oriented in any life-history stage. Fin rays are fully developed, and the body shape approaches that of juveniles at about 25 mm (Fahay 1983).

Small pelagic juveniles (<50 mm) approach the coast and begin to enter inlets in February and March. During this stage in their development, these individuals are highly pelagic and capable swimmers. Therefore, passive collecting gear fails to detect these stages, and other collection methods probably underestimate their abundance. Nevertheless, seines and trawls collect them often enough that a pattern emerges indicating an ingress from oceanic to estuarine habitats, as reported for Nova Scotia (Clay et al. 1989) and other localities to the north (e.g., Nauset Marsh; M. P. Fahay and K. W. Able unpubl. observ.). Juveniles are susceptible to capture in traps in the Great Bay–Little Egg Harbor estuary from March to June (fig. 19.4). They are seldom collected after June, perhaps because they leave estuaries as a result of higher water temperatures, which average 23 C by July (Able et al. 1992).

Growth rates estimated from the progression of length modes in Delaware Bay and coastal New Jersey fish (fig. 19.5) are consistent with those described for Nova Scotian young-of-the-year (Clay et al. 1989) and indicate that they reach lengths of 4 to 13 cm by midsummer, at a presumed age of 4 or 5 months. In Nova Scotia, age 1 reach about 20 cm, and age 2 about 30 cm on January 1 of ensuing years.

Although patterns of estuarine occupation in young-of-the-year in the central part of the Middle Atlantic Bight are similar to those from Nova Scotia and other more northern locations, there are apparent differences in the extent to which succeeding cohorts utilize inshore nursery areas. In Nova Scotia, both young-of-the-year and age 1 occur in bays and estuaries, where they are referred to as "harbor pollock" (Clay et al. 1989). In the central part of the Middle Atlantic Bight, length-frequency distributions only indicate the presence of young-of-the-year. Another difference between the two populations concerns the temporal pattern of estuarine occupation by young-of-the-year. In Nova Scotian waters, young-of-the-year spend

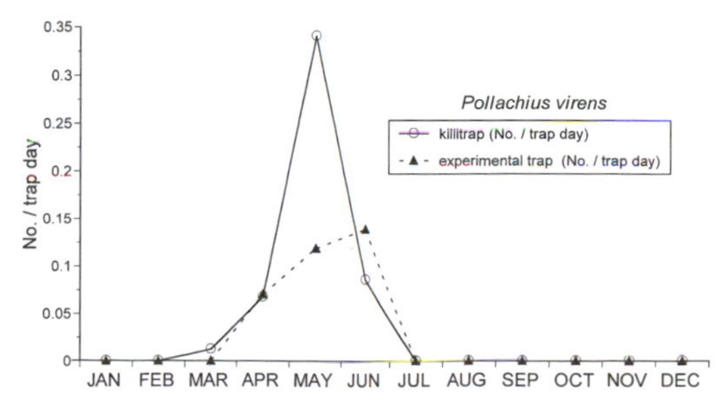

19.4 Monthly occurrences of *Pollachius virens* young-of-the-year in the Great Bay study area based on collections in killitraps 1991–1994 ($n = 176$) and experimental traps 1992–1993 ($n = 24$).

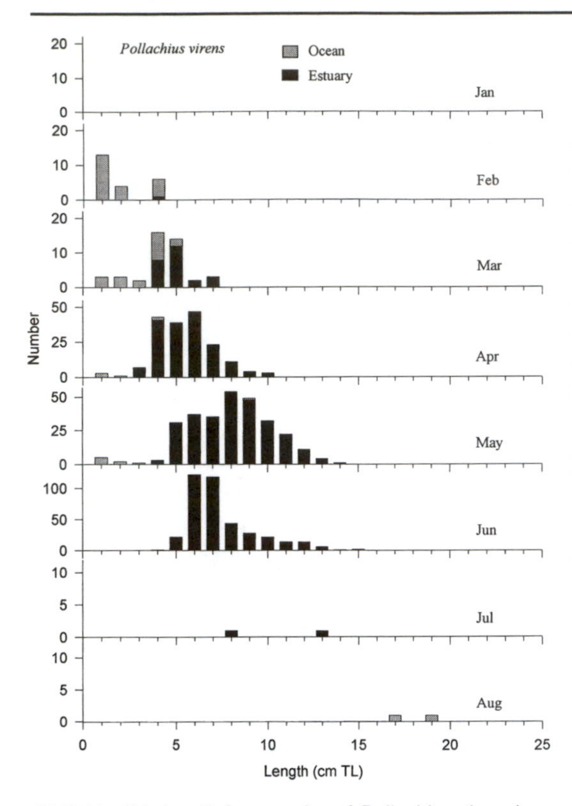

19.5 Monthly length frequencies of *Pollachius virens* larvae and young-of-the-year collected from estuarine and coastal ocean waters in the central part of the Middle Atlantic Bight. Sources: de Sylva et al. 1962; Wilk 1976; Wilk et al. 1992; MARMAP 1977–1987; ANSP (primarily IA collections); RUMFS: otter trawl ($n = 2$); 1-m plankton net ($n = 2$); killitrap ($n = 173$); experimental trap ($n = 157$); weir ($n = 123$); seine ($n = 1$); night-light ($n = 1$).

the first 6 months at sea (after spawning arbitrarily centered on January 1) and the second 6 months (July through December) within estuarine waters (Clay et al. 1989). In the central part of the Middle Atlantic Bight, young-of-the-year are only found from late winter (in the ocean) to early summer (in estuaries) (fig. 19.5).

Correlations with specific estuarine habitat types are difficult to demonstrate. However, collections elsewhere have shown that young-of-the-year feed heavily on amphipods that inhabit hydroid beds (Richards 1963). The rocky intertidal zone has been shown to be an important nursery area for juveniles in Canadian waters (Rangely and Kramer 1995a, b). On rising tides, juveniles were shown to move (as individuals or small schools) from subtidal zones into intertidal zones characterized by dense fucoid macroalgae beds. On falling tides, juveniles schooled in more open subtidal zones. As the first growing season progressed, there was a distributional shift toward greater depths in the intertidal zone or into subtidal zones exclusively. Details of processes associated with emigration from estuaries in our study area are lacking, but collections end abruptly in early July at the latest, when maximum lengths of juveniles are 10 to 12 cm SL (about 11–14 cm TL). Presumably this coincides with an offshore migration, possibly associated with rising estuarine temperatures, although this has not been demonstrated conclusively.

CHAPTER 20

Hakes of the Genus *Urophycis*

Many previous treatments of hakes in the genus *Urophycis* have confused or combined the species because of their external resemblance to each other. Reproduction, nursery use, and occurrences of young-of-the-year for three of the species in the genus overlap in the marine waters of the Middle Atlantic Bight. The central part of the Middle Atlantic Bight represents the approximate southern limit to spawning and first-year nursery use in *Urophycis tenuis,* the approximate northern limit in *U. regia,* and the approximate center of these activities in *U. chuss.* Although these species closely resemble each other and are difficult to identify at small sizes, their chronological and ontogenetic patterns of occurrences are discrete (table 20.1). Eggs of *Urophycis* sp. were collected during every month of the year in the Middle Atlantic Bight in MARMAP sampling between 1977 and 1987 (Berrien and Sibunka in press), although most occurred between June and November. Winter and spring occurrences of eggs were restricted to waters near the continental shelf edge, and recent studies in the Slope Sea suggest they are those of *U. tenuis* and *Phycis chesteri* (Fahay 1987; J. A. Hare, M. P. Fahay, and R. K. Cowen in prep.). The average monthly distributions of *Urophycis* larvae during 4 years in the bight are shown in figure 20.1. After larvae first appear in June in the northern part of the bight, they become abundant over the entire continental shelf in the central part of the bight from July through November. Although these larvae are not identified to species, recent studies indicate that they represent larvae of *U. chuss* (appearing during early summer) and *U. regia* (appearing in the fall). The larvae of both species broadly co-occur in early fall. Details of these egg and larval occurrences will be treated in the individual species accounts below.

DESCRIPTION

All three species share an early life-history pattern consisting of pelagic egg and larval stages followed by a neustonic pelagic-juvenile stage (Fahay 1983). Eggs of *U. tenuis* are 0.70 to 0.79 mm in diameter, are buoyant, pelagic, and have a single oil globule (Markle and Frost 1985). Eggs of *U. chuss* and *U. regia* range from 0.63 to 0.97 mm in diameter, have a narrow perivitelline space, smooth chorion, and homogeneous yolk. Oil globules are initially multiple and then coalesce into one with a diameter of about 0.2 mm. Larvae hatch at

Table 20.1. Comparative Biological and Temporal Characteristics in Three Species of *Urophycis* in the Central Part of the Middle Atlantic Bight

Species	Spawning months	Neustonic months*	Estuarine months	Maximum size neustonic (mm)	Modal size at 12 months (mm)
U. tenuis	Apr–May	May–Jun	May–Jun	70–80	250–300
U. chuss	May–Sep†	Jul–Nov (peak Jul–Aug)	Mar–Jun‡	23–30§	150–170
U. regia	Jul–Apr (peak Sep & Mar)	Aug–May (peak Sep & Mar)	Mar–May	25–30‖	150–200

*Based on Comyns and Grant (1993) and MARMAP collections.
†Based on GSI data (Wilk et al. 1990).
‡After an initial demersal phase involving an inquiline association with sea scallops.
§Maximum 49 mm (Steiner and Olla 1985).
‖Typically much larger (maximum 75 mm) in the South Atlantic Bight (Fahay 1975).

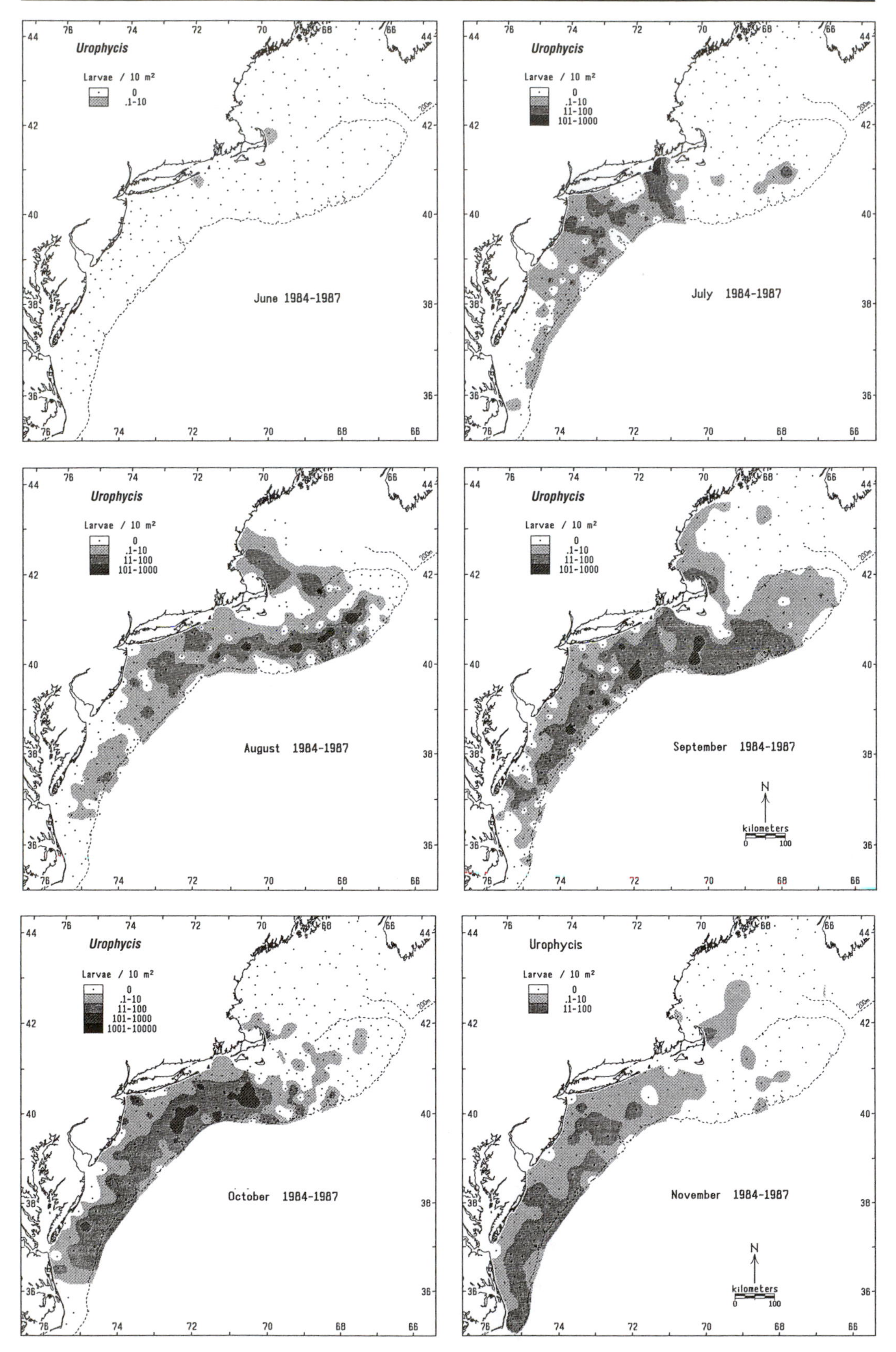
Urophycis
Larvae / 10 m²
0
.1-10
June 1984-1987
Urophycis
Larvae / 10 m²
0
.1-10
11-100
101-1000
July 1984-1987
Urophycis
Larvae / 10 m²
0
.1-10
11-100
101-1000
August 1984-1987
Urophycis
Larvae / 10 m²
0
.1-10
11-100
101-1000
September 1984-1987
N
kilometers
0
100
Urophycis
Larvae / 10 m²
0
.1-10
11-100
101-1000
1001-10000
October 1984-1987
Urophycis
Larvae / 10 m²
0
.1-10
11-100
November 1984-1987
N
kilometers
0
100
200m

Table 20.2. Meristic Characters of *Urophycis* Species Occurring in the Central Part of the Middle Atlantic Bight

		Gill	Fin Rays		
Species	Vertebrae	Rakers (Upper)	Caudal (Total)	Dorsal (1st + 2d)	Anal
U. chuss	14–15 + 32–36 (mode 15 + 34)	3	28–34	9–12 + 53–61	45–57
U. regia	13–15 + 31–35 (mode 14 + 33)	3	30–33	8–10 + 44–52	41–51
U. tenuis	15–16 + 32–35 (mode 16 + 34)	2	33–39	9–12 + 51–62	45–53

SOURCE: Fahay and Able 1989; Comyns and Grant 1993.
NOTE: Vertebrae are given as precaudal plus caudal.

lengths less than 2.0 mm, have snout-anus lengths shorter than 50% TL, rounded heads, and thin, tapering tails. Typical pigment includes a pair of bands midway between the anus and caudal tip. Larvae of *U. chuss* and *U. tenuis* develop prominent pigment at the tips of the pelvic fin rays; *U. regia* larvae lack this pigment. Larvae of *U. tenuis* resemble those of *U. chuss* but are deeper bodied (Methven 1985). Morphological specializations in the pelagic-juvenile stage include a laterally compressed body; pelagic coloration, consisting of silver sides and ventral surfaces combined with a dark blue dorsum; and three trailing, diaphanous pelvic fin rays connected by a fin membrane. Pelagic-juvenile *U. chuss* are noticeably slimmer bodied than their congeners. Maximum size attained during this stage varies among the species and coincides with settlement to habitats on the bottom (table 20.1). After settlement, the body becomes more rounded, pelvic fin rays are thickened and reduced to two, and coloration becomes more cryptic.

Juvenile hakes in their initial demersal stages are fairly elongate fishes with two distinct dorsal fins, touching at the base, and a single, continuous anal fin (see figs. 21.1, 22.1, and 23.1). The fins are composed of rays only (no spines), and the pelvic fins are reduced to two long and filamentous rays, used as sensory organs (Bardach and Case 1965). Depending on the developmental stage, there may be a single barbel under the tip of the lower jaw, which is also used as a sensory organ. Color patterns are dusky and lack strong patterns. Juveniles in our study area are best distinguished by meristic characters (table 20.2) and by the pigment pattern on the first dorsal fin (dusky in *U. chuss* and *U. tenuis* and with a strong ocellus pattern in *U. regia*).

20.1 Monthly distributions of *Urophycis* spp. larvae collected during MARMAP surveys 1984–1987. These larvae were not identified to species level. There were no samples taken in December south or west of Georges Bank during these four years. Although not shown, a few collections were made in April and May, and these were restricted to the edge of the Middle Atlantic Bight continental shelf, near the 200-m contour.

CHAPTER 21

Urophycis chuss (Walbaum) Red Hake

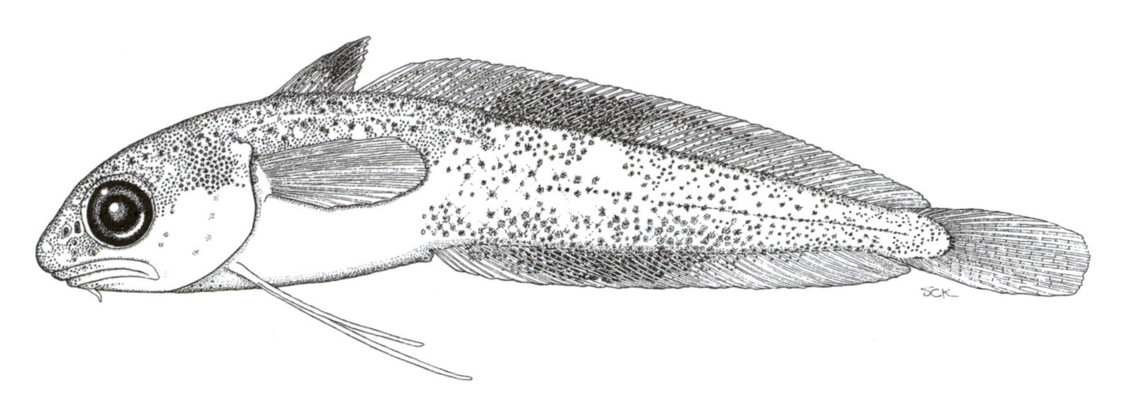

21.1 *Urophycis chuss* early demersal juvenile, 35.0 mm SL. Collected August 6, 1994, 2-m beam trawl, LEO-15 Site #1, tow #1, Beach Haven Ridge, New Jersey. ANSP 175215. Illustrated by Susan Kaiser.

DISTRIBUTION

Urophycis chuss occurs along the United States East Coast from Nova Scotia to North Carolina with a center of abundance between Georges Bank and Hudson Canyon (Anderson 1982). They are typically found on soft mud, silt, or sand bottoms (Fritz 1965; Musick 1974), but they also occur over rocky bottoms when temperatures are below 13 C (Eklund and Targett 1991). Depths of occurrence range from bays to the outer continental shelf (as deep as 550 m) and vary with age and season. They are most common in very deep water during winter and movements include an inshore migration in spring, followed by an offshore movement in summer associated with spawning (Musick 1974). Juveniles have been collected in most estuaries throughout the Middle Atlantic Bight, but eggs and larvae are more restricted to larger systems east and north of the Hudson River (table 4.2).

REPRODUCTION

Urophycis chuss reach sexual maturity in their second year at lengths of about 300 mm (Markle et al. 1982). Spawning occurs in the Middle Atlantic Bight between April and October (Musick 1969; Wilk et al. 1990). There is a strong peak in spawning activity during late June and July off Maryland and northern Virginia (Eklund and Targett 1990). Occurrences of larvae indicate spawning in the central part of the Middle Atlantic Bight probably begins in June at the earliest and continues through September, with sporadic activity continuing through October and November (MARMAP data and Comyns and Grant 1993). During this period, spawning individuals are most concentrated on the continental shelf in waters less than 110 m deep between Martha's Vineyard and Long Island (Anderson 1982), where larvae are also heavily concentrated (see chap. 20).

THE FIRST YEAR

Larvae were most abundant during August and September (when the water column was most strongly stratified) in a study conducted over the central part of the Middle Atlantic Bight continental shelf (Comyns and Grant 1993). Morphological changes, as larvae develop into pelagic juveniles and then demersal juveniles (described in chap. 20), occur during the first few months of life. After approximately 2 months in pelagic stages, descent to the bottom occurs during the late summer, when a strong thermocline is present, and fall, when the thermocline is breaking down in response to seasonal, thermal mixing. Laboratory observations suggest that pelagic juveniles descending through a thermocline require an acclimation period to adjust to the sharp, negative change in temperature (Steiner and Olla 1985). This acclimation is apparently accomplished while remaining within the thermocline for a period. Lacking acclimation, descending fishes become moribund on encountering colder bottom waters. With descent to the bottom on the continental shelf, fish initially retain pelagic-juvenile coloration and filamentous pelvic fin rays (Musick 1969; Steiner and Olla 1985). After this descent, and within a matter of hours, the body shape becomes terete, neustonic coloration changes to an adultlike pattern (fig. 21.1), and the pelvic fin rays begin to be deployed in a forward orientation (Musick 1969).

The available evidence suggests that structure of some kind is critical to survival of just settled *U. chuss* (Steiner et al. 1982). During their initial demersal stage they have been observed lying in the troughs between sand waves with their bodies curled into a C-shape (M. P. Fahay and K. W. Able unpubl. observ. from submersible). Although these fish will utilize nonliving objects, it is clear that the most common form of shelter use involves an inquiline association with the sea scallop, *Placopecten magellanicus*. This relationship, in which the initial demersal stages live within the mantles of scallops during the fall and winter, has been well described (Goode and Bean 1896; Welsh 1915; Musick 1969; Markle et al. 1982). Settlement to scallop beds begins in September, about 2 months after peak spawning in our study area, and continues through December (Steiner et al. 1982). During September, these juveniles range from 23 to 30 mm; in November, they are as large as 46 mm. Overall, juveniles from 23 to 116 mm occur in scallops, where they are most commonly found during the day, leaving this shelter to forage at night. The maximum size reported for hakes still in the scallop association is 140 mm (Musick 1974). There is evidence that the largest scallops are sought out for habitation by inquiline hakes (Markle et al. 1982), but other studies suggest that large fish seek out large scallops; smaller fish inhabit all sizes of scallops (Steiner et al. 1982).

Young-of-the-year apparently emigrate from scallop (or other) shelters between February and May, either because of their large sizes (90–116 mm), or because temperatures in the shelter habitat drop below 4 C (Musick 1974). When many large scallops are available, hakes may remain inquiline for a longer period of time. Conversely, when only small scallops are present (e.g., after overfishing), hakes must leave this relationship earlier and at smaller sizes (Musick 1974). Some of these émigrés occupy bay or estuarine habitats for a brief period after leaving the shelter of scallop beds; we have therefore included this species in the present treatment of estuarine fishes. In New Jersey waters, juveniles 7 to 20 cm (presumably following the scallop stage) have been found in Sandy Hook and Great Bays during March and April (Thomas et al. 1974; Pacheco 1983), when temperatures are typically slightly warmer than 4 C. Collection records from these studies also include juveniles lingering through May and June, but in much reduced numbers. Not all of these juveniles occur in bays, however. Studies on the inner continental shelf in our study area have also collected large numbers of these post-scallop stages from March through May (e.g., Wilk et al. 1992).

Evidence of a prolonged spawning season, variable growth rates, and successive cohorts of early demersal-stage juveniles all contribute to a large amount of size overlap between year classes of *U. chuss*. Some estimates of growth in the early stages include 15.7 mm per month while occupying scallops in our study area (Steiner et al. 1982) and a minimum average size of 100 mm in the first year off Nova Scotia (Markle et al. 1982). Laboratory studies offering large volumes of food found juveniles (92–133 mm TL) capable of growing 1.0 to 1.5 mm per day in a 24-day experiment (Luczkovich and Olla 1983). The histograms derived from sampling in our study area (fig. 21.2) show a modal length of about 100 mm at the end of fall, indications of nearly no growth through their first winter, a group of age 1 in June at about 100+ mm, and growth through their second summer and fall to lengths over 250 mm. Relatively rare, estuarine occurrences of juveniles between 100 and 200 mm are also indicated.

During May and June, *Urophycis chuss* have been reported as common in Sandy Hook

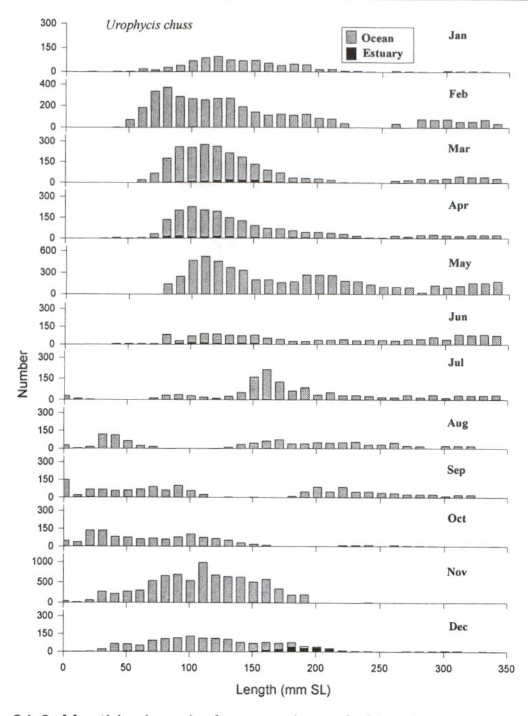

21.2 Monthly length frequencies of *Urophycis chuss* young-of-the-year collected in the central Middle Atlantic Bight study area. Sources: Clark et al. 1969; Thomas et al. 1972, 1974; Thomas and Milstein 1973; Wilk et al. 1975; MARMAP 1977–1987, Haedrich Net (uncat.); RUMFS: otter trawl ($n = 26$); 1-m beam trawl ($n = 1$); 1-m plankton net ($n = 13$); experimental trap ($n = 4$); LEO-15 Tucker trawl ($n = 1$); LEO-15 2-m beam trawl ($n = 215$); killitrap ($n = 9$); seine ($n = 2$).

Bay, where they reportedly fed voraciously on *Crangon* and other crustaceans (Nichols and Breder 1927). It is impossible, however, to determine whether these authors were referring to young-of-the-year or older fishes. Recent collections during those months include more than one year class (e.g., Pacheco 1983). All year classes prefer soft-bottomed habitats such as mud, silt, or sand and avoid rocky or shelly bottoms (Cohen et al. 1990; but see Eklund and Targett 1991).

Hypoxia (low dissolved oxygen) is a common, but unpredictable, occurrence in coastal and estuarine habitats in our study area (Garlo et al. 1979; Swanson and Sindermann 1979; Officer et al. 1984). In studies among different age groups of *U. chuss*, young-of-the-year (more so than 1- to 3-year olds) become behaviorally active in avoiding affected areas. This activity takes the form of increased swimming and orientation to near-surface layers (Bejda et al. 1987).

After the scallop-association stage, *U. chuss* young-of-the-year remain in the vicinity of scallop beds (Scott and Scott 1988), or occur coastally or within embayments. They then join older year classes in an offshore migration during their second winter at sizes greater than 250 mm.

CHAPTER 22

Urophycis regia (Walbaum)
Spotted Hake

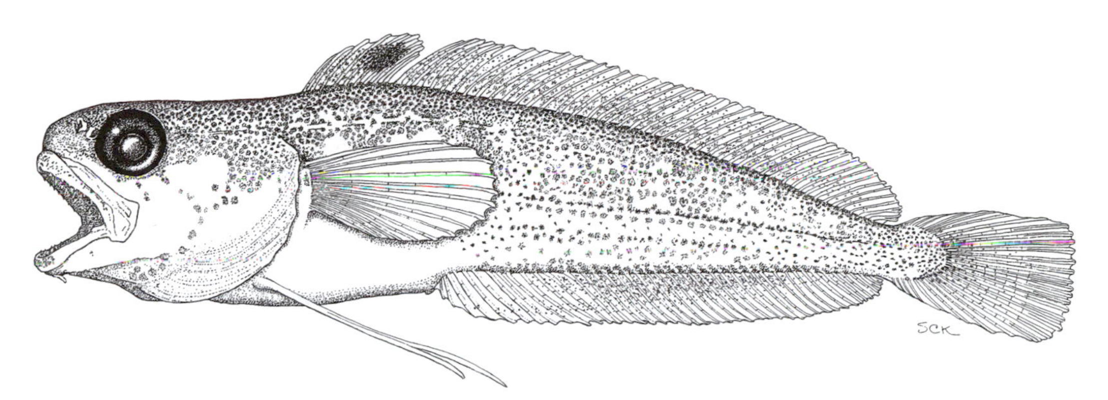

22.1 *Urophycis regia* late pelagic–early demersal juvenile, 39.0 mm SL. Collected November 3, 1993, 1-m plankton net, Little Sheepshead Creek Bridge, Great Bay, New Jersey. ANSP 175216. Illustrated by Susan Kaiser.

DISTRIBUTION

Urophycis regia (fig. 22.1) occurs most commonly in continental shelf waters between Massachusetts and the northeastern Gulf of Mexico (Hoese and Moore 1977) with a center of abundance in the Chesapeake Bay area (Bigelow and Schroeder 1953). It is rarely collected as far north as Nova Scotia (Hildebrand and Schroeder 1928; Scott and Scott 1988) and reports from as far south as Brownsville, Texas (Springer and Bullis 1956) may be based on confusion with a congener, *U. floridana* (Hoese and Moore 1977). It is a very common component of the estuarine fauna in North Carolina (Burr and Schwartz 1986). Although it was once thought to be uncommon in our study area (Nichols and Breder 1927), recent studies in Sandy Hook Bay indicate it is much less common than its congener, *U. chuss* (Pacheco 1984). In the Great Bay–Little Egg Harbor Estuary, however, it has been the most common *Urophycis* collected in recent years (Szedlmayer and Able 1996). A survey of an inner continental shelf site at the apex of the New York Bight (Wilk et al. 1992) indicated *U. regia* adults were most abundant during August through October and absent from January to April. In a study of the fauna of the upper continental slope (Haedrich et al. 1980), *U. regia* ranked high in the upper depth zone (40–264 m) by weight, but not by number, indicating the presence of larger individuals in these deep habitats. Juveniles occur in most estuaries in the Middle Atlantic Bight (table 4.2), especially those in the southern part of the bight.

REPRODUCTION

There are two temporal modes of spawning activity in our study area. Gonadosomatic studies (Wilk et al. 1990) indicate that ripe females occur from August to October (at lengths of 191–360 mm) and again (January not sampled) from February through April (at lengths of 283–396 mm). A similar August–October observation was made off Maryland and northern Virginia (Eklund and Targett 1990). In the Chesapeake Bay region, ripe females (>225 mm) were collected from August to November (Barans 1969), and ripe females have been found in December off North Carolina (Bigelow and Schroeder 1953). South of Cape Hatteras, the presence of larvae indicates most spawning activity occurs during the winter (Fahay 1975). The pattern emerges, therefore, of spawning commencing in late summer in the central part of the Middle Atlantic Bight and occurring progressively later with distance to the south. A discrete spawning event then occurs in our study area during early spring in waters near the continental shelf edge. Wilk et al. (1990) demonstrated that the cohort present (and spawning) during spring is composed of older individuals than the age 1 fish present (and spawning) during late summer and fall. Thus, the apparent bimodality in reproduction may reflect different distribution, migration, and maturation patterns of different age groups.

THE FIRST YEAR

The occurrences of larvae and pelagic juveniles off New Jersey and Virginia support the split pattern of reproduction described above (Comyns and Grant 1993). In the latter study, larvae were most abundant in October and November in mid-shelf regions but also occurred during February, March and May on stations near the edge of the shelf. Pelagic juveniles were also reported between November and May in Middle Atlantic Bight continental shelf waters (Fahay 1987). Recent collections during May (1993) on stations beyond the edge of the continental shelf off the Delmarva Peninsula (J. A. Hare, M. P. Fahay, and R. K. Cowen in prep.) yielded large numbers of neustonic larvae in an area dominated by Slope Sea water. The patterns of circulation in this area (Cowen et al. 1993) would not support a hypothesis of transport from the south but instead would suggest that these fishes were produced locally, or perhaps slightly to the north (e.g., in Slope Sea waters off New Jersey). Patterns of spring occurrences of adults presented by Barans (1969) are consistent with these larval occurrences and anecdotal observations reported therein indicate the presence of larvae at least as late as April off Chesapeake Bay (Massmann et al. 1961, 1962). The various sources of larvae summarized here, help explain the seeming multiple size cohorts observed in estuaries during spring in our study area.

There are no published observations of morphological changes associated with the development of pelagic juveniles into early demersal stages, but this change is presumably similar to that in *U. chuss*. The diagnostic black spot surrounded by a white margin on the first dorsal fin (fig. 22.1) is formed between 35 and 50 mm (in early demersal juveniles), but the series of pale spots occurring along the length of the lateral line apparently does not become visible until juveniles are about 60 mm (Hildebrand and Cable 1938).

Settlement occurs at sizes of about 25 to 30 mm, based on maximum sizes of the neustonic stage in the Middle Atlantic Bight, although rare specimens may remain in the neuston until about 75 mm in the South Atlantic Bight (Fahay 1975, 1987). Data on specific habitats where this descent occurs are lacking, however. Early demersal juveniles ($n = 43$, collected by bottom trawl) were found in various depths on the continental shelf during March between New Jersey and Cape Hatteras and ranged from 34.1 to 64.6 mm SL (Fahay 1987).

Collections in the central part of the Middle Atlantic Bight indicate an estuarine ingress occurring as early as October as well as one in February–March (fig. 22.2). Most of these collections were made in the vicinity of Little Egg Inlet and were composed of individuals in the process of descending to the early demersal stage (Witting 1995; D. A. Witting, K. W. Able, and M. P. Fahay in prep.). Most previously available data concerning estuarine occurrences of young-of-the-year *U. regia* derive from studies in Chesapeake Bay (Barans 1972). Young-of-the-year enter that bay during March, although a lower bottom temperature limit of about 6.5 C seems to impede this ingress. After entering the bay, these young-of-the-year appear to penetrate lower-salinity habitats upriver, reaching a maximum upstream limit in April and May (at a minimum salinity of 7 ppt). Growth accompanies this upstream penetration; young-of-the-year found farther upstream are larger than those remaining in lower bay habitats.

There are growth differences between young-of-the-year in Great Bay–Little Egg Harbor (data included in fig. 22.2) and those in the Chesapeake Bay study (Barans 1972). The modes in Chesapeake Bay increase from about 100 mm in March to about 200 mm in June. Comparable figures from the Great Bay–Little Egg Harbor data are about 50 mm in March and only about 75 mm in June (although a few individuals reach 200 mm). We have also observed two modes in young-of-the-year from January through March, involving both oceanic and estuarine specimens (fig. 22.2). We inter-

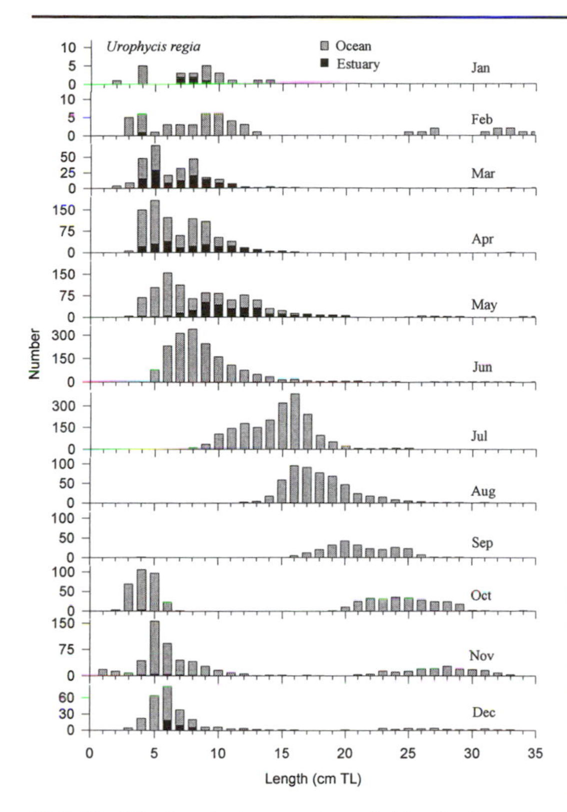

22.2 Monthly length frequencies of *Urophycis regia* young-of-the-year and 1-year-olds collected in the central Middle Atlantic Bight study area. Sources: de Sylva et al. 1962; Clark et al. 1969; Thomas et al 1972, 1974; Thomas and Milstein 1973; Wilk et al. 1975; MARMAP Haedrich Net (uncat., 1977–1987); MARMAP bongo net collections, NJBK survey area, 1984–1987; RUMFS: otter trawl (*n* = 82); 1-m beam trawl (*n* = 14); 1-m plankton net (*n* = 60); experimental trap (*n* = 17); LEO-15 Tucker trawl (*n* = 9); LEO-15 2-m beam trawl (*n* = 276); gear comparison (*n* = 5); killitrap (*n* = 63); seine (*n* = 8); weir (*n* = 39).

pret the larger of these as representing either products of November–December spawning occurring near the edge of the continental shelf off Virginia or of spawning known to occur during the fall in New Jersey waters. The smaller mode may be representative of larvae resulting from spawning that occurs in the New Jersey study area during January through April (Wilk et al. 1990; see "Reproduction"). Whatever the source for these modes, it is apparent in neuston sampling that two size classes of young-of-the-year are present in our study area. Whether the sizes of these two cohorts merge as the first year proceeds is not known. Size at age 1 must be similar in the two areas because on the first anniversary of the fall spawning event in our study area, the mode (about 20 cm) is precisely the estimate provided by Barans (1972) for age 1 fish in his Chesapeake Bay study. Estimates of sizes of this cohort after a season of estuarine growth south of Cape Hatteras are remarkably similar (Burr and Schwartz 1986).

Our observations and those of Barans (1972) indicate that not all young-of-the-year enter estuaries and that estuarine fishes are larger than their continental shelf counterparts. The estuarine components of the year class account for the largest part of the monthly histograms (fig. 22.2), while fishes apparently remaining offshore lag behind in size. There is little information on specific habitats occupied by *U. regia* young-of-the-year while in their estuarine phase, although Barans (1972) suggested they were somewhat more abundant in channels than in adjacent shallows. Observations of small *U. regia* (107–135 mm) indicate a propensity for burying in sand until the entire body is hidden except for the eyes and snout (Barans 1969). Edwards and Emory (1968) reported small *U. regia* are not associated with structures on bottom as much as *U. chuss.*

We have found no records of young-of-the-year remaining in estuarine habitats past May, and these fish apparently spend the rest of the first year in continental shelf habitats. In the Chesapeake Bay, most migrate downstream and leave the bay by June, rarely as late as July (Barans 1972). Similar results have been reported for estuaries south of Cape Hatteras (e.g., Burr and Schwartz 1986).

CHAPTER 23

Urophycis tenuis (Mitchill)
White Hake

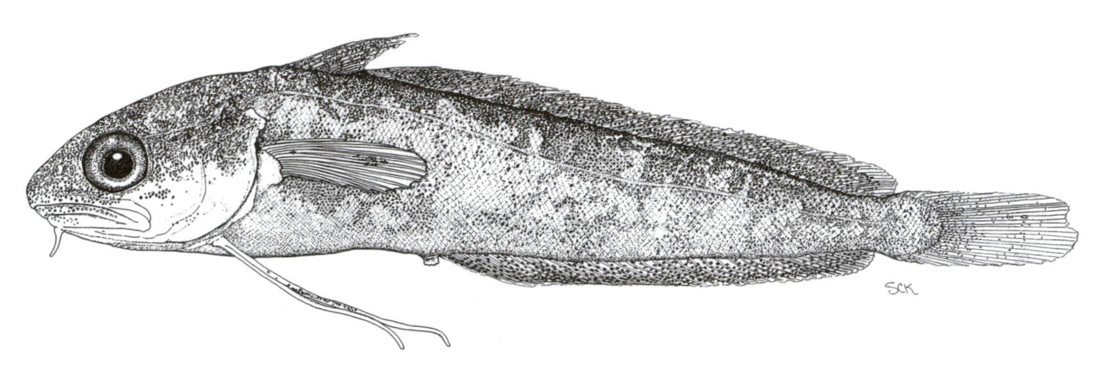

23.1 *Urophycis tenuis* early demersal juvenile, 55.0 mm SL. Collected June 7, 1988, 5.3-m otter trawl, Nauset Marsh, Cape Cod, Massachusetts. ANSP 175217. Illustrated by Susan Kaiser.

DISTRIBUTION

The overall range of *Urophycis tenuis* (fig. 23.1) is Iceland to North Carolina (Scott and Scott 1988) with occurrences in very deep water as far south as Florida (Musick 1974). Most of the population occurs deeper than 200 m (Scott and Scott 1988), where it is found over the continental shelf edge, deeper basins in the Gulf of Maine, and submarine canyons along the continental slope (Bigelow and Schroeder 1953; Cooper et al. 1987). In the central part of the Middle Atlantic Bight, *U. tenuis* is regularly collected in submarine canyon habitats, 400 to 800 m deep, near the edge of the continental shelf (Markle and Musick 1974; Haedrich et al. 1980). Many details concerning occurrences, life history, and biology are obscured by the close resemblance between this species and *U. chuss* in all life-history stages. For this reason, the reported occurrence of juveniles in only a few Middle Atlantic Bight estuaries between Massachusetts and southern New Jersey (table 4.2) may be underestimates.

REPRODUCTION

Spawning occurs in early spring in waters over the continental slope off Georges Bank and the Middle Atlantic Bight (Fahay and Able 1989; Lang et al. 1996). A separate spawning occurs during summer over the Scotian Shelf and Gulf of St. Lawrence, but progeny from that group would not be expected to occur in the Middle Atlantic Bight. Fecundity estimates are limited to Canadian studies (Nepszy 1968; Beacham and Nepszy 1980; Beacham 1983) because attempts to find ripe females in our study area or the Gulf of Maine have never been successful (Musick 1969; Burnett et al. 1984). Spawning on the Middle Atlantic Bight continental slope, therefore, has been inferred by the presence of the smallest larvae there. Subsequent collections are progressively larger on stations closer to shore (Fahay and Able 1989; Comyns and Grant 1993). More recent studies (J. A. Hare, M. P. Fahay, and R. K. Cowen in prep.) have collected very small larvae (<5.0 mm NL) at a number of locations in Slope Sea waters (fig. 2.1) off the Middle Atlantic Bight during May, providing further evidence of spawning there. These larvae were part of a ubiquitous slope assemblage, and the water masses with which they were associated suggested an origin within the Slope Sea rather than transport from other areas.

THE FIRST YEAR

In a recent ichthyoplankton study undertaken in the Slope Sea off the Middle Atlantic Bight during May 1993, the distribution of the smallest size class (<5.0 mm TL) was approximately the same as that of the largest (>56.0 mm TL) and intervening size classes, although there was a negative correlation between larval size and temperature (J. A. Hare, M. P. Fahay, and R. K. Cowen in prep.). This suggests that the smallest larvae are found in warm Slope Sea waters, but, as they develop, they are increasingly found in continental shelf waters, which are cooler during this time of year. Larvae and pelagic juveniles have been collected along two transects over the continental shelf in the central part of the Middle Atlantic Bight (Comyns and Grant 1993). In this study, larvae as small as 3 to 4 mm were collected near the shelf/slope break, and the largest sizes (40–50 mm SL) were found closest to shore, suggesting an offshore-inshore migration with growth. Available evidence indicates *U. tenuis* young-of-the-year remain in a pelagic-juvenile stage until they enter certain New England and Canadian estuaries, at which time they begin to occupy bottom habitats and their coloration and morphology change as described for *U. chuss* (Markle et al. 1982, Fahay and Able 1989). Individuals (67–76 mm TL) going through this morphological change (fig. 23.1) have also been reported from shallow water near Shark River, New Jersey during May (Nichols and Breder 1927). However, inadequate sampling for recently settled fishes in bottom habitats on the continental shelf prevents us from concluding that the entire year class migrates cross-shelf and enters estuarine areas. In view of these results, and considering those of other studies (Fahay and Able 1989; Comyns and Grant 1993), we suggest that while a portion of the year class can certainly be traced from an offshore spawning across the shelf to eventual settlement in estuarine nursery areas, another portion might simply descend to undescribed continental shelf habitats during the early demersal stage.

When young-of-the-year have been detected entering New England estuaries, they have arrived there during June and July (Fahay and Able 1989). In our central Middle Atlantic Bight study area, a small number of young-of-the-year have been collected as early as April and May (fig. 23.2). Recent intensive sampling with a variety of gears in a variety of Great Bay-Little Egg Inlet estuarine habitats (table 3.3), however, has failed to find young-of-the-year. It appears from available evidence that *U. tenuis* young-of-the-year at sizes comparable to New England estuarine inhabitants also settle in inner continental shelf sites, such as the Beach Haven Ridge near Little Egg Inlet, which was sampled intensively during the 1970s. These data were represented in an earlier study (Fahay and Able 1989), and some of the same individuals contribute here to the length frequencies portrayed in figure 23.2.

Estimates of larval (Markle et al. 1982) and pelagic-juvenile (Fahay and Able 1989) growth rates range from 10–22 to 35 mm per month. Estimates of the size at which the habitat shift from pelagic juvenile stage to the initial demersal stage occurs range from 50 to 60 mm TL (Markle et al. 1982) to more than 63.8 mm SL (Fahay and Able 1989), thereby implying that the early pelagic stages occupy about 2 months. In an earlier study, we based the rate of growth

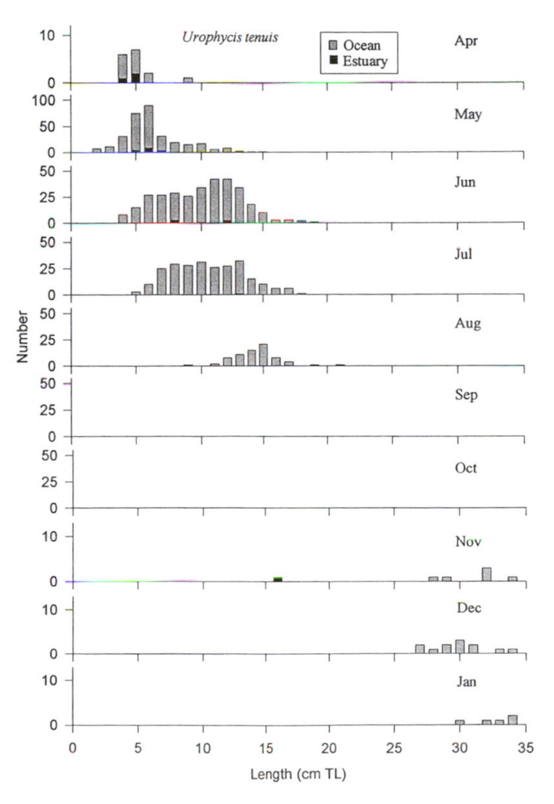

23.2 Monthly length frequencies of *Urophycis tenuis* collected in the central Middle Atlantic Bight study area. Sources: Thomas et al. 1972, 1974; Thomas and Milstein 1973; Fahay 1987; Atlantic Twin 1988 (see Lang et al. 1996); RUMFS: otter trawl ($n = 1$).

on the progression of modes and estimated that young-of-the-year in Nauset Marsh grew at the rate of about 30 mm per month (Fahay and Able 1989). A more recent study, based on an otolith analysis and covering various areas in the Middle Atlantic Bight, found a juvenile growth rate of 0.99 mm per day in June through September (Lang et al. 1996). We have too few specimens from our New Jersey study area to make equivalent monthly growth rate estimates, but the length frequencies of available material suggest that young-of-the-year reach about 30-35 cm as they begin their first winter (fig. 23.2). This large size at age 1 is a demonstration of the "get big quick" strategy employed by this species (Markle et al. 1982).

Several sources have mentioned the importance of eelgrass beds as habitat for young-of-the-year (Bigelow and Schroeder 1953; Fahay and Able 1989). In our study area, however, as well as in the Bay of Fundy (Markle et al. 1982), it is apparent that while young-of-the-year are spatially segregated from older year classes, occurring generally in shallower depths, they are not necessarily tied to eelgrass, other vegetation, or structured bottom habitats. Because of their relative scarcity in our study area, however, it is difficult to ascertain habitat associations. A singular observation of burying in sand substrates (McAllister 1960) indicates a behavior similar to that described for *U. regia* young-of-the-year. In our study area, we have no information on occurrences after the first winter. Presumably young-of-the-year leave estuarine and/or coastal habitats then and do not reoccur close to shore in subsequent years.

CHAPTER 24

Ophidion marginatum (DeKay) Striped Cusk-Eel

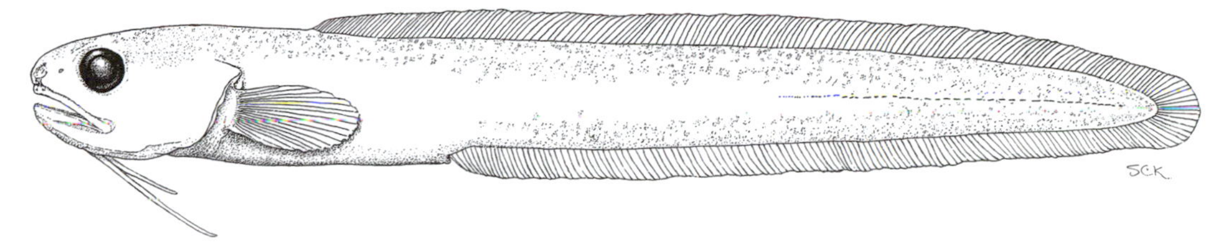

24.1 *Ophidion marginatum* early juvenile, 41.8 mm SL. Collected March 13, 1991, 1-m plankton net, Little Sheepshead Creek Bridge, Great Bay, New Jersey. ANSP 175218. Illustrated by Susan Kaiser.

DISTRIBUTION

A common ophidiid, *Ophidion marginatum* (fig. 24.1) is distributed from New York to northeastern Florida in relatively shallow bays, estuaries, and inner continental shelf waters. It burrows tail first into the substrate during daylight hours and actively forages for food at night (Allen et al. 1978; Bowers-Altman 1993). As a result of this behavior, surveys undertaken during daylight hours do not contribute to our understanding of its distribution or behavior. Winter collections are rare, and, if seasonal migrations occur, they have gone undetected. Larvae and juveniles have been reported from estuaries between Barnegat and Chesapeake Bays (table 4.2), with most reports resulting from studies that focus on inlets.

REPRODUCTION

Ophidion marginatum is a summer spawner. Observations in the laboratory indicate that courtship and spawning involve the production of sound (Mann et al. 1997) and close tandem swimming by a spawning pair (Fahay 1992; Bowers-Altman 1993). Individual females may release a small batch of eggs nightly for a period up to 2 months. Centers of reproductive activity are found off Delaware Bay and the Delmarva Peninsula (Fahay 1992) and correspond latitudinally to estuarine occurrences cited above. Larval occurrences indicate that spawning extends from June to November, with a peak off the New Jersey coast in August and September.

DESCRIPTION

Eggs are contained in an amorphous, mucilaginous sac that is positively buoyant. They are slightly off-round, and the long axis is between 0.88 and 1.02 mm. The chorion is unsculptured, and there is no oil globule (Fahay 1992). Larvae are elongate, and the preanus length is short. Pigment includes a line of melanophores along the base of the anal fin, a streak along the midline in the posterior third of the body and a scattering of fine spots on the venter of the gut. Both the dorsal and anal fins are long, and the pelvic rays form late in development. Further details of egg and larval devlopment are in Fahay (1992). Juveniles are recognized by their eel-like body shape and the presence of filamentous pelvic fin rays located far forward under the lower jaw (fig. 24.1). The dorsal, caudal, and anal fins are continuous. Their color is rather drab and plain, with at most a scattering of fine black spots along the venter. Older stages develop faint stripes along the body. Vert: 67–70; D: 138–162; A: 116–129; Pect: ca. 21; Plv: 2.

THE FIRST YEAR

Larvae hatch at about 2.0 mm after an incubation period of 36 hours at 24 to 26 C (Fahay 1992). They are pelagic and are distributed in

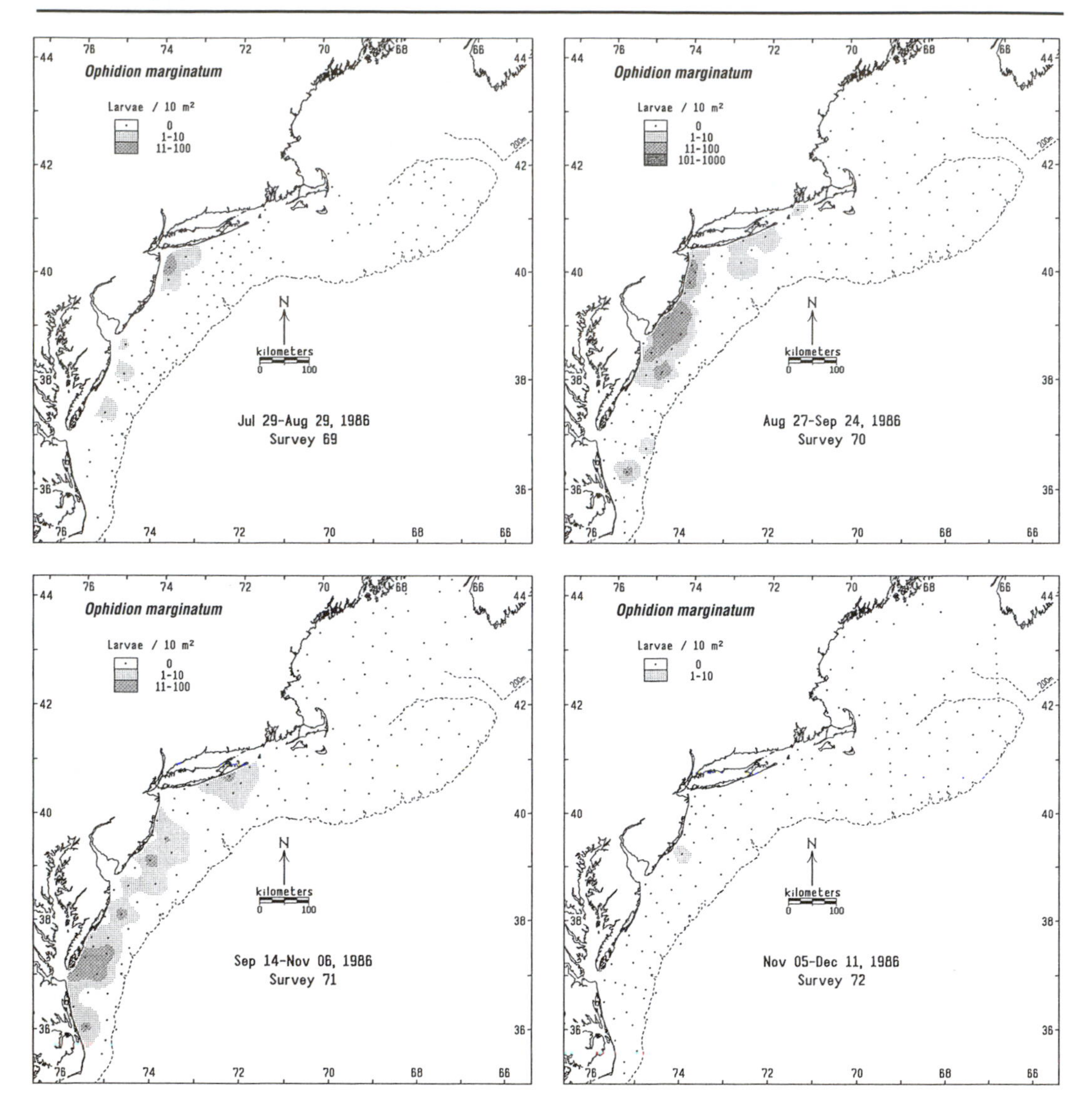

24.2 Distributions of *Ophidion marginatum* larvae during four consecutive MARMAP ichthyoplankton surveys (1986). Although spawning activity extends from June through November in the Middle Atlantic Bight, larval occurrences are most common during August–October in the central part of the bight.

coastal waters from the coast of Long Island to the Cape Hatteras area (fig. 24.2) (Fahay 1992). Preflexion and flexion stages are more common in inner continental shelf waters less than 30 m deep; postflexion stages occur more commonly between 20 and 40 m. The largest pelagic larvae collected are about 22 mm, and settlement to the bottom presumably occurs at about this size. Development in this species is gradual, and there are no abrupt morphological changes involved in the transition from pelagic larva to settled juvenile. In pelagic larvae, however, the pelvic fin rays arise from a position typical of many teleosts on the chest area near the symphysis of the cleithral bones. In juveniles, the pelvic fin basipterygium elongates, and the rays appear to originate close to the tip of the lower jaw. The development of this character, therefore, takes place after the juveniles have settled to the bottom (Fahay 1992). All available evidence suggests that this settlement occurs during the fall on the continental shelf, mostly at depths between 20 and 40 m (Fahay 1992).

Several studies sampling in the vicinity of major inlets have documented the appearance of juveniles (20–70 mm TL) from mid-March to mid-April (Pacheco and Grant 1973; ANSP voucher material examined), and this is also true in the vicinity of Little Egg Inlet (Witting 1995; D. A. Witting, K. W. Able, and M. P. Fahay in prep.). Although it is not known whether this indicates that all individuals of this length cohort are entering estuaries at this time, these are the only records available for this size range. The earliest juveniles collected in the spring (March and April) average 65.8 mm SL as they enter Little Egg Inlet (Witting 1995; D. A. Witting, K. W. Able, and M. P. Fahay in prep.), and many are approximately the same sizes as larvae settling to bottom the previous fall. Thus, inferences concerning growth rates derived from accumulated length frequencies (fig. 24.3) suggest that there is essentially no growth in just settled individuals over their first winter, but they grow at an estimated rate of about 1.0 mm per day during the following summer. One-year-olds apparently reach sizes of 100 to 200 mm by the end of the summer, with a mode at about 160 mm. The group we interpret as 2-year-olds also apparently grow little over their second winter, as they approximate this same size range during their second spring (fig. 24.3).

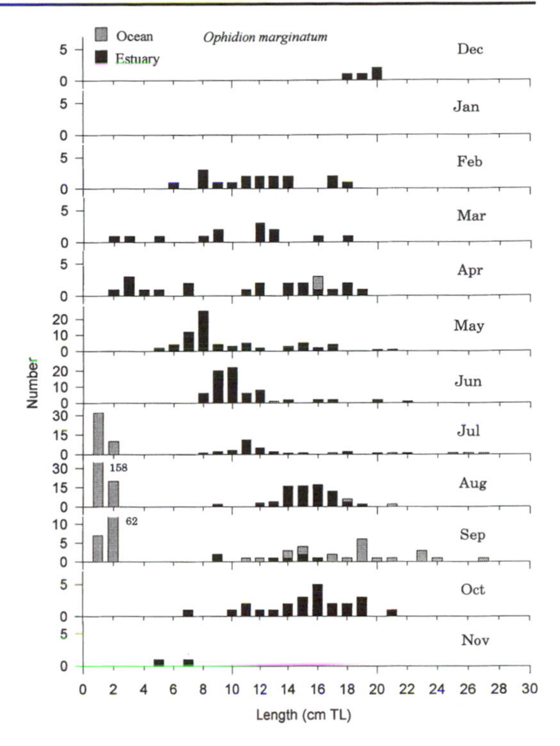

24.3 Monthly length frequencies of *Ophidion marginatum* larvae and young-of-the-year collected in estuarine and adjacent coastal ocean waters of the central Middle Atlantic Bight. Sources: Pacheco and Grant 1973; Wilk and Silverman 1976a; ANSP (uncat.); Delaware Bay Juvenile Crab Survey; Oyster Creek Power Plant Study; NMFS Bottom Trawl Survey Cruise AL8809; RUMFS: otter trawl (n = 10); 1-m plankton net (n = 98); LEO-15 Tucker trawl (n = 277); LEO-15 2-m beam trawl (n = 21); gear comparison (n = 2); seine (n = 1); weir (n = 1). Note: The few specimens collected in the estuary during December were found in the vicinity of a power plant and their occurrence may be an artifact of an artificially warmed environment.

Habitats used by this species vary with life-history stage. Pelagic young-of-the-year (larvae) occur in the central part of the Middle Atlantic Bight from July through September, but there is no evidence that they also occur in estuarine waters. Young-of-the-year occur in inlets beginning in March, when their collection in plankton nets suggests they are pelagic. They continue to occupy estuarine habitats between March and October. There is evidence that 2-year-olds are also present there in the spring, and this largest size class then presumably migrates offshore to spawn over the continental shelf. This burrowing species is usually found in sandy bottom habitats. They have also been commonly collected in marsh creeks near submerged blocks of marsh sod (Allen et al. 1978). In studies in other New Jersey estuarine habitats (Rountree and Able 1992a; Szedlmayer and Able 1992), they have gone undetected despite intensive sampling with a variety of gears designed to sample tidal migrants. An intensive seining study undertaken in Barnegat Bay (Marcellus 1972) also failed to detect *O. marginatum.* The latter study, however, restricted sampling to daylight hours. Another Barnegat Bay study found them to be abundant on power-plant intake screens and also noted their increased availability in night sampling (Tatham et al. 1984). *Ophidion marginatum,* like other members of the family, is well known to be strongly nocturnal (e.g., Greenfield 1968; Matallanas and Riba 1980), and sampling during daylight hours should not be expected to detect it.

There are no winter records of striped cusk eels from any source available to us. This lack of records indicates that they probably burrow into the substrate during the winter, but whether this occurs in estuaries, on the ocean bottom, or both, is unknown.

CHAPTER 25

Opsanus tau (Linnaeus) Oyster Toadfish

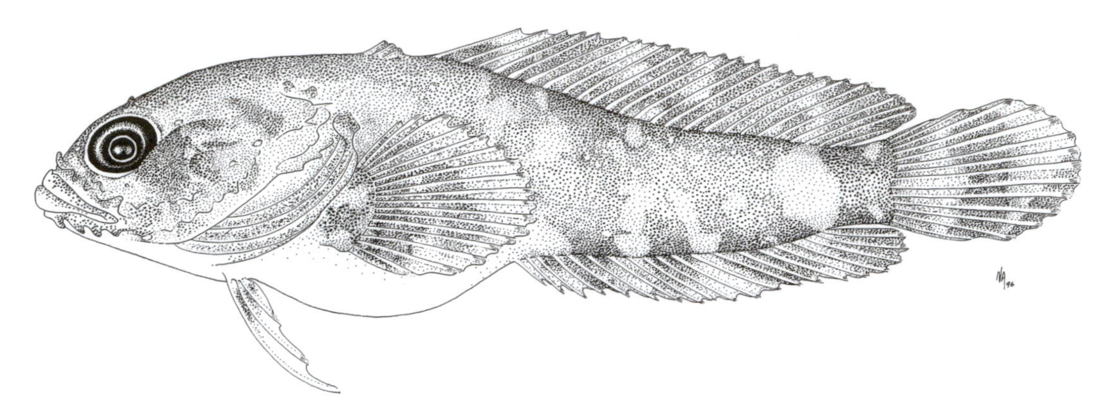

25.1 *Opsanus tau* early juvenile, 23.0 mm TL. Collected July 12, 1973, seine, Absecon Bay, New Jersey. ANSP 175219. (#FWH 73–001, NJ Mar. Ecol. Study). Illustrated by Nancy Arthur.

DISTRIBUTION

Opsanus tau (fig. 25.1) occurs in estuaries from Massachusetts to Florida (Hildebrand and Schroeder 1928), where it is found on sandy and muddy bottom and often in eelgrass (Bigelow and Schroeder 1953). In the Middle Atlantic Bight, early life-history stages are reported as common in most estuaries (table 4.2).

REPRODUCTION

Spawning has been reported from June through August in Woods Hole, Massachusetts (Gudger 1910) and in June and July in the vicinity of New York City (Nichols and Breder 1927). In Chesapeake Bay, the spawning season is from April to early August or October (Hildebrand and Schroeder 1928; Dovel 1960; Schwartz 1965b) at temperatures of 17.5–27 C (Gray and Winn 1961). Sound production, especially by the male, occurs during courtship and while defending the spawning site, eggs, and larvae (Gray and Winn 1961). The production of mating calls decreases toward the end of July in the York River (Fine 1978), coinciding with gonadal atrophy. Nests were observed in June, July, and August in North Carolina (Gudger 1910). In our observations in Great Bay–Little Egg Harbor, it appears that spawning occurs in the summer because the smallest individuals are collected in July and August (fig. 25.2).

DESCRIPTION

The large (5.0–5.5 mm in diameter) yellow eggs are spherical but flattened at their site of attachment. There are no oil globules. The larvae hatch at 6.0 to 7.4 mm. The adult complements of fin rays are complete at about 18 mm TL. Details of development are provided by Dovel (1960). The juveniles (fig. 25.2) have a single dorsal fin that is notched behind a few soft, flexible, anterior spines. The body and fins are crossed by dark bars. The mouth is large and wide, and older juveniles have a series of cirri along the lower jaw as do the adults. The caudal peduncle is thick. All life-history stages lack scales. Vert: 34–35; D: III, 24–28; A: 20–24; Pect: 19–21; Plv: I, 2–3.

THE FIRST YEAR

The embryos develop while attached to the substrate in a nest (Tracy 1959; Dovel 1960). Depending on temperature, hatching occurs in 5 to 12 days, but the young remain attached for another 6 to 19 days (Gray and Winn 1961). Development is gradual, and adult characters are complete approximately 20 days after hatching

(Dovel 1960). The young remain attached to the substrate until yolk absorption, when they are 16 to 18 mm TL, and they become free-swimming. The young stay with the male in the nest for another 5 to 18 days (Gray and Winn 1961). Young leave the nest and seek shelter on the bottom; thus, they have no major dispersal stage. In Great Bay–Little Egg Harbor, the smallest individuals are collected in July at around 2 cm TL with larger juveniles (presumably age 1) evident at sizes of 8.5 to 11 cm TL (fig. 25.2). They are most abundant from June through September, with the peak in August (fig. 25.3). By October, the young-of-the-year have reached approximately 4 to 9 cm TL. Thus, the size at age 1 in New Jersey overlaps that for Maryland (Schwartz and Dutcher 1963) and Virginia (Radtke et al. 1985). They do not appear to grow over the winter and are of a similar size in June; thus, this is the range of sizes at age 1 (Radtke et al. 1985). In a study in South Carolina (Wilson et al. 1982), the size at age 1 was 14.0 cm. This estimate is consistent with our estimate for size at age 2. Wilson et al. (1982) admitted that their size at age 1 may have been an underestimate because the first opaque zone of the otolith may have obscured an earlier annulus.

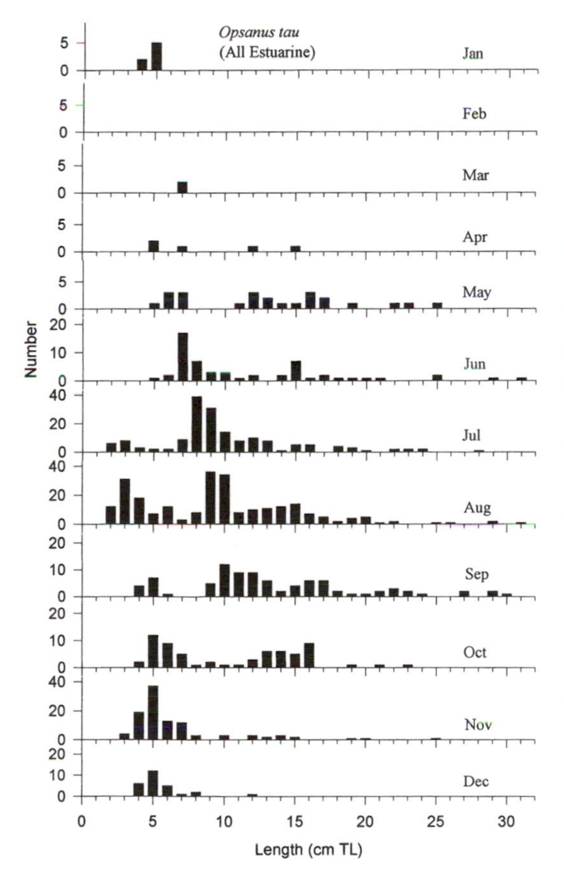

25.2 Monthly length frequencies of *Opsanus tau* in the Great Bay–Little Egg Harbor study area. Sources: RUMFS: otter trawl (*n* = 318); 1-m beam trawl (*n* = 37); 1-m plankton net (*n* = 34); experimental trap (*n* = 115); LEO-15 2-m beam trawl (*n* = 12); gear comparison (*n* = 35); killitrap (*n* = 155); seine (*n* = 26); weir (*n* = 59).

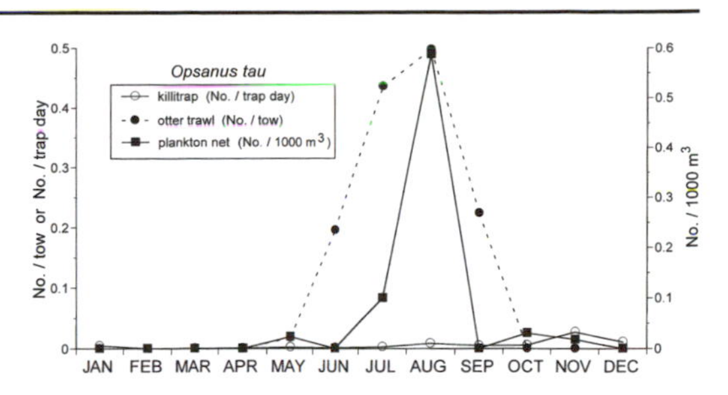

25.3 Monthly occurrences of *Opsanus tau* larvae and young-of-the-year in the Great Bay–Little Egg Harbor study area based on Little Sheepshead Creek Bridge 1-m plankton net collections 1989–1994 (*n* = 36), otter trawl collections 1988–1989 (*n* = 174), and trapping 1990–1994 (*n* = 111).

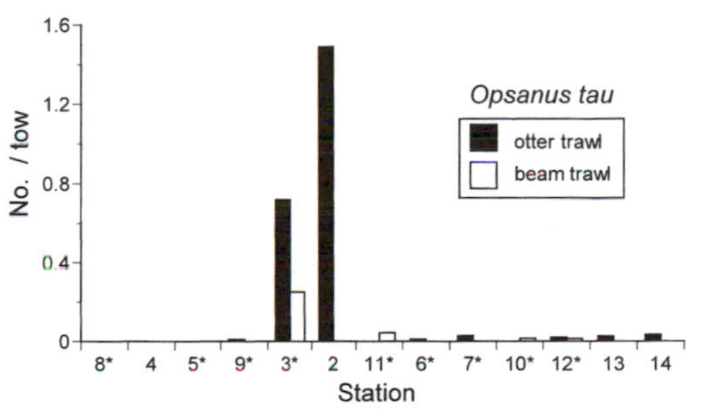

25.4 Distribution of *Opsanus tau* young-of-the-year by habitat type based on otter trawls (*n* = 213) and beam trawls (*n* = 22) deployed at regular stations 2–14 in Great Bay–Little Egg Harbor. Both gears used on stations designated by asterisk; otherwise, only the otter trawl was used. See figure 2.3 for station locations; see table 2.1 for specific habitat characteristics by station.

In the Great Bay–Little Egg Harbor study area, the young-of-the-year occurred in a variety of habitats, but they were clearly more abundant in eelgrass (fig. 25.4). During the winter, they are rarely collected, presumably because they overwinter in the mud (Gudger 1910).

CHAPTER 26

Strongylura marina (Walbaum) Atlantic Needlefish

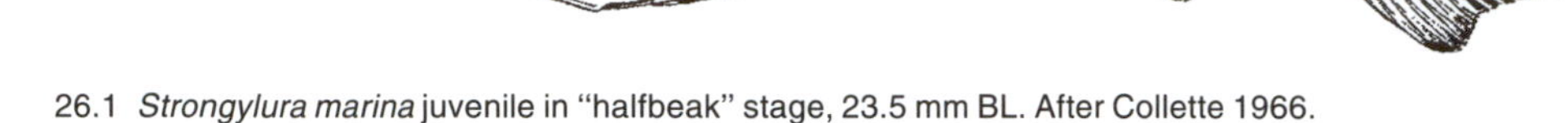

26.1 *Strongylura marina* juvenile in "halfbeak" stage, 23.5 mm BL. After Collette 1966.

DISTRIBUTION

Strongylura marina (fig. 26.1) occurs from Maine to Florida and throughout the Gulf of Mexico and south to Brazil (Kendall 1914; Collette 1968; Dahlberg 1975). Is is common in marine and estuarine waters and extends up into freshwaters (Massmann 1954; Tagatz and Dudley 1961). In the Middle Atlantic Bight, the larvae and juveniles are reported in estuaries as far north as Great South Bay (table 4.2).

REPRODUCTION

Relatively little is known about reproduction. Adults apparently move north in the spring with rising temperature. Spawning occurs in shallow estuaries and freshwaters within masses of submerged algae (Foster 1974). In Great Bay–Little Egg Harbor, spawning appears to occur in the spring, because small individuals occur as early as May (fig. 26.2). Spawning and hatching occur over a relatively long period; the smallest size classes have been collected from May through August.

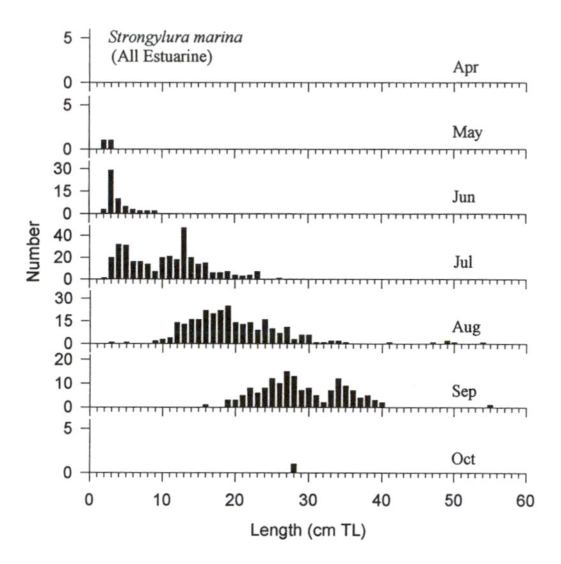

26.2 Monthly length frequencies of *Strongylura marina* young-of-the-year collected in the Great Bay–Little Egg Harbor study area. Sources: RUMFS: otter trawl (n = 2); 1-m plankton net (n = 12); gear comparison (n = 6); seine (n = 3); night-light (n = 114); weir (n = 619); dip net (daylight) boat basin (n = 46).

DESCRIPTION

The eggs are large (approximately 3.0 mm in diameter), spherical, with a narrow perivitelline space and no oil globules. They are typically attached to vegetation with numerous chorionic filaments that arise at opposite poles of the egg. The larvae are approximately 14.0 mm at hatching, densely pigmented—especially below the lateral line—and can be distinguished by their elongate shape and elongate jaws of equal length. The lower jaw grows faster than the upper (fig. 26.1) throughout the larval or "halfbeak stage" (Ryder 1882; Collette et al. 1984a). As the juveniles develop, the jaws again become similar in length at approximately 100 mm. Additional details of early development are provided by Foster (1974). Vert: 69–77; D: 13–17; A: 16–21; Pect: 10–13; Plv: 6.

THE FIRST YEAR

All of the development takes place in estuaries. During development, the jaw morphology changes so that the halfbeak form is lost and the jaws become equal in length (Boughton et al. 1991). These slender elongate fishes increase rapidly in length during the summer so that by September they range from 19–40 cm TL (fig. 26.2). The increase in modal length frequency shows that estimated growth is relatively fast, approximately 1.5 mm per day.

Our knowledge of the habitat of the young-of-the-year is limited because their habit of swimming very close to the surface and their speed probably make them inaccessible to many sampling gears. Visual observation in the RUMFS boat basin and frequent collection with small dip nets (Able et al. 1997) indicate that the habitat of the smallest individuals is open surface waters. Elsewhere in Great Bay, they have been found in subtidal creeks (Rountree and Able 1992a) and along subtidal shorelines but not in intertidal creeks or in marsh surface pools or on the adjacent continental shelf (Able et al. 1996b; K. W. Able and M. P. Fahay, unpubl. data). In Delaware Bay, the young have been collected as far upstream as the vicinity of the C&D Canal (Smith 1971) and reported as far upstream as the head of tide at Trenton (Fowler 1906). In Chesapeake Bay, the juveniles are found throughout the bay including up into freshwater (Musick 1972). This same general pattern has been observed in the Hudson River, where presumed young-of-the-year are found in marshes and weed beds from just above the Hudson Highlands to just below the Tappan Zee Bridge (Lake 1983). They are seldom collected after September, and they are not represented in winter collections presumably because they leave the estuary and retreat to the south.

CHAPTER 27

Cyprinodon variegatus Lacepède
Sheepshead Minnow

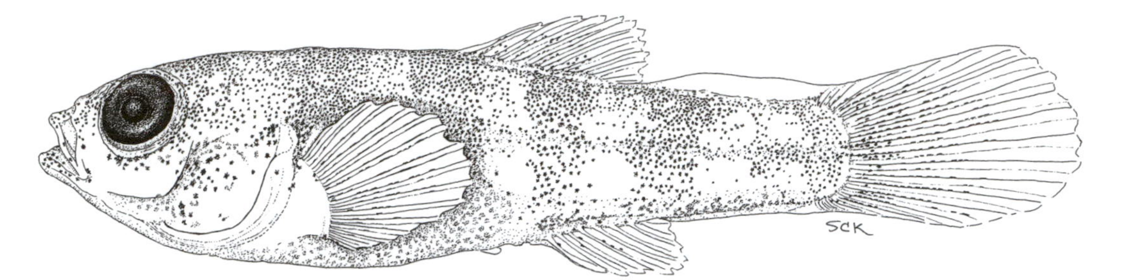

27.1 *Cyprinodon variegatus* juvenile, 10.6 mm TL. collected June 6, 1991, dip net, small marsh pool, RUMFS, Great Bay, New Jersey. ANSP 175220. Illustrated by Susan Kaiser.

DISTRIBUTION

Cyprinodon variegatus (fig. 27.1) occurs in very shallow estuarine waters, typically less than 0.5 m, from Cape Cod to Mexico (Duggins et al. 1983). They are common in protected coves, bays, ponds, and are particularly abundant in salt marshes. Early life-history stages occur in all of the estuaries in the Middle Atlantic Bight (table 4.2).

REPRODUCTION

Ovarian development and the occurrence of small larvae from May through September indicate that spawning occurs at age 1+ during spring and summer (fig. 27.2). This pattern is consistent with that for previous reports from New Jersey (Talbot and Able 1984; Able 1990; Smith 1995), Delaware Bay (Wang and Kernehan 1979), and North Carolina (Hildebrand 1919). The primary spawning sites are shallow areas, particularly marsh surface pools with vegetation, where the smallest larvae are collected (Talbot and Able 1984; Smith 1995).

DESCRIPTION

The demersal eggs are spherical (1.2–1.4 mm in diameter) with a narrow perivitelline space and one large (approximately 0.24 mm) and many minute oil globules. The chorion surface has many filaments. The larvae hatch at approximately 4.2 to 5.2 mm TL and have a short, plump body. Larvae eventually develop six or more dark vertical bands across the body. Fin rays are complete by 12.0 mm. By this size, it is clear that the anal fin originates posterior to the dorsal fin. Details of development are summarized by Kuntz (1914a) and Wang and Kernehan (1979). The juveniles (fig. 27.1) retain these characteristics. Vert: 25–27; D: 9–13; A: 9–12; Pect: 14–17; Plv: 5–7.

THE FIRST YEAR

The larvae hatch in 5 to 13 days (Kuntz 1916; Wang and Kernehan 1979). On the basis of monthly increases in modal length frequencies in Great Bay salt marshes, the young-of-the-year appear to grow slowly (0.3 mm per day). Individuals reach 19 to 49 mm TL (Able 1990) or perhaps up to 60 mm (fig. 27.2) by the end of the first summer. Growth may be faster in vegetated pools than in pools without vegetation in southern New Jersey marshes (Smith 1995). Most of the population in the fall is composed of young-of-the-year, as has been reported for other New Jersey populations (Pyle 1964). There is essentially no growth over the winter so that at the end of the first year they are 3 to 6 cm TL. It was suggested that some members of a Delaware population may live up to 3 years (Warlen 1963).

The habitat is broad and variable with all life-history stages occurring in the same habitats. The demersal larvae of *C. variegatus* (Seligman 1951) were less likely to be collected in pit traps on the marsh surface than were *Fundulus* spp. (Talbot and Able 1984), apparently because the former move less with flooding tides and prefer to stay in deeper pools and ditches. Some leave marsh pools and move onto the marsh surface during high tides (Smith 1995). Young-of-the-year and older ages occur in a variety of shallow water habitat types, including intertidal and subtidal marsh creeks and pools on the marsh surface in New Jersey (Talbot and Able 1984; Rountree and Able 1992a; Smith 1995; Able et al. 1996b). In direct comparisons between vegetated (*Ruppia maritima*) and unvegetated pools on the marsh surface, young-of-the-year were more abundant in vegetated pools (K. J. Smith and K. W. Able unpubl. data). During the summer, they appear to be limited to intertidal and shallow subtidal habitats because they are not important components of the fauna in deeper portions (1–8 m) in the same bay (Szedlmayer and Able 1996). However, they occur deeper (1–3 m), adjacent to a marsh creek (RUMFS boat basin) in November and December in Great Bay (fig. 27.3), suggesting a movement into deeper water for the winter. This broad range of habitats for young-of-the-year is not surprising because they are extremely tolerant of a range of environmental conditions. When exposed to low levels of dissolved oxygen similar to that observed in natural pools, they survived at levels as low as 0.5 ppm (K. J. Smith and K. W. Able unpubl. data). Individuals initially responded to a lowering of dissolved oxygen by resting on the bottom, then changed to aquatic surface respiration. Similar broad ranges are reported for salinity (e.g., 4–35 ppt, de Sylva et al. 1962) and temperature (Hildebrand and Schroeder 1928). Individuals that spend the winter on the marsh surface in ditches or pools may be susceptible to mortality during especially cold winters, at least in New Jersey (K. W. Able pers. observ.).

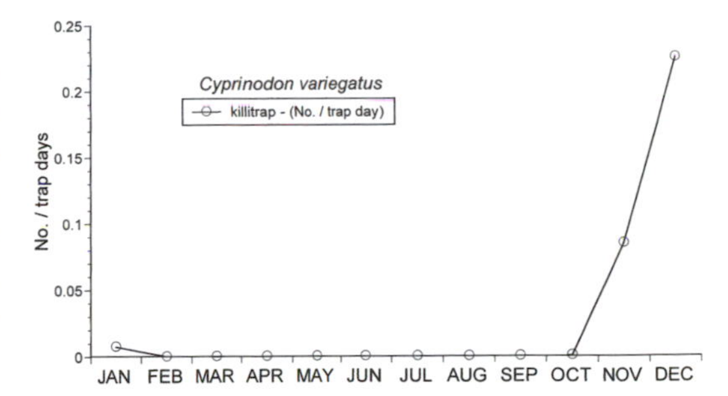

27.3 Monthly abundance of *Cyprinodon variegatus* larvae and young-of-the-year in the Great Bay–Little Egg Harbor study area based on trap collections in the RUMFS boat basin (n = 566).

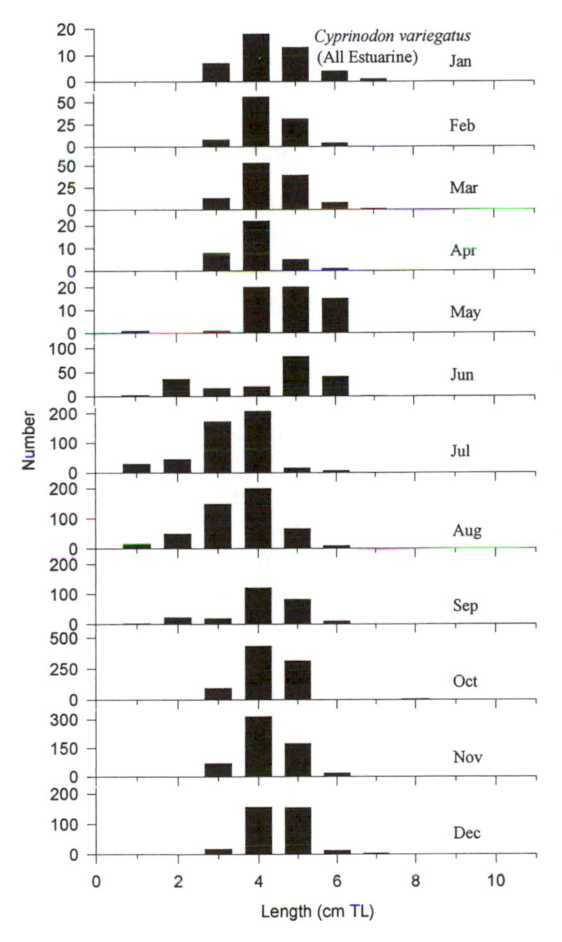

27.2 Monthly length frequencies of *Cyprinodon variegatus* collected in the Great Bay–Little Egg Harbor study area and other New Jersey salt marshes. Sources: Able 1990; RUMFS: otter trawl (n = 2); 1-m plankton net (n = 12); experimental trap (n = 2); gear comparison (n = 130); killitrap (n = 253); seine (n = 217); night-light (n = 5); weir (n = 1,198).

CHAPTER 28

Fundulus heteroclitus (Linnaeus) Mummichog

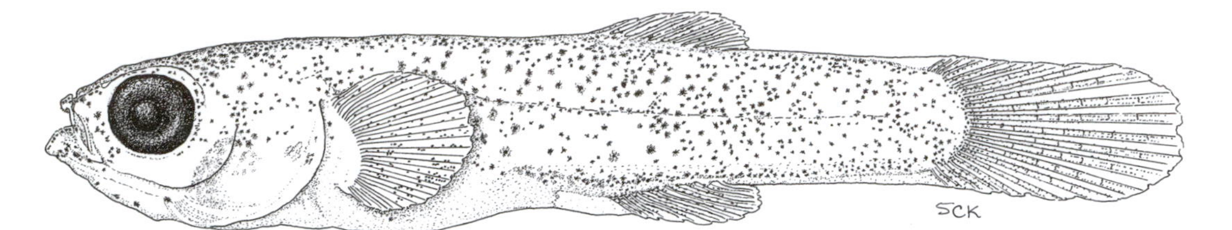

28.1 *Fundulus heteroclitus* juvenile, 11.0 mm TL prior to formation of vertical bars on the body. Laboratory-raised, April 27, 1991. ANSP 175221. Illustrated by Susan Kaiser.

DISTRIBUTION

Fundulus heteroclitus (fig. 28.1) is euryhaline and occurs in shallow estuarine waters from southwestern Newfoundland and Prince Edward Island to northern Florida (Relyea 1983; Scott and Scott 1988). Freshwater populations also occur (Hastings and Good 1977; Denoncourt et al. 1978; Samaritan and Schmidt 1982). Some authors recognize two subspecies (Able and Felley 1986). *Fundulus h. heteroclitus* occurs from northern New Jersey south to Florida including lower Chesapeake and Delaware Bays whereas *F. h. macrolepidotus* is distributed from Connecticut north to Newfoundland with disjunct populations in upper Chesapeake and Delaware Bays. Intergrade zones occur where the two morphs come together. In the Middle Atlantic Bight, intergrade populations occur in northern New Jersey (Raritan Bay, Arthur Kill, and Kill van Kull) and on Long Island and in upper Chesapeake Bay including the Potomac River (Morin and Able 1985; Marteinsdottir and Able 1988; Gonzalez-Villasenor and Powers 1990; Smith et al. 1992; Powers et al. 1993). Intensive mark-recapture studies in a Delaware marsh found a home range of approximately 18 m (Lotrich 1975), while movements up to 1 km occurred over the winter (Fritz et al. 1975). Early life-history stages of this species occur in every estuarine system in the Middle Atlantic Bight (table 4.2). Movements of adults in salt marshes appear to be restricted during the summer.

REPRODUCTION

Spawning occurs for many populations at age 1, in early spring through the summer, but the season varies in response to local temperatures. Reproduction occurs in Massachusetts from May to July (Wallace and Selman 1981), in New Jersey from late April through August (Talbot and Able 1984; Able 1990), and in North Carolina from March to August (Kneib and Stiven 1978).

All populations appear to spawn intertidally. Some populations, primarily those identified as *F. h. heteroclitus,* exhibit pronounced lunar periodicity. This has been demonstrated most clearly for a population in lower Delaware Bay (Taylor et al. 1977, 1979; DiMichele and Taylor 1980), where spawning occurs during the 3 to 5 days around lunar spring tides. The same lunar periodicity is evident in Great Bay–Little Egg Harbor (K. W. Able pers. observ.), North Carolina (Kneib and Stiven 1978), South Carolina (Radtke and Dean 1982), and Georgia (Kneib 1986). Populations of *F. h. macrolepidotus,* although not as well documented, may spawn daily during the reproductive season in Massachusetts (Taylor et al. 1982) and Long Island (Conover and Kynard 1984). As further evidence of subspecific differences in spawning periodicity, in New Jersey, a *F. h. heteroclitus* population from lower Delaware Bay had a pronounced peak in spawning on spring tides, but a *F. h. macrolepidotus* population from the lower Delaware River spawned almost daily

(Marteinsdottir 1991). On the other hand, a presumed *F. h. macrolepidotus* in Chesapeake Bay had distinct semimonthly cycles, which were usually in phase with the lunar cycle (Hines et al. 1985).

Subspecific differences may be responsible for the site of egg deposition as well (Able 1984a). Females, regardless of subspecies, deposit eggs in cracks, crevices, and small interstices, a behavior that is facilitated by the anal sheath (ovipositor) of the female (Able and Castagna 1975). Populations of *F. h. heteroclitus* are known to deposit eggs between the empty valves of the ribbed mussel, *Geukeusia demissa,* (Able and Castagna 1975) and the basal leaves of *Spartina alterniflora* (Taylor and DiMichele 1983), whereas populations of *F. h. macrolepidotus* deposit their eggs in the substrate or in mats of vegetation. Spawning over mud substrates has been reported for Delaware populations (Wang and Kernehan 1979).

DESCRIPTION

Egg morphology varies between populations and was the original basis for resurrecting subspecific designations (Morin and Able 1983). In addition, egg morphology between populations appears to be generally concordant with spawning site preference and spawning periodicity (Able and Felley 1986). Populations of *F. h. heteroclitus* have large eggs (2.0–2.2 mm in diameter), with numerous small-diameter short filaments on the surface of the chorion and numerous oil droplets, whereas populations of *F. h. macrolepidotus* have small eggs (1.6–1.9 mm), with few large-diameter, long chorionic filaments and relatively few oil droplets (Morin and Able 1983). Eggs intermediate in these characters are found in intergrade zones (Able and Felley 1986; Marteinsdottir and Able 1988). Embryonic development of a *F. h. macrolepidotus* population has been described in detail (Armstrong and Child 1965). Larval size at hatching also varies between the subspecies; *F. h. heteroclitus* has larger eggs and larger larvae (approximately 5.5 mm TL), whereas *F. h. macrolepidotus* has smaller eggs and smaller larvae (approximately 4.8–5.1 mm TL; Marteinsdottir and Able 1992). For all populations, the larvae (fig. 28.1) have dense melanophores concentrated on the dorsal surface of the head and a double row of melanophores on the middorsal and midventral regions. Larval development is described in greater detail by Lippson and Moran (1974) and Richards and McBean (1966). Larger juveniles have vertical bars on the lateral surfaces of the body and a dense patch of melanophores anterior to the dorsal fin as well as a plump body and a single, short-based dorsal fin consisting of soft rays only. Vert: 32–35; D: 10–14; A: 9–12; Pect: 16–20; Pelv: 6–7.

THE FIRST YEAR

The eggs are deposited intertidally and, as a result, are often exposed to the air. Hatching normally occurs in 9 to 18 days but is a function of temperature, dissolved-oxygen level, and elevation in the intertidal spawning site, because hatching is cued to emersion after the eggs are exposed at low tide (DiMichele and Taylor 1980). In addition, the intertidal site may cause hatching to be delayed for longer periods if emersion does not occur because of lower than normal tides (Taylor et al. 1977; DiMichele and Powers 1982). Thus, larvae from delayed hatching are more advanced developmentally (fin ray formation, pigmentation; K. W. Able pers. observ.).

Growth of young-of-the-year is typically slow; in Great Bay, it averaged 0.26 mm per day based on the slope of length regressions (Rountree 1992). Young-of-the-year attain lengths of about 3 to 7 cm TL by late fall (fig. 28.2; Able 1990). Growth of young-of-the-year varies latitudinally, in a countergradient manner, with the greatest growth in the northern part of the range (Schultz et al. 1996). There is little evidence of growth during the winter, and they are 4 to 8 cm TL by the end of the first 12 months. These sizes are in approximate agreement with the values for age-1 fish in North Carolina (Kneib and Stiven 1978). In a Delaware salt marsh, young-of-the-year with similar growth rates produced at least 78% of the annual production for all year classes and experienced 99.5% mortality (Meredith and Lotrich 1979). Given the abundance of this species in many salt marsh systems, these parameters indicate the ecological significance of the population dynamics of this species to salt marsh ecosystems.

Habitat use varies with life-history stage.

After hatching on flooding tides, the larvae are typically carried into the high marsh where they are found in shallow pools, ponds, ditches, and depressions on the marsh surface (Kneib and Stiven 1978; Wang and Kernehan 1979; Smith 1995). In New Jersey marshes, the larvae and juveniles were the most abundant form found across a variety of high marsh habitat types (Talbot and Able 1984). On rare occasions, the larvae have been collected in plankton tows in deeper estuarine waters (Croker 1965; Himchak 1982b; Witting 1995; D. A. Witting, K. W. Able, and M. P. Fahay in prep.). Transforming larvae and small juveniles have been captured from mid-May through September in high marsh pools, ponds, and ditches but were most abundant in these habitats during June and July (Talbot and Able 1984; Able 1990). During the same period, other young-of-the-year are abundant along a variety of intertidal and subtidal shorelines (fig. 28.3; Able et al. 1996). Many of the larger juveniles apparently remain in marsh pools during the first summer (Smith 1995). In Great Bay and other southern New Jersey estuaries, this species occurs in a diversity of shallow habitats, particularly in salt marshes and associated creeks (Clymer 1978; Talbot and Able 1984; Rountree and Able 1992a), but can be collected in almost any shallow habitat from eelgrass beds (Sogard and Able 1991) to open shorelines over substrate that range from sand to silt (Able et al. 1996b).

At least some portion of the population in Great Bay, including young-of-the-year, move into pools on the marsh surface during the fall (fig. 28.3) and remain through the winter (Smith and Able 1994). During exceptionally cold winters, at least in New Jersey, individuals in marsh surface pools and ditches are susceptible to increased mortality (K. W. Able pers. observ.).

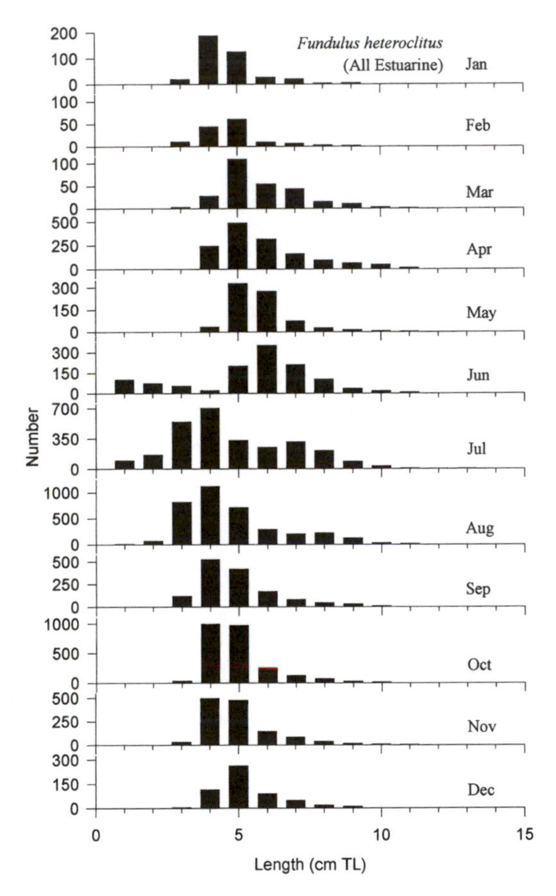

28.2 Monthly length frequencies of *Fundulus heteroclitus* collected in the Great Bay–Little Egg Harbor study area and other New Jersey salt marshes. Sources: Able 1990: RUMFS: otter trawl (n = 101); 1-m beam trawl (n = 32); 1-m plankton net (n = 193); experimental trap (n = 393); gear comparison (n = 464); killitrap (n = 957); seine (n = 1,374); night-light (n = 2); weir (n = 10,262).

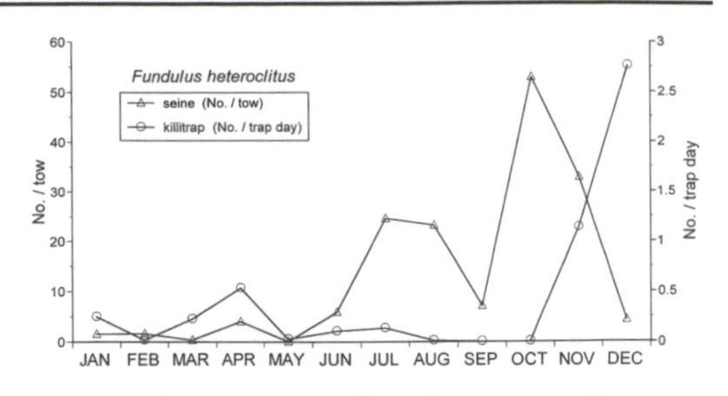

28.3 Monthly abundance of *Fundulus heteroclitus* larvae and young-of-the-year in the Great Bay–Little Egg Harbor study area based on killitrapping in the RUMFS boat basin during 1990–1994 (n = 11,560) and seine collections along a variety of intertidal and shallow subtidal shorelines during 1990–1991 (n = 7,827).

CHAPTER 29

Fundulus luciae (Baird) Spotfin Killifish

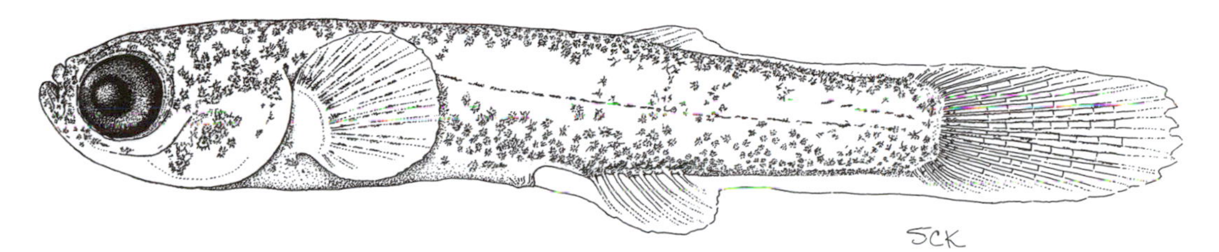

29.1 *Fundulus luciae* juvenile, 6.9 mm SL. Collected June 23, 1995, dip net, small marsh pool, RUMFS, Great Bay, New Jersey. ANSP 175222. Illustrated by Susan Kaiser.

DISTRIBUTION

Fundulus luciae (fig. 29.1) is found along the East Coast of the United States from Long Island and Connecticut to Georgia, and along a wide range of salinities (Lee et al. 1980). It is probably much more common than is recognized because it is likely to be confused with *F. heteroclitus* (Byrne 1978). Where efforts have been made to specifically collect this species, it is common but never abundant (Able et al. 1983). Early life-history stages are reported primarily from the southern part of the Middle Atlantic Bight (table 4.2).

REPRODUCTION

Maturation occurs at age 1 when males are 24 to 27 mm TL and females are 28 to 30 mm TL (Byrne 1978). On the basis of gonadal development and the occurrence of larvae, spawning in New Jersey occurs in salt marshes from May through September (Talbot and Able 1984; Able 1990). In Virginia, spawning may begin as early as April and extend to August (Byrne 1978); in North Carolina, it has been reported from April to October (Hildebrand and Schroeder 1928).

DESCRIPTION

The demersal eggs are spherical (1.7–2.2 mm in diameter) with 5 to 58 oil globules and filaments and distinct pits on the surface of the chorion (Byrne 1978; Able 1984b). The larvae have numerous small melanophores on the dorsal surface of the head and body, along the base of the finfolds, on the developing fins, and on the yolk sac. Further details of larval development are available from Byrne (1978). Juveniles are recognized as fundulids by the dorsally oriented mouth and the single short-based dorsal fin consisting only of soft rays (fig. 29.1). They can be distinguished from other juvenile killifish by the prominent line of melanophores extending from the dorsal surface of the head to the origin of the dorsal fin (Byrne 1978). Other killifishes have an aggregation of melanophores at the origin of the dorsal fin or lack both the aggregation and line. Vert: 31–33; D: 8–9; A: 10–11; Pect: 16–18; Plv: 6.

THE FIRST YEAR

Hatching occurs from June to September in New Jersey. Incubation time determines length at hatching. Those hatching in 12 to 16 days are 5.3 to 6.0 mm TL and have large yolk sacs and undeveloped fin rays; those in which hatching is delayed are 6.0 to 6.3 mm and have small yolk sacs with partially formed anal fin rays (Byrne 1978). Remnants of finfolds are evident at 11 to 12 mm TL, but all larger individuals have completely formed fins. Scales are first observed at 11 to 12 mm TL and scalation is completed when the hatchling is larger than

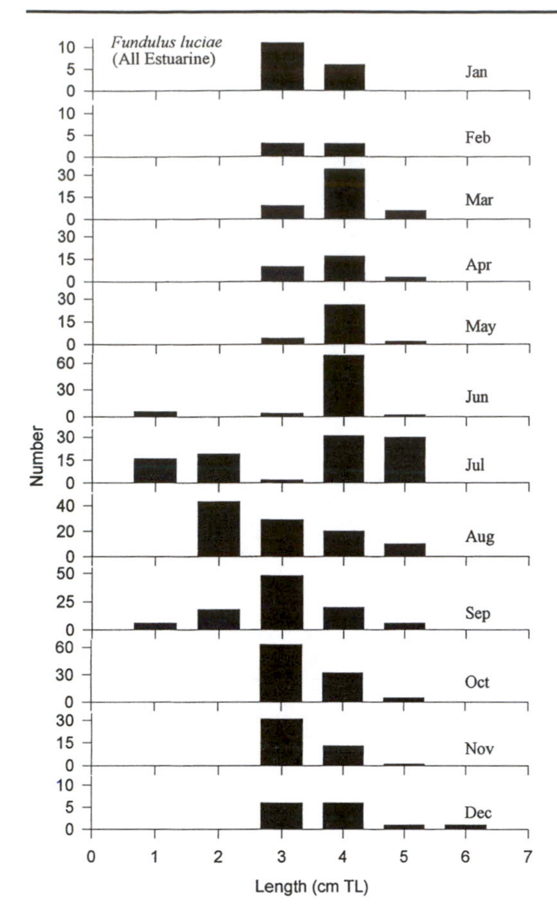

29.2 Monthly length frequencies of *Fundulus luciae* collected in the Great Bay–Little Egg Harbor study area and other New Jersey salt marshes. Sources: Able 1990: RUMFS: weir ($n = 86$).

17 mm TL. Based on modal length frequency progression, the juveniles grow an average of 0.2 mm TL per day between June and October and reach 3.0 to 6.0 cm TL by the fall (fig. 29.2). There appears to be little growth over the winter and they are the same size at the end of the first year.

Larvae, juveniles, and adults share habitat. They occur in shallow bodies of water on the marsh surface, where they experience a broad range of temperatures and salinities. They have also been collected in New Jersey at 0 to 46 ppt and 0.6 to 34.5 C (Able et al. 1983; Talbot and Able 1984), and similar broad ranges have been reported from Virginia (Byrne 1978). In other sampling, they have been collected in intertidal creeks (Able et al. 1996b). They are also tolerant of a variety of marsh alterations for mosquito control, which creates shallow pools on the marsh surface, but they do not occur in large expanses of shallow standing water such as impoundments (Talbot et al. 1986). The young-of-the-year were more abundant in pools without vegetation (*Ruppia maritima*) and where ammonia levels were high (Smith 1995). Although little is known about winter habitat, it is assumed to be the marsh surface because there is no evidence of immigration to deeper habitats (Byrne 1978).

CHAPTER 30

Fundulus majalis (Walbaum) Striped Killifish

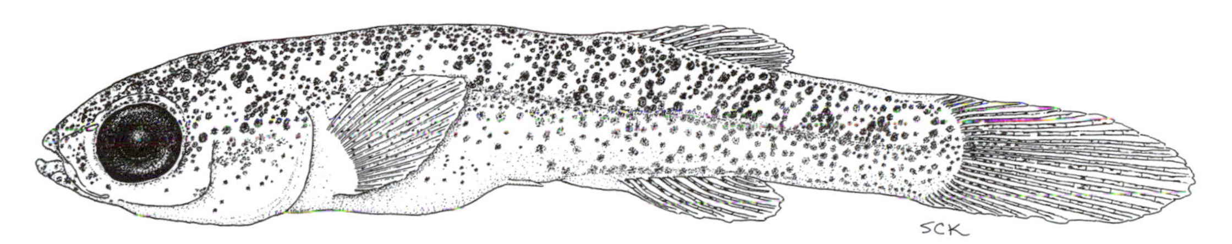

30.1 *Fundulus majalis* juvenile, 14.5 mm TL. Collected July 7, 1991, push-net, cove near RUMFS, Great Bay, New Jersey. ANSP 175223. Illustrated by Susan Kaiser.

DISTRIBUTION

Fundulus majalis (fig. 30.1) occurs from New Hampshire to the Gulf of Mexico (Jackson 1953; Relyea 1983). It is found in a variety of shallow habitats from open beaches to coves and bays and occasionally marsh creeks (Rountree and Able 1992a). In the Middle Atlantic Bight, early life-history stages occur in most estuaries but are less frequently reported in the northern part of the bight (table 4.2).

REPRODUCTION

Spawning in New Jersey at Corson's Inlet occurs as early as late May or early June through July. This information is based on a study in which adults (55–135 mm TL; fig. 30.2) in reproductive condition (males in spawning coloration, males and females running ripe, females with well-developed oviducts) and post-reproductive condition (males and females with torn anal fins, females emaciated) were collected. Spawning occurs in shallow, intertidal pools over sand substrates. Others have reported that the eggs are buried in substrates as much as 7 to 10 cm deep near the water's edge (Newman 1909; Sumner et al. 1913). Recently hatched larvae (<25 mm TL) occurred in early June and July. The presence of multiple cohorts (figs. 30.2, 30.3; late June, July, August) among recently hatched individuals suggests two or three spawning peaks, perhaps associated with spring tides. In Chesapeake Bay, spawning occurs from April to September at sizes as small as 75 and 63 mm for females and males, respectively (Hildebrand and Schroeder 1928).

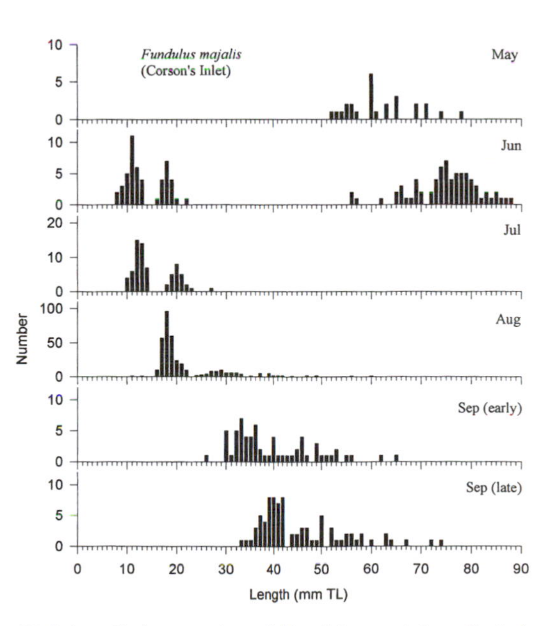

30.2 Length frequencies of *Fundulus majalis* collected with seine on several dates between May and September 1993 at Corson's Inlet, New Jersey (n = 744).

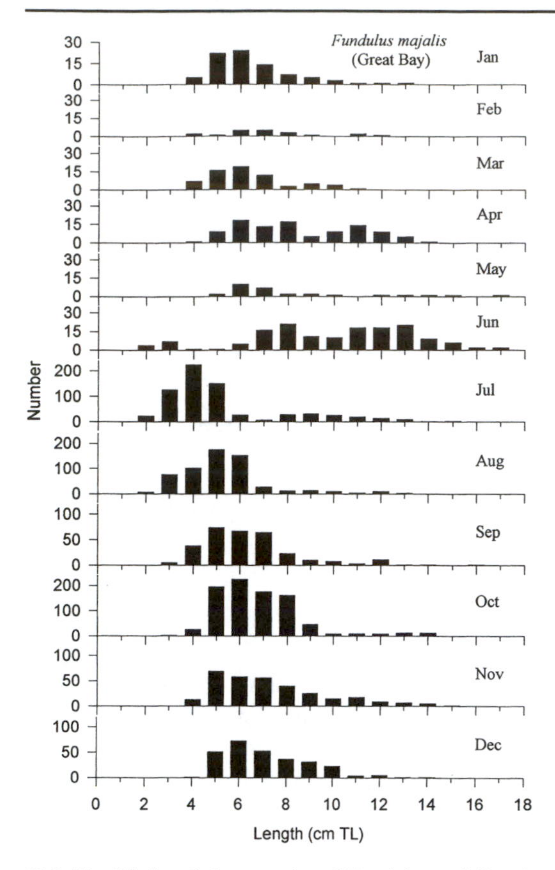

30.3 Monthly length frequencies of *Fundulus majalis* collected in the Great Bay–Little Egg Harbor study area. Sources: RUMFS: otter trawl (*n* = 16); 1-m beam trawl (*n* = 1); 1-m plankton net (*n* = 31); experimental trap (*n* = 68); gear comparison (*n* = 374); killitrap (*n* = 277); seine (*n* = 1,259); night-light (*n* = 22); weir (*n* = 1,485).

DESCRIPTION

The demersal eggs are spherical with a narrow perivitelline space. The diameter is 2.0 to 3.0 mm, and there are approximately 50 oil globules. Large filaments are present on the surface of the chorion, and microfilaments occur on the larger filaments and on the surface of the chorion (Able 1984b). Hatching occurs after 22 to 23 days at approximately 7.0 mm (Newman 1908, 1914). The larvae have numerous, large melanophores on the dorsal surface of the head and body and along the bases of the finfolds; there is a distinct aggregation at the origin of the dorsal fin. All other available descriptions and illustrations are summarized by Hardy (1978a). The juveniles have a typical killifish shape with a rather plump body and a single, short-based dorsal fin of soft rays (fig. 30.1). They have only five branchiostegal rays, whereas the similar *F. heteroclitus* have six (Richards and McBean 1966). Vert: 14–20; D: 11–16; A: 9–13; Pect: 18–21; Plv: 6.

THE FIRST YEAR

Hatching occurs after an incubation period of 22 to 23 days (Newman 1908, 1914). Temperature and degree of crowding of the eggs influence the meristic characters of embryos and the subsequent larvae (Fahy 1978, 1979, 1980, 1982). The larvae develop into juveniles with typical vertical bars within a few weeks of hatching. Larvae occur in shallow (<3 cm at low tide), intertidal sand-bottom pools, at least at the Corson's Inlet study site. Multiple cohorts of young-of-the-year were present during several sampling events (fig. 30.2). These were evident during late June, July, and early August but, by September, the length differences between cohorts were not obvious. The earliest spawned, or fastest growing, reached approximately 75 mm TL by the end of September. In May, the majority of individuals, presumably representing the same year class, were 49 to 78 mm TL. This suggests that very little or no growth occurs during the winter. By June, when these individuals were age 1, they had grown to approximately 57 to 98 mm TL, although it is difficult to separate the larger individuals from those 1 year older. The same general pattern is evident in collections from Great Bay–Little Egg Harbor (fig. 30.3). Scales have been used to age a Chesapeake Bay population (Clemmer and Schwartz 1964), with the

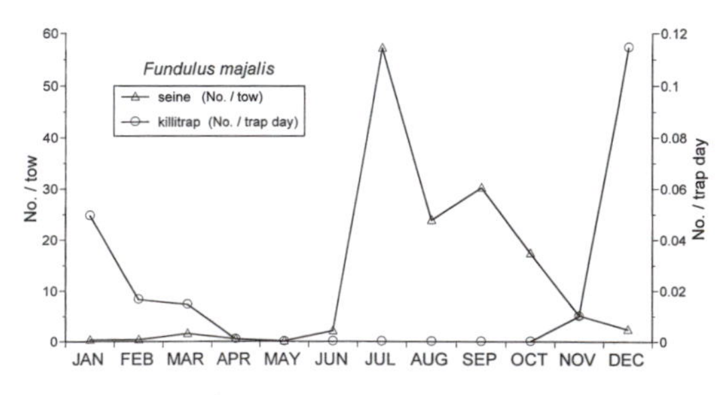

30.4 Monthly occurrences of *Fundulus majalis* larvae and young-of-the-year in the Great Bay–Little Egg Harbor study area based on killitrapping in the RUMFS boat basin during 1991–1994 (*n* = 311) and seine collections along a variety of intertidal and subtidal shorelines during 1990–1991 (*n* = 6,512).

resulting interpretation that males reached sizes of approximately 65 mm and females 61 mm at the end of the first year.

Habitat use varies with life-history stage and season during the first year. Generally, the young-of-the-year are found in a variety of shallow habitats along estuarine shores (Able et al. 1996b) from open beaches to coves and bays. They can be abundant in marsh creeks (Werme 1981; Rountree and Able 1992a, 1993) and, from there, move to the marsh surface (Werme 1981), although the species is never resident there (Weisberg 1986; Smith and Able unpubl data). The observations of Briggs and O'Connor (1971) indicate that this species may prefer sandy sediments. It has not been captured at greater depths (<8 m) in open waters of the adjacent bay (Szedlmayer and Able 1996). It is usually most abundant at higher estuarine salinities (Foster 1967; Weinstein 1979; Weisberg 1986), but has been collected from salinities of 1.0 to 37.0 ppt (Griffith 1974). Temperature-salinity experiments for young-of-the-year (8–69 mm TL) indicate that survival time and salinity tolerance vary primarily with temperature, with the greatest survival at 20 C (Schmelz 1970). Larvae are less resistant to temperature and salinity extremes (Schmelz 1970). The seasonal change from shallow shoreline habitats to deeper water in the winter (fig. 30.4) probably reflects the movement into a more stable temperature regime.

CHAPTER 31

Lucania parva (Baird and Girard) Rainwater Killifish

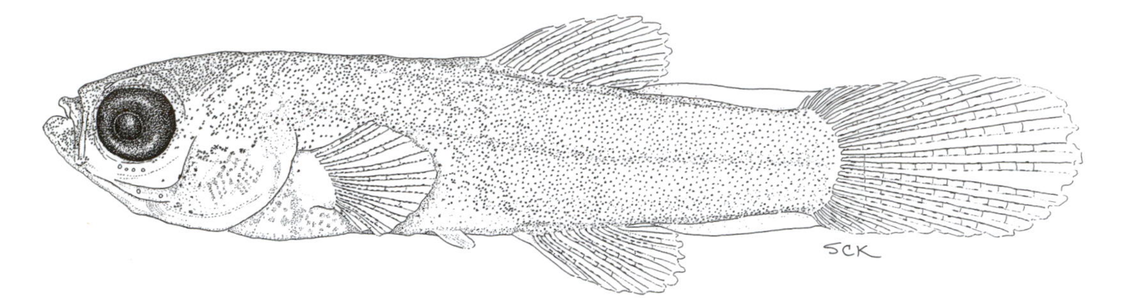

31.1 *Lucania parva* juvenile, 11.1 mm TL. Collected June 12, 1991, throw trap, upper marsh creek, heavily vegetated pool near RUMFS, Great Bay, New Jersey. ANSP 175224. Illustrated by Susan Kaiser.

DISTRIBUTION

A euryhaline species, *Lucania parva* (fig. 31.1) occurs in shallow, vegetated habitats or along open shores in coves, bays, and creeks from Cape Cod to Mexico (Hubbs and Miller 1965; Duggins et al. 1983). Early life-history stages are found in shallow estuaries in the southern portion of the Middle Atlantic Bight but are less frequently encountered from Long Island north (table 4.2).

REPRODUCTION

This species matures at about 25 mm TL (Hildebrand and Schroeder 1928), presumably at age 1. On the basis of gonadosomatic indices (Able 1990) and the occurrence of the smallest larvae, spawning lasts from June through August (fig. 31.2). This is consistent with other reports from New Jersey (Talbot and Able 1984), but spawning may occur from May through July in Delaware (Wang and Kernehan 1979). Spawning has been reported to occur in freshwater in Delaware (Wang and Kernehan 1979) and in the Gulf of Mexico (Gunter 1945, 1950). It also occurs at higher salinities based on the occurrence of recently hatched larvae in polyhaline marshes in New Jersey (Smith 1995).

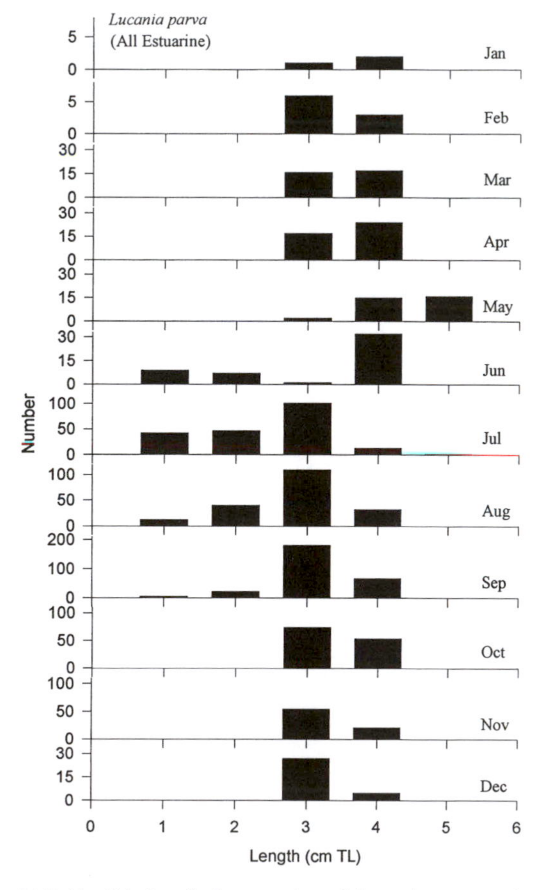

31.2 Monthly length frequencies of *Lucania parva* collected in the Great Bay–Little Egg Harbor study area and other New Jersey salt marshes. Sources: Able 1990: RUMFS: beam trawl (*n* = 1); experimental trap (*n* = 1); gear comparison (*n* = 2); killitrap (*n* = 3); weir (*n* = 79).

DESCRIPTION

The demersal eggs are large (1.0–1.3 mm in diameter) and spherical with a narrow perivitelline space and approximately 8 to 12 large (0.3–0.4 mm) oil globules (Able 1984b). The surface filaments occur in a tuft. The larvae hatch at 4.0 to 5.5 mm TL (Wang and Kernehan 1979) and are uniformly covered with small melanophores. Available descriptions and illustrations during larval development have been summarized by Hardy (1978a). The juveniles have a laterally flattened body with a single, short-based dorsal fin (fig. 31.1). The dorsal fin originates anterior to the anal fin but not as markedly as in *Cyprinodon variegatus.* The larger juveniles lack the vertical bars typical of larger *Fundulus* spp. This is one of the smallest killifishes (maximum size 43 mm TL; Able 1990). Vert: 25–30; D: 9–14; A: 8–13; Pect: 10–15; Plv: 4–7.

THE FIRST YEAR

Hatching takes place after 11 days at 25 C in the laboratory (Wang and Kernehan 1979) but has also been reported to occur in as little as 5 to 6 days at "summer" temperatures (Hildebrand and Schroeder 1928). In New Jersey, larvae first appear in June and may be found through September (fig. 31.2). A comparison of monthly length frequencies indicates that the young-of-the-year are 7 to 35 mm TL by the end of the first summer (Able 1990). A comparison of the modal lengths between June and September (fig. 31.2) suggests that larvae grow approximately 0.3 mm per day during the summer. There is little apparent growth during the winter, and the length at the end of the first year is 30 to 40 mm TL.

Most of the life history occurs on the marsh surface. The larvae are found primarily in shallow (0.3–0.6 m), marsh surface pools with submerged vegetation (*Ruppia maritima*) (Talbot and Able 1984; Smith 1995). The larger juveniles are found in the same habitats. Through the winter, the juveniles do not appear to use the intertidal marsh surface during most high tides but remain in pools or subtidal ditches (Talbot and Able 1984; Smith 1995). When exposed to low dissolved oxygen, as occurs in nature, laboratory populations of young-of-the-year survived levels as low as 0.5 ppm by switching from midwater swimming to aquatic surface respiration (Smith 1995). There is little evidence of movements out of these marsh pool habitats. However, a mass downstream migration of juveniles and adults (>270,000 individuals in a few hours) was observed during October in the York River, Virginia (Beck and Massmann 1951).

CHAPTER 32

Gambusia holbrooki Girard Eastern Mosquitofish

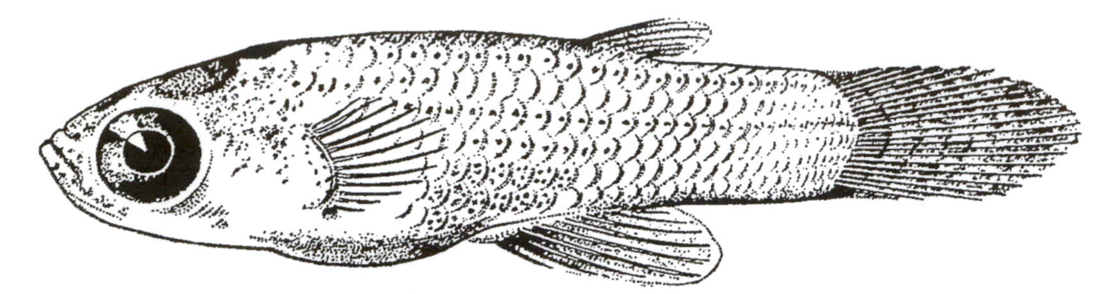

32.1 *Gambusia holbrooki* embryo, 6.3 mm TL, shortly before hatching. Source: Kuntz 1914b. Species designation presumably correct based on date and North Carolina collection location. Questions remain because of the widespread introduction of *G. affinis* for mosquito control.

DISTRIBUTION

In the Middle Atlantic Bight, *Gambusia holbrooki* (fig. 32.1) extends south from the Delaware shores of Delaware Bay (Wang and Kernehan 1979). The natural occurrence of this species in New Jersey has been questioned. The first records, from the turn of the century, coincide with the introduction of *Gambusia* spp. for mosquito control (Gooley and Lesser 1977). Some of the introductions have been from other populations in the central United States that are now recognized as *G. affinis* (Wooten et al. 1988). It is typically found in quiet, shallow, fresh, and brackish waters.

REPRODUCTION

Eastern mosquitofish mature in 4 to 6 weeks after birth (Krumholz 1948) at sizes of approximately 20 mm (Hildebrand and Schroeder 1928). On the basis of the occurrence of young, reproduction of New Jersey populations of *Gambusia* spp. occurs in late spring and summer (fig. 32.2). In Chesapeake Bay, reproduction occurs from May through September (Hildebrand and Schroeder 1928). Fertilization and development of eggs is internal.

DESCRIPTION

The taxonomy of the forms in the study area is confounded by potential introductions of several species for mosquito control. In the eastern United States, two naturally occurring forms have been recognized and designated as either species or subspecies. Most recently, genetic and other data suggest a degree of isolation such that they are now recognized as separate species (i.e., *G. affinis* and *G. holbrooki*) (Wooten et al. 1988). *Gambusia holbrooki* is the coastal species, extending from central Alabama east into Florida and along the Atlantic coast drainages to New Jersey (Rivas 1963; Rosen and Bailey 1963). Both of these forms have been introduced into numerous locations for mosquito control and, as a result, it is difficult to determine which species has been collected in the New Jersey study site.

The young of these livebearers are born at an advanced stage of development (Kuntz 1914b) at sizes of 8 to 10 mm (Hildebrand and Schroeder 1928) (fig. 32.1). They superficially resemble fundulids in body shape but differ in that almost all the fin rays are completely formed at birth and the number of dorsal fin rays is low (8). In addition, the origin of the anal fin is markedly anterior to the dorsal fin. Scales form at 7 mm. The pelvic fins are not formed at birth but are complete when the fish is larger than 10 mm. Sexual dimorphism in anal fins (intromittent organ) is evident at about 20 mm. D: 7–8; A: 9.

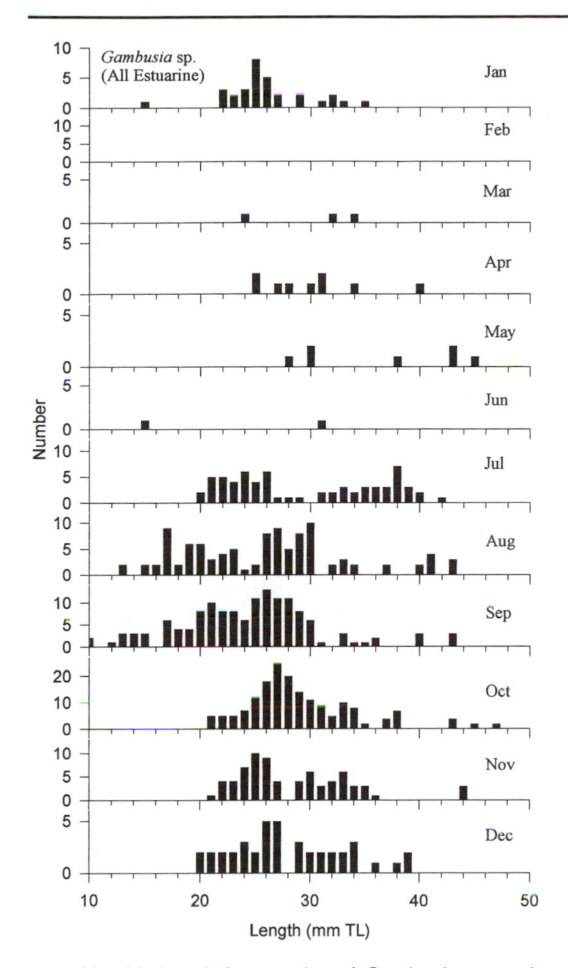

32.2 Monthly length frequencies of *Gambusia* spp. collected from Manahawkin Bay between 1978 and 1979. Source: Talbot et al. 1980.

THE FIRST YEAR

The young-of-the-year occur in some New Jersey locations from June through September (Talbot et al. 1980). During this time, modal length frequency progression indicates that they grow approximately 0.1 mm per day. They reach 15 to 40 mm TL by the following October (fig. 32.2), but there is no evidence of growth during the winter. Young-of-the-year typically occur in shallow estuarine and freshwater habitats and can be very abundant in man-made impoundments (Talbot et al. 1980), especially where there is aquatic vegetation (Wang and Kernehan 1979). Many may succumb to low temperatures in the winter (Talbot et al. 1980, 1986).

Membras martinica (Valenciennes) Rough Silverside

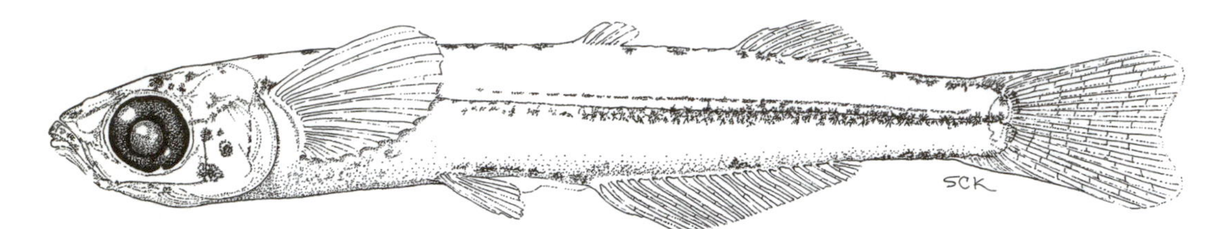

33.1 *Membras martinica* juvenile, 16.0 mm SL. Collected October 12, 1991, dip net (night-light), RUMFS boat basin, Great Bay, New Jersey. ANSP 175225. Illustrated by Susan Kaiser.

DISTRIBUTION

Membras martinica (fig. 33.1) occurs from New York to Mexico but is most abundant in the southern part of its range (Robins and Ray 1986). Over this range, it is found in open, shallow water, usually along exposed shorelines and beaches with little vegetation. In the Middle Atlantic Bight, early life-history stages have been reported from estuaries in New York, New Jersey, and south (table 4.2).

REPRODUCTION

Little is known about reproduction of this species. Spawning occurs from late spring through most of the summer in polyhaline to oligohaline waters of Delaware Bay at 20–30 C (Wang and Kernehan 1979). A large collection of eggs in Delaware Bay suggested that many individuals had spawned simultaneously (Wang and Kernehan 1979).

DESCRIPTION

The demersal eggs of this atherinid are small (0.7–0.8 mm), spherical, and have one to three large filaments on the surface of the chorion. The larvae have a single row of elongate melanophores along the lateral line and other melanophores on the dorsal surface of the head. Further details of development are available (Kuntz 1916; Lippson and Moran 1974). Juveniles have dorsal pigment restricted to a few, large, widely spaced spots along the midline (fig. 33.1). The upper sides are usually clear of pigment. The anal fin has a long base, beginning with a single feeble spine. There are two distinct dorsal fins, both with short bases. Note anal fin ray count relative to other atherinids. Vert: 40–44; D: II-VII, 6–9; A: I, 15–23; Pect: 11–15; Plv: I, 5.

THE FIRST YEAR

Eggs are attached to submerged vegetation, algae, and other debris in shallow waters. In Delaware Bay, larvae and juveniles were most abundant throughout the summer and early fall (Wang and Kernehan 1979). At Sandy Hook, New Jersey, the occurrence of young-of-the-year was enigmatic, perhaps reflecting that they are near the northern limit of their range. In regular collections in September of several years, they were abundant in one year (1979) but absent in most (1980–1982). When synoptic collections occurred in Sandy Hook Bay and in the ocean during 1979, they appeared to represent a single year class (young-of-the-year) that ranged from 30 to 110 mm TL with a mean size of 49 mm TL (fig. 33.2).

In Delaware Bay, this species is common along beaches with hard sand or mud substrate

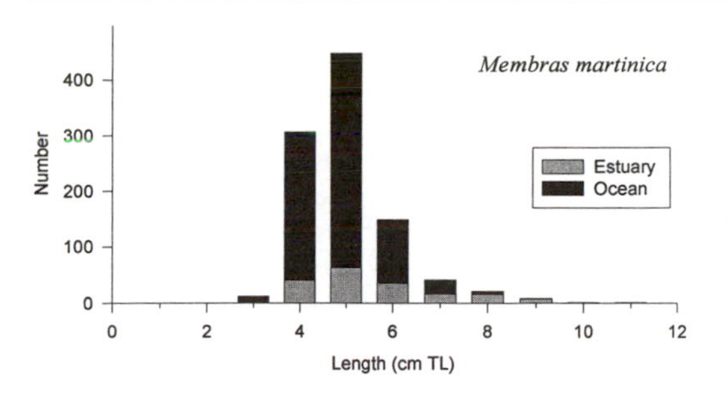

33.2 Length frequencies of *Membras martinica* collected October 2, 1979, from the estuary and ocean at Sandy Hook, New Jersey ($n = 986$).

in polyhaline to oligohaline shallow waters (Wang and Kernehan 1979). The collections in New Jersey, both in the estuary and ocean, were over sand substrates. The species is apparently most abundant at higher salinities (>25 ppt, Shuster 1959). In Chesapeake Bay, it has been reported as mesohaline in its distribution and common over grass flats and channel edges (Musick 1972). During the winter, this species is less abundant, indicating movement into deeper waters or possibly offshore (Wang and Kernehan 1979).

CHAPTER 34

Menidia beryllina (Cope)
Inland Silverside

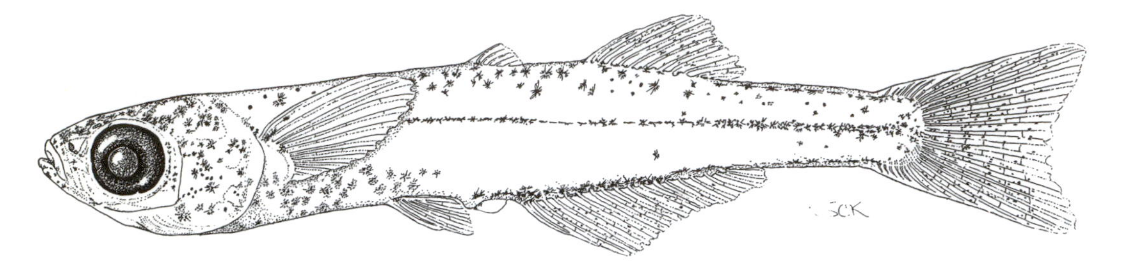

34.1 *Menidia beryllina* juvenile, 10.3 mm SL. Collected June 25, 1993, dip net, Bass River Marina, Bass River, New Jersey. ANSP 175226. Illustrated by Susan Kaiser.

DISTRIBUTION

Menidia beryllina (fig. 34.1) inhabits shallow estuarine and freshwater marshes from Cape Cod, Massachusetts to Vera Cruz, Mexico (Gosline 1948; Robbins 1969; Johnson 1974), where it is found commonly in tidal channels, particularly in the vicinity of submerged vegetation. In New Jersey, it enters marshes during the spring, spends the summer, and then moves out of the marshes into deeper water during the fall (Coorey et al. 1985). Early life-history stages are distributed in estuaries in the Middle Atlantic Bight up to Long Island and then less consistently farther north (table 4.2).

REPRODUCTION

Spawning in New Jersey occurs at sizes as small as 42 mm TL (Coorey et al. 1985). Duration of the spawning season varies from north to south: it is relatively short (June–July) in Rhode Island (Bengston 1984) and longer in Delaware (May–August) (Wang and Kernehan 1979). On the basis of a female gonadosomatic index, variation in ova diameters, and the occurrence of eggs, larvae, and young-of-the-year, spawning occurred from May to August in a Barnegat Bay, New Jersey salt marsh (Coorey et al. 1985). Most spawning was reported from oligohaline waters, which is consistent with other reports from Delaware Bay (Wang and Kernehan 1979) and Chesapeake Bay (Dovel 1971).

DESCRIPTION

Eggs are small (0.8–0.9 mm in diameter), spherical, and have a narrow perivitelline space and one to three large oil globules. Chorionic filaments consist of one large and five or more thin filaments. Larvae hatch at 3.5 to 4.0 mm TL and have a few elongate melanophores along the lateral line and more numerous melanophores on the dorsal surface of the head and along the ventral surface. Juveniles have small pigment spots scattered along the dorsal midline and extending down onto the upper sides (fig. 34.1). Eventually, a silvery stripe extends the length of the side. The anal fin has a long base, although not as long as, and with fewer rays than that of *Menidia menidia.* The anal fin originates with a single feeble spine. There are two distinct dorsal fins, both short-based. Vert: 41–42; D: V, 6–8; A: I, 16–19; Pect: 12–14; Plv: I, 5.

THE FIRST YEAR

The eggs are often attached to vegetation during spawning (Coorey et al. 1985; Wang and Kernehan 1979). Hatching occurs after 8 days at 17 to 25 C (Wang and Kernehan 1979). Mean instantaneous growth of larvae (7–14 days post hatch) in mesocosms in a Rhode Island estuary was 0.12 to 0.13 per day (Gleason and Bengston 1996). Mortality experiments in the same mesocosms suggested that size-selective predation was species specific and, in nature, likely to vary with the local suite of predators (Gleason and Bengston 1996). Young-of-the-year (4.0–20.0 mm TL) were present at a Barnegat Bay study site from mid-May through early August but were most abundant during May and June (Coorey et al. 1985). Growth of juveniles in New Jersey marsh systems, based on modal length frequency progression, averaged 0.2 mm per day (Marcellus 1972; Coorey et al. 1985). Young-of-the-year reached lengths of approximately 4 to 6 cm TL by the end of the first summer in New Jersey (fig. 34.2) and Rhode Island and showed little growth over the winter (Bengston 1984; Coorey et al. 1985).

Recent intensive sampling on estuarine shores in the Great Bay–Little Egg Harbor Estuary during the summer found all life-history stages across the gradient from marsh surface pools to intertidal creeks to shallow subtidal creeks (Able et al. 1996b) but not in the deeper portions (<1–8 m) of the estuary (Szedlmayer and Able 1996). The larvae are most abundant at low salinities (<10 ppt) (Dovel 1971; Wang and Kernehan 1979; Coorey et al. 1985). Dissolved-oxygen levels appear to influence habitat use and behavior of young-of-the-year. When exposed to low levels of dissolved oxygen that occur naturally in marsh pool habitats, the young-of-the-year survived at levels as low as 0.5 ppm by using aquatic surface respiration (Smith 1995). This ability to tolerate low dissolved oxygen allows them to use marsh surface pools as a nursery, where the larvae and juveniles are often collected (Coorey et al. 1985; K. J. Smith and K. W. Able unpubl. data).

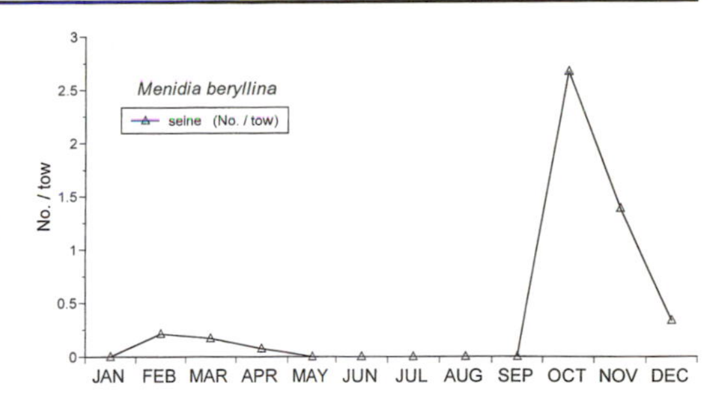

34.3 Monthly occurrences of *Menidia beryllina* young-of-the-year in the Great Bay–Little Egg Harbor study area based on seine collections along a variety of intertidal and shallow subtidal shorelines during 1990–1991 (n = 252).

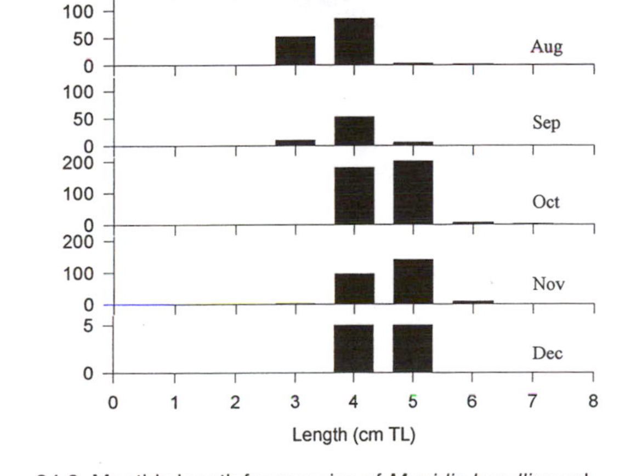

34.2 Monthly length frequencies of *Menidia beryllina* collected in the Great Bay–Little Egg Harbor study area. Sources: RUMFS: otter trawl (n = 7); 1-m plankton net (n = 17); gear comparison (n = 1); seine (n = 127); night-light (n = 24); weir (n = 960).

Seasonal variation in abundance in Barnegat Bay indicates that the young-of-the-year move out of salt marshes during late fall and winter, presumably into deeper waters of the estuary, and reenter during the spring (Coorey et al. 1985). The same pattern is evident in Great Bay–Little Egg Harbor, where seasonal sampling along shallow polyhaline subtidal areas indicates that they are rare in the summer, when they are in marshes, but become increasingly abundant in the fall and can be found in low numbers through the winter and early spring (fig. 34.3). In Delaware Bay, young-of-the-year and adults reportedly migrate to higher-salinity waters in the lower bay during late fall when temperatures are declining (Wang and Kernehan 1979).

CHAPTER 35

Menidia menidia (Linnaeus) Atlantic Silverside

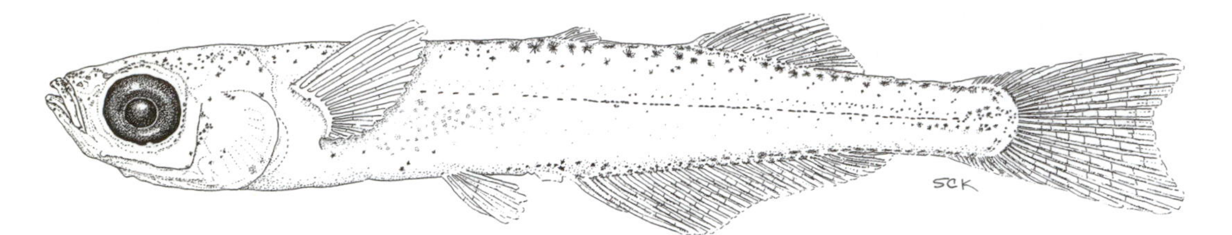

35.1 *Menidia menidia* juvenile, 17.4 mm SL. Collected June 13, 1987, dip net (night-light), RUMFS boat basin, Great Bay, New Jersey. ANSP 175227. Illustrated by Susan Kaiser.

DISTRIBUTION

Menidia menidia (fig 35.1) is found in a variety of shallow estuarine and marine habitats from the southern Gulf of St. Lawrence to Florida (Johnson 1974), but it is most abundant from Cape Cod to South Carolina, where it is common on open beaches over sandy and gravely bottoms and in tidal creeks. This is one of the most abundant fishes throughout the Middle Atlantic Bight, and early life-history stages have been collected in all estuaries we surveyed (table 4.2). Seasonal movements, especially in the northern portion of the bight, are common. Atlantic silverside typically moves offshore and into the ocean in the fall, returning to estuaries and coastal waters in the spring (Conover and Murawski 1982).

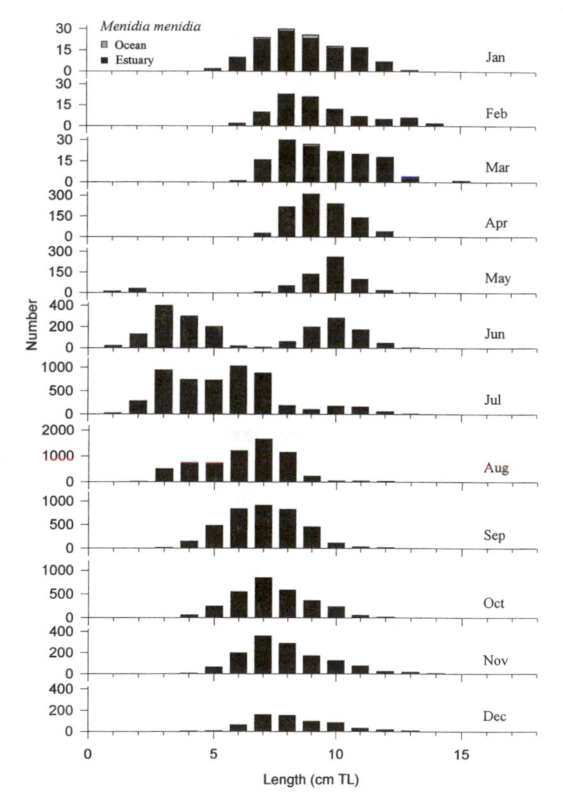

35.2 Monthly length frequencies of *Menidia menidia* young-of-the-year collected in the Great Bay–Little Egg Harbor study area. Sources: RUMFS: otter trawl (*n* = 2,402); 1-m beam trawl (*n* = 686); 1-m plankton net (*n* = 872); experimental trap (*n* = 142); LEO-15 2-m beam trawl (*n* = 7); gear comparison (*n* = 1,822); killitrap (*n* = 497); seine (*n* = 2,752); night-light (*n* = 641); weir (*n* = 14,343).

REPRODUCTION

This species spawns in estuaries in the spring only during daytime high tides in intertidal areas, including marsh creeks (Middaugh and Takita 1983). Spawning occurs from April to June in Massachusetts (Conover and Kynard 1984), late April to early July in Rhode Island (Bengston 1984). In New Jersey small, recently hatched individuals occurred from May through July (fig. 35.2).

DESCRIPTION

The benthic eggs are relatively large (1.0–1.5 mm in diameter) and spherical and have 5 to 12 large oil droplets and numerous small ones. The chorion has numerous long filaments that form a tuft. The larvae have a double row of melanophores on the dorsal surface from the nape to the origin of the second dorsal fin. The juveniles (fig. 35.1) have a silvery stripe that extends along the side and smooth cycloid scales. The anal fin has a long base with a single feeble spine and a greater number of fin rays than other atherinids in the region. V: 37–47; D: III–VII, 7–11; A: I, 19–29; Pect: 12–16; Plv: I, 5.

THE FIRST YEAR

The eggs hatch in 8 days at 22 to 29 C (Wang and Kernehan 1979). Larval development is gradual with no pronounced changes in morphology. In Great Bay–Little Egg Harbor, the larvae, which are dominant components of the ichthyoplankton, reach a peak in abundance in June (fig. 35.3; Witting 1995; D. A. Witting, K. W. Able, and M. P. Fahay in prep.). The juveniles reach their peak in July, but they are abundant through the summer and fall and then decline until December (fig. 35.3). Thereafter, they are less abundant, especially from January to March. This seasonal pattern is similar to that reported for Massachusetts, except that all individuals left inshore waters during the winter (Conover and Kynard 1984) and were most abundant in the ocean at that time (Conover and Murawski 1982). In Great Bay–Little Egg Harbor, relative abundance of larvae and juveniles, as captured in weekly, nighttime plankton samples was variable between years (fig. 35.4).

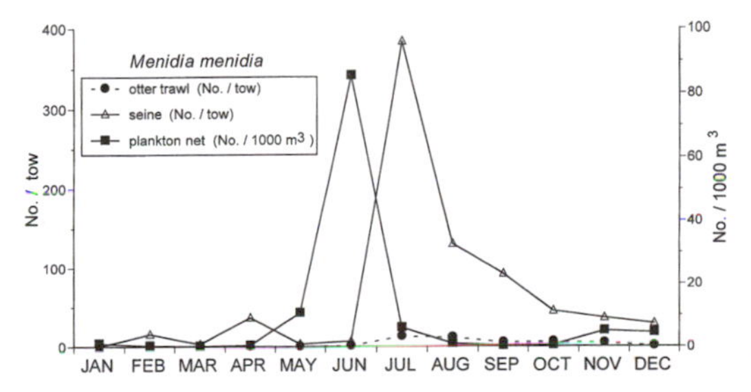

35.3 Monthly occurrences of *Menidia menidia* larvae and young-of-the-year in the Great Bay–Little Egg Harbor study area based on seine collections 1990–1991 (n = 34,242), otter trawl collections 1988–1989 (n = 4,812), and Little Sheepshead Creek Bridge 1-m plankton net collections 1989–1994 (n = 4,486).

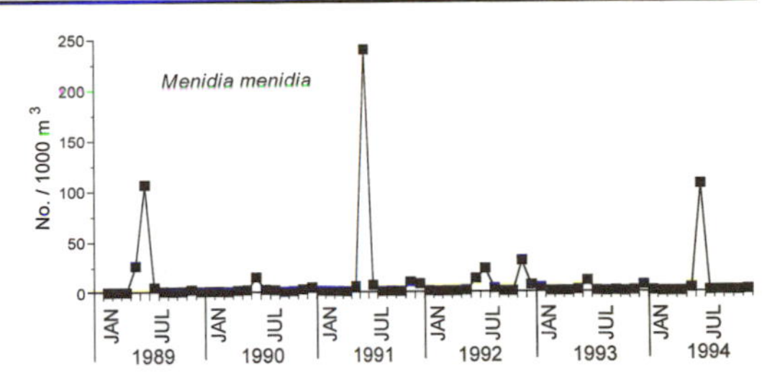

35.4 Annual variation in abundance of *Menidia menidia* larvae in Little Sheepshead Creek Bridge 1-m plankton net collections 1989–1994 (n = 4,486).

The juveniles grow quickly during the summer. Estimates from otolith daily growth rings suggest rates of 0.84 mm per day in Rhode Island (Barkman et al. 1981), and these growth rates appear to be influenced by temperature (Bengtson and Barkman 1981). In New Jersey, the young-of-the-year grow to 4 to 13 cm by September (fig. 35.2), and this is consistent with the pattern of countergradient growth relative to latitude as suggested by Conover (1992). There is little evidence of growth during the winter in New Jersey (fig. 35.2) or Rhode Island (Bengston 1984). In New Jersey, during May, at approximately age 1, these individuals are 7 to 13 cm. This general pattern of length distribution by month is consistent for other areas of the Middle Atlantic Bight including the Hudson River (Perlmutter et al. 1967) and Delaware Bay (Schuler 1974) as well as Great Bay (Thomas et al. 1972).

Young-of-the-year are abundant along a variety of intertidal and subtidal habitats on estuarine shores and in deeper waters and this is well documented for Great Bay (Able et al. 1996b; Szedlmayer and Able 1996). An examination of collections, primarily from otter trawls, indicates that they were generally more abundant at shallower depths (1–3 m), over sandier sediments (< 20% silt/clay), and where the structural portions of the habitat included eelgrass, amphipod tubes, shell, and sea lettuce (fig. 35.5). Intensive sampling in the intertidal and subtidal portions of marsh creeks indicated ontogenetic patterns in habitat use that were largely influenced by diel differences between different young-of-the-year cohorts (Rountree and Able 1993). The smaller size cohorts make limited use of intertidal creeks during the day.

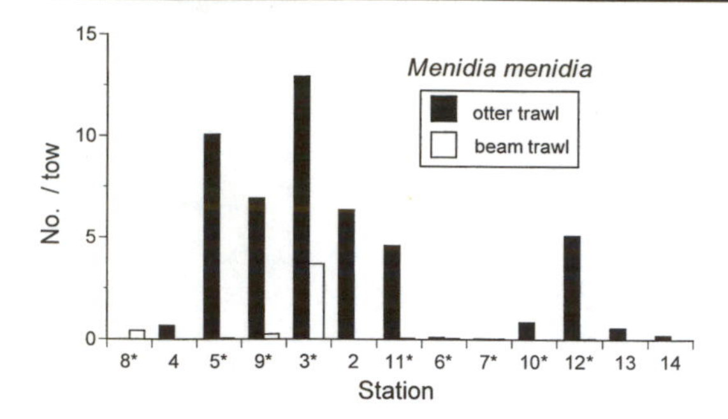

35.5 Distribution of *Menidia menidia* young-of-the-year by habitat type based on otter trawls (n = 4,715) and beam trawls (n = 308) deployed at regular stations 2–14 in Great Bay and Little Egg Harbor. Both gears used on stations designated by asterisk; otherwise, only the otter trawl was used. See figure 2.3 for station locations; table 2.1 for specific habitat characteristics by station.

However, the larger size cohorts are abundant in intertidal and subtidal creeks at night, presumably as the result of nocturnal movements into marsh creeks. Ontogenetic differences in habitat use have also been reported for Great South Bay, New York, where smaller individuals appear to prefer eelgrass (Briggs and O'Connor 1971).

The young-of-the-year are the least tolerant of low dissolved-oxygen conditions of any of the fundulids, cyprinodontids, and atherinids found in marsh surface pools (Smith 1995). When exposed to low dissolved oxygen, they first show signs of stress (aquatic surface respiration) at approximately 2.6 ppm, thus they are much less tolerant than the cooccurring congener *M. beryllina.* This relative intolerance may explain why the larvae and juveniles are only present in marsh pools until July. Thereafter, the nocturnal periods of low dissolved oxygen are more consistent and of longer duration (Smith 1995).

The diel movements in and out of marsh creeks by young-of-the-year during the summer continue until they reach 60 to 80 mm TL in early August, at which time they move into the deeper waters of the estuary (Rountree and Able 1992a). Most young-of-the-year leave the estuary during the late summer and early fall to move offshore for the winter and then return the following spring to spawn (Conover and Murawski 1982). This movement out of the estuary may occur as late as December because in several years small peaks in collections have occurred at that time (fig. 35.4).

CHAPTER 36

Apeltes quadracus (Mitchill) Fourspine Stickleback

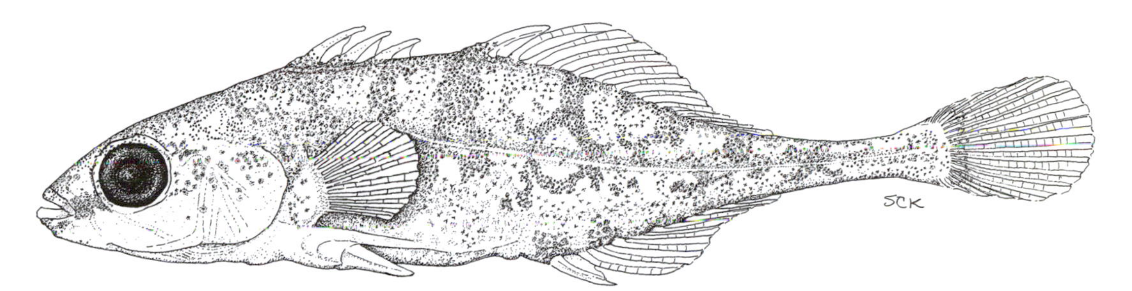

36.1 *Apeltes quadracus* juvenile, 17.9 mm SL. Collected May 30, 1991, 1-m beam trawl, small cove, Holgate, New Jersey. ANSP 175228. Illustrated by Susan Kaiser.

DISTRIBUTION

Apeltes quadracus (fig. 36.1) is found in polyhaline to freshwater areas from Virginia to the Gulf of St. Lawrence (Bigelow and Schroeder 1953). It can also be found across a variety of shallow habitats, including tide pools, ponds, and salt marshes, and appears to be most abundant in areas with submerged vegetation. Early life-history stages are common in most estuaries in the Middle Atlantic Bight (table 4.2).

REPRODUCTION

Courtship and spawning occur in the spring. Nests are constructed of vegetation glued together by secretions of the kidney (Breder 1936). The eggs and larvae are defended by the male (Reisman 1963; Scott and Crossman 1973). Based on the occurrence of the smallest individuals, most spawning occurs during April and May in Delaware Bay (Wang and Kernehan 1979), Chesapeake Bay (Dovel 1971), and Great Bay–Little Egg Harbor (fig. 36.2).

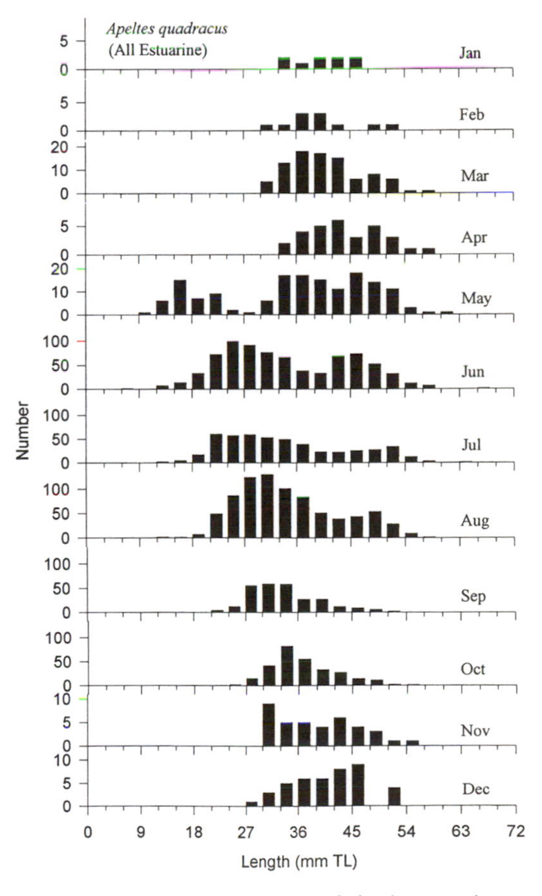

36.2 Monthly length frequencies of *Apeltes quadracus* collected in the Great Bay–Little Egg Harbor study area. Sources: RUMFS: otter trawl (n = 1,234); 1-m beam trawl (n = 1,099); 1-m plankton net (n = 9); experimental trap (n = 9); gear comparison (n = 525); killitrap (n = 81); seine (n = 42); weir (n = 3).

DESCRIPTION

The eggs are spherical, large (1.3–1.6 mm in diameter), and adhesive, and lack a perivitelline space early in development. A few small (0.1 mm in diameter) oil globules are present. The larvae hatch at 4.2 to 4.5 mm TL. They are darkly pigmented except for the finfolds and alternating dark and light areas along the middorsal surface. Fin rays are complete by 16.0 mm. Further details of development are presented by Wang and Kernehan (1979). Juveniles (fig. 36.1) have a very thin caudal peduncle but lack the overlapping bony plates along the body as in Middle Atlantic Bight populations of *Gasterosteus aculeatus*. There are typically four prominent, isolated dorsal spines, the last two more widely spaced and the last associated with the second dorsal fin. A prominent spine also precedes the anal fin and each pelvic fin. The color pattern is blotchy. Vert: 29–34; D: II–IV, 9–14; A: I, 7–11; Pect: 11–12; Plv: I, 2.

THE FIRST YEAR

The eggs hatch in 7 to 8 days at 21 C (Wang and Kernehan 1979). The demersal larvae are guarded by the male parent. This restriction to the area of the nest may account for the absence of larvae in plankton collections. They grow quickly after hatching and by 8 days are 7 mm (Kuntz and Radcliffe 1917). The young-of-the-year first appear in May at sizes of approximately 9 to 27 mm TL (fig. 36.2). By October, the young-of-the-year are approximately 27 to 48 mm TL. Estimates based on the increases in modal length frequency over this period indicate a growth rate of approximately 0.1 mm TL per day. There appears to be little growth during the winter so that, by May, this same year class is 33 to 60 mm TL. At this time, the bimodal length frequencies representing both the young-of-the-year and age-1 individuals are apparent (fig. 36.2). These estimates of size at age are greater than others from Chesapeake Bay (Schwartz 1965a), where males, based on examination of annuli in vertebrae, were estimated to be 36.0 to 41.0 mm at age 1, while females were on average 41.4 mm, and some females reach 2 to 3 years of age.

In the Great Bay–Little Egg Harbor estuary, in a comparison across a variety of habitat types at different depths (<1–8 m), juveniles were consistently most abundant in the shallow (<1 m) *Zostera marina* habitat (fig. 36.3; Able et al. 1989; Szedlmayer and Able 1996). This same pattern had been observed in earlier years.

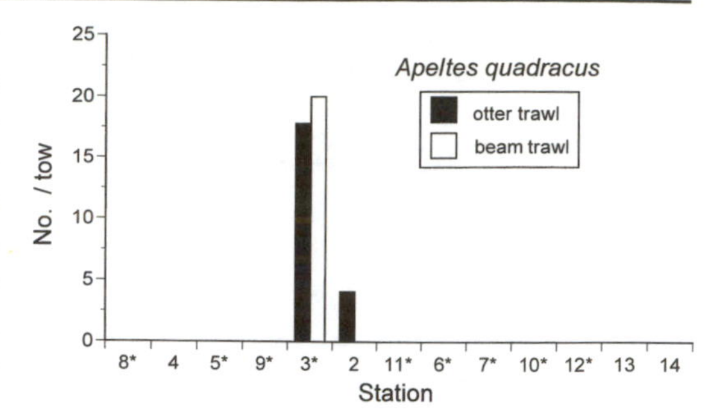

36.3 Distribution of *Apeltes quadracus* young-of-the-year by habitat type based on otter trawls (n = 2,075) and beam trawls (n = 1,362) deployed on regular stations 2–14 in Great Bay and Little Egg Harbor. Both gears used on stations designated by asterisk; otherwise, only the otter trawl was used. See figure 2.3 for station locations; table 2.1 for specific habitat characteristics by station.

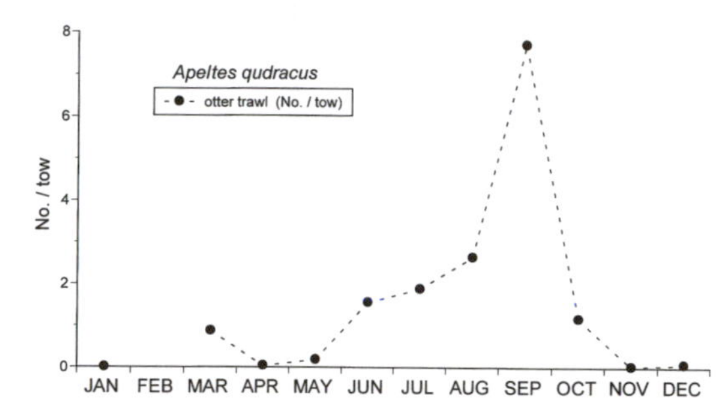

36.4 Monthly occurrences of *Apeltes quadracus* young-of-the-year in the Great Bay–Little Egg Harbor study area based on otter trawl collections 1988–1989 (n = 1,840).

This species also occurs rarely or occasionally in macroalgae or marsh creeks or pools on the marsh surface and along subtidal shorelines in the same system (Sogard and Able 1991; Rountree and Able 1992a; Able et al. 1996; K. J. Smith and K. W. Able unpubl. data). Elsewhere, they have been collected in low salinity and freshwater habitats (Scott and Scott 1988).

In Great Bay–Little Egg Harbor, the young-of-the-year of this estuarine resident have been collected in every month (fig. 36.2), making them one of the most consistent components of the estuarine fauna. They are most abundant in late summer and fall (fig. 36.4). During the winter they have been found as deep as 31 m in Chesapeake Bay (Hildebrand and Schroeder 1928).

CHAPTER 37

Gasterosteus aculeatus Linnaeus
Threespine Stickleback

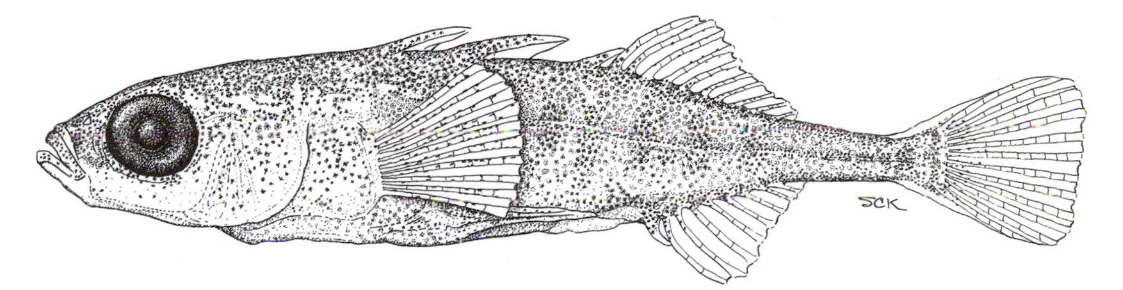

37.1 *Gasterosteus aculeatus* juvenile, 18.6 mm TL. Collected May 17, 1987, dip net (night-light), RUMFS boat basin, Great Bay, New Jersey. ANSP 175229. Illustrated by Susan Kaiser.

DISTRIBUTION

Gasterosteus aculeatus (fig. 37.1) is found along the East Coast of North America from Labrador to Virginia (Musick 1972; Scott and Scott 1988). It occurs from oceanic waters, where it is pelagic, to a variety of shallow estuarine and freshwater habitats. In the central part of the Middle Atlantic Bight, it has been collected pelagically over the continental shelf (Cowen et al. 1991). Seasonal movements are pronounced, with adults moving into estuaries in late winter and early spring and then disappearing from inshore waters during the rest of the year. Early life-history stages are recorded from estuaries throughout the bight (table 4.2).

REPRODUCTION

Many populations of this widely distributed species are anadromous (Wooten 1976) including those from Virginia (Musick 1972). Recent studies have demonstrated that adults occur in the ocean (Quinn and Light 1989; Williams and Delbeck 1989; Cowen et al. 1991) and that they migrate into estuaries or freshwater (Smith 1971) to spawn, although exact locations are generally unknown for the Middle Atlantic Bight. In the estuary behind Hereford Inlet (Allen et al. 1978), Delaware Bay (Wang and Kernehan 1979), and Great Bay–Little Egg Harbor, pelagic adults (>36 mm TL) typically enter estuaries during the winter, as early as November, and remain as late as June (fig. 37.2, 37.3). Reproduction in Great Bay occurs March through May. Based on the examination of numerous females (56–71 mm TL, n = 748), the gonadosomatic index increased during the winter and spring (Jan. = 8.9, Feb. = 13.2, Mar. = 20.6, Apr. = 26.5, May = 34.7). By May, very few females were collected, suggesting that most spawning occurred during March and April. Otolith increment counts from juveniles collected in the ocean indicated they were 29 to 56 days old and thus had a birth date in mid-April to mid-May (Cowen et al. 1991). This agrees with the estimated spawning and hatching period based on the gonadosomatic index from Great Bay–Little Egg Harbor and other observations on Long Island (Perlmutter 1963; Monteleone 1992). We suspect that nest building, courtship, and spawning occurs in shallow marsh pools in the polyhaline portion of Great Bay–Little Egg Harbor estuary because adult females with high gonadosomatic indices are abundant in these pools during the spring and small young-of-the-year are collected later from these pools (see below). Thus, at least some portion of the populations in New Jersey are not anadromous but spawn in high-salinity marshes. In the western North Atlantic, spawning in the littoral zone has also been reported

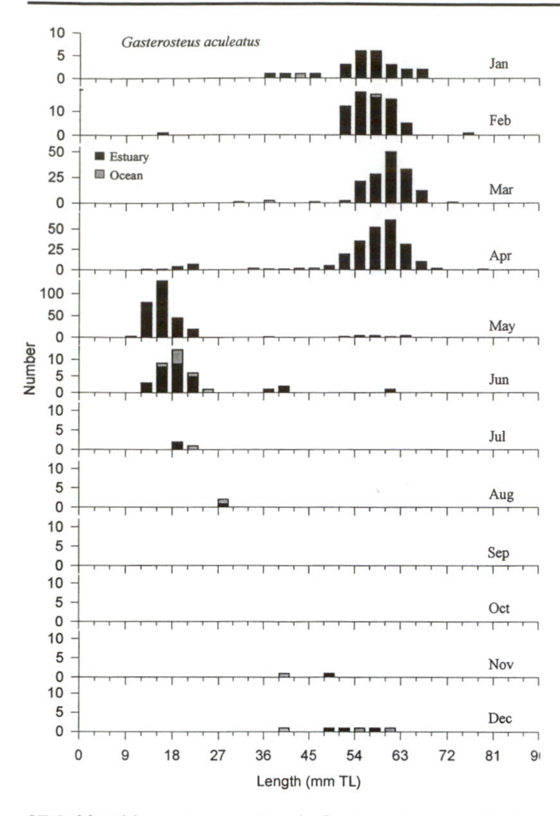

37.2 Monthly occurrences of *Gasterosteus aculeatus* young-of-the-year and adults in the Great Bay–Little Egg Harbor study area based on Little Sheepshead Creek Bridge 1-m plankton net collections 1989–1994 ($n = 346$) and otter trawl collections 1988–1989 ($n = 411$).

for Virginia (Richards and Castagna 1970), Long Island (Perlmutter 1963), and the St. Lawrence estuary (Picard et al. 1990).

DESCRIPTION

The eggs are adhesive, spherical and large (1.5–1.9 mm in diameter) and have numerous (≥25) oil droplets (largest 0.66–0.73 mm in diameter). The larvae hatch at 4.2 to 4.5 mm TL (Kuntz and Radcliffe 1917). After hatching, the larvae have melanophores concentrated on the dorsal surface of the head and middorsal and dorsolateral surfaces of the body. All fin rays are complete by 15 mm. Additional details of development are described by Kuntz and Radcliffe (1917). Juveniles are slender and laterally compressed and have a narrow caudal peduncle with lateral keels (fig. 37.1). Three (rarely four or five) prominent dorsal spines are evident anterior to the dorsal fin. A strong spine is also associated with each pelvic fin. Vert: 39–33; D: II–V, 6–13; A: I, 6–13; Pect: 9–11; Plv: I.

THE FIRST YEAR

Embryonic development occurs over 6 to 10 days while eggs are tended by the male in the nest (Bigelow and Schroeder 1953). After hatching, the demersal larvae are guarded by the parent. Eventually the juveniles form lateral plates and become silvery but not countercolored as is typical of oceanic forms. These individuals then leave Little Egg Inlet after June (figs. 37.2, 37.3). Late larvae and early juveniles (14–28 mm SL) that had completed lateral plate formation and were countercolored have been collected in the ocean off Great Bay during May and June (Cowen et al. 1991). It is not clear if the adults that appear in the winter all belong to the same year class. It has been reported, based on otoliths, that young-of-the-year reach 14 to 44 mm by the end of the first summer, 18 to 48 mm by the end of the first year, and 31 to 54 mm by the end of the second summer (Jones and Hynes 1950). If these ages apply to the Great Bay populations, then they are 2 to 3 years old when they reproduce. It seems just as likely that spawners could be age 1 given that they have already grown to 12 to 27 mm in early summer and they still had several months to grow during the summer and thus could easily reach size at maturation by the following spring.

While in the estuary, the various life-history stages use a variety of habitats including pools

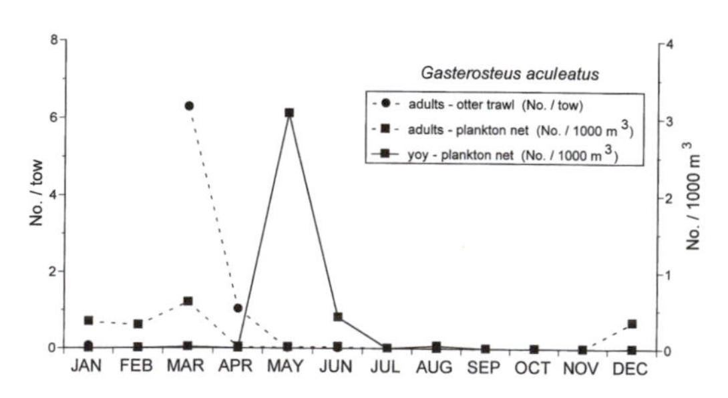

37.3 Monthly length frequencies of *Gasterosteus aculeatus* collected in the Great Bay–Little Egg Harbor study area. Sources: RUMFS: otter trawl ($n = 142$); 1-m beam trawl ($n = 7$); 1-m plankton net ($n = 226$); experimental trap ($n = 13$); killitrap ($n = 60$); seine ($n = 28$); night-light ($n = 14$); weir ($n = 312$).

and shallow depressions on the marsh surface and intertidal and subtidal creeks (Talbot and Able 1984; Rountree and Able 1992a). During 1992, when larval fish traps were deployed in marsh pools from May 6 to July 29, large numbers of small juveniles were collected in May (n = 106, $\bar{x}$ = 24.8 mm TL, range 17 to 29 mm TL) and June (n = 28, $\bar{x}$ = 22.4 mm TL, 19 to 27 mm TL). The adults migrating into the estuary and juveniles migrating out have also been collected in larger, deeper thoroughfares through the marsh, such as Little Sheepshead Creek (fig. 37.3; Witting 1995; D. A. Witting, K. W. Able and M. P. Fahay in prep.).

CHAPTER 38

Hippocampus erectus Perry
Lined Seahorse

DISTRIBUTION

Hippocampus erectus (fig. 38.1) occurs in estuaries and the inner continental shelf from Nova Scotia to Uruguay (Vari 1982). In estuaries, it is typically found in eelgrass or other types of vegetation, oyster beds, salt marshes, and deep channels. It is occasionally found at the surface, even over deep water. It is reported to make annual inshore-offshore migrations. The early life-history stages are common in most Middle Atlantic Bight estuaries (table 4.2).

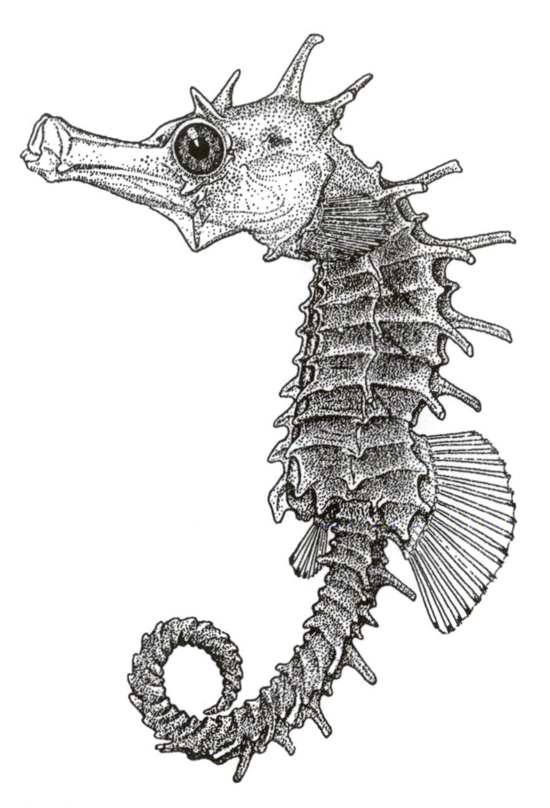

38.1 *Hippocampus erectus* juvenile, 3.3 mm TL. After Lippson and Moran 1974.

REPRODUCTION

This species, like many other syngnathids, has reversed sex roles (Vincent et al. 1992), with the male bearing the fertilized embryos and larvae in a "placentalike" brood pouch (Linton and Soloff 1964). Mating and birth occur over a protracted period in the Middle Atlantic Bight. In Chesapeake Bay, the gonadosomatic index for adult females (>70 mm TL) was highest from May through August or October (Teixeira 1995). In southern New Jersey, gonadosomatic index values for females (n = 37) began to increase in May, reached a peak in July, and stayed relatively high through September (J. Morrison and K. W. Able unpubl. data). The males examined (63–127 mm TL, n = 39) had as many as 1,515 embryos in the pouch at sizes as small as 63 mm TL, but brood size increased with the size of the male.

DESCRIPTION

Fertilized eggs are pear-shaped and 3.1 to 3.9 mm long. They are deposited into the male brood pouch where they are fertilized and complete incubation. Hatching occurs in the pouch after not less than 12 to 14 days when larvae are 3.3 to 4.8 mm TL. At this size, they are recognizable by their unique body shape, although the snout length is less than in adults. These have the full fin ray complement, and they are distinctive in that the large head is held at right angles to the rest of the body (fig. 38.1). The tail is prehensile and lacks a caudal fin. Details of development are summarized in Hardy (1978a). Vert: 49–51; D: 17–20; A: 3–4; Pect: 15–19; Plv: 4.

THE FIRST YEAR

Small, recently born individuals (1–2 cm TL) were present in the vicinity of Great Bay–Little Egg Harbor from June through October (fig. 38.2). The pattern of length frequency distribution is confusing for these collections. The

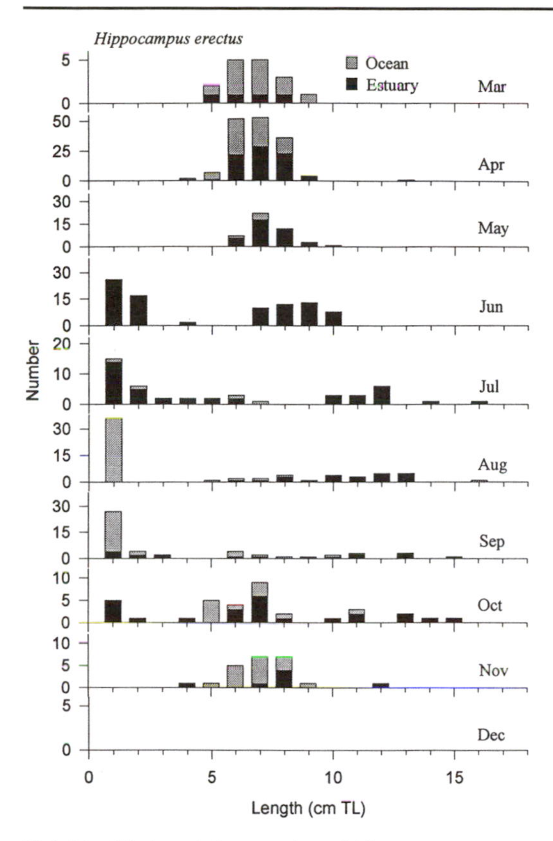

38.2 Monthly length frequencies of *Hippocampus erectus* collected in the Great Bay–Little Egg Harbor study area. Sources: ANSP (uncat.); RUMFS: otter trawl (n = 49); 1-m beam trawl (n = 17); 1-m plankton net (n = 97); experimental trap (n = 4); LEO-15 Tucker trawl (n = 61); LEO-15 2-m beam trawl (n = 16); gear comparison (n = 13); killitrap (n = 10); seine (n = 11); night-light (n = 4).

young-of-the-year are apparent in June at 1 to 2 cm TL and may reach 6 to 7 cm TL in July. By November, most are 4 to 9 cm TL, although they are not well represented in August and September when the collections are dominated by recently born individuals. On the basis of the sizes collected during March and April, they do not appear to grow over the winter (fig. 38.2). By approximately 12 months they are 7 to 10 cm TL and clearly differentiated from the recently born young-of-the-year.

The pattern of habitat use is either variable, or we simply lack sufficient data to interpret it for our study area. In early summer (June and July), most individuals are in the Great Bay–Little Egg Harbor study area. During August and September, when collections are dominated by recently born individuals, most were collected from the inner continental shelf near Beach Haven Ridge. The abundance of recently born individuals in the ocean during August and September is inconsistent with seasonal inshore-offshore migrations, but perhaps occurrences at Beach Haven Ridge (ocean) at this time are the result of estuarine outwelling of these small individuals. A variety of sources suggest they move offshore for the winter. By November, most are collected from the ocean, and this remains true for the collections the following spring (March). During the winter, at a temperature of 10.6 C, individuals have been observed by divers on the inner continental shelf off Long Island, where they lay motionless on the substrate (Wicklund et al. 1968). Several individuals, size unreported, have also been found in the guts of *Urophycis chuss* and *Merluccius bilinearis* outside of Delaware Bay (46 m) in the early spring (Allen et al. 1978). In deeper, more southern estuaries, such as Chesapeake Bay, they may overwinter in channels (Musick 1972).

CHAPTER 39

Syngnathus fuscus Storer
Northern Pipefish

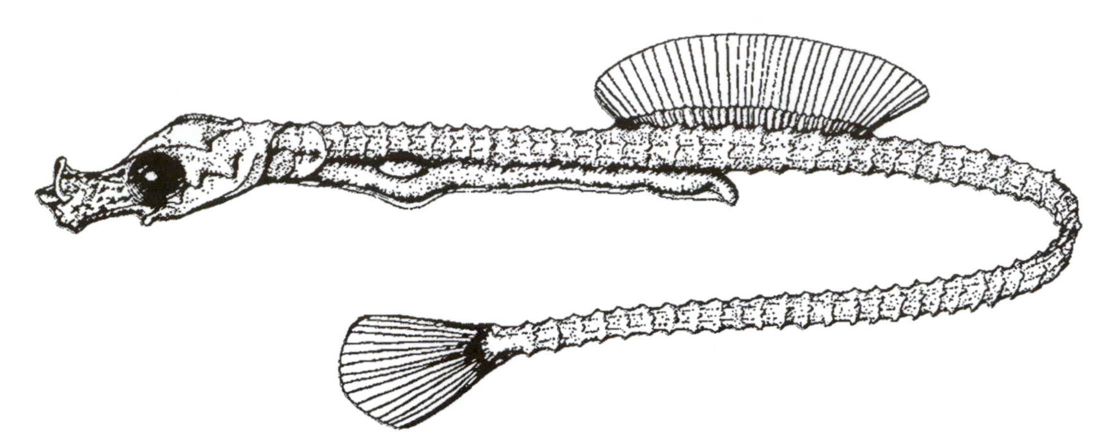

39.1 *Syngnathus fuscus* juvenile, 13.2 mm TL. After Lippson and Moran 1974.

DISTRIBUTION

Syngnathus fuscus (fig. 39.1) is distributed from the Gulf of St. Lawrence to Florida (Dawson 1982). It occurs in shallow bays, harbors, rivers, creeks, and salt marshes and is frequently associated with submerged vegetation. It migrates from the estuary into the coastal ocean for the winter, at least in the northern Middle Atlantic Bight (Lazzari and Able 1990). In the Middle Atlantic Bight, early life-history stages occur in every estuary (table 4.2).

REPRODUCTION

The mode of reproduction for this species, as for other pipefishes, is one with reversed sex roles, in which the female deposits the eggs in the male brood pouch and the male carries the fertilized eggs throughout the gestation period (Vincent et al. 1992; Teixeira 1995). Males as small as 83 mm may carry a brood, but most are at least 90 to 100 mm (Dawson 1982). In Great Bay–Little Egg Harbor, reproduction occurs from spring through late summer as indicated by the timing of the maturation of the gonads, the presence of embryos in the male pouch, and the occurrence of small, recently born larvae (Campbell and Able in press). Brood size is variable, ranging from 45 to 1,380 embryos in males ranging from 119 to 222 mm TL. Each brood appears to be from a single mating because all of the embryos are in the same developmental stage (Campbell and Able in press). A similar pattern occurs in Chesapeake Bay (Teixeira 1995).

DESCRIPTION

The eggs (0.75–1.0 mm in diameter) develop in the brood pouch of the male until birth at 10 to 12 mm TL (Dawson 1982). The larvae resemble juveniles except that they lack pigmentation (Lippson and Moran 1974). At any size, the larvae and juveniles are distinctive and unmistakable (fig. 39.1). The extremely elongate body ends in a small, rounded caudal fin. The snout is elongate and the tiny mouth is located at its tip. There is no anal fin. D: 35–43; Pect: 12–16; Pelv: none.

THE FIRST YEAR

Incubation lasts approximately 10 days (Bigelow and Schroeder 1953), after which, the larvae are released from the pouch. The abundance of small planktonic individuals peaks in June (fig. 39.2). A midsummer peak in planktonic individuals occurred every summer from 1989 to 1994 (fig. 39.3 and Witting 1995; D. A. Witting, K. W. Able, and M. P. Fahay in prep.). They apparently remain planktonic until

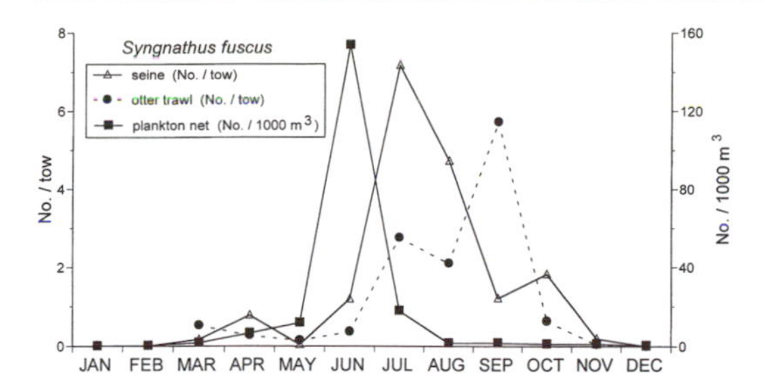

39.2 Monthly occurrences of *Syngnathus fuscus* young-of-the-year in the Great Bay–Little Egg Harbor study area based on otter trawl collections 1988–1989 (n = 1,441), seine collections 1990–1991 (n = 804), and Little Sheepshead Creek Bridge 1-m plankton net collections 1989–1994 (n = 7,056).

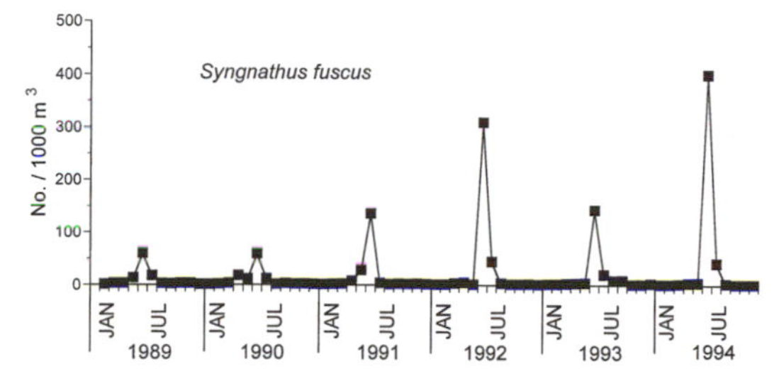

39.3 Annual variation in abundance of *Syngnathus fuscus* larvae in Little Sheepshead Creek Bridge 1-m plankton net collections 1989–1994 (n = 7,056).

they are approximately 40 mm TL (Campbell and Able in press). In Great Bay–Little Egg Harbor, young-of-the-year occurred from May through October. Peaks in demersal individuals (>40 mm TL), collected in otter trawls, occurred in July and September (fig. 39.2). The young-of-the-year are extremely variable in size by the end of the first growing season (November), ranging from approximately 5 to 20 cm TL (fig. 39.4). This broad range is due in part to the long period of reproduction and release of larvae (Campbell and Able in press). There is probably little growth during the winter. By the following spring, at approximately 12 months of age, the largest young-of-the-year are 8 to 24 cm TL and are large enough to reproduce, as suggested by Bigelow and Welsh (1925). Thus, the population may be dominated by young-of-the-year individuals. Others have suggested that some individuals may reach 2 years of age and 28 to 30 cm (Mercer 1973; Dawson 1982; Tatham et al. 1984; Warfel and Merriman 1944) but definitive studies of the age and growth need to be conducted.

Young-of-the-year are most abundant during the summer in shallow, vegetated habitats, including eelgrass (*Zostera marina*) and sea lettuce (*Ulva lactuca*) (Hildebrand and Schroeder 1928; Briggs and O'Connor 1971; Tatham et al. 1984). They also occur in unvegetated areas adjacent to these habitats, in marsh creeks (Sogard and Able 1991; Rountree and Able 1992a) and a variety of shallow estuarine shoreline habitats (Able et al. 1996b). In sampling in slightly deeper (<1–8 m) waters, they have also been collected across a variety of habitat types but are most abundant in the shallowest vegetated habitats (fig. 39.5; Szedlmayer and Able 1996). The relatively high catch rates over

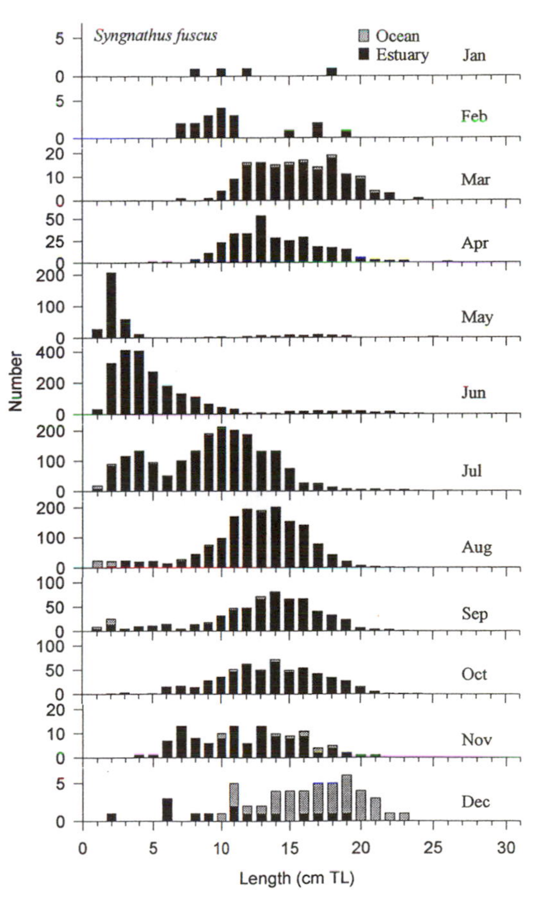

39.4 Monthly length frequencies of *Syngnathus fuscus* young-of-the-year collected in the Great Bay–Little Egg Harbor study area. Sources: RUMFS: otter trawl (n = 1,399); 1-m beam trawl (n = 1,082); 1-m plankton net (n = 2,772); experimental trap (n = 168); LEO-15 Tucker trawl (n = 92); LEO-15 2-m beam trawl (n = 134); gear comparison (n = 972); killitrap (n = 291); seine (n = 623); night-light (n = 252); weir (n = 244).

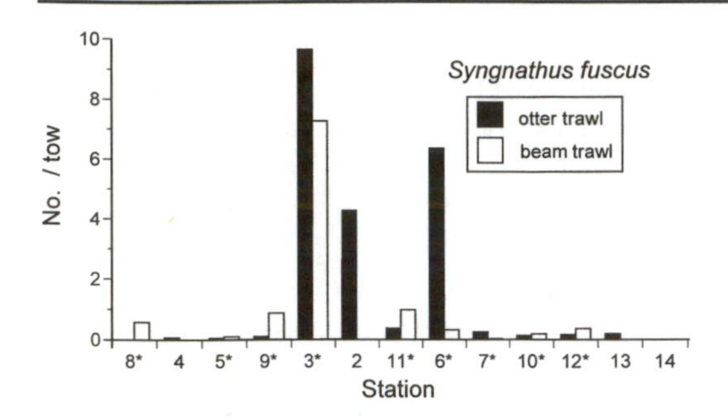

39.5 Distribution of *Syngnathus fuscus* young-of-the-year by habitat type based on otter trawls (n = 2,026) and beam trawls (n = 721) deployed on regular stations 2–14 in Great Bay and Little Egg Harbor. Both gears used on stations designated by asterisk; otherwise, only the otter trawl was used. See figure 2.3 for station locations; table 2.1 for specific habitat characteristics by station.

sandy substrate with hydroids (fig. 39.5, Sta. 6) resulted from a single large collection (559 individuals in a single 2-minute tow). In experiments with artificial eelgrass in the same estuary (Sogard 1989; Sogard and Able 1994), young-of-the-year readily colonized these habitats, even if they had to cross broad expanses of sand to do so. Colonization occurred during both day and night, but most small individuals (<70 mm TL) colonized at night. Some evidence suggests movements between habitats may be gender-based, with males that are carrying embryos less likely to leave protective habitats than are females (Roelke and Sogard 1993).

Seasonal migrations in the northern Middle Atlantic Bight are common, with most individuals leaving the estuaries to spend the winter on the inner continental shelf (Lazzari and Able 1990). They move into the ocean during November and December and return to the estuary in March and April (fig. 39.4). During the winter, they are found in the ocean within 10 km of shore in depths of 2 to 19 m (Lazzari and Able 1990). Other observations in the winter off Long Island in depths of 14 to 17 m found torpid individuals partially buried in the sand with only the head and caudal fin exposed (Wicklund et al. 1968). The observations off both New Jersey and Long Island suggest these movements into the ocean are in response to low winter temperatures. Comparison of length frequencies between the fall (November) and spring (April) indicate that individuals smaller than 7 cm do not appear to survive the winter (fig. 39.4).

CHAPTER 40

Searobins of the Genus *Prionotus*

Although triglid larvae are often abundant constituents of temperate ichthyoplankton surveys over continental shelves, they lack distinctive characters and are usually not identified to species (e.g., Smith et al. 1979; Morse et al. 1987). Juveniles are also not well described. Consequently, little has been reported concerning their early life histories. Much of the following summary is based on recently published or ongoing studies in the central part of the Middle Atlantic Bight (e.g., McBride 1994; McBride and Able 1994; McBride et al. in press, in review).

DESCRIPTIONS

Eggs and early larvae of *Prionotus carolinus* and *P. evolans* have been reared in the laboratory and described (Yuschak and Lund 1984; Yuschak 1985). We extrapolated from those descriptions and augmented them with our own observations of larvae and juveniles collected in the Middle Atlantic Bight. On the basis of those observations, we have established useful criteria for identifying the various life-history stages (table 40.1).

Both species are recognized as triglids because of the lower three rays of the pectoral fin, which are feelerlike and separate from the remainder of the fin. The head is ornamented with ridges, spines, or bumps. In juveniles of *P. carolinus* (see fig. 41.1), the body pigment includes two prominent saddles of dark pigment on the sides below the second dorsal fin. The pectoral fin reaches to (or just past) the origin of the anal fin and is darkly pigmented with a pale border. Note counts of anal and pectoral fin rays. Vert: 26; D: X, 13–14; A: 12; Pect: 14 + 3; Plv: I, 5. In juveniles of *P. evolans* (see fig. 42.1), the body is darkly pigmented, but the caudal peduncle is noticeably lighter in color. Spinous scales cover the body. The pectoral fin reaches well past the anal fin origin and is darkly pigmented. Note counts of anal and pectoral fin rays. Vert: 26; D: X, 13–14; A: 11; Pect: 13 + 3; Plv: I, 5.

DISTRIBUTIONS

Prionotus carolinus occurs from the Gulf of Maine to South Carolina, where it may be found from estuaries to the edge of the continental shelf, depending on the time of year. It prefers sand-bottomed substrates in coastal waters between May and October and spends the winter months at mid- to outer shelf depths (Roberts 1978). This species is generally more abundant than its congener *P. evolans* in the Middle Atlantic Bight (McBride and Able 1994; McBride et al. in press), and juveniles and adults are generally more common in estuaries in the northern part of the Middle Atlantic Bight than in the southern part (Stone et al. 1994). It is among the four most abundant species in Sandy Hook Bay during the summer (Wilk and Silverman 1976b), but it is less common in the shallower Great Bay–Little Egg Harbor (Szedlmayer and Able 1996). Eggs, larvae, and juveniles have been reported from most estuaries in the Middle Atlantic Bight (table 4.2).

Prionotus evolans occurs between Cape Cod, Massachusetts and South Carolina, from estuaries to the continental shelf edge. It prefers sand bottoms in coastal waters between May and November and spends the winter months at outer continental shelf depths (Roberts-Goodwin 1981; McBride et al. in press). An analysis of seasonal distributions found that while the two species often co-occur during summer, *P. evolans* tends to occur in warmer, less oxygenated, and more turbid habitats than *P. carolinus,* and also arrives at, and departs from, coastal areas later than its congener (McBride and Able 1994). It is among the four most abundant species in Sandy Hook Bay during the summer (Wilk and Silverman 1976b),

Table 40.1. Summary of Early Life-History Characters in Two Species of *Prionotus* in the Middle Atlantic Bight

Character	*Prionotus carolinus*	*Prionotus evolans*
	Egg stages	
Diameter (mm)	0.86–0.97 1.0–1.5*	1.05–1.25
Chorion	Unpigmented, lightly sculptured	Same
Oil globules	11–37, in 1 hemisphere 10–20*	16–37, in 1 hemisphere
Perivitelline space	Very narrow	Same
Incubation (hrs @ temp.)	60 at 15–22 C 120–155 at 15 C	80–90 at 19–20 C
Size at hatch	3.0 mm NL (SD = 0.09)	2.8 mm NL (SD = 1.77)
	Preflexion-flexion stages	
Nuchal spine†	Forms early (ca. 4.0 mm SL), then double spines (ca. 6.5 mm SL)	Forms later (ca. 6.0 mm SL), then develops stout ridge (>6.5 mm SL) with supporting struts
Teeth formed	Early	Later
Depth at cleithrum	Deep	Very deep
Occiput shape	Concave	Strongly concave
Early pectoral fin rays	Long	Very elongate
Size at flexion	6.0–7.0 mm	6.0–7.0 mm
	Flexion-postflexion stages	
Pectoral fin rays	Pigment on upper rays	Darkly pigmented overall
Pectoral fin ray number	14 + 3‡	13 + 3‡
Anal fin ray number	12	11
Parietal spiny ridge	Absent	Present (anterior to nuchal spine)
Anterior teeth	Slightly expressed	Pronounced
	Late postflexion-juvenile stages	
Pectoral fins	Extend to anal fin origin	Extend beyond anal fin origin
Head spines	Slightly developed	Well-developed
Body pigment	2 saddles dorsolaterally	Dark with lighter caudal peduncle
Subocular spine§	Present (forms ca. 7.0 mm SL)	Absent
Nasal spine‖	Absent	Present (forms ca. 7.0 mm SL)
Spinous scales	None	Present over body, >10 mm
(also see anal and pectoral fin ray numbers above)		

NOTE: Egg data after Yuschak and Lund (1984) or Yuschak (1985) unless indicated by footnote. SD = standard deviation. SL = standard length; NL = notochord length.
*Kuntz and Radcliffe (1917).
†Head spine above and behind each eye, usually one of the earliest to form.
‡Count of fin rays in main part of fin plus 3 lower rays, usually fleshy and separate from rest of fin.
§Spine on cheek area, under posterior part of each eye.
‖Spines (paired) on snout, behind tip of upper jaw.

Table 40.2. Comparative Biological and Ecological Characteristics in *Prionotus carolinus* and *Prionotus evolans* in the Middle Atlantic Bight

	Prionotus carolinus	*Prionotus evolans*
Spawn	June–September	June–August
Flexion size	5.4–6.8 mm SL	6.7–7.5 mm SL
Flexion age (begin)	18 days	13 days
Largest larvae	9.8 mm SL	11.9 mm SL
Settlement size	7–12 mm SL	7–12 mm SL
Settlement age	18–19 days	18–19 days
YOY in estuaries†	April–June	July–December
Modal size @ 12 months	ca. 7.5 cm	ca. 9.0 cm

SOURCE: Most data from McBride et al. in review.
*SL = standard length.
†YOY = young-of-the-year.

but it is less common in Great Bay–Little Egg Harbor (Szedlmayer and Able 1996). Although both species winter on the mid- or outer-continental shelf, *P. carolinus* is then found at latitudes farther north than *P. evolans,* which is restricted to offshore areas south of Delaware (McBride and Able 1994; McBride et al. in press). Young-of-the-year of both species spend their first winter with adults, judging from bottom-trawl survey results (e.g., Edwards et al. 1962). Juveniles have been reported from several estuaries in the Middle Atlantic Bight, but eggs and larvae have been reported from only a few (table 4.2), possibly because of the inability of investigators to discriminate species in early stages. To further facilitate the separation of early life-history stages of these species, we have compared biological and ecological characteristics during their first year (table 40.2).

CHAPTER 41

Prionotus carolinus (Linnaeus) Northern Searobin

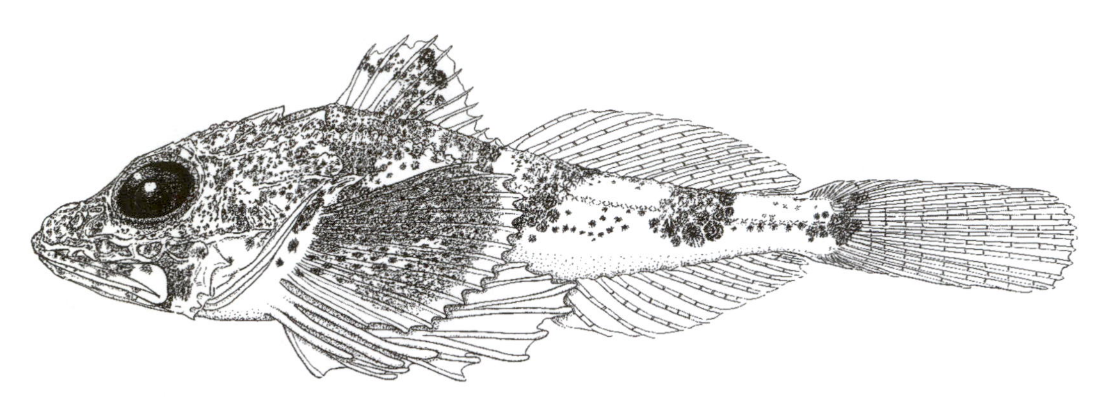

41.1 *Prionotus carolinus* early juvenile, 12.0 mm SL. Collected September 23, 1991, 2-m beam trawl, LEO-15 study site, Beach Haven Ridge, New Jersey. ANSP 175230. Illustrated by Susan Kaiser.

REPRODUCTION

In *Prionotus carolinus* (fig. 41.1), spawning occurs during the summer in continental shelf and estuarine waters throughout the Middle Atlantic Bight. Gonadosomatic index levels indicate females are ripe from May through September and peak in July in the New York Bight (Wilk et al. 1990). Ichthyoplankton evidence shows that spawning occurs simultaneously throughout this range. A study in Peconic Bays, New York, found most spawning occurred in the evening or at night (Ferraro 1980). At an inner continental shelf site off the coast of New Jersey, eggs (identified only to the genus *Prionotus*) were present from May to October, and their abundance peaked during July or August (McBride and Able 1994). Eggs also occurred in Great Bay, New Jersey, with an early peak in June, although these were not identified to species (McBride 1994). In Sandy Hook Bay, eggs of *Prionotus* sp. were the most abundant taxon collected in May and June (Croker 1965), but larvae were not collected. Eggs occurred between June and August in Narragansett Bay, where they were fifth or sixth in abundance, although larvae were rare (Herman 1963; Bourne and Govoni 1988). A study in the Hereford Inlet, New Jersey area that used immunochemical methods to discriminate the eggs of *P. carolinus* from those of *P. evolans* (Keirans et al. 1986) detected an initial peak in spawning intensity of *P. carolinus* in June or July, followed by a second peak in August (1973) or September (1974). Larvae are most abundant in Middle Atlantic Bight continental shelf waters from July through October (fig. 41.2). At the Beach Haven Ridge study site off New Jersey, larvae only occurred during September, and a few also then occurred in plankton sampling near Little Egg Inlet (fig. 41.3).

THE FIRST YEAR

The pelagic larvae are found most commonly in that part of the bight south of the drowned Hudson River Valley (e.g., New Jersey to Cape Hatteras). Settlement from pelagic stages to bottom stages occurs on the continental shelf. In a recent study, the largest planktonic larvae collected were 9.8 mm SL, and settlement occurred at sizes of 7 to 12 mm SL and ages of 18 to 19 days (McBride et al. in review). The lower three rays of the pectoral fin become separated from the rest of the fin at about 12 mm SL in specimens we have examined (fig. 41.1). Thus this occurs shortly after settlement at 8 to 9 mm (McBride et al. in review). It is not known when scalation begins or the size at which the body is fully scaled.

The generally late peak in spawning activity (August) and the slow growth rate of about

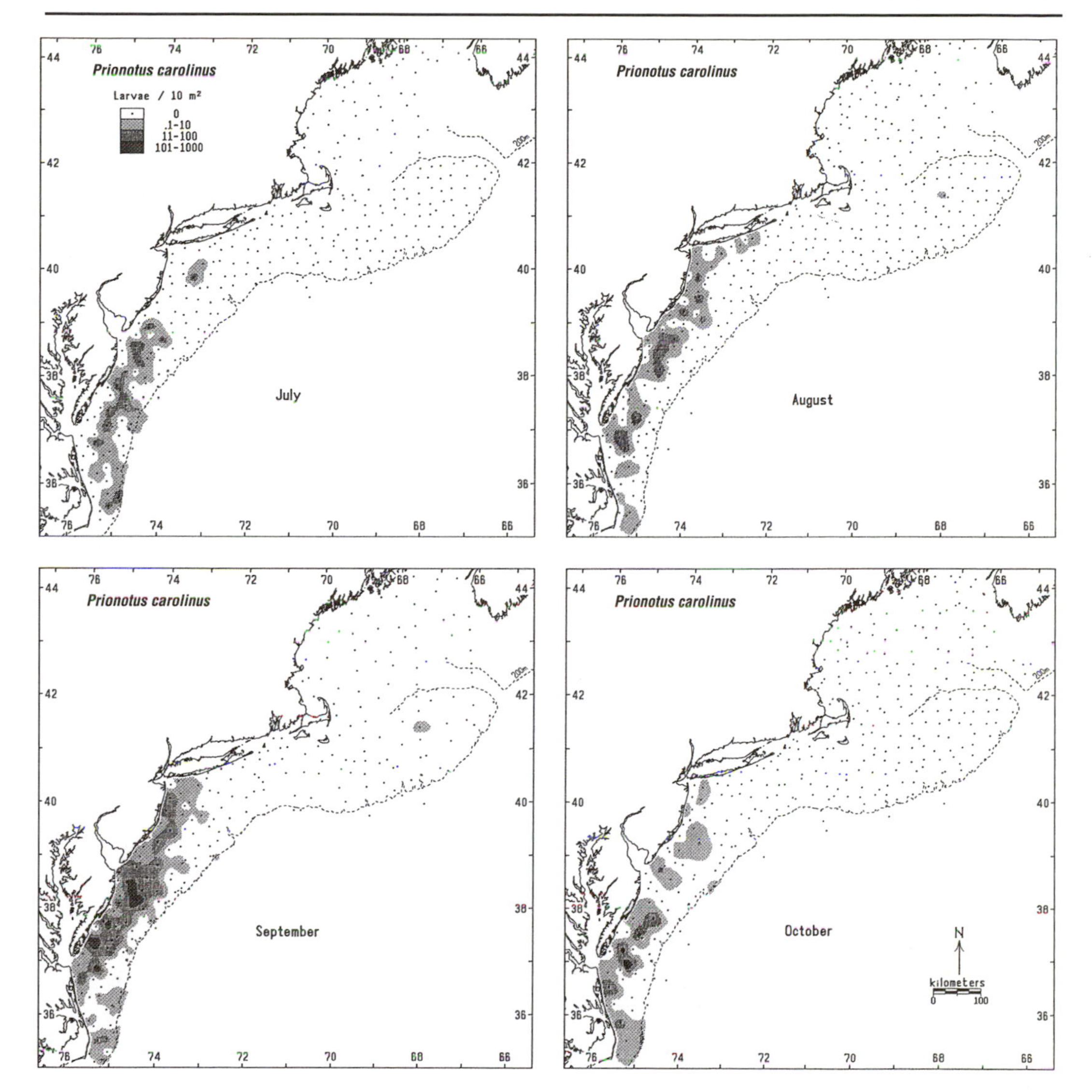

41.2 Monthly distributions of *Prionotus carolinus* larvae during MARMAP surveys, 1977–1987.

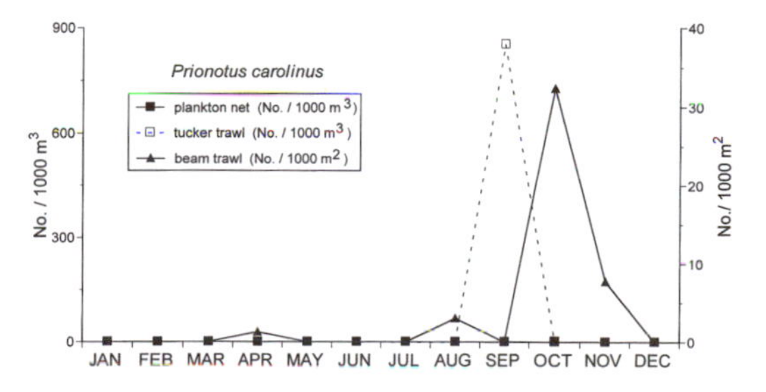

41.3 Monthly occurrences of *Prionotus carolinus* larvae and young-of-the-year at the LEO-15 study site and the Great Bay–Little Egg Harbor study area based on 1-m plankton net collections 1989–1994 (n = 24), Tucker trawl collections 1991–1992 (n = 1,241), and 2-m beam trawl collections 1991–1992 (n = 76).

0.5 mm per day (McBride et al. in review) combine to produce relatively small juveniles (2–3 cm TL) as the winter begins (fig. 41.4). These data are based on sampling with a 2-m beam trawl in the vicinity of Little Egg Inlet. Somewhat larger sizes were reported from the southern New England and Long Island Sound areas (table 41.1) in fishes collected by otter trawl. A recent study has explained this discrepancy as being based on size selectivity of these two gears (McBride et al. in press); thus, fish entering their first winter range from 2 to 10 cm TL.

Early life-history stages use the continental shelf as a nursery. Eggs and larvae occur in the

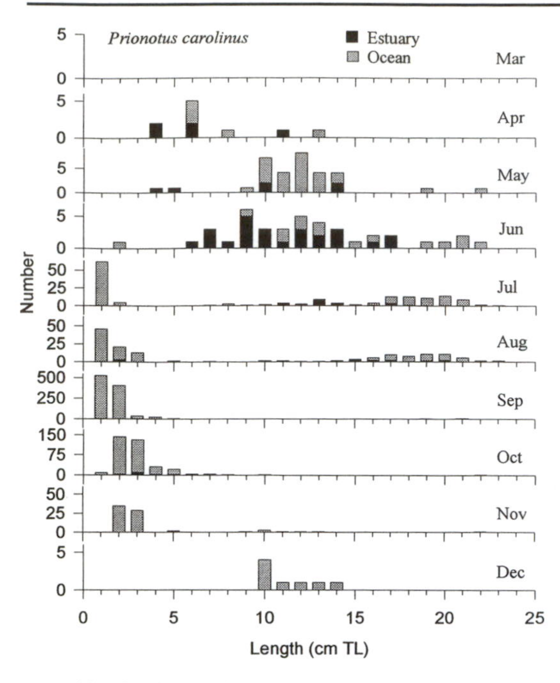

41.4 Monthly length frequencies of *Prionotus carolinus* young-of-the-year and older stages collected in estuarine and adjacent coastal ocean waters of the central Middle Atlantic Bight. Sources: Wilk et al. 1992; RUMFS: otter trawl (n = 76); 1-m beam trawl (n = 6); 1-m plankton net (n = 23); LEO-15 Tucker trawl (n = 682); LEO-15 2-m beam trawl (n = 978); gear comparison (n = 11); seine (n = 24); night-light (n = 1).

water column there and recently settled juveniles are found on the bottom in maximum densities >10/100 m² (McBride et al. in review). Any use of estuaries (and inner continental shelf sites) is made during the spring when young-of-the-year are just less than 12 months old and at sizes between 4 and 15 cm, presumably after they have spent their first winter in bottom habitats offshore.

Patterns of estuarine use vary throughout the Middle Atlantic Bight. In the central part of the Middle Atlantic Bight, young-of-the-year and early age-1 fish leave estuaries during the warmer parts of summer and move offshore to cooler temperatures. However, through the summer, they occur in southern New England coastal areas and then emigrate in October or November in response to dropping temperatures (Lux and Nichy 1971). Young-of-the-year generally emigrate from Long Island Sound during the fall, but occasional specimens remain through the winter (Richards et al. 1979). In the Chesapeake Bay area, young-of-the-year between 3 and 8 cm TL have been collected in the deeper portions of Chesapeake Bay between December and May, indicating they sometimes spend the winter there (Hildebrand and Schroeder 1928).

Table 41.1. Mean Lengths (cm TL) of Young-of-the-Year *Prionotus carolinus* Collected in Long Island Sound and Near Woods Hole, Massachusetts

Long Island Sound*			Woods Hole, MA†	
Date	1961	1962	Date	1943–1977
July 16–31		3.5		
August 1–15	3.7			
August 16–31	5.1		August	5.0
September 1–15	4.8	5.5		
September 16–30	5.9	6.0	September	6.6
October 1–15	6.5	6.8		
October 16–31	6.7	6.8	October	10.3
November 1–15	6.4			
November 16–30	7.3		November	10.3

NOTE: Lengths converted from reported fork lengths and standard lengths.
*Semimonthly collections; Lux and Nichy 1971.
†Monthly collections; Richards et al. 1979.

CHAPTER 42

Prionotus evolans (Linnaeus)
Striped Searobin

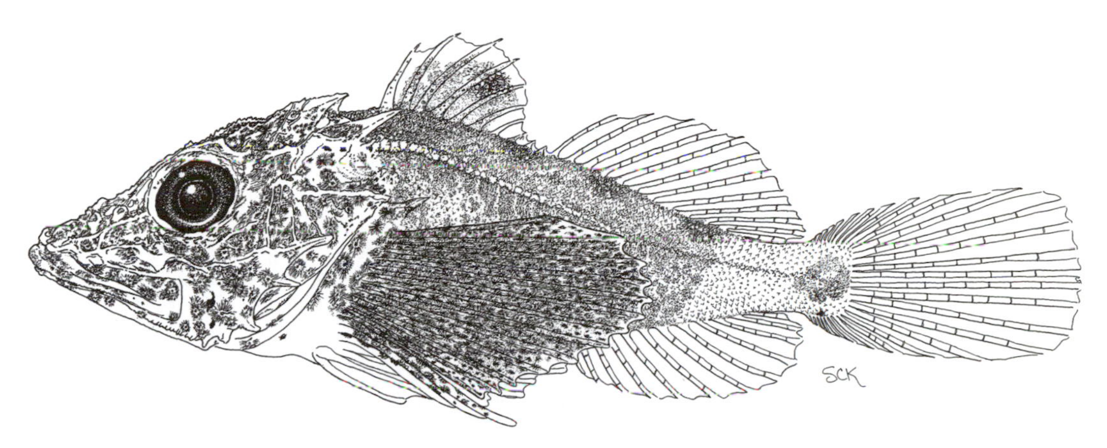

42.1 *Prionotus evolans* early juvenile, 12.5 mm SL. Collected August 6, 1986, dip net (night-light), boat basin, Shark River, New Jersey. ANSP 175231. Illustrated by Susan Kaiser.

REPRODUCTION

In the central part of the Middle Atlantic Bight, gonadosomatic evidence indicates that spawning in *Prionotus evolans* (fig. 42.1) occurs between May and October with a peak in June (Wilk et al. 1990), but the occurrence of eggs or larvae indicates that spawning peaks in July or August (McBride and Able 1994). Most spawning occurs in inner continental shelf waters, although eggs and larvae have also been collected near inlets and in estuaries. During 4 years of MARMAP sampling (1984–1987), only 24 larvae were collected in the entire Middle Atlantic Bight, and these occurred between March and September and were scattered between Long Island and Cape Hatteras (McBride et al. in review). A study in the New York Bight collected larvae between July and August, and these occurred mostly within 50 km of shore and southwest of the Hudson Shelf Valley (McBride et al. in review). Eggs (identified by immunochemical techniques) were collected near Hereford Inlet, New Jersey, between May and September with the strongest peak in the latter month (Keirans et al. 1986).

THE FIRST YEAR

Larvae and settled juveniles occurred during August and September at the Beach Haven Ridge study site off Little Egg Inlet (fig. 42.2). Pelagic larvae 8 to 9 mm settle to the bottom at ages of 18 to 19 days (McBride et al. in review). The lower three rays of the pectoral fin become separated from the rest of the fin at about 10 to 15 mm (Morrill 1895). At this size, spinous scales cover the body, but it is unknown how long they are retained by juveniles (fig. 42.1).

The growth rate is slow in young-of-the-year, only about 0.5 mm/day during the first year (McBride et al. in review). Young-of-the-year 23 to 88 mm SL have been reported from New Haven Harbor, Connecticut, in August and September (Warfel and Merriman 1944), and juve-

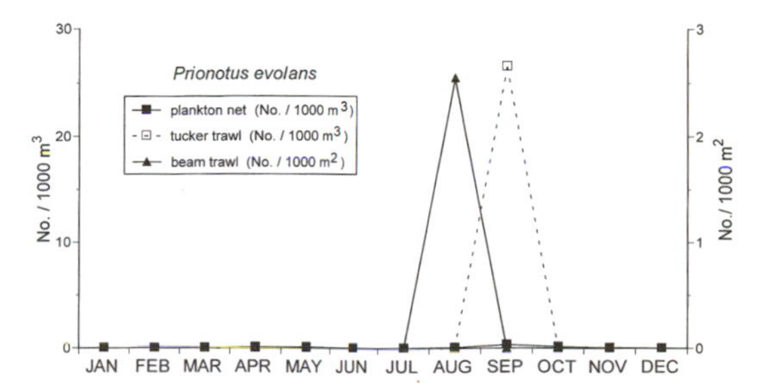

42.2 Monthly occurrences of *Prionotus evolans* larvae and young-of-the-year at the LEO-15 study site and the Great Bay–Little Egg Harbor study area based on 1-m plankton net collections 1989–1994 (n = 69), Tucker trawl collections 1991–1992 (n = 40) and 2-m beam trawl collections 1991–1992 (n = 7).

niles have been reported from Sandy Hook Bay, where they averaged 55 mm SL in August and 70 mm SL by October (Nichols and Breder 1927). During the fall (October–December), we have observed a clear separation of young-of-the-year in estuaries and 1-year-olds in the ocean (fig. 42.3). Their slow growth, combined with a late peak in spawning, results in fish entering their first winter at sizes between 2 and 8 cm TL (fig. 42.3). There is apparently little or no growth over the winter, and juveniles are observed the following spring in the same size range. Because we have few observations 12 months after spawning, we cannot estimate the size at age 1. Other studies have given the size at age 1 as 151 or 173 mm FL (in two different years) (McEachran and Davis 1970), or 173 mm SL (Richards et al. 1979).

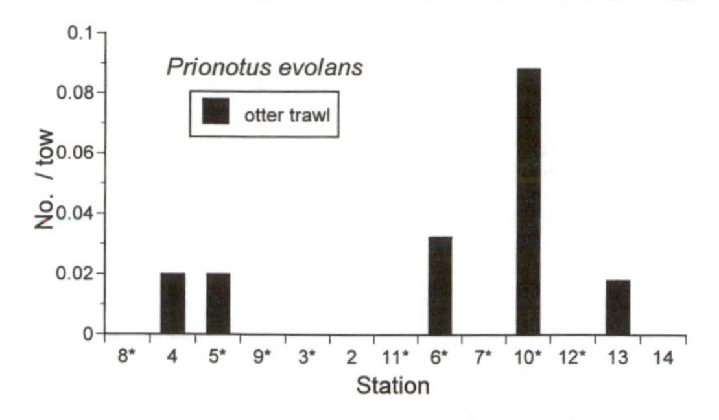

42.4 Distribution of *Prionotus evolans* young-of-the-year by habitat type based on otter trawls (n = 17) deployed on regular stations 2–14 in Great Bay and Little Egg Harbor. Both otter trawl and beam trawl used on stations designated by asterisk; otherwise, only the otter trawl was used. See figure 2.3 for station locations; table 2.1 for specific habitat characteristics by station.

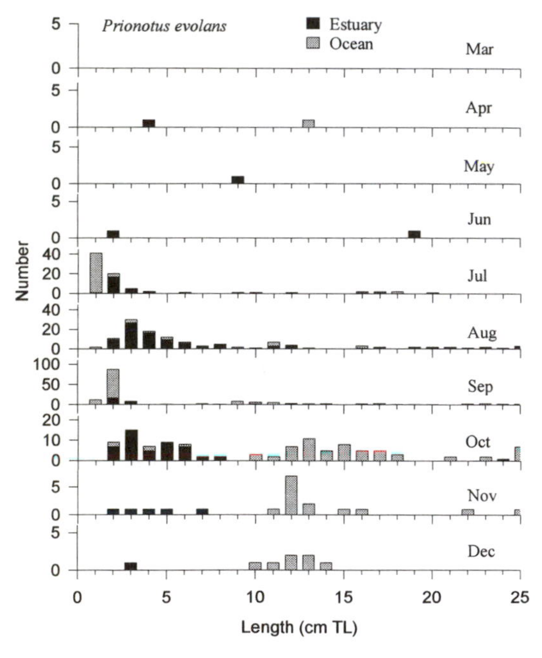

42.3 Monthly length frequencies of *Prionotus evolans* young-of-the-year and older stages collected in estuarine and adjacent coastal ocean waters of the central Middle Atlantic Bight. Sources: Wilk et al. 1992; RUMFS: otter trawl (n = 48); 1-m beam trawl (n = 10); 1-m plankton net (n = 65); LEO-15 Tucker trawl (n = 95); LEO-15 2-m beam trawl (n = 81); gear comparison (n = 77); seine (n = 23); night-light (n = 6); weir (n = 18).

Habitat use varies with life-history stage. Larvae occur most frequently over the continental shelf, although a few have been collected in plankton sampling inside Little Egg Inlet (Witting 1995; D. A. Witting, K. W. Able, and M. P. Fahay in prep). Most recently settled juveniles also occur on the shelf. In the few months after settlement, some young-of-the-year are also found in estuaries over a wide range of substrates at depths greater than 2 m (fig. 42.4). Unlike its congener, *P. carolinus, P. evolans* young-of-the-year are more likely to be collected in estuaries between spawning and their first winter. Where they are distributed on their first birthday (summer) remains enigmatic.

Young-of-the-year as well as older year classes move to the outer continental shelf during the late fall and remain there through the winter. This species spends the winter on the continental shelf in areas farther south than *P. carolinus* (Edwards et al. 1962; McBride and Able 1994). Occasional young-of-the-year specimens have been found in deep water in Long Island Sound as late as February (Richards et al. 1979).

CHAPTER 43

Myoxocephalus aenaeus (Mitchill)
Grubby

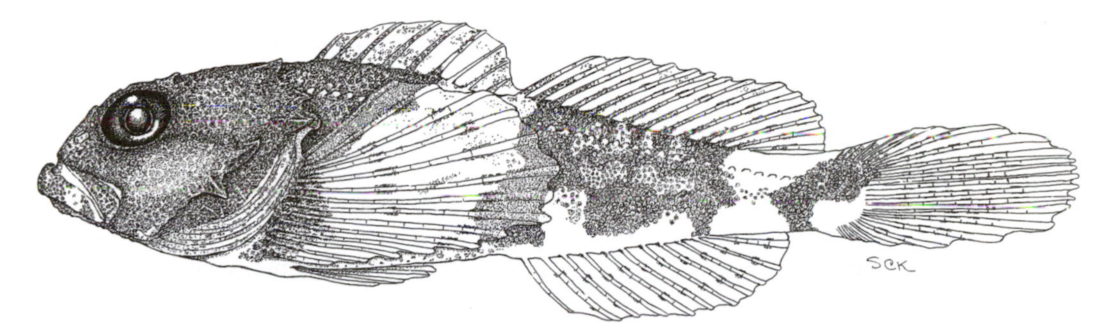

43.1 *Myoxocephalus aenaeus* juvenile, 19.4 mm SL. Collected May 12, 1989, 1-m plankton net, Little Sheepshead Creek Bridge, Great Bay, New Jersey. ANSP 175232. Illustrated by Susan Kaiser.

DISTRIBUTION

Myoxocephalus aenaeus (fig. 43.1) is found from the Gulf of St. Lawrence and Newfoundland to New Jersey (Bigelow and Schroeder 1953; Ennis 1969). Within this range, it inhabits coastal waters from the low-tide mark to depths of about 27 m (Sumner et al. 1913), but it has also been reported from as deep as 130 m (Robins and Ray 1986). Early life-history stages have been reported from several estuaries between Cape Cod and Delaware Bay (table 4.2).

REPRODUCTION

On the basis of literature accounts and recent ichthyoplankton sampling throughout the Middle Atlantic Bight, it appears that spawning begins in coastal waters and ends somewhat later in offshore oceanic waters. Numerous accounts indicate coastal spawning occurs in winter (Sumner et al. 1913; Morrow 1951; Bigelow and Schroeder 1953) or early spring (Richards 1959; Smith 1985). Winter spawning in a Cape Cod estuary was verified by gonadosomatic analysis, presence of egg masses, and increased reproductive coloration of males (Lazzari et al. 1989). In Newfoundland, spawning may extend from the fall into winter (Ennis 1969). On the basis of larval occurrences in the Middle Atlantic Bight between 1977 and 1987, spawning occurred between Georges Bank (where it has not been previously reported) and New Jersey coastal waters (fig. 43.2). This spawning occurred between March and June, but in the central part of the Middle Atlantic Bight it was limited to March through May (fig. 43.3). In this area, it is not known whether spawning occurs in estuaries, coastal ocean waters, or both.

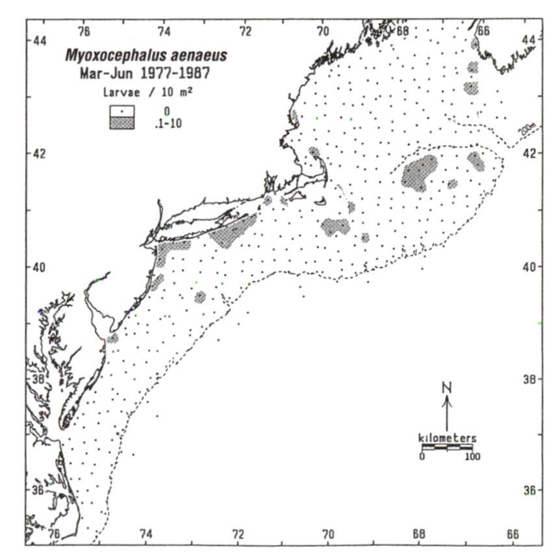

43.2 Occurrences of *Myoxocephalus aenaeus* larvae in MARMAP sampling 1977–1987. Average densities are combined for four months, March through June.

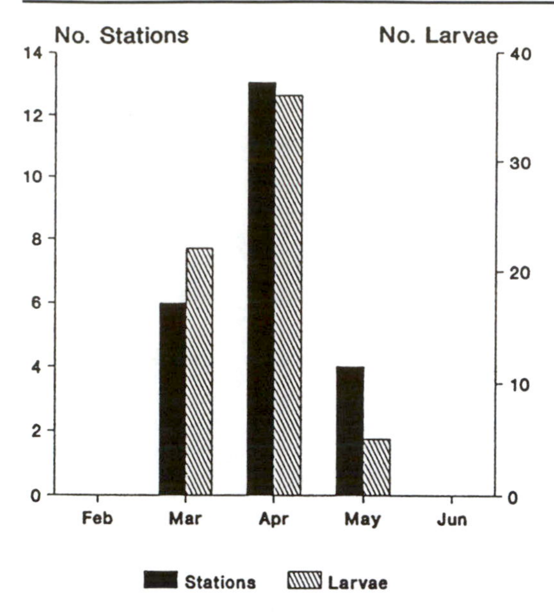

43.3 Monthly occurrences of *Myoxocephalus aenaeus* larvae collected from the central part of the Middle Atlantic Bight, 1977–1987.

DESCRIPTION

Eggs are spherical, transparent, and adhesive, and their diameters range from 1.5 to 1.7 mm. The color varies among yellow, green, red, and white, although color is consistent in individual females. Two large and several smaller oil globules are present, and the perivitelline space is narrow. Larvae hatch at a mean size of 5.4 mm TL (Lund and Marcy 1975). They have bulbous, heavily pigmented guts, and the tail has a row of melanophores along the ventral edge. Dorsal (parietal) head spines and preopercle spines are visible at sizes as small as 5.0 mm. There is a row of melanophores along the midventral line of the gut. Hatching, flexion (at about 6–8 mm), and relative development of spines and fin rays all occur at sizes smaller than in congeners (Fahay 1983). See Lund and Marcy (1975) or Fahay (1983) for further details of larval development. In juveniles (fig. 43.1), the body tapers from a thick pectoral region to a slim caudal peduncle. The head is covered with bumps, ridges, or spines, but the body lacks bony, lateral plates. Body pigment includes two prominent saddle-shaped blotches. The lower three pectoral fin rays are continuous with the rest of the fin, unlike the pectoral fin of searobins, *Prionotus* spp. Dorsal fin spines are present, but the anal fin lacks spines. Vert: 30–35; D: VIII–XI, 12–17; A: 8–14 (usually 10–11); Pect: 14–17; Plv: I, 3.

THE FIRST YEAR

Larvae hatch after 40 to 57 days incubation at a mean size of 5.4 mm TL (Lund and Marcy 1975). Larvae transform into juveniles at an age of about 55 days and at lengths about 9 to 10 mm. Most adult characters are developed at this size, although the urostyle tip is not yet resorbed and scales have not yet formed (Lund and Marcy 1975). Lengths of larvae occurring over the continental shelf range from 4 to 5 mm (corresponding to the size at hatching) to about 15 mm, but most are smaller than 10 mm (fig. 43.4). The size at settlement can be estimated at about 10 to 15 mm, since this represents the largest size reached in planktonic individuals and the smallest collected in bottom-sampling in estuaries. Although young-of-the-year of this species are never abundant in the central part of the Middle Atlantic Bight, an influx of settlement-sized individuals occurs in estuaries in April (Witting 1995; D. A. Witting, K. W. Able, and M. P. Fahay in prep.). In Nauset Marsh, larvae were abundant in April, and juveniles appeared in summer collections (Lazzari et al. 1989).

Otolith analysis shows that juveniles grow to 60 to 65 mm SL in the first year (Lazzari et al. 1989). Observations on growth in the Great

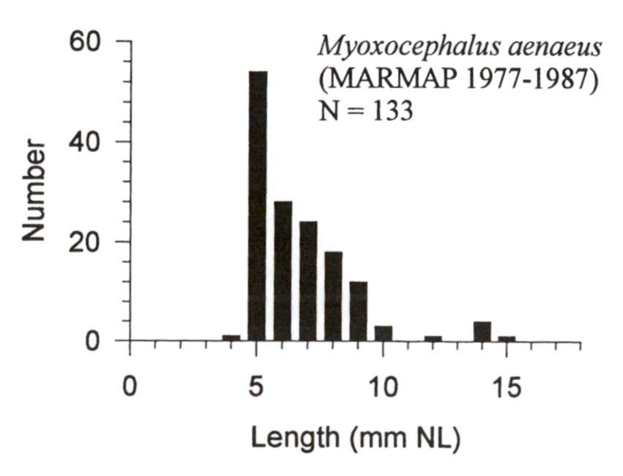

43.4 Lengths of *Myoxocephalus aenaeus* larvae collected from the central part of the Middle Atlantic Bight, 1977–1987.

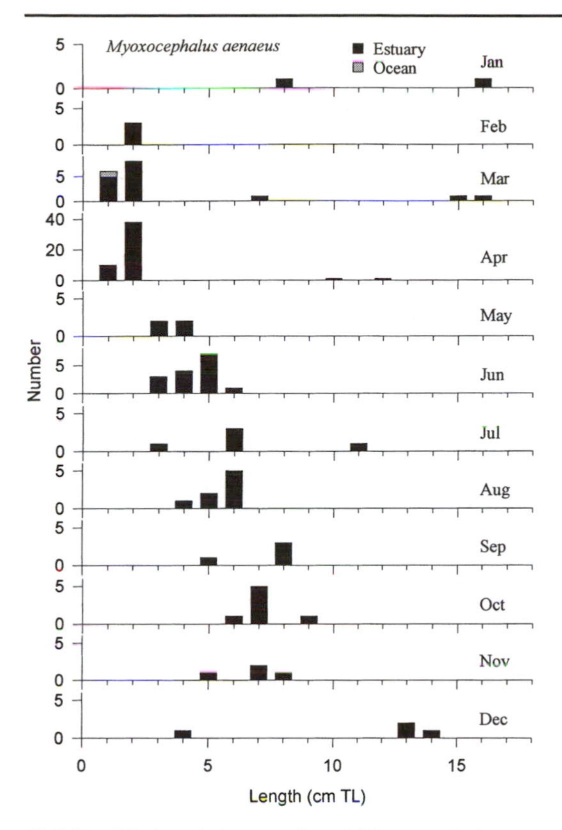

43.5 Monthly length frequencies of *Myoxocephalus aenaeus* young-of-the-year collected in estuarine and adjacent coastal ocean waters of the central Middle Atlantic Bight. Sources: RUMFS: otter trawl (n = 11); 1-m beam trawl (n = 15); 1-m plankton net (n = 39); experimental trap (n = 10); LEO-15 Tucker trawl (n = 1); seine (n = 15); gear comparison (n = 2); killitrap (n = 5); night-light (n = 25).

Bay–Little Egg Harbor study area (fig. 43.5) indicate that young-of-the-year reach lengths from 5 to 10 cm by their first fall. The similarity in sizes the following spring indicates little or no growth through the winter. A summer growth rate, calculated on the progression of these observed modes, would therefore be about 0.3 mm per day between April and October. Although our records are few, it appears that the maximum size of about 15 cm (Scott and Scott 1988) is reached by fish beginning their second winter.

Both estuarine and continental shelf habitats are used during the first year. Planktonic larvae have been collected from oceanic and estuarine waters (Lazzari et al. 1989). Young-of-the-year, as well as older stages, occurred most commonly in eelgrass beds in Nauset Marsh (Lazzari et al. 1989), and this habitat apparently is also most commonly used in the area around Woods Hole, Massachusetts (Bigelow and Schroeder 1953). All sizes were collected in similar habitats in the Bay of Fundy (Huntsman 1922). Young-of-the-year may also undertake diel movements between habitats. In a Great Bay–Little Egg Harbor study, these movements (into artificial seagrass beds) were greatest at night (Sogard and Able 1994). Their abundance in some estuaries may be underestimated as they have been observed by divers to be common in channels among blocks of peat sloughed off the marsh surface, and this habitat is not readily sampled by traditional gears (Able et al. 1988; Lazzari et al. 1989).

In the central part of the Middle Atlantic Bight, it is not known whether winter and summer distributions differ, therefore nothing is known about migrations. Although we collected relatively few specimens, they occurred during all months, indicating that they may reside in the estuary year-round (fig. 43.5).

CHAPTER 44

Morone americana (Gmelin)
White Perch

44.1 *Morone americana* early juvenile, 28.0 mm TL. After Mansueti 1964.

DISTRIBUTION

The range of *Morone americana* (fig. 44.1) extends from the upper St. Lawrence River to South Carolina (Robins and Ray 1986; Scott and Scott 1988). Within this range, it is endemic to estuaries, although landlocked freshwater populations also occur (Stanley and Danie 1983). All life-history stages are semi-anadromous, with seasonal migrations occurring between deeper, highly saline areas of estuaries during winter to shallow, brackish, or freshwater areas from early spring through summer (Beck 1995). It is one of the most abundant resident species in Delaware Bay (O'Herron et al. 1994), Chesapeake Bay (Setzler-Hamilton 1991), and parts of the Hudson River Estuary (Bath and O'Connor 1982). Its range in the Delaware River extends from Marcus Hook, Pennsylvania (river km 0) to Long Eddy, New York (river km 364) (Ashton et al. 1975), but it is most common in the tidal portion of the river south of Trenton. In the Hudson River, their 250-km range extends from Manhattan to Albany (Bath and O'Connor 1982). Genetically distinct subpopulations may occur in certain larger tributaries to the lower Delaware Bay, as they do in the Chesapeake Bay (Mulligan and Chapman 1989). Early life-history stages occur in most Middle Atlantic Bight estuaries, although eggs and larvae are not found in systems lacking tidal freshwaters, such as Nauset Marsh and Great South Bay (table 4.2).

REPRODUCTION

Spawning occurs in brackish (<4 ppt) and freshwater areas during the early spring (late March through early June) after a migration from deeper, wintering habitats (Hardy 1978b; Setzler-Hamilton 1991; Beck 1995). Most spawning occurs in the lower reaches of large coastal rivers. In the Delaware Bay area, most activity occurs when temperatures are between 14 and 18 C (Wang and Kernehan 1979). Spawning also occurs in the C&D Canal and in most larger tributaries to Delaware Bay (Wang and Kernehan 1979; Weisberg and Burton 1993). In the Chesapeake Bay, spawning reaches a peak at temperatures of 10 to 16 C and usually occurs in fresh or brackish water over beds of fine gravel or sand (Setzler-Hamilton 1991). Spawning in the Hudson River occurs in shallow flats, embayments, and tidal creeks, primarily between river km 138 and 198. Spawning begins in this river in late April when temperatures are 10 to 12 C and reaches a peak in late May when they are 16 to 20 C (Klauda et al. 1988).

DESCRIPTION

Eggs are spherical, demersal, and have a flattened, adhesive disk. Diameters range from 0.75 to 1.04 mm, and adhesive properties are lost after water hardening (Mansueti 1964). Larvae hatch at about 2.6 mm TL and have unpigmented eyes and unformed mouth parts. They strongly resemble larvae of *Morone saxatilis*. Details for separating early stages of the two species can be found in Hardy (1978b) and Olney et al. (1983). In juveniles (fig. 44.1), fins have both sharp spines and soft rays. The second dorsal fin base is only slightly longer than the anal fin base. The second anal fin spine is thicker than the first and third. Vertical bars of pigment may be present on the body at sizes between 25 and 75 mm TL (Lippson and Moran 1974). The body is deeper than that of comparably sized *M. saxatilis*. The most reliable method for separating juveniles of the two species of *Morone* in our study area involves determining patterns of interdigitation of the dorsal fin pterygiophores and neural spines of adjacent vertebrae (Olney et al. 1983). Vert: 25; D: VII–XI, I, 11–13; A: III, 9–10; Pect: 10–18; Plv: I, 5.

THE FIRST YEAR

The demersal eggs are attached to the substrate or objects on the bottom in still water but may drift with currents. Hatching occurs after 2 to 6 days, depending on temperature (Hardy 1978b). Larvae complete the transformation to the juvenile stage at about 20 to 30 mm after about 6 weeks of passive, planktonic existence in near-surface waters (Wang and Kernehan 1979; Public Service Electric and Gas 1984). Body depth relative to length increases during this size interval. Fin rays are complete by 30 mm TL, and scalation begins at sizes between 16 and 35 mm, depending on geographic area (Marcy and Richards 1974). Scales were complete over most of the body in 100% of those young-of-the-year larger than 23 mm FL in a Delaware Bay study (Wallace 1971). Juveniles become demersal as body depth increases and fin rays and scales develop, and

Table 44.1. Mean Sizes at End of the First Year in *Morone americana* from Several Locations and Years (where known) within Middle Atlantic Bight Estuaries

Location, study years	Length (mm FL)	(mm TL)	Source
Thames River	—	91.0	Whitworth et al. 1975
Connecticut River, 1959–1965	—	88.1	Marcy and Richards 1974
Hudson River, 1963–1969	86.5	90.6	Bath and O'Connor 1982
Hudson River, 1979	—	72.0	Klauda et al. 1988
Hudson River	72.0	76.5	Lawler et al. 1982
Delaware River, 1960–1967	83.5	88.5	Wallace 1971
Delaware River	80.0	84.8	Public Service Electric & Gas 1984
C & D Canal	80.0	84.8	Horseman and Shirey 1974
York River, 1960–1968	79.0	83.8	St. Pierre and Davis 1972
James River, 1960–1968	76.0	80.7	St. Pierre and Davis 1972
Patuxent River, 1942–1954	—	93.0	Mansueti 1961
Albemarle Sound	69.0	73.4	Conover 1958

NOTE: Sizes converted to Fork Length (FL) or Total Length (TL) from those given in original studies according to conversion formulas presented in Appendix B. Lengths are averages of male and female lengths if sexes were distinguished in original study.

they then move inshore to relatively shallow areas that they use as nurseries through the summer (O'Herron et al. 1994).

In general, growth is fastest during the first summer, and 37% to 40% of the ultimate maximum size is reached after the first year (Wallace 1971; Bath and O'Connor 1982). Growth rates of young-of-the-year are influenced by environmental and habitat factors and differ strongly between study areas (Setzler-Hamilton 1991) and in different temperatures (Klauda et al. 1988). Populations in freshwater impoundments grow faster than those in rivers (Bath and O'Connor 1982). Sizes after the first year in several study areas range from 72 to 93 mm TL (table 44.1). Young-of-the-year grow through the summer and fall and reach these sizes in November, while older year classes stop growing in September of each year (Wallace 1971). A positive correlation between mean river temperatures and annual variation in growth rate has been demonstrated for larvae and early juveniles (< 25 mm TL) in the Hudson River (table 44.2), but rates during older juvenile stages (25 to 60 mm TL) were not correlated with temperatures, freshwater flow rate, or juvenile abundance (Klauda et al. 1988). Results of all the previously cited studies indicate that growth in all year classes is nearly nil through the winter. The lengths presented here from the Delaware River and Bay (fig. 44.2) demonstrate some of these trends. Young-of-the-year exhibit a modal length of 7 cm FL (about 7.4 cm TL) as they enter their first winter. Modes are about the same the following spring, and at age 1

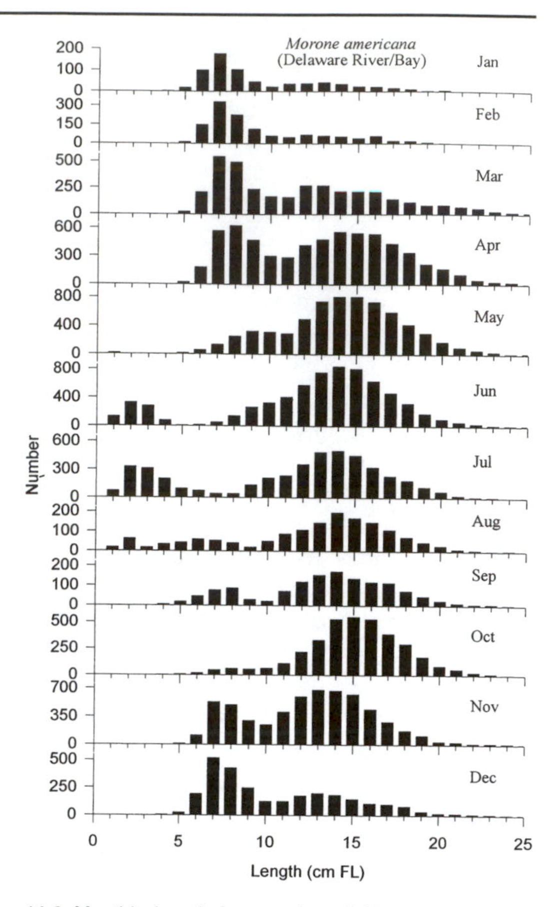

44.2 Monthly length frequencies of *Morone americana* from Upper Delaware Bay and tidal portion of Delaware River (1970–1981). Data courtesy of Public Service Electric & Gas and Environmental Consulting Services, Inc.

they have grown to about 9 to 10 cm. Growth of a single year class from the Hudson River (fig. 44.3) is comparable to young-of-the-year from the Delaware region when a mode of 8 cm TL is apparent at the end of the first summer (fig. 44.2).

Table 44.2. Growth Rates of Larval and Early Juvenile (< 25 mm TL) *Morone americana* from the Hudson River during Several Years

Sampling period	Mean water temperature (C)	Growth rate (mm per day)
Jun 13–Jul 10, 1973	21.9	0.82
May 22–Jul 10, 1974	18.9	0.45
May 24–Jul 26, 1975	20.6	0.67
Jun 12–Jul 8, 1976	22.7	0.85
May 24–Jul 1, 1977	19.9	0.60
May 31–Jun 28, 1978	21.1	0.79
May 16–Jun 28, 1979	19.2	0.50

SOURCE: Klauda et al. 1988.

This species has been studied intensively enough that aspects of its habitat use, especially for the earliest life-history stages, are available. Eggs incubate in temperatures between 10 and 20 C, with the optimum at about 14 C (Morgan and Rasin 1982). Temperatures near the high part of this range may be lethal, as are sudden drops in temperatures during incubation (Setzler-Hamilton 1991). Optimum survival and growth of larvae is attained at temperatures between 15 and 20 C (Marguiles 1988). All early stages have a high tolerance to salinity differences, ranging from 0 to 13 ppt

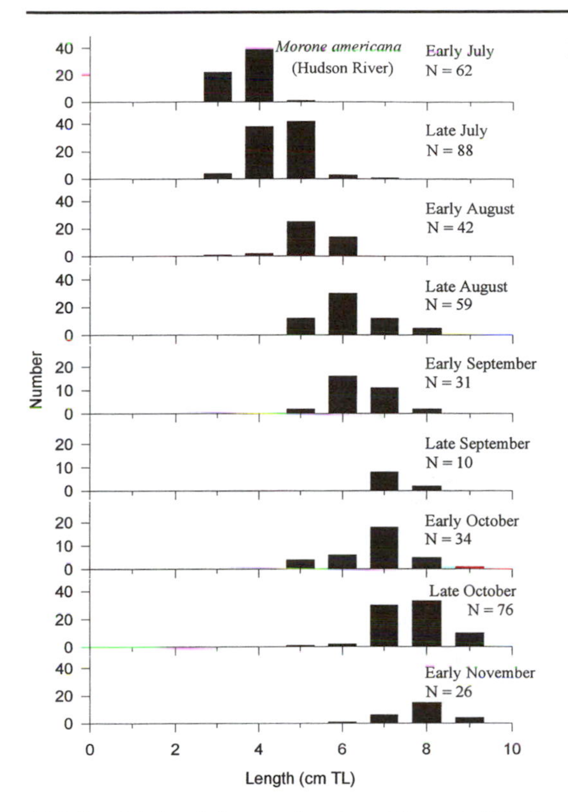

44.3 Monthly length frequencies of a single year-class of *Morone americana* collected from the Hudson River Estuary (Kahnle and Hattala 1988).

(Dovel 1971). Improvements to water quality in the Delaware River, especially between Philadelphia and Wilmington, have resulted in enhanced survival of a congener, *M. saxatilis,* but may have had less effect on *M. americana* production, presumably because young-of-the-year of the latter species complete their development upstream from the formerly impacted area (Weisberg and Burton 1993). The latter study, however, did find increased abundances of larvae in the impacted part of the river compared to studies undertaken before the improvements. Juvenile nursery areas during the first summer tend to be located downstream from spawning areas. Young-of-the-year occur most frequently in relatively shallow, brackish to freshwater areas including tidal creeks. They appear to prefer level bottoms of compact silt, mud, sand or clay with little or no cover and to avoid soft muck, highly organic substrate, and gravelly or rocky areas (Stanley and Danie 1983; Setzler-Hamilton 1991). Vegetation or other structural components are not essential habitats (Hardy 1978b; Stanley and Danie 1983).

In the Delaware Bay area, young-of-the-year leave shallow nursery areas during the fall and spend their first winter in deeper parts of the lower river and upper bay (Beck 1995). In the Hudson River, older juveniles begin moving from shore and shoal areas into deep midriver areas in October, and by mid-December a majority of this age class is found in deep areas (Klauda et al. 1988). In the Chesapeake Bay area, they overwinter in deep channel areas with depths between 12 and 18 m (maximum 42 m) and in temperatures between 2 and 5 C (Setzler-Hamilton 1991).

CHAPTER 45

Morone saxatilis (Walbaum) Striped Bass

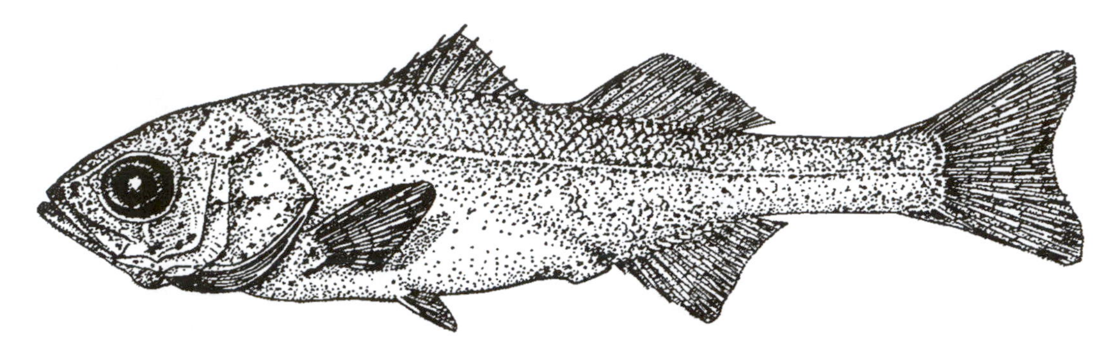

45.1 *Morone saxatilis* early juvenile, 29.0 mm TL. After Mansueti 1958.

DISTRIBUTION

Morone saxatilis (fig. 45.1) occurs along the Atlantic coast between the St. Lawrence River, Canada and St. Johns River, Florida (Smith and Wells 1977). Within this range, different populations occur, many migratory, some nonmigratory, and some well established in freshwater lakes. Several principal anadromous stocks have been recognized in the Middle Atlantic Bight, centered in Roanoke River (North Carolina); Chesapeake Bay and tributaries; Delaware River; and Hudson River. Recent tagging studies indicate that Hudson River fish migrate seasonally and range between the Bay of Fundy and Cape Hatteras, with the smallest individuals occurring nearest the estuary and the largest wandering farther from the Hudson (Waldman et al. 1990). Members of all populations usually occur very close to the coast, often in the surf zone, although they also occur in shallow areas in bays and estuaries. They frequent sandy bottom areas where tidal currents are strong. They are less abundant in smaller, shallower estuarine bays in the central part of the Middle Atlantic Bight but are common to abundant in larger, deeper systems (Stone et al. 1994). Juveniles occur in most estuaries throughout the Middle Atlantic Bight, but eggs and larvae have only been reported from the Hudson River and Chesapeake Bay and Delaware Bay (table 4.2).

REPRODUCTION

Major spawning areas include tributaries of the Chesapeake Bay (e.g., Potomac, James, York, and many smaller rivers along the eastern shore of Maryland) but significant reproduction also occurs in tributaries to Albemarle Sound, the C&D Canal, and certain tributaries to, and mainstems of, the Delaware and Hudson Rivers. It may also occur in freshwater and low-salinity estuarine rivers and streams along the eastern shore of Chesapeake Bay. Spawning is generally reported to occur in the first 40 km upstream from saltwater intrusion (Rathjen and Miller 1957). In the Delaware River, spawning activity occurs between river km 94 and 201 (Murawski 1969), where salinity is less than 3.0 ppt (Wang and Kernehan 1979). Eggs are also abundant in the C&D Canal, where concentrations as high as 110 eggs and 95 larvae/m^3 have been collected (Kernehan et al. 1977). Most spawning in the Hudson River occurs between West Point and Kingston (Smith 1985), and recent collections of eggs have spanned the 90 to 153 km and 37 to 164 km areas above salt water (Texas Instruments 1973, cited in Dovel 1981). Collections of eggs in a weekly survey of the Hudson River in 1972 found occurrences only above Haverstraw Bay (about river km 64) (Dovel 1981). Limited spawning activity has also been reported in the Mullica River in southern New Jersey (Hoff 1976). In the Chesapeake and Hudson systems, spawning begins when temperatures reach 11 C (Dovel 1981). In the Delaware Bay region, spawning begins when temperatures reach 14 C (Fay et al. 1983) from early April through mid-June (Wang and Kernehan 1979).

DESCRIPTION

Eggs are spherical, nonadhesive, and transparent and have a greenish yolk. There is a single oil globule ranging from 0.40 to 0.85 mm, and the perivitelline space is very wide (65%–85% of diameter). Egg diameter varies widely (1.3–4.6 mm), and size may be inversely related to salinity (Murawski 1969; Wang and Kernehan 1979). Mean egg diameters from various locations in our study area are: North Carolina, 2.43–2.63 mm; Patuxent River, Maryland, 2.18–2.23 mm; Delaware River tributaries, 1.65 mm; Delaware River, 2.90 mm. Eggs have been described as buoyant or semibuoyant and found at various levels within the water column. (For a compilation of details regarding characters of eggs reported by a variety of authors, see Hardy 1978b.) Larvae hatch at 2.0 to 3.7 mm TL. The body is slender, and the snout-anus length is greater than 50% TL. There are 23 to 27 myomeres (mean 25), of which 11 to 13 are preanal. Major pigment accumulations occur over the air bladder, along the postanal ventral edge, on the top of the head, and on the opercle. Fin spines and rays begin forming at about 7.0 mm TL, and full complements are present at 20 mm TL. The fins of juveniles (fig. 45.1), have both sharp spines and soft rays. The second dorsal fin base is only slightly longer than the anal fin base. The three anal fin spines are of three discrete lengths (first, short; second, midsized; third, long) but are about the same thickness. (In similar-sized *M. americana,* the second anal fin spine is about the same length as the third and is noticeably thicker than the first and third.) The body is not as deep as in juveniles of *M. americana.* The most reliable method for separating juveniles of the two species of *Morone* in our study area involves determining patterns of interdigitation of the dorsal fin pterygiophores and neural spines of adjacent vertebrae (Olney et al. 1983). Vert: 24–25; D: VIII-IX, I, 9–14; A: III, 7–13; Pect: 13–19; Plv: I, 5.

THE FIRST YEAR

Hatching occurs after 48 hours incubation at temperatures of about 17 to 18 C but ranges between 29 hours (at 24 C) and 80 hours (at 12 C) (Hardy 1978b). Scales begin forming in juveniles at about 16 mm and are complete at about 20 to 30 mm (Hardy 1978b).

A two-year study in the Pamunkey River, Virginia showed that temperature can affect the timing of hatching and larval development (McGovern and Olney 1996). When temperatures were lower, food densities were also lower, and the duration of development was extended, thus exposing larvae to elevated numbers of predators for a longer period in this study. When temperatures were higher, development took a shorter period of time and coincided with an increase in food items in a field of reduced predator density. Year-class success was greater in the year with higher temperatures during the time of peak egg production.

Growth of young-of-the-year appears to be consistent across a number of locations. In the Delaware River population, growth continues from spring through summer and the year class reaches a mode of about 10 cm by late fall (fig. 45.2). The same pattern was evident for Hudson River young-of-the-year combined for six year classes (fig. 45.3) (Kahnle and Hattala 1988). In another study (McKown 1991), mean lengths at age and growth rates were found to be consistent among five Hudson River year classes (1983–1987). This study also found a significant difference between lengths of the age 1 size class in the lower Hudson River and in western Long Island embayments, with the latter between 39 and 57 mm larger than the former. Growth apparently ceases during the winter and a similar mode is found the following April, when the year class reaches age 1.

Habitat use during the first year varies with life-history stage. After spawning, there is a net movement of young-of-the-year from upstream locations to those in lower, tidal reaches. Late larvae and early juveniles favor shallow waters with sluggish currents and sand or gravel bottoms for nursery areas (Wang and Kernehan

1979; Boynton et al. 1987). Tidal creeks with similar substrates are also utilized by young-of-the-year (Smith 1971). Young-of-the-year of the Delaware River population migrate downstream from spawning locations and spend their first summer within the tidal portions of the river. Most young-of-the-year from the Hudson River also move downstream from spawning areas and spend their first summer and winter in the lower Hudson River (McKown 1991). In this river, young-of-the-year have been found in and around pier pilings (Stoecker et al. 1992; Able and Studholme 1994) and in the lower Hudson River Estuary, appear to be more abundant in deep interpier habitats than in shallow ones during late summer (Cantelmo and Wahtola 1992). After young-of-the-year overwinter in the lower Hudson River Estuary,

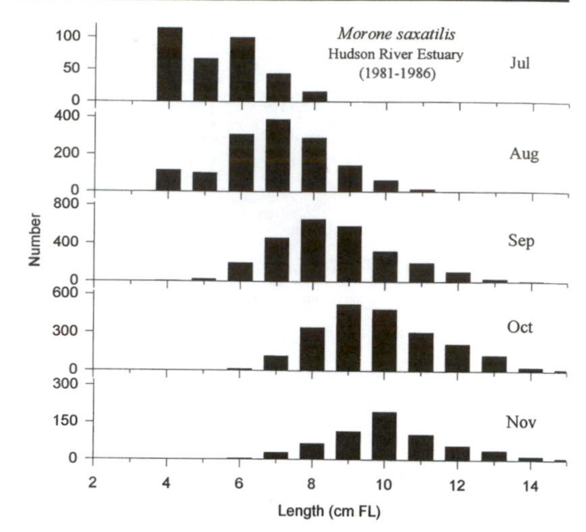

45.3 Monthly length frequencies of *Morone saxatilis* young-of-the-year collected from the Hudson River Estuary during bottom-trawl survey of juvenile fishes, 1981–1986 (Kahnle and Hattala 1988). Data from all years combined. Lengths <5.0 cm arbitrarily combined as "4.0 cm FL."

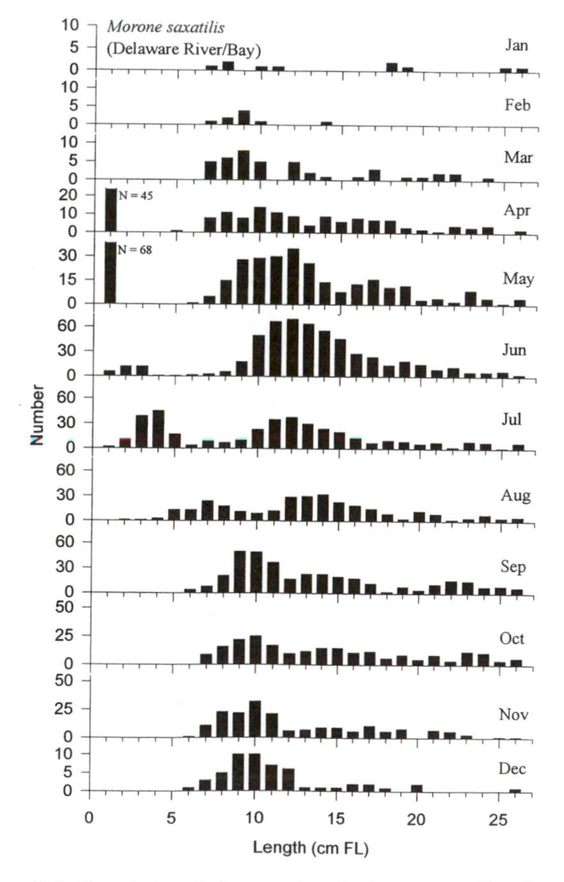

45.2 Monthly length frequencies of *Morone saxatilis* collected in Upper Delaware Bay (1970–1980) and tidal portion of Delaware River (1979–1981). All river collections indicated as 1-cm size class during April and May. Data courtesy of Public Service Electric & Gas and Environmental Consulting Services, Inc.

M. saxatilis of age 1 are found the following spring both in the lower Hudson and in various bays on the north and south shores of western Long Island (McKown 1991), especially Jamaica Bay, Little Neck Bay, Manhasset Bay, and, in some years, Bellport Bay. In years with strong year classes, young-of-the-year are found beyond the lower Hudson River nursery area and are included in the catches of age 1 (and older) in certain of these embayments (McKown 1991).

The range of acceptable environmental conditions is relatively well-known for this species. An important requirement for the spawning area is a current strong enough to keep the eggs suspended in the water column, lest they settle to the bottom and be smothered by silt (Bigelow and Schroeder 1953). The upper lethal temperature for developing embryos is 27 C (Morgan and Rasin 1982). Eggs and larvae are less tolerant of salinity fluctuations than adults, and their survival is enhanced in low salinities (2–10 ppt) (Fay et al. 1983). Turbidity also adversely affects larvae's ability to capture prey (Fay et al. 1983). Optimum temperatures for larval growth and survival are 15 to 22 C (Fay et al. 1983; Funderburk et al. 1991) and rapidly changing temperatures can be detrimental to this life-history stage (Hollis

1967). The optimum range for juveniles is 18 to 23 C (Fay et al. 1983). For dissolved oxygen, the optimum levels for all life-history stages is 6 to 12 mg/liter (Fay et al. 1983). Levels below 2.4 mg/liter are lethal for larvae, and those below 3 mg/liter are lethal for juveniles (Fay et al. 1983).

Pollution and resulting low oxygen levels prevented successful reproduction in the lower Delaware River for a time (Chittenden 1971), but improved water quality in the region over the period 1980 to 1993 has resulted in increased reproduction and enhanced production and survival of juveniles (Weisberg and Burton 1993). The latter authors cite a thousandfold increase in the abundance of juveniles during the decade studied and correlate this increase with improved water quality, not with an overall increase in the population. In particular, this improvement has resulted in reductions of anoxic conditions that once formed a block to migratory species, especially in that part of the river downstream from Philadelphia during late summer. Thus, historically important spawning and nursery areas that had been lost to the population (Chittenden 1971) have been restored to a condition allowing for its use.

In the Delaware River system, young-of-the-year and ensuing year classes may spend 2 years or more within the estuary before migrating offshore and joining the migratory population (J. P. Miller 1995). Most overwinter in deeper portions of the estuary, but some also overwinter in tidal creeks (Smith 1971). Studies in western Long Island embayments indicate that Hudson River progeny may spend up to 3 years in these estuaries before migrating offshore with the adults. Furthermore, tagging studies of young-of-the-year, age 1 and age 2 suggest that these year classes do not move far from these nursery areas (McKown 1991).

CHAPTER 46

Centropristis striata (Linnaeus)
Black Sea Bass

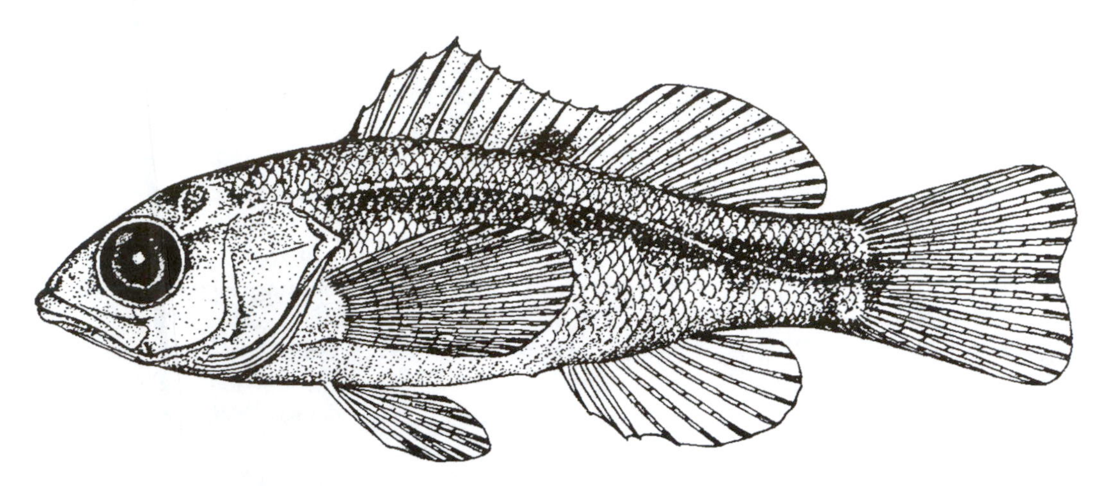

46.1 *Centropristis striata* juvenile, 22.2 mm TL. After Hardy 1978b.

DISTRIBUTION

Black sea bass (*Centropristis striata*) (fig. 46.1) is found from Massachusetts to Florida (Miller 1959), but it is most abundant in the Middle Atlantic Bight (Kendall and Mercer 1982). The Middle Atlantic Bight population is considered a separate stock from populations south of Cape Hatteras (Mercer 1978; but see Bowen and Avise 1990). This interpretation is supported by studies that show seasonal inshore–offshore and north–south movements in the Middle Atlantic Bight but only localized movements in the South Atlantic Bight (see Able et al. 1995a). Morphological studies have suggested that stock structure may not be homogenous in the Middle Atlantic Bight (Shepherd 1991). Black sea bass occurs from the edge of the continental shelf to shallow and deep drowned river valleys, in rocky areas along the coast, and barrier island estuaries. In all these areas, it is typically associated with hard-bottom or structured habitats such as reefs, wrecks, etc., both as juveniles (Able et al. 1995a) and as adults (Bigelow and Schroeder 1953). The larvae and juveniles are found in most estuaries in the Middle Atlantic Bight (table 4.2).

REPRODUCTION

This species is a sequential hermaphrodite that matures first as a female, then changes to a male at ages of 1 to 8 years (Lavenda 1949; Mercer 1978; Wenner et al. 1986). Collections of eggs over several years (Berrien and Sibunka in press) indicate that spawning occurs on the Middle Atlantic Bight continental shelf from April through October. Spawning begins between Cape Hatteras and Chesapeake Bay in April or May and expands northward, especially along the inner continental shelf, and reaches a peak in August when eggs are most concentrated between Cape Hatteras and northern New Jersey. Subsequently, egg abundance declines, and the distribution of eggs becomes patchy. Off New Jersey, spawning occurs from July through October (Able et al. 1995a) as indicated by small larvae on the inner and middle continental shelf (fig. 46.2).

DESCRIPTION

Eggs are pelagic, spherical, and 0.8 to 1.0 mm in diameter. The perivitelline space is narrow, and there is a single oil globule, 0.13 to 0.19 mm in diameter. Hatching occurs at 1.5 to 2.0 mm. The larval body is stocky (body

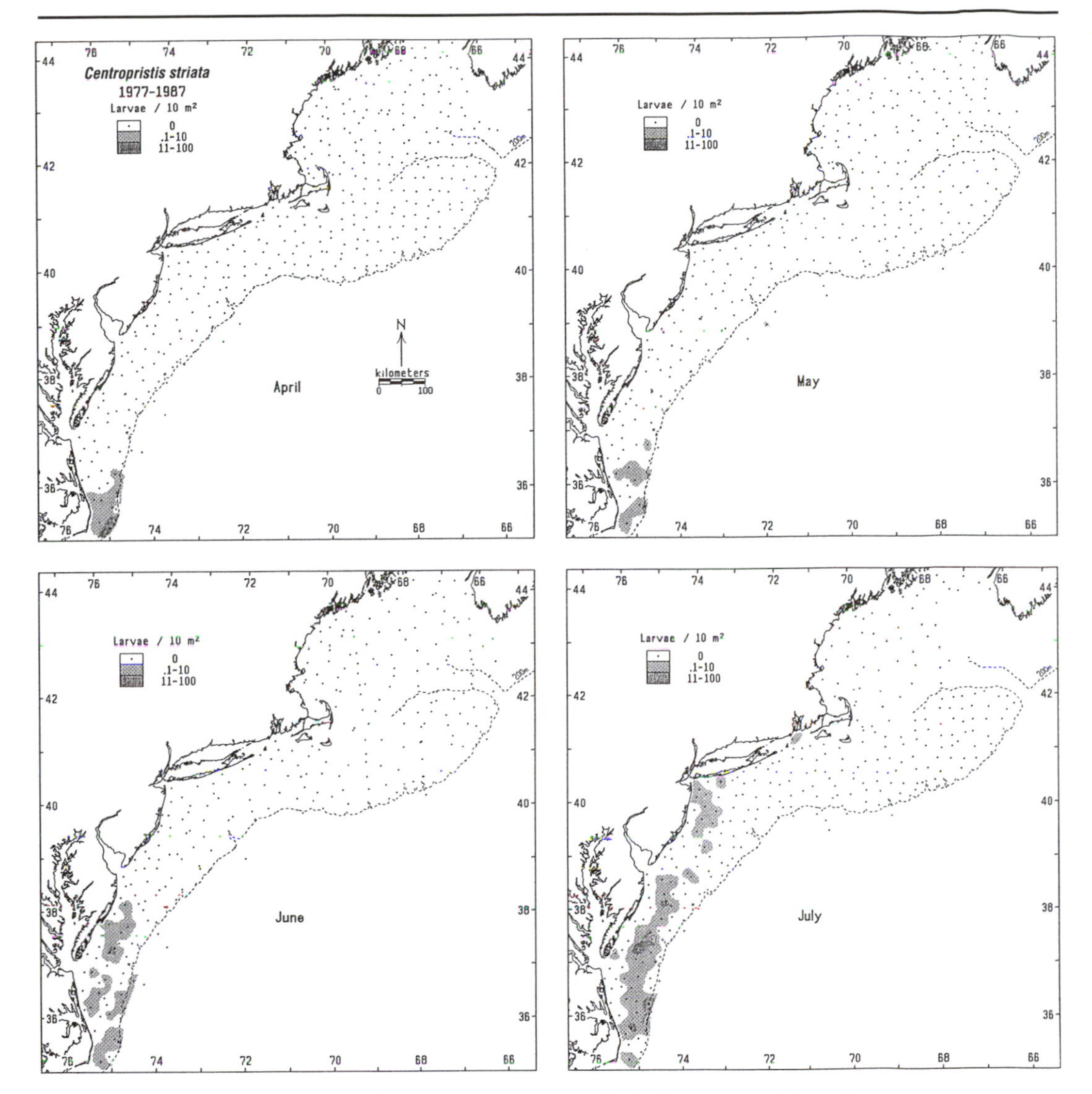

46.2 Monthly distributions of *Centropristis striata* larvae during MARMAP surveys, 1977–1987.

depth = 25%–27% of SL), and the head is large (33% of SL). Snout-to-anus length increases from 50% SL at 5 mm to 65% SL in juveniles. There are a few spines along the edge of the opercle and preopercle. Pigmentation includes prominent melanophores along the venter posterior to the anal fin and a series of spots between the anus and the cleithral symphysis. Fin spines and rays are complete at about 9.0 mm (Kendall 1972, 1979). Demersal juveniles have a single, continuous dorsal fin, with at most a slight notch between the spines and rays (fig. 46.1). A prominent melanophore occurs on the posterior part of the spiny dorsal fin, and a vague bar extends anteriorly and posteriorly from it. A dark midlateral stripe extends the length of the midbody. The maxilla reaches the midpoint of the eye or beyond. Very small *Tautogolabrus adspersus* juveniles are similarly pigmented but have a much smaller mouth. Vert: 23–24; D: X, 11; A: III, 7; Pect: 18–20; Plv: I, 5.

THE FIRST YEAR

The pelagic eggs hatch in 35 to 75 hours depending on temperature (Wilson 1891; Tucker

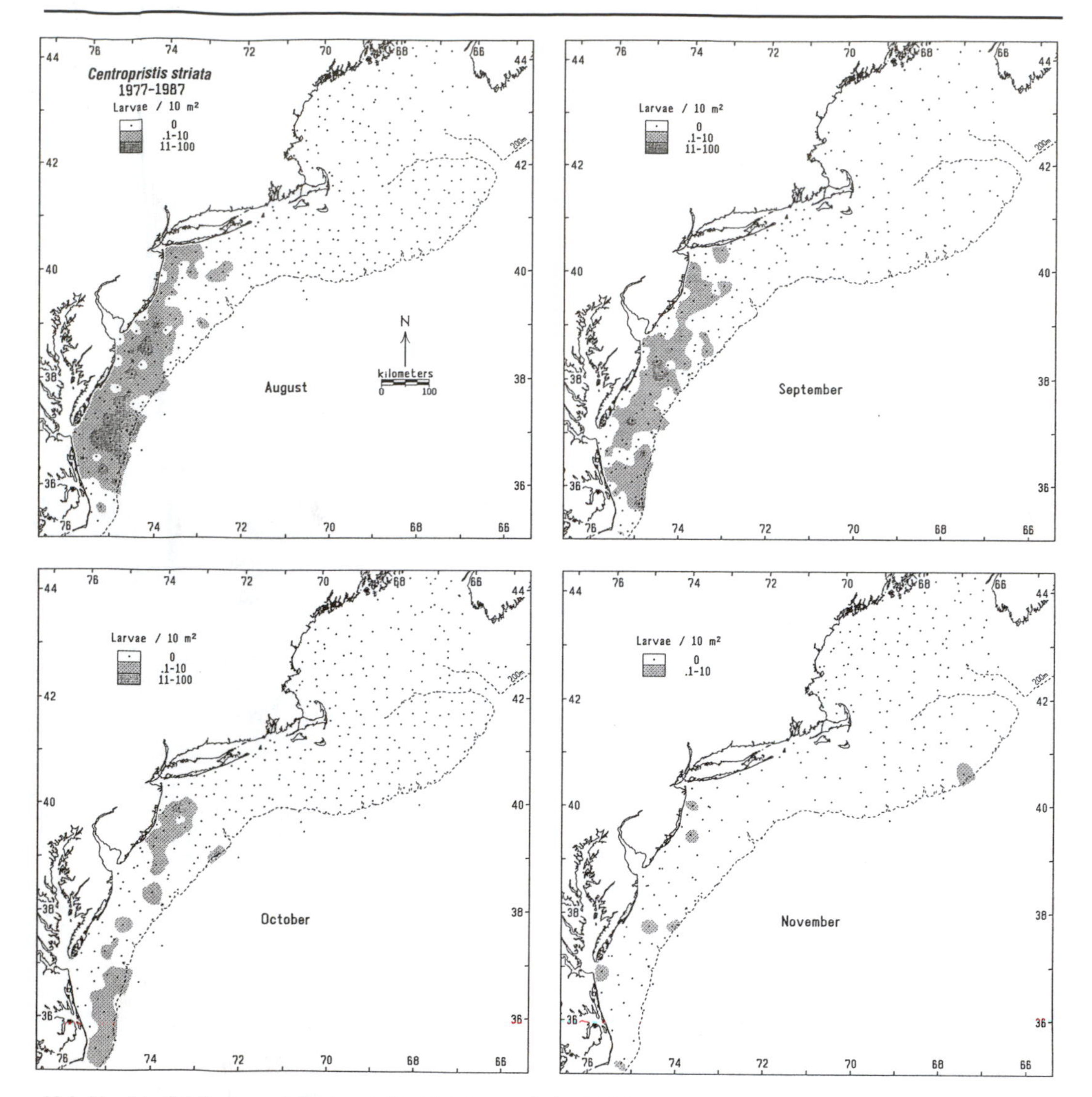

46.2 Monthly distributions of *Centropristis striata* larvae during MARMAP surveys, 1977–1987.

1989). Development of the pelagic larvae occurs in continental shelf waters in the Middle Atlantic Bight (Kendall 1972), but the developing larvae (3–13 mm SL) are most abundant in the southern portion of the bight, partly because spawning occurs over a longer period in that region (Able et al. 1995a). At Beach Haven Ridge on the inner continental shelf off southern New Jersey, small larvae (3–10 mm NL; K. W. Able and M. P. Fahay unpubl. data) were collected in July and August. Settlement occurs at sizes of 10 to 16 mm TL (Kendall 1972; Able et al. 1995a) on the inner continental shelf. Movement of recently settled individuals from the continental shelf spawning area into estuarine nurseries (Able et al. 1995a) is implied because no larvae have been collected in the Great Bay estuary over 6 years of weekly ichthyoplankton collections (Witting 1995; D. A. Witting, K. W. Able, and M. P. Fahay in prep.), yet small juveniles occur there. It is unclear how this movement into the estuary occurs. Small individuals (<20 mm TL) were collected from the Great Bay–Little Egg Harbor study area July through October (fig. 46.3). The seasonal peaks in abundance in the estuary reflect the occurrence of young-of-the-year at two different periods. The first peak in the spring is due

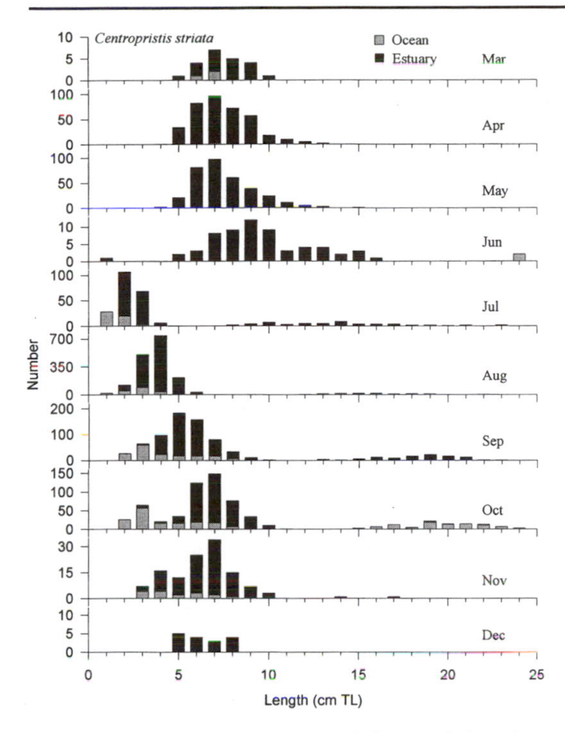

46.3 Monthly length frequencies of *Centropristis striata* collected in estuarine and coastal ocean waters in the vicinity of the Great Bay–Little Egg Harbor study area. Sources: Thomas et al. 1975; RUMFS: otter trawl (n = 87); 1-m beam trawl (n = 98); 1-m plankton net (n = 13); experimental trap (n = 1,894); LEO-15 Tucker trawl (n = 58); LEO-15 2-m beam trawl (n = 650); gear comparison (n = 28); killitrap (n = 1,253); seine (n = 2); weir (n = 5).

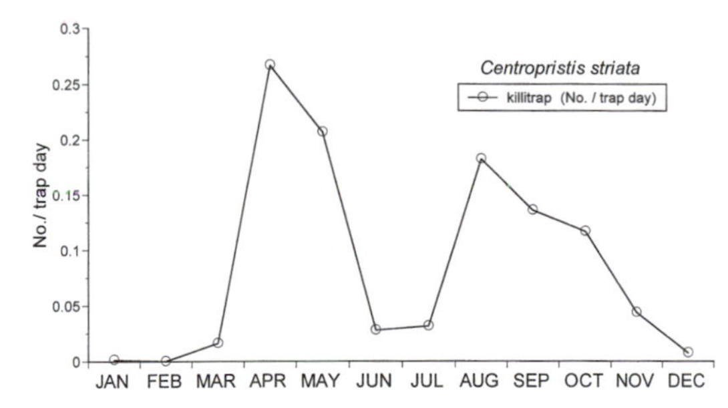

46.4 Monthly occurrences of *Centropristis striata* young-of-the-year in the Great Bay–Little Egg Harbor study area based on killitrap collections 1991–1994 (n = 1,334).

to fish returning to the estuary, and the second peak represents recently settled individuals (fig. 46.4). The decline in catches in June in these small traps is probably due to the reduced occurrences of larger individuals (see below).

Young-of-the-year from the estuary and the inner continental shelf grow to 18 to 91 mm TL by October (fig. 46.3; Able et al. 1995a). Calculations of growth, based on the progression of length modes, indicate rates of 0.42 to 0.45 mm per day. In a separate study, growth of marked and recaptured fish (0+ age) was similar (0.43 mm per day) (Able and Hales 1997). There appears to be little growth during the winter, but it resumes at a relatively fast rate (0.77 mm per day) the following spring and summer so that by midsummer, approximately 1 year after hatching, they are 78 to 175 mm TL (fig. 46.3). By the fall, at an approximate age of 14 to 17 months, they are 134 to 225 mm TL (Able et al. 1995a). This growth pattern was consistent with data from the early 1990s and the early 1970s in southern New Jersey both in the estuary and ocean (Able et al. 1995a). Laboratory studies have indicated that low dissolved-oxygen levels may negatively influence somatic growth rate of young-of-the-year (Hales and Able 1995). In multiple experiments between 2.2 and 5.8 ppm dissolved oxygen, somatic growth was positively related to oxygen level, with growth rates varying from 0% to 0.30% per day.

Young-of-the-year have been collected and observed in a variety of structured habitats both in the estuary and on the inner continental shelf off southern New Jersey (Able et al. 1995a). Average summer densities (0.33 individuals/m^2) were similar in these two habitats. This similarity and the similarities in growth rates suggest that both areas have similar habitat value as nurseries, and thus this species is not strictly estuarine-dependent (Able et al. 1995a). In the ocean, young-of-the-year are more frequently collected in catches with large amounts of shell hash, particularly valves of the surf clam, *Spisula solidissima*. In the estuary, they occur at a variety of sites including those with shell, amphipod tubes (*Ampelisca abdita*), and deep channels with rubble (fig. 46.5; Able et al. 1995a). They are especially abundant in the RUMFS boat basin, which is an embayment off a marsh creek (Able and Hales 1997) and in deep holes in that and other marsh creeks (see below). The young-of-the-year have also been captured around pier pilings and in open water in the Hudson River (Able et al. 1995b). The importance of structured habitats for young-of-the-year is further substantiated by the increase

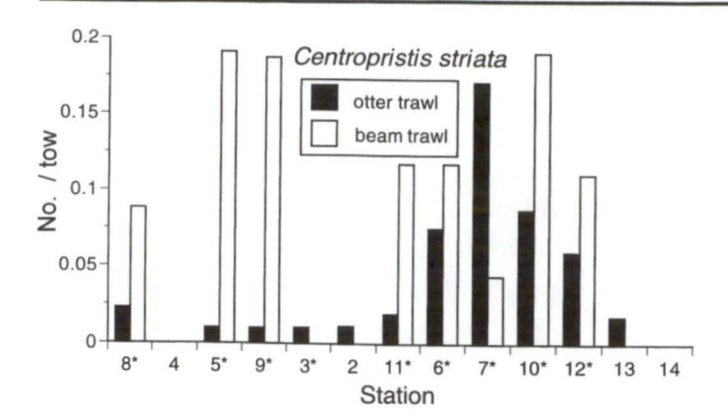

46.5 Distribution of *Centropristis striata* young-of-the-year by habitat type based on otter trawls (n = 49) and beam trawls (n = 71) deployed on regular stations 2–14 in Great Bay and Little Egg Harbor. Both gears used on stations designated by asterisk; otherwise, only the otter trawl was used. See figure 2.3 for station locations; table 2.1 for specific habitat characteristics by station.

in the catch rate of juveniles after mollusc shell was experimentally added to featureless estuarine substrate in Chesapeake Bay (Arve 1960).

Laboratory observations of individually tagged young-of-the-year (38–64 mm TL, n = 12) indicate a preference for oyster shells over barren sand substrate (2 cm deep) in large tanks (250 cm x 115 cm) (K. W. Able unpubl. data). Under these conditions, young-of-the-year spent most of their time resting during the day under or against shells. They seldom shared these habitats and were often distributed evenly with one individual per shell. When aggression occurred, it was during the day and was associated with defense or an attempt to take over the shell from a prior occupant. At night, most of these individuals left their shells, schooled together, and continuously circled the tank. This behavior might be interpreted as migratory because, during the period of the experiments (22 September–3 October 1991), the juveniles are leaving the estuary and moving offshore (Able et al. 1995a).

There seems to be a high degree of habitat fidelity during the summer in the estuary. In preliminary experiments in the summer of 1991, juveniles (n = 34) were captured in deep (2–3 m) holes in polyhaline marsh creeks in Great Bay and the RUMFS boat basin. These individuals were marked with injections of acrylic dyes and released at the same locations (K. W. Able unpubl. data). In several instances (n = 11), they were collected in the same locations after 4 to 90 days. In 1992, this same pattern of habitat fidelity occurred in a much larger tagging study in the same area (Able and Hales 1997). In that study, age 0+ (n = 337, 34–106 mm TL) and age 1+ (n = 367, 43–111 mm TL) individuals were tagged with 225 recaptures of 180 fish up to 121 days after initial release (Able and Hales 1997). Recapture frequency was 20% for age 0+ fish released in summer and fall and 30% for age 1+ fish released in spring and summer. Almost all recaptures (99%) were made within 30 m of the release site. Of 35 fish that were recaptured twice, 46% were recaptured at the original site. All of the above indicate a high degree of habitat fidelity during the summer and fall.

Temperature and oxygen seem to be especially important components of the habitat. In intensive laboratory observations of 0+ age individuals under ambient estuarine seasonal temperatures, there was a strong response to low winter temperatures (Hales and Able 1995). When temperatures fell to 6 C, they occasionally buried in sand substrate. Below 4 C, they ceased feeding. At 2 to 3 C, mortality increased sharply, and no individuals survived the entire winter.

The young-of-the-year are migratory during some portions of the first year. They migrate out of the estuary and away from inner continental shelf nursery areas during the fall as water temperatures drop (Able et al. 1995a; Able and Hales 1997). They move into deeper water on the continental shelf for the winter and then back into estuaries and the inner continental shelf by the following spring and summer (Able et al. 1995a). It would be useful to know if they return to the natal estuary.

CHAPTER 47

Pomatomus saltatrix (Linnaeus)
Bluefish

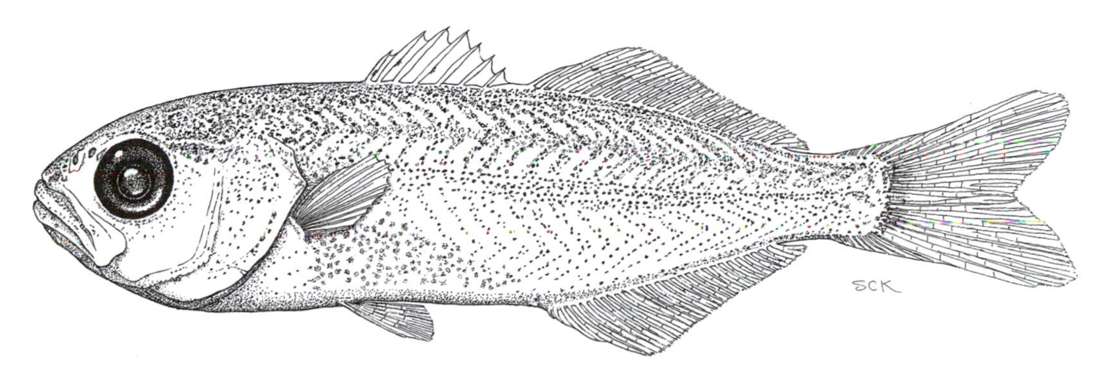

47.1 *Pomatomus saltatrix* juvenile, 24.3 mm SL. Collected August 4, 1993, 7-m seine, Great Bay, New Jersey. ANSP 175233. Illustrated by Susan Kaiser.

DISTRIBUTION

In the western Atlantic, *Pomatomus saltatrix,* a coastal, pelagic species, ranges from Nova Scotia and Bermuda to Argentina, but it is rare between southern Florida and northern South America (Robins and Ray 1986). They undertake seasonal migrations, traveling in schools of like-sized individuals in temperate and semitropical ocean waters worldwide (Briggs 1960; Juanes et al. 1996). They migrate into the Middle Atlantic Bight during the spring and south and/or offshore during fall. In our study area, they occur in large bays and estuaries as well as across the entire continental shelf. Juveniles have been reported from all estuaries surveyed within the Middle Atlantic Bight, but eggs are rare and larvae have been recorded from only a few (table 4.2).

REPRODUCTION

A seminal study, based largely on the distribution of eggs and larvae, concluded that there were two discrete spawning events in western Atlantic *P. saltatrix.* The first occurs from March to May near the edge of the continental shelf of the South Atlantic Bight. The second occurs between June and August in the Middle Atlantic Bight (Kendall and Walford 1979). Recent studies have reexamined this conclusion and refined our knowledge of a complex reproductive pattern, supporting the concept of a single, migratory spawning stock (Hare and Cowen 1993; Smith et al. 1994).

Sexual maturity and gonad ripening occur in early spring off Florida, early summer off North Carolina, and late summer off New York (Hare and Cowen 1993). In the New York Bight, gonadosomatic studies indicate both sexes are ripe or ripening between June and September with a strong peak in July (Chiarella and Conover 1990). Larvae reoccur in the South Atlantic Bight in the fall (Collins and Stender 1987), and there are also indications that gonads reach a second peak in ripeness in fishes off Florida in September.

Evidence from an intensive study of the distribution of eggs and larvae in the Middle Atlantic Bight supports the suggestion that the spawning season is a single, protracted one that begins in May off North Carolina (Smith et al. 1994; Berrien and Sibunka in press). Larval occurrences then progress northward as far as Cape Cod where they peak in July. Eggs (Berrien and Sibunka in press) and larvae (fig. 47.2) occur in central Middle Atlantic Bight continental shelf waters between early July and mid-August. During some years, they are concentrated over the inner shelf; in other years, they

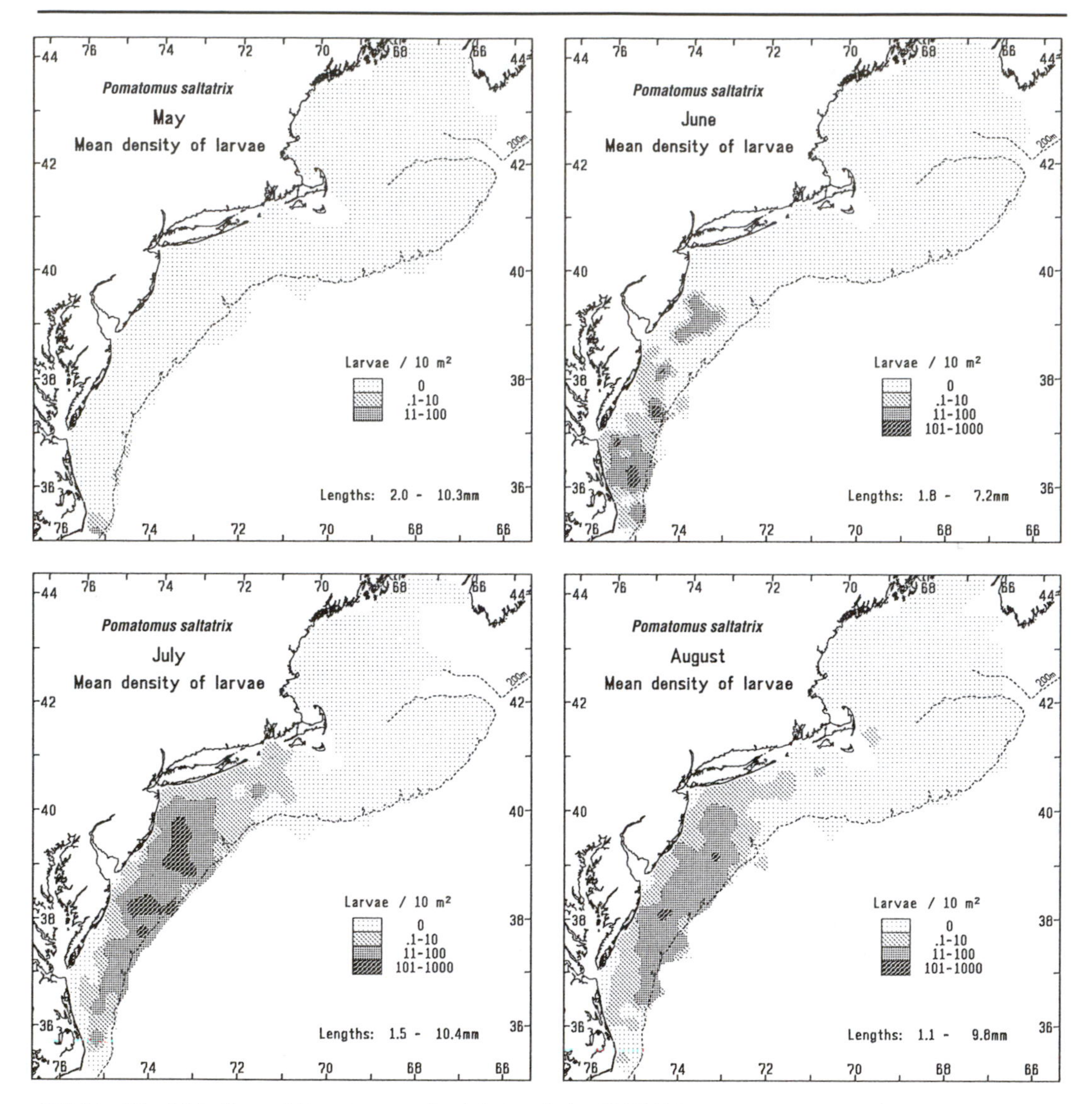

47.2 Monthly distributions of *Pomatomus saltatrix* larvae during MARMAP surveys 1977–1987. Size ranges of larvae collected are indicated in the lower right corner of each panel (after Smith et al. 1994).

are more widely distributed across the entire shelf (Morse et al. 1987; Smith et al. 1994; Berrien and Sibunka in press).

DESCRIPTION

The pelagic eggs are spherical and range from 0.90 to 1.20 mm in diameter. The yolk is homogeneous, the perivitelline space is narrow, and there is a single oil globule 0.26 to 0.29 mm in diameter. Larvae hatch at 2.0 to 2.4 mm with unpigmented eyes and a nonfunctional mouth. Characteristic pigment includes parallel rows of melanophores aligned with the dorsal fin base, anal fin base, and midline of the body. All fin rays are ossified by 14.0 mm. See Deuel et al. (1966) and Norcross et al. (1974) for further details of larval development. Juveniles have a usual fish shape without unusual features (fig. 47.1). The caudal fin is forked, and the body is somewhat compressed laterally, with a silvery unpatterned color. The mouth is large and oblique. All fin spines are strong. Two distinct dorsal fins touch at their bases. The second dorsal fin base is about the same length as the anal fin base. Vert: 26; D: IX, 24–25; A: III, 26–28; Pect: 18; Plv: I, 5.

THE FIRST YEAR

Embryonic and larval development occur in the upper levels (primarily between the surface and 15 m) of the water column in oceanic waters (Kendall and Naplin 1981). After completion of fin ray development, they go through a pelagic-juvenile stage (fig. 47.1) characterized by a silvery, laterally compressed body (Deuel et al. 1966; Norcross et al. 1974). This transition occurs at ages of 18 to 25 days and at sizes between 10 and 12 mm SL (Hare and Cowen 1994). Scales begin to form at about 12 mm on the posterior part of the lateral line region, then proceed forward, until the head is completely scaled at about 37 mm (Silverman 1975).

During the spring, oceanographic mechanisms providing for the transport of these developing larvae and pelagic juveniles from southern waters have been discussed in several papers (Kendall and Walford 1979; Cowen et al. 1993; Hare and Cowen 1993). A recent study has described in detail each step of this transport (Hare and Cowen 1996). After spawning on the outer shelf of the South Atlantic Bight from March through May (Collins and Stender 1987), some larvae are retained there and enter estuaries south of Cape Hatteras (McBride et al. 1993), while others are entrained northeastward by the Gulf Stream. The most developed of these, after entering slope waters off the Middle Atlantic Bight (J. A. Hare, M. P. Fahay, and R. K. Cowen in prep.), actively swim across the Slope Sea until they reach continental shelf waters. Less developed larvae are entrained in warm-water filaments or streamers associated with warm-core rings, and this also results in their introduction into continental shelf waters. Evidence accrued from neuston sampling indicates that pelagic juveniles mass in outer continental shelf waters in the central part of the Middle Atlantic Bight before actively crossing the shelf toward nursery areas in bays and estuaries of the region (Shima 1989; Cowen et al. 1993) after the shelf/slope temperature front dissipates in late spring or early summer (Hare and Cowen 1996).

In the New York Bight, there are two episodes of ingress by pelagic juveniles into estuarine habitats (table 47.1). The first occurs between May 28 and June 15 (Cowen et al. 1993). Studies of the otoliths of these juveniles indicate they are at least 60 days old when they first enter estuaries (McBride and Conover 1991). Therefore, they are presumably the result of a spawning event occurring during late March or early April in waters of the South Atlantic Bight. This initial ingress apparently occurs abruptly, with most of these pelagic juveniles appearing in estuaries simultaneously (Nyman and Conover 1988). Coincident with this migration, diets change from zooplankton to young-of-the-year fishes of several species (Friedland and Haas 1988; Marks and Conover 1993; Juanes et al. 1993, 1994). The second pulse of pelagic-juveniles occurs during mid- to late August (McBride and Conover 1991). Presumably, these are the result of summer spawning in the New York Bight. Some authors have

Table 47.1. Differences between Spring- and Summer-Spawned Cohorts of *Pomatomus saltatrix* Recruiting into Estuaries of the New York Bight, 1987–1988

	Spring-spawned	Summer-spawned
When spawned	March/April	July
Where spawned	South Atlantic Bight	Middle Atlantic Bight
Age at recruitment to estuaries	60–76 days	33–47 days
Size at recruitment to estuaries	< 100 mm (mean = 60 mm)	< 75 mm (mean = 46 mm)
Date of ingress to estuaries	Late May/mid-June	Mid–late August
Growth rate before estuarine ingress	0.71–0.85 mm/day	0.91–1.2 mm/day
Growth rate during first summer	1.17–1.35 mm/day	0.57–1.47 mm/day
Approximate size on October 1	180–200 mm FL	70–100 mm FL

SOURCE: Data from McBride and Conover 1991.

Table 47.2. Size at Ingress into Estuaries for Spring- and Summer-Spawned Cohorts of *Pomatomus saltatrix* from a Variety of Studies in the Middle Atlantic Bight

Range	Length	Mode	Location	Date	Source
			Spring-spawned cohorts		
21–60	mm FL	~40	North Carolina estuaries	Mar/Apr 1979–1990	McBride et al. 1993
24–36	—	—	Middle Atlantic Bight	—	Wilk 1977
65–75	mm TL	70	Horseshoe Cove, NJ	1983/1984	Friedland and Haas 1988
< 110	mm TL	—	Hudson River	1992/1993	Juanes et al. 1994
78–?	mm TL	101	Hudson River	1989	Juanes et al. 1993
40–60	mm FL	45?	South shore Long Island	1985/1986	Nyman & Conover 1988
45–65	mm FL	50–60	South shore Long Island & New Jersey	1987/1988	McBride & Conover 1991 (lean years for bluefish)
78–82	mm TL	80	South shore Long Island & New Jersey	1981	McBride & Conover 1991 (lean years for bluefish)
<90	mm TL	—	Great South Bay, NY	1988/1989	Juanes et al. 1994
60–140	mm FL	100	Narragansett Bay, RI	July 1988	McBride et al. 1995
50–140	mm FL	70	Narragansett Bay, RI	July 1989	McBride et al. 1995
50–100	mm FL	90	Narragansett Bay, RI	July 1990	McBride et al. 1995
30–110	mm FL	80	Narragansett Bay, RI	July 1991	McBride et al. 1995
			Summer-spawned cohorts		
30–37	—	—	New Jersey	Mid/late Aug	Bean 1887; Tracy 1910
47-?	mm TL	—	Hudson River	Aug 1989	Juanes et al. 1993
<75	mm FL	—	Long Island & New Jersey	Aug 1987/1988	McBride and Conover 1991
40–80	mm FL	40	Narragansett Bay	Aug 1988	McBride et al. 1995

suggested that this cohort (or a major part of it) undergoes juvenile development in inner continental shelf waters rather than entering estuaries (Kendall and Walford 1979). In the late 1950s, the contribution of these two cohorts to the overall population was observed to be approximately equal. However, studies in the New York Bight during the 1980s demonstrated that the spring-spawned cohort strongly dominated (e.g., Chiarella and Conover 1990).

Growth rates vary between life-history stages and cohorts. Before larvae enter estuaries, growth ranges from 0.3 to 0.8 mm per day (Deuel et al. 1966; Hare and Cowen 1994). In comparisons between spring- and summer-spawned cohorts from several years, a summer (1988) group was found to be the slowest growing through the juvenile stage (J. Hare pers. comm.). A separate study, comparing different groups, found a summer cohort to be the fastest growing among a different set of cohorts (McBride and Conover 1991). Juveniles entering estuaries during the first (spring) recruitment range from about 30 to 70 mm in length (Juanes et al. 1996), but this size varies somewhat between years and study sites throughout the Middle Atlantic Bight (table 47.2). Sizes of fish from the summer spawn are somewhat smaller when they appear in estuaries.

Length-frequency histograms resulting from recent sampling in the Great Bay–Little Egg Harbor and Hudson River study areas clearly demonstrate the ingress of both cohorts (fig. 47.3). Growth rates accelerate after the shift from zooplankton to fish prey in these entering juveniles, with much of the increase expressed in weight (Friedland and Haas 1988). Juveniles grow at the rate of 0.9 to 2.1 mm per day (McBride and Conover 1991; McBride et al. 1995; Juanes et al. 1993, 1996). Members of the spring-spawned cohort are much larger than those of the summer-spawned cohort at the

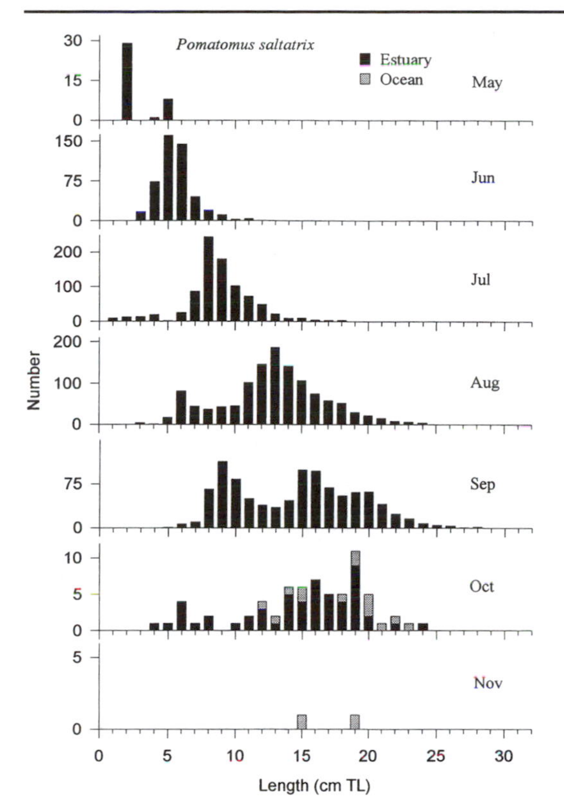

47.3 Monthly length frequencies of *Pomatomus saltatrix* young-of-the-year collected from the Hudson River Estuary, Great Bay and Little Egg Inlet. Sources: Thomas et al. 1974; Kahnle 1987; RUMFS: otter trawl (n = 66); 1-m beam trawl (n = 2); 1-m plankton net (n = 35); experimental trap (n = 3); LEO-15 2-m beam trawl (n = 2); seine (n = 24); night-light (n = 32); gear comparison (n = 1); weir (n = 1,810).

onset of fall migration (McBride and Conover 1991) and continue to be larger at ages 1 through 4 (Lassiter 1962). Size modes of the first (spring) cohort reach about 13 to 14 cm by late August in Long Island waters (Nyman and Conover 1988) or 15 to 20 cm in New England (McBride et al. 1995). By September in Great Bay–Little Egg Harbor, these cohorts are discernible as size modes at about 9 and 15 cm TL.

Little information is available on specific habitats where *P. saltatrix* young-of-the-year occur. Most studies of growth and feeding habits, however, have made collections with beach seines, indicating occurrence in relatively shallow estuarine habitats. Other studies have used small otter trawls in slightly deeper water to advantage. There are some indications that young-of-the-year undertake diel and tidal movements between marsh creeks and open bay habitats (Rountree and Able 1993).

Very little is known about fall migration. Our data indicate that young-of-the-year leave estuaries and begin to be collected in the ocean in October (fig. 47.3). After emigration from estuaries, the limited evidence suggests they move south in coastal waters. Year-round occurrences of all life-history stages have been reported from the South Atlantic Bight (Anderson 1968), suggesting a north–south seasonal migration. Tagging studies also support a north–south coastal migration (Miller 1969), but there are also suggestions that certain size classes (at least) overwinter in very deep waters near the edge of the continental shelf in the Middle Atlantic Bight (Hamer 1959; Miller 1969; Wilk 1982), which implies an inshore–offshore seasonal movement. During most years, the spring-spawned cohort dominates in these emigrating young-of-the-year. One theory, which has not been supported or refuted, says that young-of-the-year spawned in summer in the Middle Atlantic Bight: (1) use continental shelf waters as a nursery; (2) migrate in the fall to offshore waters south of Cape Hatteras where they spend the winter; and (3) move into estuaries in North Carolina the following spring (Wilk 1982). Because they leave the Middle Atlantic Bight at lengths of 50 to 75 mm and are reportedly 230 mm FL (mode) the following spring and only 290 to 300 mm the following fall (Wilk 1982), they would be required to achieve fast growth over the winter and very little growth during their summer in North Carolina. Both of these requirements are unsupported and unlikely, but details of winter distributions of all life-history stages remain enigmatic.

CHAPTER 48

Caranx hippos (Linnaeus) Crevalle Jack

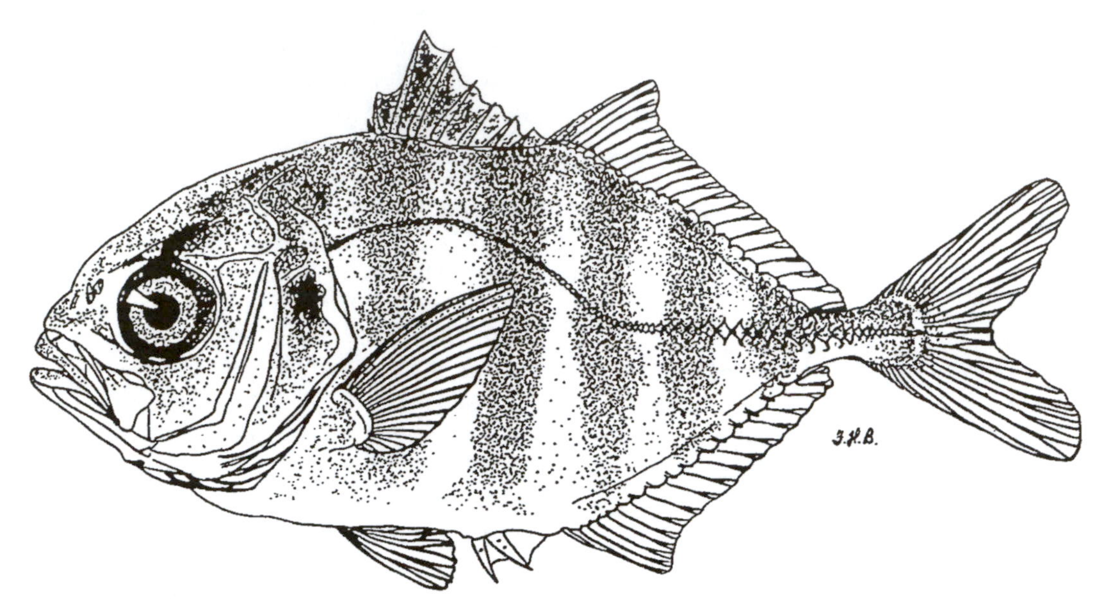

48.1 *Caranx hippos* juvenile, 32.6 mm SL. After Berry 1959.

DISTRIBUTION

Caranx hippos (fig. 48.1) is found from Nova Scotia to Florida and throughout the Gulf of Mexico and south to Uruguay (Berry 1959; Scott and Scott 1988). In the South Atlantic Bight, adults occur in inshore waters during the summer and fall and then move to offshore waters near the Gulf Stream in the winter. The young are common in inshore and estuarine waters and up into freshwater. In the Middle Atlantic Bight, juveniles have been found in a number of estuaries (table 4.2). McBride (1995) provides a complete listing of all locations including museum material.

REPRODUCTION

Spawning occurs in offshore, subtropical waters based on the occurrence of small young (18–21 mm), but actual locations are unknown (Berry 1959).

DESCRIPTION

In general, carrangid eggs are small (0.7–1.5 mm in diameter), spherical, and pelagic (Laroche et al. 1984). For this species from the Indian Ocean (Chacko 1949; Subrahmanyam 1964), eggs are spherical, 0.7 to 0.9 mm in diameter, with segmented yolk and a single yellowish oil globule (approximately 0.2 mm in diameter). The larvae hatch at 1.7 mm. Little else is known of development. Juveniles are laterally compressed with scutes along the posterior portion of the lateral line (fig. 48.1). Pigmentation includes five to seven vertical bars on sides of body, and the posteriormost three extend to the base of the anal fin in juveniles that are 15 to 60 mm. Pigment covers the entire first dorsal fin in juveniles smaller than 45 mm (Berry 1959). Vert: 24; D: VIII–IX, 15–21; A: II–III, 16–17; Pect: 19–20; Plv: I, 5.

THE FIRST YEAR

Pelagic juveniles (12.2–22.5 mm) have been collected from surface waters of the South Atlantic Bight from May through August (Fahay 1975). They are reported in Delaware estuaries from June to October at sizes greater than 20 mm (de Sylva et al. 1962; Wang and Kernehan 1979) and from July to November in the Hudson River and Haverstraw Bay and Ja-

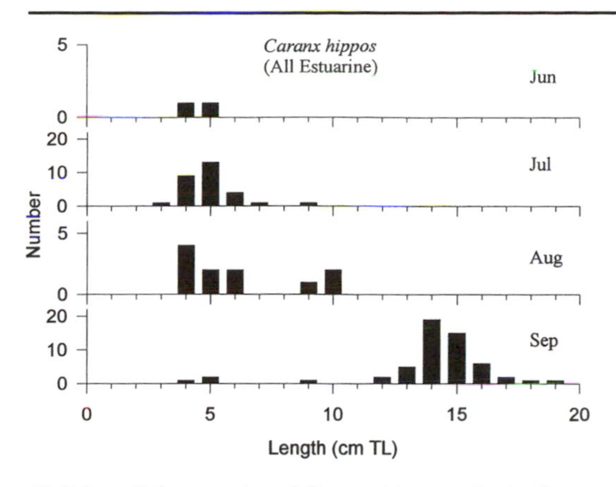

48.2 Length frequencies of *Caranx hippos* collected from the Great Bay–Little Egg Harbor study area. Sources: RUMFS: otter trawl (n = 12); gear comparison (n = 1); killitrap (n = 1); night-light (n = 1); weir (n = 82).

maica Bay in every year sampled from 1986 to 1993 (McBride 1995). In the Hudson River and Jamaica Bay and Haverstraw Bay, as well as Delaware Bay, small individuals occurred into the late summer (September–October), suggesting either a protracted spawning period or long delivery time to the estuary. In the Hudson River and Jamaica Bay, they were approximately 25 to 75 mm FL in July and ranged from 35 to 175 mm FL by September. Based on modal length-frequency progression in the study area, they grew at approximately 1.3 mm per day (fig. 48.2).

The young-of-the-year seem to occupy a wide range of estuarine environmental conditions. In the Hudson River and Jamaica Bay, they were collected across a wide range of bottom temperatures (9–30 C), salinities (1.3–32.0 ppt), and dissolved oxygen (2.0–13.6 ppm) (McBride 1995). In Delaware Bay, the young-of-the-year were found across a wide range of salinities extending up into brackish waters (Wang and Kernehan 1979). Details of their distribution in specific habitat types are poorly known, although they have been collected in marsh creeks (Rountree and Able 1993) and along shallow estuarine shorelines (Able et al. 1996b) in Great Bay–Little Egg Harbor.

They apparently leave estuaries, at least in the northern Middle Atlantic Bight, in the fall. In Jamaica Bay and Haverstraw Bay, both the number and the average size of fish declined by October and November. They probably cannot tolerate winter temperatures in Middle Atlantic Bight estuaries. Mortalities at 7.4 to 9.0 C in October have been reported from Massachusetts (Hoff 1971) and at 6 C in Florida (Gilmore et al. 1978). Additional details of the sources of Middle Atlantic Bight young-of-the-year and their ultimate fate after the summer would improve our understanding of how these, and other species that presumably spawn in the South Atlantic Bight, depend on Middle Atlantic Bight estuaries as nurseries.

CHAPTER 49

Lutjanus griseus (Linnaeus)
Gray Snapper

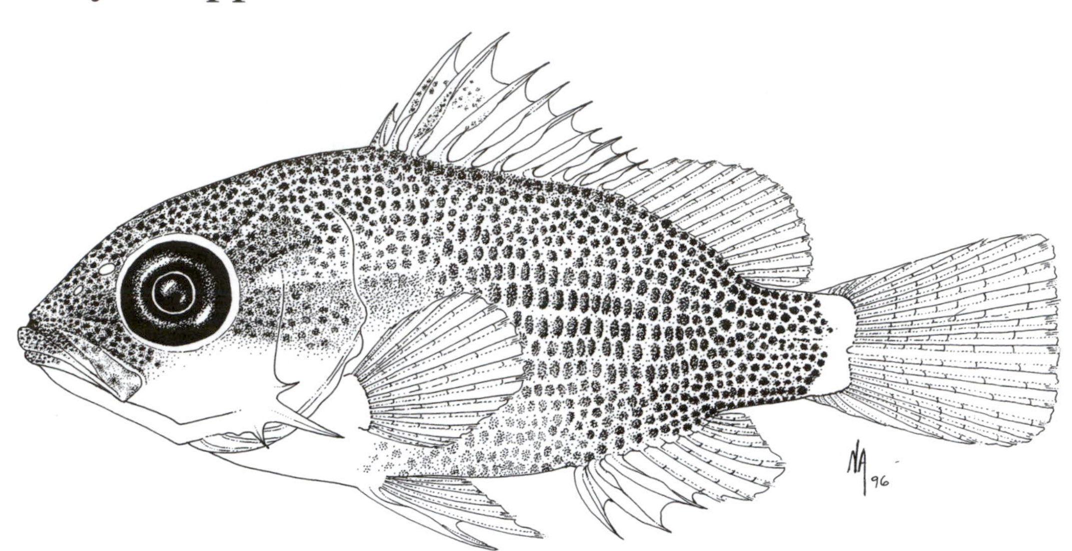

49.1 *Lutjanus griseus* juvenile, 15.8 mm TL. Collected August 27, 1973, Absecon Inlet, New Jersey. ANSP 130329. Illustrated by Nancy Arthur.

DISTRIBUTION

Lutjanus griseus (fig. 49.1) occurs from Massachusetts, Bermuda, and northern Gulf of Mexico to southeastern Brazil (Hildebrand and Schroeder 1928; Hoese and Moore 1977). Over this range, it has been found in a variety of habitats including shallow, grassy flats, deep reefs, wharves, and pilings. In the Middle Atlantic Bight, the juveniles are reported sporadically from estuaries from Long Island and south (table 4.2).

REPRODUCTION

Spawning occurs, with eggs produced in multiple batches, in offshore waters of the South Atlantic Bight from June through August (Starck 1964, 1971). There is no evidence that spawning occurs in the Middle Atlantic Bight.

DESCRIPTION

The pelagic eggs are small (0.70–0.85 mm in diameter), with a clear and homogeneous yolk and a single oil globule (0.12–0.18 mm in diameter) (Richards and Saksena 1980). The larvae have melanophores along the postanal ventral body margin, on the developing pelvic fin, and in a characteristic pattern on the head and opercle. Flexion occurs between 4 and 6 mm. Diagnostic serrations occur on the pelvic fin spine and first six dorsal spines (Richards and Saksena 1980). In juveniles (fig. 49.1), a blue line extends from the snout to below the eye. The body is heavily pigmented with dark bars across the eyes and dark horizontal stripes along the sides of the body. The mouth is fairly large with the maxilla reaching nearly the midpoint of the eye or beyond. There is a single continuous dorsal fin, with at most a shallow notch between the spines and rays. Vert: 24; D: X, 14; A: III, 7–9; Pect: 15–17; Plv: I, 5.

THE FIRST YEAR

Hatching occurs after 17 to 20 hours at 27 to 30 C. Settlement occurs at 10 to 15 mm SL (Richards and Saksena 1980). Little is known of its distribution or ecology in the Middle Atlantic Bight, probably because it is an infrequent visitor. Presumed young-of-the-year

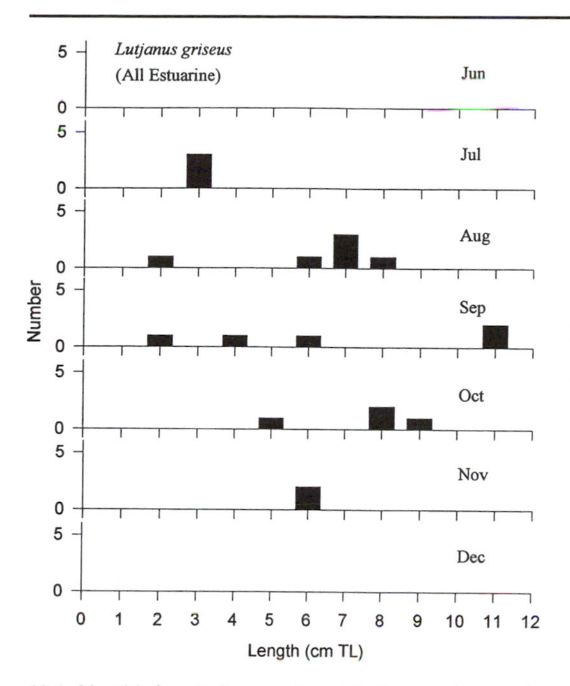

49.2 Monthly length frequencies of *Lutjanus griseus* collected from the Great Bay–Little Egg Harbor study area. Sources: RUMFS: otter trawl (n = 6); 1-m beam trawl (n = 3); 1-m plankton net (n = 2); gear comparison (n = 3); killitrap (n = 5); seine (n = 1).

(105–111 mm) are known from Chesapeake Bay as strays (Hildebrand and Schroeder 1928), and the same applies to Delaware Bay, where individuals 14 to 72 mm have been collected (de Sylva et al. 1962; Smith 1971; Wang and Kernehan 1979). A similar size range has been reported in New Jersey (14–39 mm) (Milstein and Thomas 1976b). More recent collections of presumed young-of-the-year in the Great Bay–Little Egg Harbor area occurred from July to November and ranged from 20 to 110 mm TL (fig. 49.2).

CHAPTER 50

Stenotomus chrysops (Linnaeus)
Scup

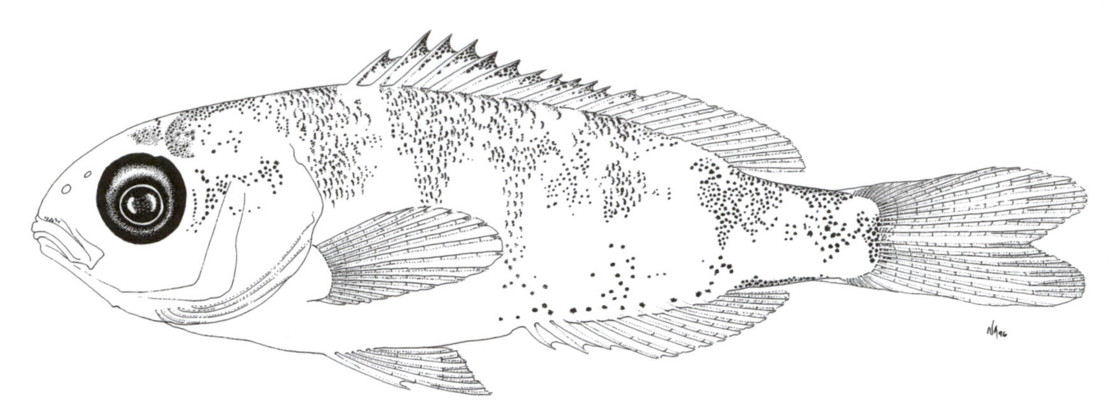

50.1 *Stenotomus chrysops* juvenile, 26.0 mm TL. Collected August 14, 1929, off Broadkill Beach, Delaware. ANSP 128648. Illustrated by Nancy Arthur.

DISTRIBUTION

Stenotomus chrysops (fig. 50.1) is found between Nova Scotia and South Carolina but is uncommon north of Cape Cod (Hildebrand and Schroeder 1928; Bigelow and Schroeder 1953; Scott and Scott 1988). It is more abundant in larger and deeper estuaries, such as Delaware Bay and Narragansett Bay, than in smaller ones, such as the Lower Hudson–Raritan Estuary and Great Bay (Berg and Levinton 1985). Tagging studies and meristic analyses suggest that two stocks occur within the Middle Atlantic Bight, one during summer in southern New England, and the other off New Jersey and the southern part of the bight (Hamer 1970). However, recent assessments have concluded that the two-stock concept is based on relatively weak data, and this species is presently managed as a single stock (G. Shepherd pers. comm.).

In the northern Middle Atlantic Bight, schools move inshore during April and May and spend the summer in bays and coastal waters within 10 km of the coast, where they prefer hard bottoms and structured habitats (Perlmutter 1939; Bigelow and Schroeder 1953; Wheatland 1956). Fish in the central part of the Middle Atlantic Bight move inshore during early May (Morse 1982) and June (Eklund and Targett 1991) beginning with the larger size classes. In this part of the bight, fishes 4 years and older tend to remain in coastal ocean waters or near mouths of larger bays and tend not to penetrate estuaries. Younger fish are found inside bays and sounds but do not penetrate to low-salinity areas (Morse 1982). Similar observations have been made near the mouth of Delaware Bay (de Sylva et al. 1962). Juveniles have been reported from all estuaries between Buzzards Bay and Chesapeake Bay (table 4.2), but eggs and larvae have only been found in those east and north of the Hudson River estuary. In the Great Bay–Little Egg Harbor study area, a single larva was collected in a 6-year weekly plankton-sampling program (Witting 1995; D. A. Witting, K. W. Able and M. P. Fahay in prep.).

REPRODUCTION

Sexual maturity is reached at age 2 (Nichols and Breder 1927; Finkelstein 1969b; Morse 1982). The precise location of spawning, and of concentrations of eggs and larvae, has remained enigmatic. Spawning appears to take place in larger bodies of water, such as Long Island Sound (especially the more-saline eastern part; Richards 1959), Sandy Hook Bay (possibly near the bay/ocean interface), Narragansett Bay, and coastal ocean waters between those systems (Morse 1982). Gardiner's and Peconic Bays have also been characterized as important spawning areas (Perlmutter 1939). However, eggs and larvae are rare or absent in other well-studied systems such as Block Island Sound, Great South Bay, the Hudson River estuary, and Great Bay (Merriman and Sclar 1952; Swiecicki and Tatham 1977; Kahnle and Hattala 1988; Monteleone 1992; Witting 1995; D. A. Witting, K. W. Able and M. P. Fahay in prep.). It has been suggested that spawning occurs over sandy and weed-covered bottoms (Morse 1978). Spawning presumably does not occur over the continental shelf, because during 11 years (1977–1987), the MARMAP surveys (table 3.3) collected only 14 larvae on three sampling stations near the mouth of Narragansett Bay. Spawning occurs between May and August and peaks during June, as shown in several studies between Sandy Hook, New Jersey and Vineyard Sound, Massachusetts (Nichols and Breder 1927; Perlmutter 1939; Wheatland 1956; Finkelstein 1969a; Sisson 1974; O'Brien et al. 1993). All of these studies indicate a single peak in abundance of eggs and larvae; thus, there is apparently only one spawning per season. Most *S. chrysops* occupying Sandy Hook Bay during summer are age 1 (Wilk and Silverman 1976b, fig. 17) and presumably are too young to spawn. Recent ichthyoplankton studies restricted to coastal or estuarine waters have not added much to our knowledge. For example, in Narragansett Bay, thorough sampling failed to collect larvae, although eggs (possibly confused with those of *Cynoscion* sp.) were fairly common (Bourne and Govoni 1988). It is interesting that the most recent study we can find reporting the presence of eggs and/or larvae is that of Sisson (1974).

DESCRIPTION

The transparent eggs are spherical, ranging from 0.8 to 1.0 mm in diameter. There is a single, pigmented oil globule, 0.17 to 0.21 mm in diameter. Larvae hatch at about 2.0 mm with undeveloped mouths and unpigmented eyes. Later larvae have snout-anus lengths less than 50% TL and a prominent, ventral row of melanophores from the anus to the caudal fin. Fin rays begin forming at about 4 mm and complete their formation at about 13 mm SL. See Griswold and McKenney (1984) and Fahay (1983) for more details of early development. Early juveniles have a rounded snout and strong, sharp dorsal fin spines (fig. 50.1). The dorsal fin is continuous, with a slight notch between spines and rays. The anal fin base is shorter than the dorsal fin base. The body is generally silvery (with vague, vertical bars) and is laterally compressed. Vert: 24; D: XII, 12; A: III, 11–12; Plv: I, 5.

THE FIRST YEAR

The embryos incubate for 70 to 75 hours at 18 C and 44 to 54 hours at 21 C. Larvae hatch at about 2.0 mm, and their yolk is absorbed after 2 or 3 days when active feeding begins (Griswold and McKenney 1984). Full complements of fin rays are formed by about 25 mm. The deep-bodied, sharp-snouted adult morphology is acquired by 40 to 60 mm (Nichols and Breder 1927). At about 15 to 30 mm, early juveniles descend to the bottom (Lux and Nichy 1971). It is unknown whether pelagic larvae and early demersal stages occur in different areas, necessitating a migration to nursery habitats. The smallest young-of-the-year appear in estuaries during June, and they are only found there through September (fig. 50.2).

Growth is relatively slow during the first year, based on combined collections made in the Great Bay–Little Egg Harbor estuary, Long Island embayments, and waters within 24 km of shore in the central part of the Middle Atlantic Bight (fig. 50.3). By October, most young-of-the-year are 8 to 11 cm TL. These sizes are somewhat larger than those reported for New

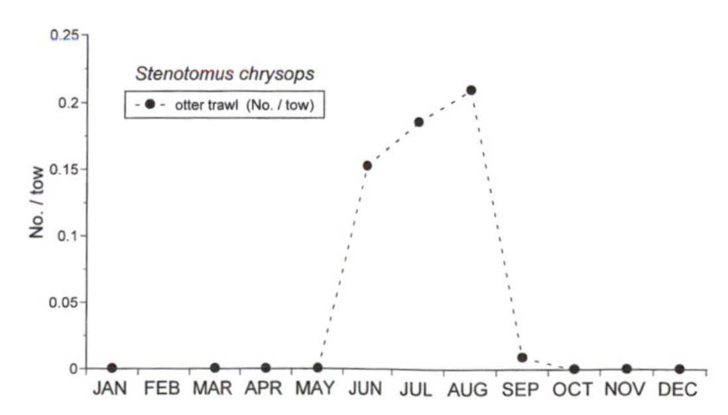

50.2 Monthly occurrences of *Stenotomus chrysops* young-of-the-year based on otter trawl collections in Great Bay–Little Egg Harbor 1988–1989 ($n = 70$).

England young-of-the-year (Lux and Nichy 1971). Growth over the winter continues to be slow, and 0+ fishes returning in the spring are 11 to 15 cm TL (Wilk et al. 1992). The average length of New England fish at age 1 is 10.6 cm (Morse 1982).

Little information is available regarding specific habitats favored by juveniles. In Great Bay–Little Egg Harbor sampling covering a variety of habitats, they occurred over a variety of substrates but were most abundant over a bottom with no structure, and in depths ranging from 3 to 5 m (fig. 50.4). They have been reported as "abundant" in the highest salinity portion (e.g., >25.0 ppt) of the Hudson River–Raritan Bay estuary, and "rare" in the mixed-salinity zone (0.5–25.0 ppt) (Stone et al. 1994). In the same report, juveniles are regarded as "highly abundant" in the higher salinity portions of Long Island Sound. Details of other physical attributes of these habitats are lacking, however.

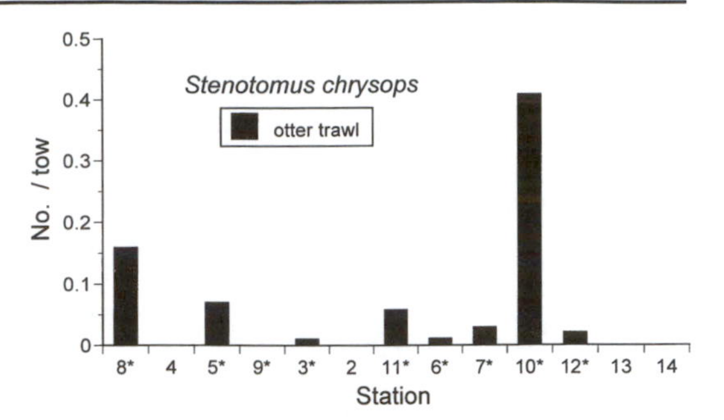

50.4 Distribution of *Stenotomus chrysops* young-of-the-year by habitat type based on otter trawls (n = 71) deployed on regular stations 2–14 in Great Bay and Little Egg Harbor. Both otter trawl and beam trawl used on stations designated by asterisk; otherwise, only the otter trawl was used. See figure 2.3 for station locations; table 2.1 for specific habitat characteristics by station.

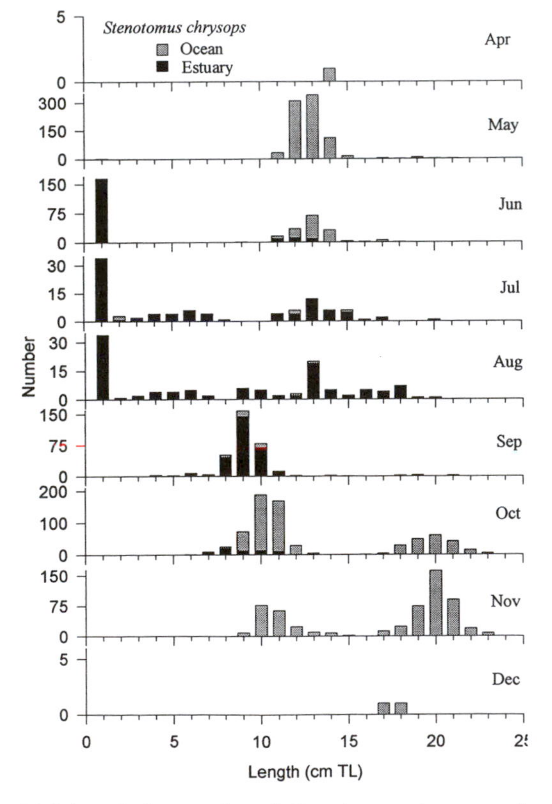

50.3 Length frequencies of *Stenotomus chrysops* collected from estuarine and coastal ocean waters of the central Middle Atlantic Bight. Sources: Perlmutter 1939; Wilk et al. 1992; RUMFS: otter trawl (n = 112); 1-m plankton net (n = 1); experimental trap (n = 1); LEO-15 Tucker trawl (n = 1); LEO-15 2-m beam trawl (n = 5); gear comparison (n = 24); killitrap (n = 9); weir (n = 6). (The Perlmutter fish collected in "July–August" are here arbitrarily divided between the July and August histograms.)

The emigration of young-of-the-year from estuaries and bays begins in September and lasts through November (fig. 50.3). During this period, fish 6 to 13 cm TL (along with larger size classes) are frequently collected by trawl in coastal ocean waters in our study area (Wilk et al. 1992). The fall migration proceeds southward inside the 10-m isobath and is then followed by an offshore movement (Hamer 1970). The population then spends the winter in depths between 37 and 146 m, mostly on the outer continental shelf between New Jersey and Cape Hatteras (Neville and Talbot 1964; Morse 1982; Shepherd and Terceiro 1994). No evidence suggests that young-of-the-year overwinter in areas separate from older year classes. Fish from New England also spend the winter on the outer continental shelf, but in areas slightly farther to the north and east than those from the southern Middle Atlantic Bight (Hamer 1970). Winter distributions appear to be determined largely by temperature because the species occurs mostly in bottom waters at temperatures above 7.3 C (the lower temperature tolerance limit), and the location of that isotherm can vary annually (Neville and Talbot 1964).

CHAPTER 51

Bairdiella chrysoura (Lacepède) Silver Perch

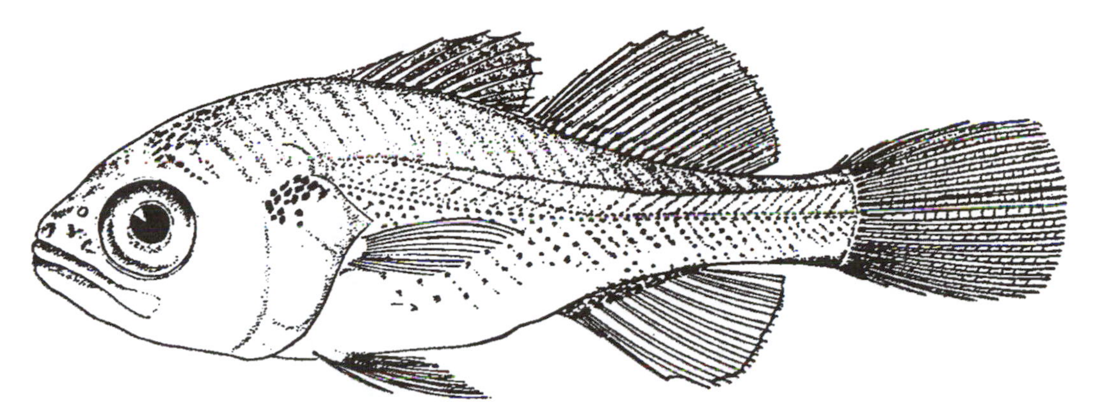

51.1 *Bairdiella chrysoura* juvenile, 30.0 mm TL. After Kuntz 1914a.

DISTRIBUTION

Bairdiella chrysoura (fig. 51.1) occurs along the East Coast of the United States from New York and south into the Gulf of Mexico (Robins and Ray 1986). Throughout its range, it is often associated with marshes and seagrass beds. It reaches its northern limit in the Middle Atlantic Bight, where early life-history stages are common in estuaries from the southern part of the bight and as far north as Great South Bay and Long Island Sound (table 4.2).

REPRODUCTION

Individuals mature in the third year (Welsh and Breder 1923). Spawning in northern New Jersey has been reported from June to August with a peak in June (Welsh and Breder 1923). In Great Bay, we have collected running ripe individuals (204–245 cm TL) in late June. In Delaware Bay, most spawning occurs in shallow areas from June through August with peaks in June and July (Thomas 1971). Spawning is reported in late spring and early summer in Chesapeake Bay (Hildebrand and Schroeder 1928). At Beaufort, North Carolina, spawning was reported to have occurred within the harbor, the estuaries, and sounds and also some distance at sea (Hildebrand and Cable 1930).

DESCRIPTION

Eggs are pelagic and spherical (0.66–0.88 mm in diameter), with one or two sparsely pigmented oil globules (0.6–0.22 mm in diameter). The chorion is thin and horny. Hatching occurs at 1.5 to 1.9 mm. Through larval development, the snout-anus length increases from 40% to 58% of the SL, body depth increases from 32% to 37% SL, and the head length increases from less than 30% to 38% SL. Spines are present on the lateral surface and margin of the preopercle. Characteristic pigment includes a swatch of internal and external melanophores from the nape through the cleithral symphysis. All fin rays are complete by 8.8 mm SL. See Kuntz (1914a) and Powles (1980) for more details of larval development. Juveniles (fig. 51.1) lack obvious diagnostic characters and are best identified by counts and shape and location of the fins. The body is somewhat compressed laterally, is silvery gray overall, and there is no prominent spot behind the gill cover. There are no barbels on the snout or lower jaw. The caudal fin is slightly rounded. The dorsal and anal fins are composed of both spines and rays. The separate dorsal fins touch at the base and the dorsal spines are at least as high as the rays. Vert: 25; D: X–XI, 19–23; A: II, 8–10; Pect: 15–17; Plv: I, 5.

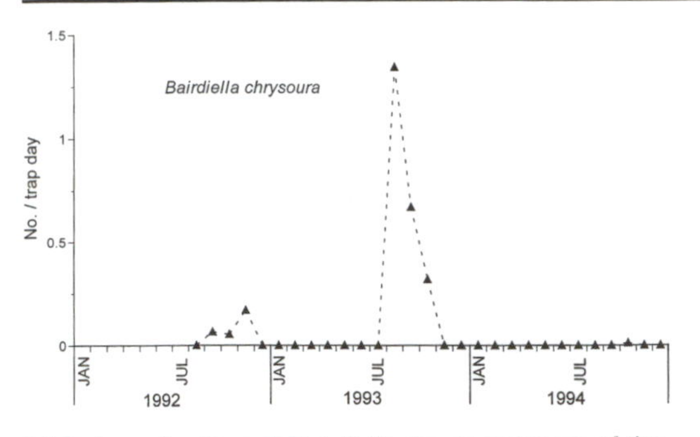

51.2 Annual pattern of *Bairdiella chrysoura* young-of-the-year in trap collections in the RUMFS boat basin in the Great Bay–Little Egg Harbor study area.

THE FIRST YEAR

Larvae were almost never collected ($n = 12$ individuals) in the Middle Atlantic Bight during the MARMAP surveys (1977–1987). The occurrence of young-of-the-year in the central part of the Middle Atlantic Bight varies annually, with large numbers in some years and very few in other years. As an example, in Great Bay–Little Egg Harbor, young-of-the-year were present in intensive trap collections in 1992, reached very high levels in 1993 and were virtually absent in 1994 (fig. 51.2). They first appear in the estuary as larvae and small juveniles in June and July at sizes of approximately 1 to 2 cm (fig. 51.3). A single size mode of young-of-the-year is evident through the summer. In most years, they reach greatest abundance in August and decline thereafter (fig. 51.4). By October, the young-of-the-year range from 4 to 15 cm TL or 6 to 14 cm as found by Welsh and Breder (1923). The modal progression in lengths in Great Bay–Little Egg Harbor indicate that they grow an average of 0.75 mm per day during this period. By the following spring, the presumed age-1 individuals are 15 to 20 cm TL (fig. 51.3), which is in close agreement with the size (13–14 cm) at the end of the first year as determined by analysis of scales (Thomas 1971).

The range of habitats occupied may vary as the result of seasonal migrations. In Great Bay–Little Egg Harbor, most of the young-of-the-year collected in otter trawls were found in eelgrass beds (fig. 51.5), but they have also been reported from marsh creeks (Rountree and Able 1992a) and estuarine shorelines (Able et al.

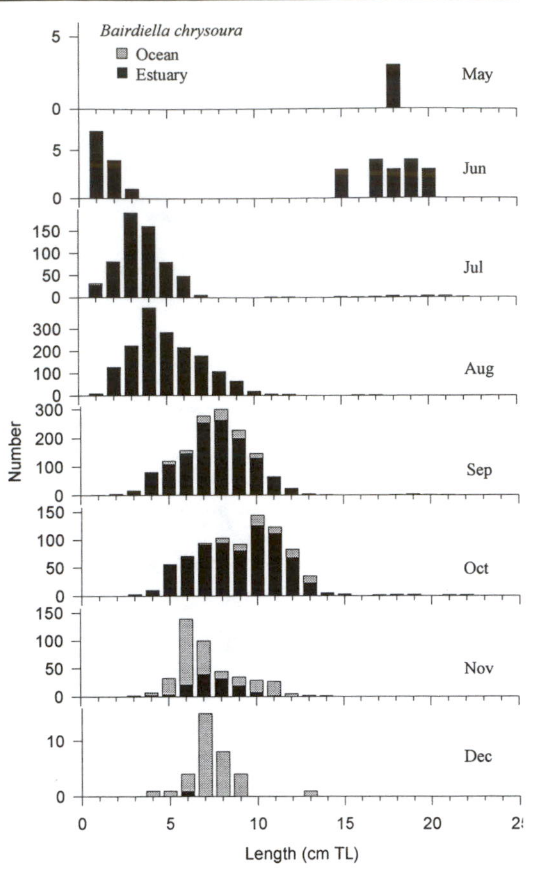

51.3 Monthly length frequencies of *Bairdiella chrysoura* in estuarine and coastal ocean waters of the central Middle Atlantic Bight. Sources: Perlmutter 1939; de Sylva et al. 1962; Clark et al. 1969; Thomas 1971; Thomas et al. 1972, 1974; Thomas and Milstein 1973. RUMFS: otter trawl ($n = 62$); 1-m beam trawl ($n = 46$); 1-m plankton net ($n = 7$); experimental trap ($n = 539$); LEO-15 Tucker trawl ($n = 3$); LEO-15 2-m beam trawl ($n = 7$); gear comparison ($n = 119$); killitrap ($n = 508$); seine ($n = 3$); weir ($n = 14$).

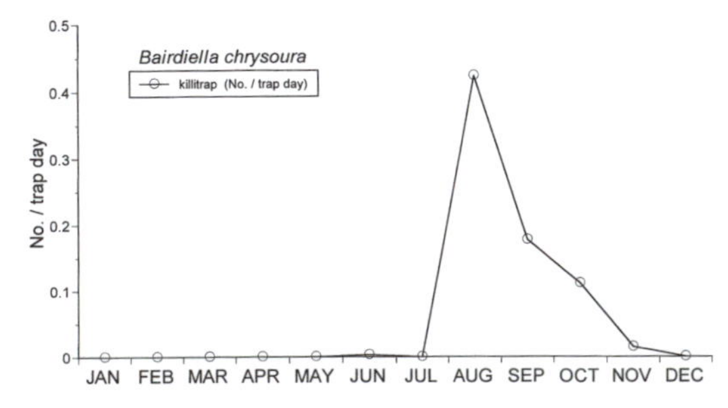

51.4 Monthly occurrences of *Bairdiella chrysoura* young-of-the-year based on killitrap collections in the Great Bay–Little Egg Harbor study area 1991–1994 ($n = 1,059$).

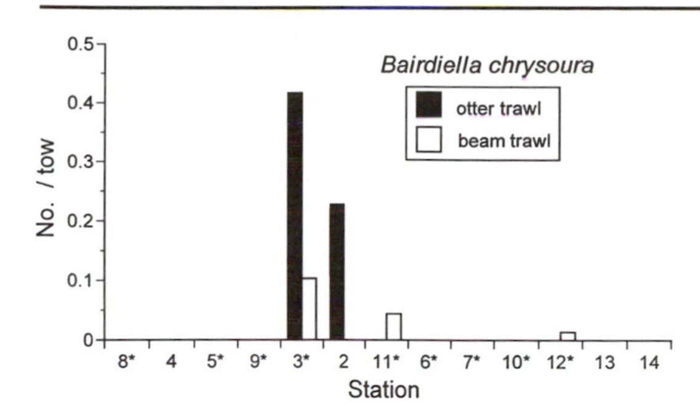

51.5 Distribution of *Bairdiella chrysoura* young-of-the-year by habitat type based on otter trawls (n = 60) and beam trawls (n = 11) deployed at regular stations 2–14 in Great Bay–Little Egg Harbor. Both otter trawl and beam trawl used on stations designated by asterisk; otherwise, only the otter trawl was used. See figure 2.3 for station locations; table 2.1 for specific habitat characteristics by station.

1996b). However, some were collected in the ocean in September and October, and by November and December, most of the collections were from the ocean (Beach Haven Ridge; fig. 51.3). The average size of young-of-the-year decreases in November and December, perhaps because the larger individuals are moving out of the estuary, into the ocean, and beyond our study area (fig. 51.3). In Delaware Bay, the young-of-the-year are most abundant in shallow waters, typically in the lower portions of creeks and ditches and along the bay in protected areas where they are found over a bottom of mud or mud and sand where detritus is abundant (Thomas 1971). They extend up the bay as far as the C&D Canal.

In Delaware Bay, most specimens leave the upper part of the bay by October and move into the deeper portions, where they have been collected in otter trawls (de Sylva et al. 1962). Movements out of estuaries and into the ocean may be precipitated by declining temperature because, at ambient winter estuarine temperatures, young-of-the-year experienced 100% mortality in a laboratory study (M. C. Curran and K. W. Able unpubl. data). However, some young-of-the-year are reported to overwinter in the deeper waters of Chesapeake Bay (Hildebrand and Schroeder 1928).

CHAPTER 52

Cynoscion regalis (Bloch and Schneider) Weakfish

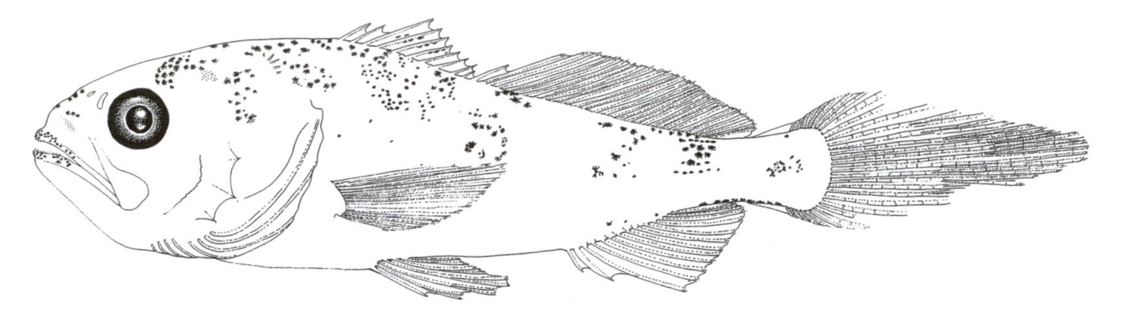

52.1 *Cynoscion regalis* juvenile, 31.0 mm TL. Collected July 16, 1996, mouth of Dennis Creek, Delaware Bay, 4.2-m otter trawl. ANSP 175234. Illustrated by Nancy Arthur.

DISTRIBUTION

Cynoscion regalis (fig. 52.1) occurs between Nova Scotia and the Cape Canaveral region of Florida. Its center of abundance lies between New York and North Carolina (Shepherd and Grimes 1983). It is a common member of the ichthyofauna in our study area, migrating northward and inshore in spring and summer, offshore and southward in fall and winter (Wilk 1976). Adults are found in a variety of estuarine habitats but seem to favor shallow waters with sand bottoms. During some years, *C. regalis* ranks among the top three species landed in New Jersey, both by commercial and sport fishermen (Thomas 1971). As with other sciaenid species, the population has declined precipitously in the last half century, and there are wide annual fluctuations in local abundances. Eggs, larvae, and juvenile stages are found in most Middle Atlantic Bight estuaries (table 4.2).

REPRODUCTION

Most spawning activity occurs close to the coast, near major inlets, or within bays or estuaries from March to August, with peak production from May through July (Welsh and Breder 1923; Hildebrand and Schroeder 1928; Pearson 1941; Daiber 1957; Massmann et al. 1958; Harmic 1958; Merriner 1976; Lowerre-Barbieri et al. 1996a; Berrien and Sibunka in press). Establishing the temporal distribution of spawning in discrete estuarine systems may be confounded by different groups (possibly composed of same-aged individuals) migrating into and out of those systems and spawning at different times of the year (see above citations). For example, there are reports of two peaks of activity in the Delaware Bay region (Daiber 1957; Goshorn and Epifanio 1991). In Chesapeake Bay, all mature fish begin spawning together, but the cessation of spawning is asynchronous, varying among adults, and the duration (or resumption after an initial peak) of spawning varies between years, contributing to this variable pattern (Lowerre-Barbieri et al. 1996a,b). Variation in spawning time has also been found to be a function of the size of adults (Shepherd and Grimes 1984). Judging from MARMAP collections from 1977 to 1987, spawning (based on egg and larval occurrences) reaches a peak in July along the central part of the Middle Atlantic Bight coast. Eleven-year average occurrences of eggs (Berrien and Sibunka in press) and occasional dense concentrations of larvae (fig. 52.2) indicate that the inner continental shelf area adjacent to central New Jersey is an important spawning area.

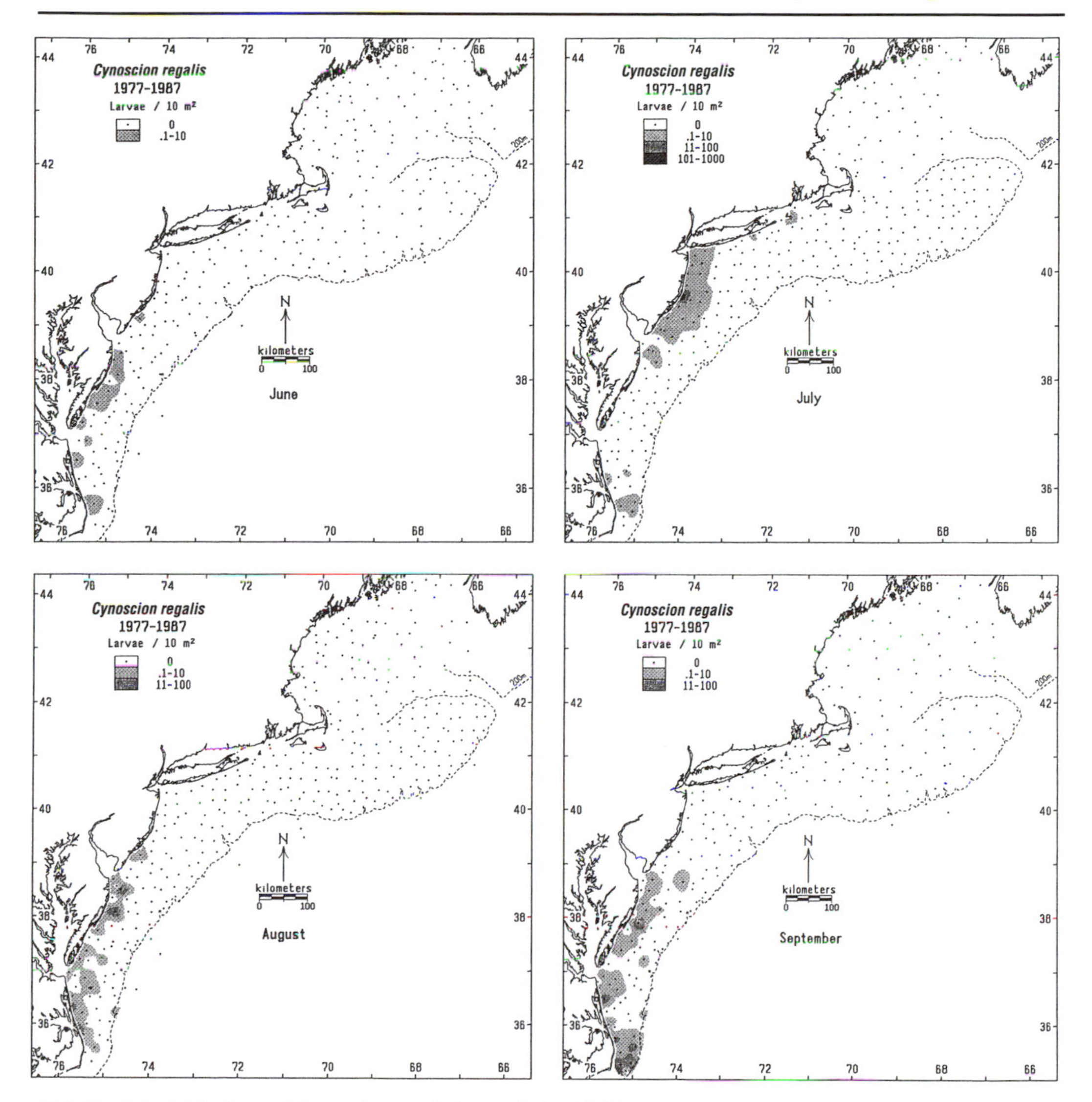

52.2 Monthly distributions of *Cynoscion regalis* larvae during MARMAP surveys 1977–1987.

DESCRIPTION

The eggs of *C. regalis* are pelagic and spherical and range from 0.75 to 0.87 mm in diameter (Fahay 1983). A Chesapeake Bay study, using genetic techniques to compare characters of sciaenid eggs, found diameters to range from 0.82 to 0.98 mm (Daniel and Graves 1994). Oil globules are sometimes multiple (up to 6) but usually single, with a diameter of 0.20 to 0.25 mm. The larvae hatch at about 1.5 to 1.7 mm SL. The snout-anus length is less than 50% SL in early larvae. A characteristic melanophore occurs in the gap between the anus and origin of the developing anal fin. Other melanophores occur along the venter, between the opercle and caudal fin, and there is internal pigment in the nape and on the anterior surface of the gut. Second dorsal and anal fin rays are complete at 5.0 mm, and all fin rays are complete by 10.0 mm. See Pearson (1941), Powles and Stender (1978), Fahay (1983), and Ditty (1989) for further details of larval development. In juveniles, the caudal fin is symmetrical and the central rays are the longest (fig. 52.1). The lower jaw tip protrudes beyond the snout tip, and there are no barbels at the tip. The second dorsal fin base is about twice the length of the anal fin base. The body has faint, dusky vertical bars against a silvery gray background. Vert: 24–25; D: X, I, 24–29; A: II, 10–12; Plv: I, 5.

THE FIRST YEAR

Optimum incubation temperatures are 18 to 24 C, and hatching occurs after 1,000 degree hours. Thus, at 20 C, hatching occurs 50 hours after fertilization (Harmic 1958). A report that newly hatched larvae sink to the bottom (Harmic 1958) is contrary to recent laboratory observations, in which larvae were found to be buoyant (J. Duffy pers. comm.). During a 2-year study in the Beach Haven Ridge area, larvae were commonly collected during July, with lingering, sparse occurrences into September. They were also commonly collected in the adjacent Great Bay–Little Egg Harbor estuary (Witting 1995; D. A. Witting, K. W. Able, and M. P. Fahay in prep.). In that study, larvae reached a peak in abundance during July, but annual abundances varied, and they were not present every year (fig. 52.3). Because spawning occurs near or within major inlets, or in estuaries themselves (Olney 1983; Bourne and Govoni 1988; Lankford and Targett 1994), most larvae are not required to undergo an active cross-shelf migration from oceanic areas through inlets. Recent studies suggest larvae utilize selective tidal stream transport in estuarine habitats such as Delaware Bay, whereby larvae migrate into the upper water column during flood tides and descend to near bottom during ebb, thus effecting retention (and net up-estuary transport) (Rowe and Epifanio 1994a,b). Older stages move upstream to nurseries from spawning sites in higher salinities. It has been assumed that these young-of-the-year use a net up-estuary flow in deeper layers of channels to do this (Thomas 1971). Other authors have remarked that young-of-the-year are more common in channel habitats than in adjacent grass beds, at least in lower bay (higher-salinity) areas (Olney and Boehlert 1988), thus substantiating the possible use of channels for upstream migration.

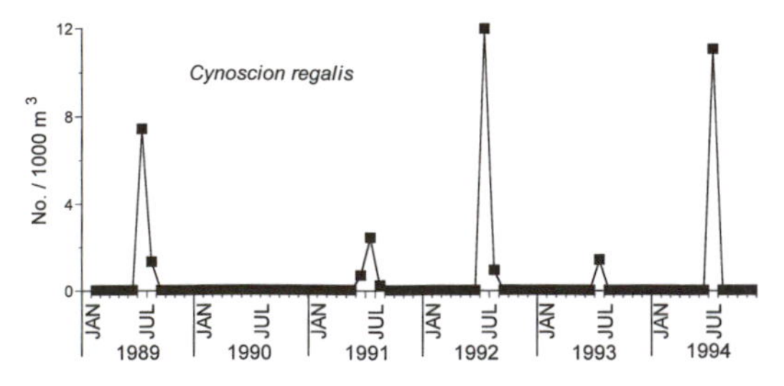

52.3 Annual variation in abundance of *Cynoscion regalis* larvae collected in the Great Bay–Little Egg Harbor study area based on Little Sheepshead Creek Bridge 1-m plankton net collections 1989–1994 (n = 203).

Development from larvae to juveniles is gradual, with no sudden morphological changes occurring at transformation. Snout-anus length increases to about 68% in juveniles. During their first summer, young-of-the-year are characterized by elongate central rays in the caudal fin (fig. 52.1). Fishes larger than about 170 mm lack this character. Estimated size and age when scales are first formed are 14.3 mm SL and 26 days, respectively (Szedlmayer et al. 1991).

Growth rates during the first year vary greatly between areas, and estimates are confounded by season, year, subpopulation, and investigative method (Wilk 1976). According to several combined collections from the central part of the Middle Atlantic Bight, multiple young-of-the-year cohorts are apparent (fig. 52.4). For example, the bimodal distributions shown in August through October are most likely due to the incursion of progeny from more than one spawning group, as has been described for Delaware Bay (Thomas 1971). During some years, this phenomenon is not apparent, and instead a single cohort dominates (Thomas 1971). In early fall, young-of-the-year range from about 4 to 16 cm TL (fig. 52.4). In early winter, the mode is 5 to 10 cm TL, possibly because the larger members of the cohort migrate out of the study area first. Several authors have reported potential lengths at age 1 of 180 to 200 mm (Welsh and Breder 1923; Pearson 1941; Thomas 1971; Shepherd and Grimes 1983). A discrete cohort is more evident in recent Delaware Bay collections (fig. 52.5), where a single mode can be traced through summer and fall months. A mode between 10 and 15 cm FL is reached in the fall, and similar-sized fish appear the following spring, indicating little or no growth through the first winter. In Delaware Bay, growth rates

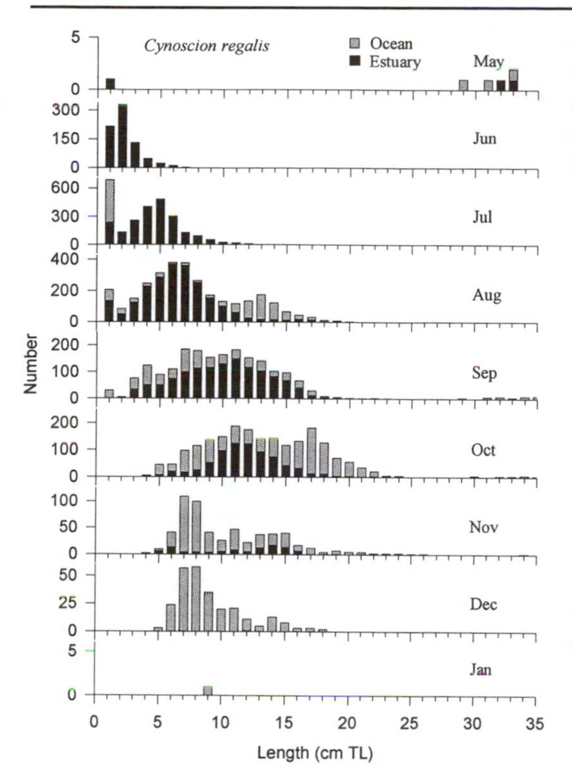

52.4 Monthly length frequencies of *Cynoscion regalis* collected in estuarine and coastal ocean waters of the central Middle Atlantic Bight. Sources: Perlmutter 1939; de Sylva et al. 1962; Clark et al. 1969; David's Island Report (unpubl.); Thomas 1971; Thomas et al. 1972, 1974; Pacheco 1983; Kahnle 1987; RUMFS: otter trawl (n = 291); 1-m beam trawl (n = 24); 1-m plankton net (n = 188); experimental trap (n = 9); LEO-15 Tucker trawl (n = 284); gear comparison (n = 215); killitrap (n = 4); weir (n = 41).

increase with increasing temperatures and vary from 0.29 mm per day at 20 C and 19 or 26 ppt to 1.49 mm per day at 28 C and 19 ppt (Lankford and Targett 1994).

Habitat use is variable because of extensive movements of young-of-the-year. Eggs and larvae occur pelagically in both ocean and estuarine environments. Both stages have been found in Delaware Bay waters in temperatures of 17.0 to 26.5 C and salinities of 12.1 to 31.3 ppt (Harmic 1958). We have found young-of-the-year at Beach Haven Ridge during all months and suggest that the coastal ocean may also provide nursery habitat (fig. 52.4). In general, young-of-the-year are most abundant in estuarine waters soon after spawning peaks (fig. 52.6); they then penetrate as far upstream as freshwater nursery areas during summer and return downstream to the ocean in the fall (e.g., Massmann et al. 1958). Deeper channels have been found to harbor most young-of-the-year by several authors (e.g., Thomas 1971; Olney

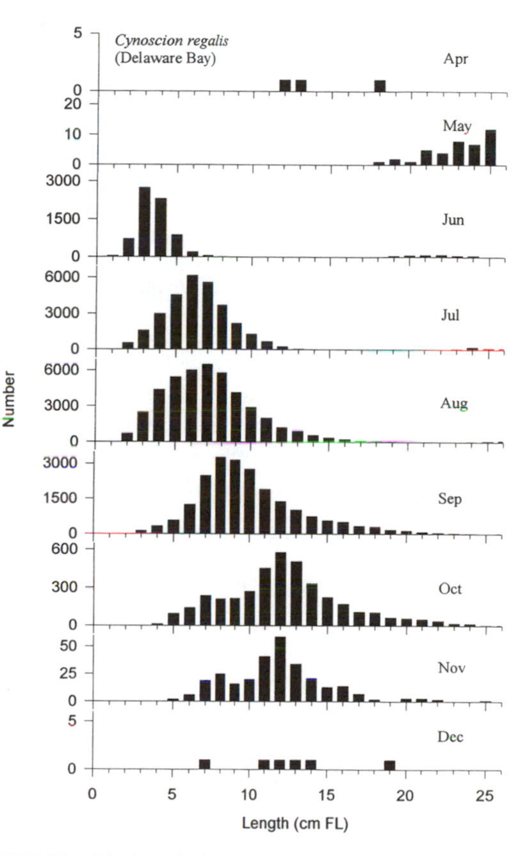

52.5 Monthly length frequencies of *Cynoscion regalis* collected 1970–1980 in Delaware Bay. Data courtesy of Public Service Electric & Gas and Environmental Consulting Services, Inc.

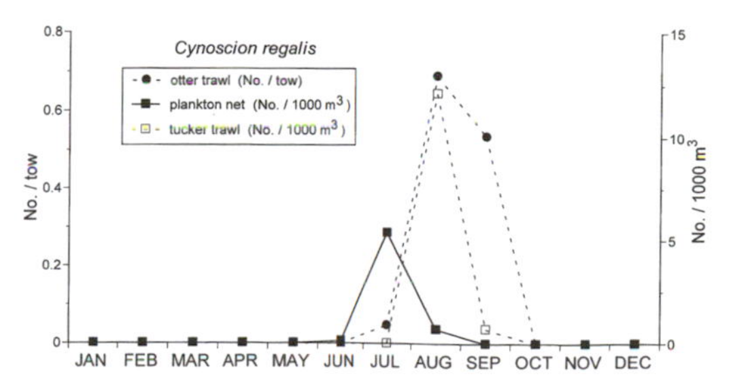

52.6 Monthly occurrences of *Cynoscion regalis* larvae and young-of-the-year in the Great Bay and adjacent coastal ocean study area based on 1-m plankton net collections 1991–1994 (n = 203), otter trawl collections 1988–1989 (n = 180), and Tucker trawl collections 1991–1992 (n = 12).

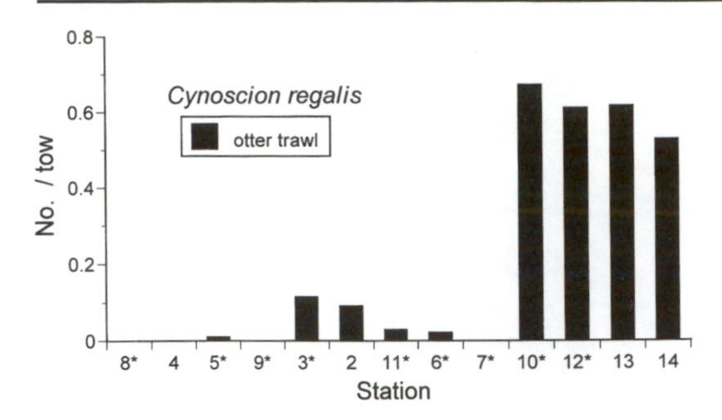

52.7 Distribution of *Cynoscion regalis* young-of-the-year by habitat type based on otter trawls (n = 262) deployed on regular stations 2–14 in Great Bay and Little Egg Harbor. Both otter trawl and beam trawl used on stations designated by asterisk; otherwise, only the otter trawl was used. See figure 2.3 for station locations; table 2.1 for specific habitat characteristics by station.

and Boehlert 1988). Young-of-the-year feed by sight in the upper to middle layers of the water column, whereas their confamilials are more oriented to the bottom for food resources (Chao and Musick 1977). In habitat-specific sampling in Great Bay–Little Egg Harbor, young-of-the-year were collected mostly from stations where the substrate was silty and depths were 2 to 3 m (fig. 52.7). Structural components at these locations varied from none to shell, rubble, or peat deposits. In Chesapeake Bay during summer, juveniles have been collected at temperatures of 21.8 to 28.4 C and salinities of 32.9 to 36.8 ppt (Cowan and Birdsong 1985) and have also been reported to occur in freshwater habitats (e.g., Massmann 1954). In Delaware Bay, they are found as far upstream as the intrusion of salt water (Thomas 1971). During several years of intensive sampling of shallow sub- and intertidal marsh creeks (where salinities ranged from 22 to 32 ppt) in the Great Bay–Little Egg Harbor system, Rountree et al. (1992) collected no young-of-the-year, but a confamilial, *Leiostomus xanthurus,* was abundant. Thomas (1971), however, reported *C. regalis* from all four low-salinity tidal creeks he sampled.

Recent habitat studies have found that growth rates of young-of-the-year within Delaware and Chesapeake Bays can vary between specific nursery zones located along a gradient of salinities and temperatures (Szedlmayer et al. 1990; Lankford and Targett 1994). Young-of-the-year in mesohaline areas in Delaware Bay experienced increased growth rates and growth efficiencies over those in oligohaline areas. Young-of-the-year initially occur in low-salinity nurseries, then move to higher salinities with growth (Chao and Musick 1977). Szedlmayer et al. (1990), however, found the opposite behavior in smaller larvae, with four of five cohorts moving upstream (into lower salinities) with growth. Paperno (1991) found no net movement with growth in the Delaware Bay. Given that the potential for growth was better in mesohaline habitats, the fact that Delaware Bay young-of-the-year migrated into oligohaline areas might imply an early life-history trade-off, sacrificing enhanced growth for more protection from predation (Lankford and Targett 1994).

Young-of-the-year begin to leave estuarine and coastal ocean nursery areas and migrate offshore during October. In deeper parts of estuaries, however, young-of-the-year may remain until winter in our study area, although our data indicate these occurrences are irregular. At times they have been found to be among the most abundant components of trawl hauls in Delaware Bay as late as December (Abbe 1967). After leaving the estuary, young-of-the-year undertake a longer winter migration than do older fishes (Wilk 1976; Mercer 1983), eventually spending the colder months south of Cape Hatteras, while older fishes may remain in the Cape Hatteras area for the winter.

CHAPTER 53

Leiostomus xanthurus Lacepède
Spot

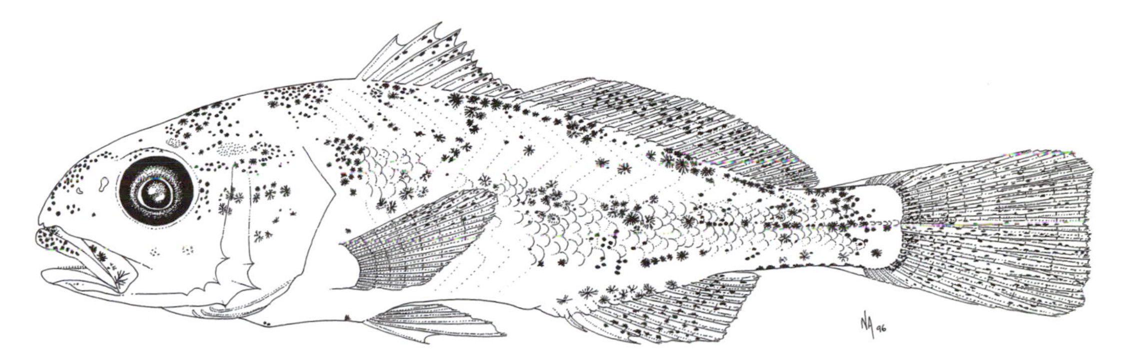

53.1 *Leiostomus xanthurus* juvenile, 23.5 mm SL. Collected June 13, 1995, seine, New Cove, Great Bay, New Jersey. ANSP 175235. Illustrated by Nancy Arthur.

DISTRIBUTION

Leiostomus xanthurus (fig. 53.1) occurs from Massachusetts Bay to Campeche Bay, Mexico and is most abundant between North Carolina and Chesapeake Bay (Hildebrand and Schroeder 1928; Thomas 1971). It is a euryhaline species that commonly occurs in fresh or low-salinity water (e.g., Massmann 1954). Populations in the central part of the Middle Atlantic Bight demonstrate wide annual fluctuations in abundance. In Delaware Bay, they ranked 12th in abundance in 1967, did not rank at all in 1968, and ranked 3rd in 1969 (Daiber and Smith 1970). In a 2-year (1988–1989) Great Bay study, nearly all *L. xanthurus* were taken during 1 year (1988) when they were abundant (Rountree and Able 1992a). Likewise, nearly all larvae collected in the Indian River Inlet, Delaware during a 3-year study were collected during a single year (Pacheco and Grant 1973). Larvae entering Little Egg Inlet were collected during 3 of 6 years studied (Witting 1995; D. A. Witting, K. W. Able, and M. P. Fahay in prep.). Larvae and juveniles occur in Middle Atlantic Bight estuaries between the Hudson River and Cape Hatteras, while in most estuaries north and east of the Hudson River, only juveniles have been found (table 4.2). We have been unable to document the occurrences of eggs and larvae in Narragansett Bay as reported by Stone et al. (1994); because synoptic surveys of that bay failed to collect these stages (e.g., Herman 1963; Bourne and Govoni 1988), these records should be viewed as suspect.

REPRODUCTION

Spawning (by 2- or 3-year-old adults) occurs in continental shelf waters from winter through early spring (Mercer 1987). Spawning activity is most intense in relatively warm water on the outer shelf south of Cape Hatteras (Norcross and Bodolus 1991), especially during winter in the proximity of the Gulf Stream front (Govoni 1993). In North Carolina, spawning is protracted from mid-October to mid-March, but 90% of larvae were produced in the 2-month period from mid-November to mid-January (Flores-Coto and Warlen 1993). Distributional data support the hypothesis that larvae spawned south of Cape Hatteras contribute to recruitment in Chesapeake Bay (Flores-Coto and Warlen 1993), but limited production also occurs in the area between the Chesapeake Bay mouth and Cape Hatteras. Between 1977 and 1987, larvae were infrequently collected in the Middle Atlantic Bight, but when larvae were present, they were restricted to the southern part of the bight (fig. 53.2) and occurred primarily during January (fig. 53.3). During some years, there is no evident spawning within the bight (table 53.1).

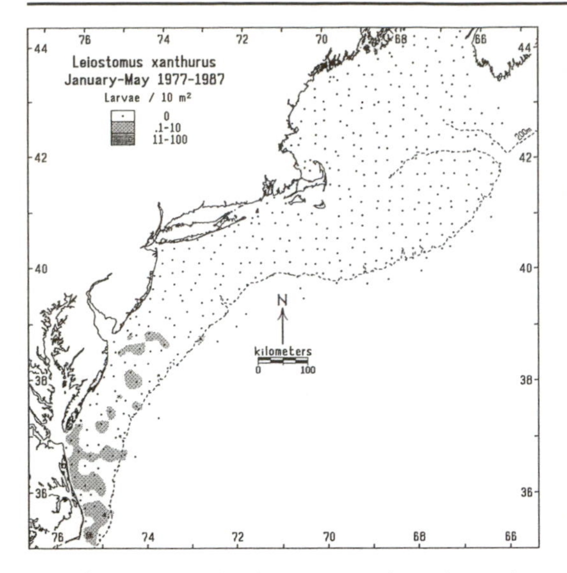

53.2 Occurrences of *Leiostomus xanthurus* larvae between January and May (inclusive) during MARMAP surveys 1977–1987.

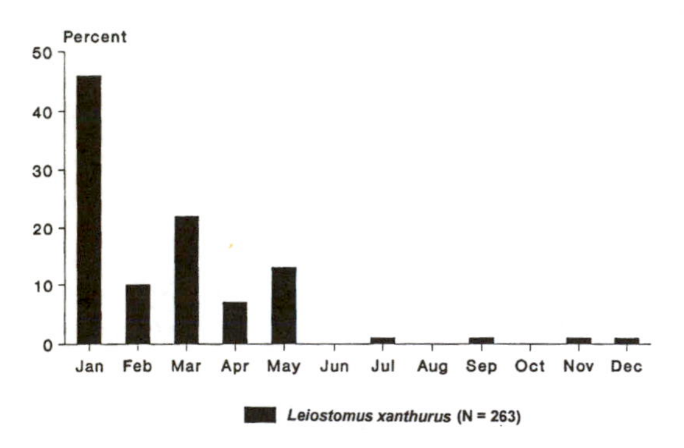

53.3 Monthly occurrences of *Leiostomus xanthurus* larvae collected during MARMAP surveys 1977–1987.

DESCRIPTION

The pelagic eggs are spherical, transparent, and range from 0.72 to 0.87 mm in diameter. The multiple oil globules coalesce into a single one, 0.18 to 0.28 mm in diameter. Larvae are 1.6 to 1.7 mm NL at hatching, lack functional mouth parts, and have unpigmented eyes. Flexion occurs when larvae are 3.8 to 5.3 mm NL. The snout-anus length increases from 20% NL in yolk-sac larvae to about 50% NL at 8 to 14 mm NL. Larvae are relatively slim, and the body depth increases from about 20% NL at flexion to about 30% NL in later larvae. Characteristic pigment includes a row of melanophores along the ventral edge of the tail, and three melanophores arranged in a triangle on the venter anterior to the anus. More details of larval development, including comparisons with other sciaenids, are in Powell and Gordy (1980), Powles and Stender (1978), Fahay (1983), and Ditty (1989). In juveniles (fig. 53.1), the eye diameter is about equal to the snout length. The second dorsal fin base is about twice the length of the anal fin base. The caudal fin is squared off or very slightly forked. There is a dark spot on the side behind the gill cover in older juveniles. There are no barbels at the tip of lower jaw. Vert: 25; D: XI–XII, 29–32; A: II, 12–13; Pect: 18–24; Plv: I, 5.

THE FIRST YEAR

Larval development occurs during winter and is gradual, with no sudden changes in acquisition of adult characters. Fin rays are fully developed at about 11 mm SL (Fahay 1983) while larvae occur pelagically in continental shelf waters. Scales begin to form in juveniles after they enter estuaries but while remnants of larval pigmentation remain. Forty-eight newly settled juveniles seined from New Cove, Little Egg Harbor, New Jersey, on June 13, 1995, were in the process of acquiring scales (fig. 53.4). In this collection, scales began forming at about 23 mm TL and were complete in fishes greater than 25 mm TL. Previous studies have suggested scales are complete by 30 mm (Hildebrand and Cable 1930). The juvenile

Table 53.1. Summary of Larval *Leiostomus xanthurus* Collections during MARMAP Surveys in the Middle Atlantic Bight, 1977–1987

Year	1977	1978	1979	1980	1981	1982	1983	1984	1985	1986	1987
Larvae	25	0	26	12	16	13	12	5	52	144	30
Pos. stn.	7	0	5	5	3	2	1	4	4	26	12
N	1158	867	910	958	971	870	963	1131	1151	1161	1298

NOTE: *N* = number of stations sampled. Pos. stn. = number of stations where larvae occurred.

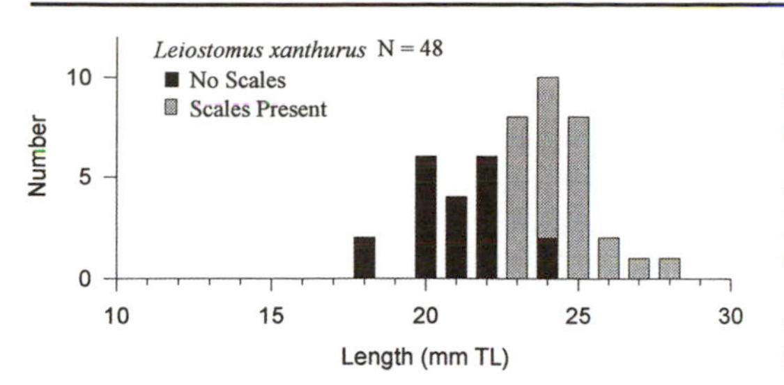

53.4 Onset of scale formation in juvenile *Leiostomus xanthurus* collected by M. C. Curran, June 13, 1995, New Cove, Great Bay, New Jersey.

morphology, consisting of a relatively large eye, broad, truncate caudal fin, and inferior mouth, indicates that *L. xanthurus* feeds in the near-bottom part of the water column by sight and olfaction (Chao and Musick 1977). Dentary teeth are lost as the juvenile diet shifts from planktivory to benthic organisms (Govoni 1987).

After larval development in continental shelf waters, larvae enter Middle Atlantic Bight estuaries during the spring at sizes of about 9 to 20 mm (de Sylva et al. 1962; Thomas 1971; Pacheco and Grant 1973; Hettler and Barker 1993). This compares with a mean size of 17.2 mm and a mean age of 82 days in a North Carolina study (Flores-Coto and Warlen 1993). In the Indian River Inlet–Delaware Bay–Little Egg Inlet regions, ingress reaches a peak in March (Pacheco and Grant 1973; Witting 1995; D. A. Witting, K. W. Able, and M. P. Fahay in prep.), April, or May (de Sylva et al. 1962; Thomas 1971). This ingress is followed by migration into areas of the upper estuary. Although larvae were not collected in the 2 years of collecting at Beach Haven Ridge and were only rarely found in central Middle Atlantic Bight continental shelf waters during the 11-year MARMAP study, young-of-the-year sometimes occur in abundance, both in upper Delaware Bay habitats (Thomas 1971) and in Great Bay (Rountree and Able 1992a; Witting 1995; D. A. Witting, K. W. Able, and M. P. Fahay in prep.). In Delaware Bay, it has been observed that young-of-the-year also commonly occur in the C&D Canal. The possibility exists, therefore, that upper Delaware Bay young-of-the-year arrive there from the upper Chesapeake Bay via that canal.

Growth of young-of-the-year has been well studied. Larvae collected off North Carolina increased from 1.2 mm at hatching to 16.1 mm in 80 days (Flores-Coto and Warlen 1993). The smallest young-of-the-year collected in New Jersey estuaries are between 15 and 20 mm TL as they enter from the ocean in early spring. After growing through the first summer in Great Bay–Little Egg Harbor, young-of-the-year reach a mode of about 10 to 11 cm TL by the fall (fig. 53.5), and this mode is sustained as larger fishes apparently migrate out of the system first. Growth rates are comparable in the upper Delaware Bay and lower Delaware River area (fig. 53.6), and the apparent autumn emigration of larger fishes first is also indicated. A slightly different pattern is apparent in the data presented by Thomas (1971), who sampled

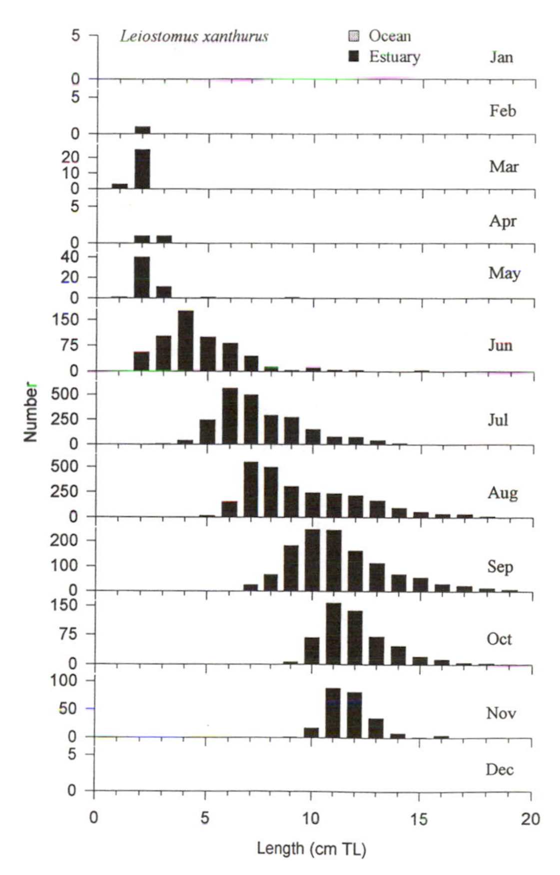

53.5 Monthly length frequencies of *Leiostomus xanthurus* young-of-the-year collected in estuarine and adjacent coastal ocean waters of the Great Bay–Little Egg Harbor study area. Sources: Thomas et al. 1974; RUMFS: otter trawl (*n* = 1,403); 1-m beam trawl (*n* = 208); 1-m plankton net (*n* = 64); LEO-15 2-m beam trawl (*n* = 1); gear comparison (*n* = 118); seine (*n* = 57); weir (*n* = 2,941).

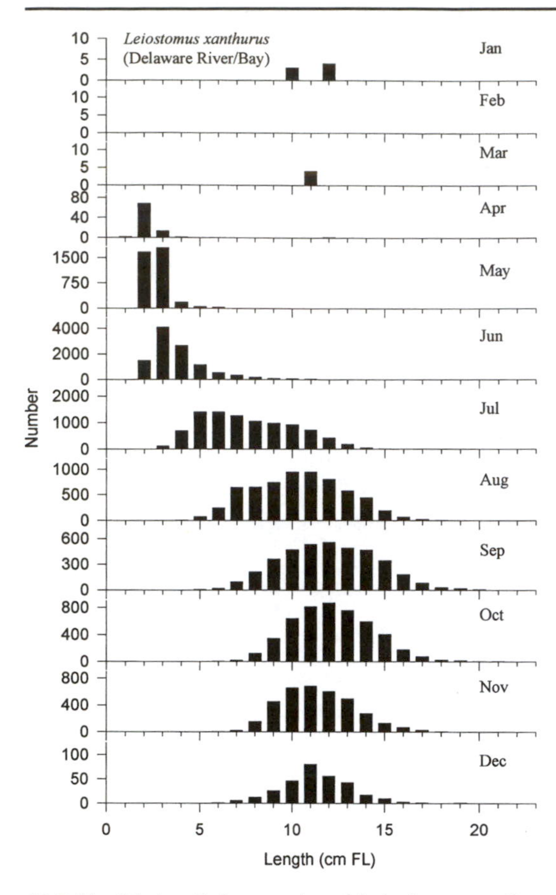

53.6 Monthly length frequencies of *Leiostomus xanthurus* young-of-the-year collected in upper Delaware Bay. Fork Length in this species approximates Total Length (Appendix B). Data courtesy Public Service Electric & Gas and Environmental Consulting Services, Inc.

fishes near the mouth of Delaware Bay and in the lower Delaware River (fig. 53.7). The smallest July and August cohorts in this study correspond well with the previous two studies, but in both months a larger cohort also appears. The smaller fishes were collected near the mouth of the bay, while the larger were collected there or from the C&D Canal. Without age data, it is not known whether the single mode during September and October represents a blend of both cohorts or survivors of the larger cohort (fig. 53.7). Presumably, movements of young-of-the-year within Chesapeake Bay are capable of affecting interpretations of growth patterns in Delaware Bay, especially in the case of this species, which uses nursery areas very far upstream to and including freshwater habitats or the C&D Canal (Massmann 1954).

In a James River (Chesapeake Bay) nursery, young-of-the-year grew 11.3 mm per month during spring and summer, but growth then leveled off during fall (McCambridge and Alden 1984). Winter growth in North Carolina estuaries has been estimated at 0.14 to 0.16 mm per day (Weinstein and Walters 1981). Previous conclusions concerning size at age are equivocal. Estimates of the sizes attained at age 1 range from 8 to 10 cm in New Jersey (Welsh and Breder 1923) to an average of 15 to 17 cm in Chesapeake Bay (McCambridge and Alden 1984; Pacheco 1962). Our own data from Great Bay–Little Egg Harbor would indicate young-of-the-year reach between 10 and 15 cm TL as they begin their first winter (October), whereas in North Carolina, they range from 5 to 11 cm SL (6–14 cm TL) in October (Weinstein and Walters 1981).

Juveniles occupy a variety of estuarine habitats during their first year. In Great Bay–Little Egg Harbor, they are somewhat more common in subtidal creeks than in intertidal habitats (Rountree and Able 1992a). In a Virginia study, they occurred in tidal creeks but moved between habitats, with individuals remaining within single creeks for an average of 81 days (Weinstein et al. 1984). In a trawl survey covering several habitats other than tidal creeks, they were most abundant at a relatively deep (5-m) station characterized by a shell, sponge, and peat substrate (fig. 53.8). They are also especially abundant over beds of thick, loose

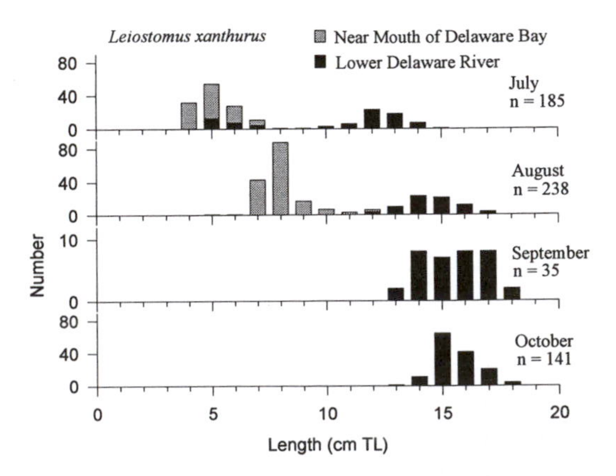

53.7 Monthly length frequencies of *Leiostomus xanthurus* young-of-the-year collected from sampling sites near mouth of Delaware Bay and lower Delaware River (Thomas 1971).

mud (Reid 1955; Cowan and Birdsong 1985.) Juveniles occur in a wide range of salinities and can survive prolonged periods in freshwater (Massmann 1954). In 23 collections in Delaware Bay between August and October, young-of-the-year were collected in salinities ranging from 0 to 10 ppt, temperatures from 14.0 to 27.2 C, and dissolved oxygen from 3.8 to 10.8 ppm (Thomas 1971). Juveniles have also been collected at sites with salinities as high as 35.2 ppt (Cowan and Birdsong 1985). Olney and Boehlert (1988) observed that larvae and juveniles became the most abundant taxa in their lower Chesapeake Bay grass-bed study site in the spring as the submerged vegetation was beginning to grow.

Habitat quality may affect growth rates. Weight-length relationships were compared among young-of-the-year from several locations between Chesapeake Bay and Texas, and the former fish were found to be heavier at length than fish from other nurseries, suggesting that habitat quality was greater in the Chesapeake Bay system (McCambridge and Alden 1984). These authors also summarized monthly growth rates during the first year from a number of studies, presented here for comparison (table 53.2).

As indicated in the length-frequency histograms (figs. 53.5, 53.6), young-of-the-year begin to vacate estuaries in the central part of the Middle Atlantic Bight in October or November, concurrent with fall cooling. They were most commonly collected during November in a trawl survey of Great Bay–Little Egg Harbor, presumably as they were moving from brackish habitats, through the estuary, to the ocean (fig. 53.9). Larger individuals (e.g., >15 cm) apparently leave estuarine habitats earliest, and all have migrated to oceanic habitats by December. There is no evidence that young-of-the-year overwinter in estuaries in the central part of the Middle Atlantic Bight, as they sometimes do in Chesapeake Bay. Tagging studies indicate a fall and winter migration (along with older year classes) to wintering grounds south of Cape Hatteras (Pearson 1932; Pacheco 1962).

Table 53.2. Monthly Growth Rates of *Leiostomus xanthurus* during the First Year and Selected Months, Reported from a Number of Studies along the Atlantic coast between Virginia and Florida

Growth Rate (mm/mo)	Period	Location
10.5	First year	Chesapeake Bay
11.7	First year	Beaufort, NC
11.8	First year	Florida
16.3	First year	York River, VA
12.8	First year	Bears Bluff, SC
18.5	August	Beaufort, NC
19.1	August	Chesapeake Bay
15.8	July	Florida
15.2	August	James River, VA.

SOURCE: McCambridge and Alden 1984.

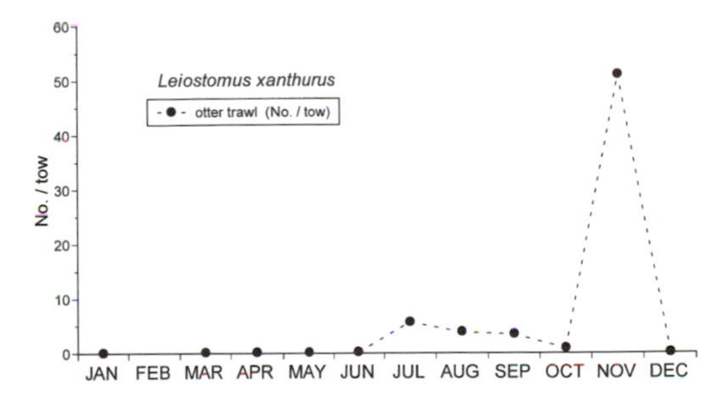

53.9 Monthly occurrences of *Leiostomus xanthurus* young-of-the-year in the Great Bay–Little Egg Harbor study area based on otter trawl collections 1988–1989 ($n = 3,916$).

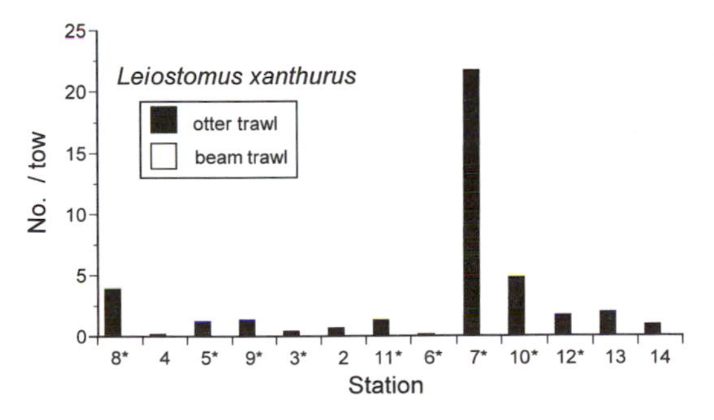

53.8 Distribution of *Leiostomus xanthurus* young-of-the-year by habitat type based on otter trawls ($n = 4,091$) and beam trawls ($n = 4$) deployed on regular stations 2–14 in Great Bay and Little Egg Harbor. Both otter trawl and beam trawl used on stations designated by asterisk; otherwise, only the otter trawl was used. See figure 2.3 for station locations; table 2.1 for specific habitat characteristics by station.

CHAPTER 54

Menticirrhus saxatilis (Bloch and Schneider) Northern Kingfish

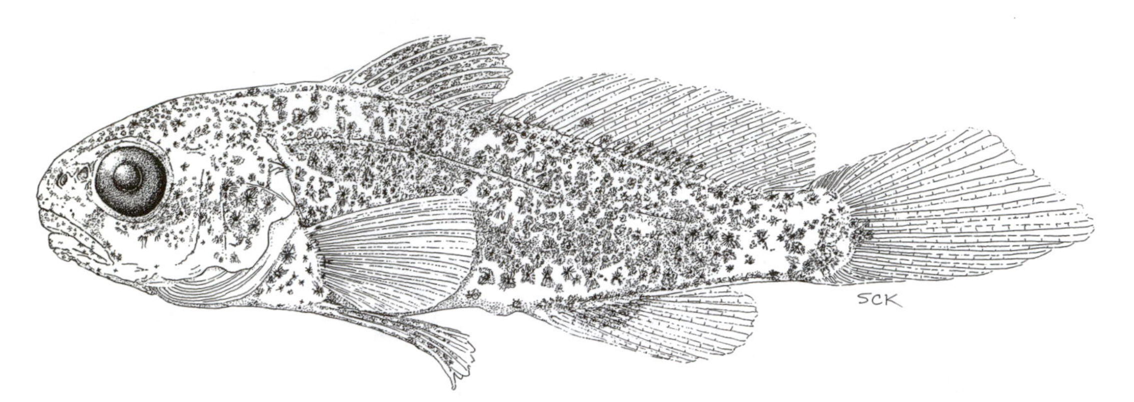

54.1 *Menticirrhus saxatilis* juvenile, 16.4 mm TL. Collected July 7, 1991, seine, intertidal pool, Corson's Inlet, New Jersey. ANSP 175236. Illustrated by Susan Kaiser.

DISTRIBUTION

Menticirrhus saxatilis (fig. 54.1) occurs in inner continental shelf and estuarine waters from Maine to Florida but is most abundant in the Middle Atlantic Bight (Bigelow and Schroeder 1953). It typically occurs over sandy bottom in the surf zone and in sandy channels near inlets. The early life-history stages are distributed in most estuaries in the bight, but curiously, the eggs are only reported from estuaries in New Jersey and north (table 4.2). It is common in the central part of the Middle Atlantic Bight from late spring through fall (Phillips 1914; Welsh and Breder 1923; Nichols and Breder 1927).

REPRODUCTION

The occurrence of larvae in May through September indicates that spawning, primarily by age-1 individuals (Schaefer 1965), occurs in the Middle Atlantic Bight from spring through late summer (fig. 54.2). The finding of eggs in every portion of the bay shows that spawning occurs throughout Narragansett Bay (Bourne and Govoni 1988).

DESCRIPTION

The pelagic eggs are small (0.7–0.9 mm in diameter), with numerous oil globules (1–18) of various sizes (Welsh and Breder 1923) and a narrow perivitelline space. Larvae hatch at 2.0 to 2.5 mm. Larvae are densely pigmented over the head and body, including a midline stripe and melanophores over the gut surface. Fin rays are complete at about 8.0 mm SL (Fahay 1983; B. H. Comyns pers. comm.). Further develop-

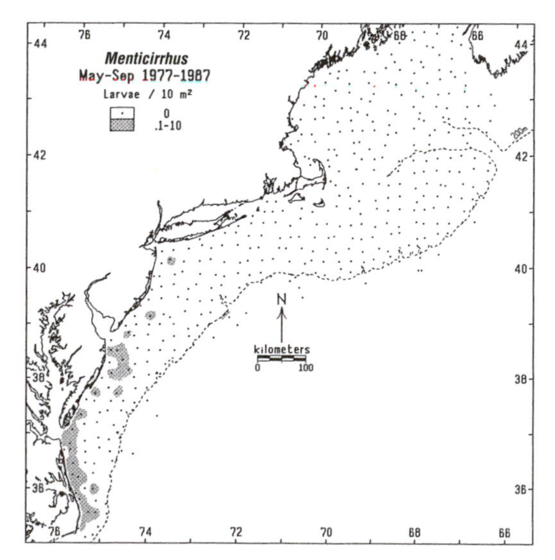

54.2 Occurrences of *Menticirrhus* spp. larvae between May and September (inclusive) during MARMAP surveys 1977–1987. Mean density, in shaded areas, was 3 per 10 m² of sea surface.

ment is described by Perlmutter (1939) and Hildebrand and Cable (1934). Juveniles have two distinct dorsal fins, touching at their bases (fig. 54.1). Dorsal and anal fins have both sharp spines and soft rays and the second dorsal fin is about twice the length of the anal fin. A single barbel occurs at the tip of the lower jaw. Note the shape of the caudal fin. The body is dusky to dark, with or without prominent bars. Vert: 25; D: XI, 23–35; A: II, 7–8; Pect: 18–21; Plv: I, 5.

THE FIRST YEAR

The incubation period for eggs is 46 to 50 hours at a temperature of 20 to 21 C (Welsh and Breder 1923). Larvae identified as *Menticirrhus* sp., have been only rarely collected from the inner continental shelf in the Middle Atlantic Bight from Cape Hatteras to as far north as northern New Jersey, but they are most consistently found south of Delaware Bay from May through September (fig. 54.2). This period corresponds roughly to the June through August period described for New Jersey by Welsh and Breder (1923). In samples taken from 1989 to 1993 in the Great Bay–Little Egg Harbor estuary, larvae (n = 50) were collected primarily in late June and July but occurred through late September (fig. 54.3). In the same estuary, juveniles were most abundant during July (fig. 54.4). In lower Delaware Bay, young-of-the-year occurred from July through October (de Sylva et al. 1962).

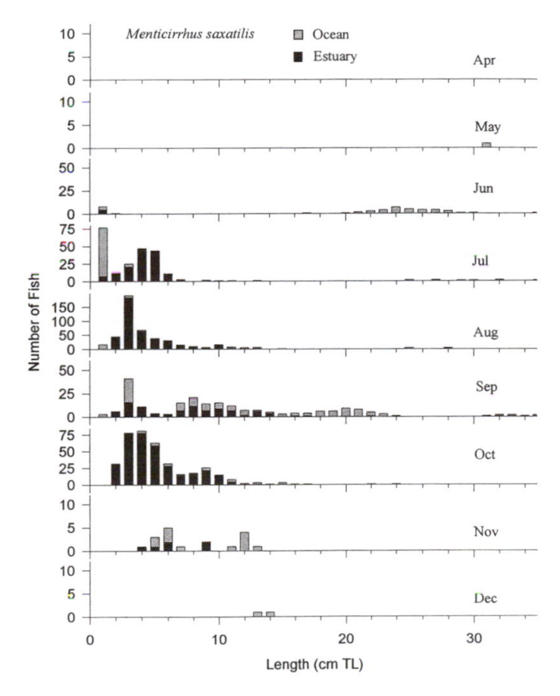

54.3 Monthly length frequencies of *Menticirrhus saxatilis* collected from estuarine and coastal ocean waters of the central Middle Atlantic Bight. Sources: de Sylva et al. 1962; Thomas et al. 1972, 1974; Pacheco and Grant 1973; Wilk and Silverman 1976a; Corson's Inlet Seine Survey (unpubl.); RUMFS: otter trawl (n = 30); Tucker trawl (n = 59); 1-m beam trawl (n = 16); 1-m plankton net (n = 15); LEO-15 2-m beam trawl (n = 4); gear comparison (n = 85); seine (n = 43); night-light (n = 12); weir (n = 3).

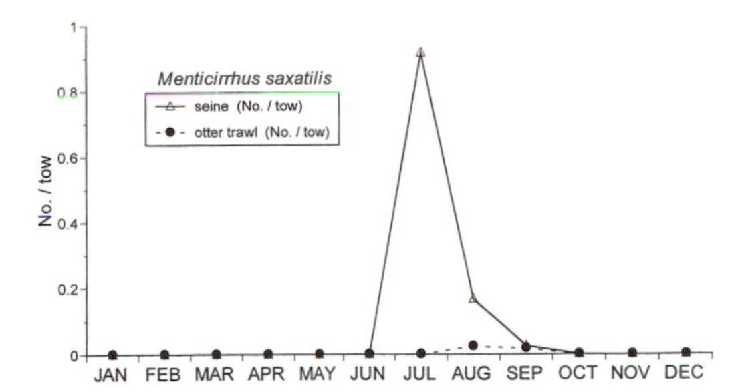

54.4 Monthly occurrences of *Menticirrhus saxatilis* young-of-the-year collected in the Great Bay–Little Egg Harbor study area based on otter trawl, 1988–1989 (n = 6) and seine, 1990–1991 (n = 52) collections.

The monthly length-frequency distributions from New Jersey waters suggest that at least two year classes may be represented (fig. 54.3), although these interpretations are open to question. Collections in June suggest that young-of-the-year are represented by those smaller than 3 cm while age 1+ are indicated by those between 20 and 30 cm. This interpretation assumes that young-of-the-year, from both the estuary and the ocean, are quite variable in size and may attain up to 24 cm by September (fig. 54.3). This interpretation of relatively fast growth is supported by similar estimates from Long Island where extensive collections found young-of-the-year ranging from 2.5 to 31.5 cm TL, with a modal length of 25 cm in October (Schaefer 1965). These estimates were confirmed from back calculations of annuli from scales.

The habitats of young-of-the-year range from the estuary to the ocean. All sizes have been collected from sandy beaches in the vicinity of Little Egg Inlet (D. Haroski and K. W.

Able unpubl. data). Our recent observations of young-of-the-year in Corson's Inlet, New Jersey, duplicate those of Phillip's (1914) from the same location some 90 years earlier, which suggests that their occurrence is regular. The preference for ocean beach and inlets in southern New Jersey is supported by the lack of young-of-the-year in extensive collections in salt marsh creeks (Rountree and Able 1992a) and shallow muddy shorelines (Able et al. 1996b) throughout the estuary behind Little Egg Inlet. In polyhaline waters of lower Delaware Bay, the young-of-the-year typically occur at salinities greater than 8.0 ppt, although some have been collected as low as 5.2 ppt (Thomas 1971). Abundance in the estuary declines through the fall (fig. 54.4), and by November relatively few individuals are represented in our collections (fig. 54.3), presumably because they emigrate.

CHAPTER 55

Micropogonias undulatus (Linnaeus) Atlantic Croaker

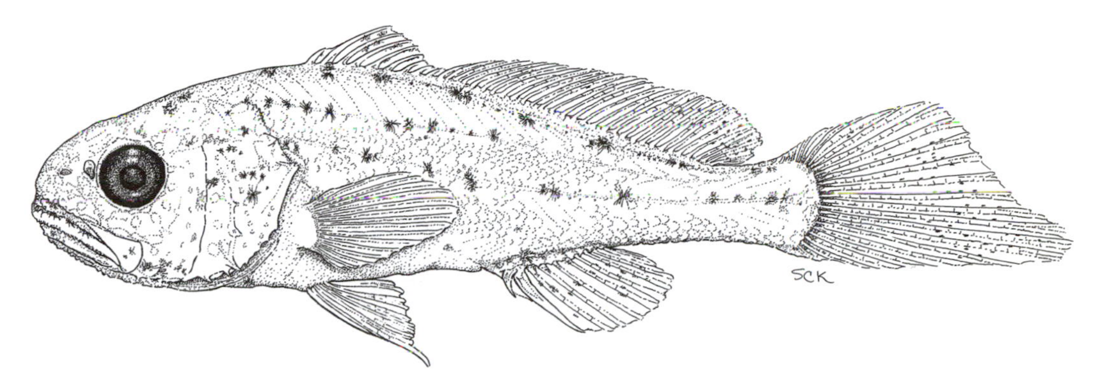

55.1 *Micropogonias undulatus* juvenile, 19.0 mm SL. Collected October 22, 1992, experimental trap, RUMFS boat basin, Great Bay, New Jersey. ANSP 175237. Illustrated by Susan Kaiser.

DISTRIBUTION

Micropogonias undulatus (fig. 55.1) is found in the western Atlantic Ocean between Massachusetts and the Gulf of Mexico (Robins and Ray 1986). It is uncommon north of New Jersey, but farther south it is a common (sometimes abundant) bottom-feeding fish in coastal waters and embayments. It is strongly euryhaline and can be found in salinities ranging from freshwater to 70 ppt (Simmons 1957). During winter in the Middle Atlantic Bight, adults move offshore and south, perhaps as far as the South Atlantic Bight. It was the most common species collected by A. E. Parr and associates from 1929 to 1933 in the ocean and bays of southern New Jersey but failed to rank among the top 20 in a similar effort in 1972 (Thomas and Milstein 1974). It was once a common constituent of seine collections in the Delaware Bay (Shuster 1959), but the population has apparently decreased drastically since the 1950s. Annual fluctuations in abundance were apparent between 1959 and 1961, when 98% of the 3,641 larvae collected were taken during one of three study years (Pacheco and Grant 1973). Larvae and juveniles occur in Middle Atlantic Bight estuaries from the Hudson River to Cape Hatteras (table 4.2) with only rare occurrences north and east of the Hudson River.

REPRODUCTION

A gonadosomatic study suggests spawning in the Middle Atlantic Bight begins in early September, peaks in October, and ends in late December (Morse 1980). On the basis of larval occurrences, spawning activity in the Middle Atlantic Bight from 1977 to 1987 was almost totally restricted to the continental shelf between Delaware Bay and Cape Hatteras (fig. 55.2).

DESCRIPTION

Eggs are undescribed. In larvae, snout-anus length increases from 50% SL in early stages to about 60% SL in larger larvae. Body depth remains less than 30% SL through development. Flexion occurs at lengths of 4 to 5 mm SL. All fin rays are ossified by 12 mm SL. Typical pigment includes a line of melanophores from the anus to the notochord tip and a series of three melanophores in line between the anus and cleithral symphysis. Internal pigment on the anterior surface of the gut, present in larvae of most sciaenid species, is absent in this species. More details of larval development may be found in Holt et al. (1981), Fahay (1983), and Ditty (1989). Early juveniles have a rounded snout, an anal fin base less than half as long as the second dorsal fin base, and a caudal fin with elongate central rays (fig. 55.1). The

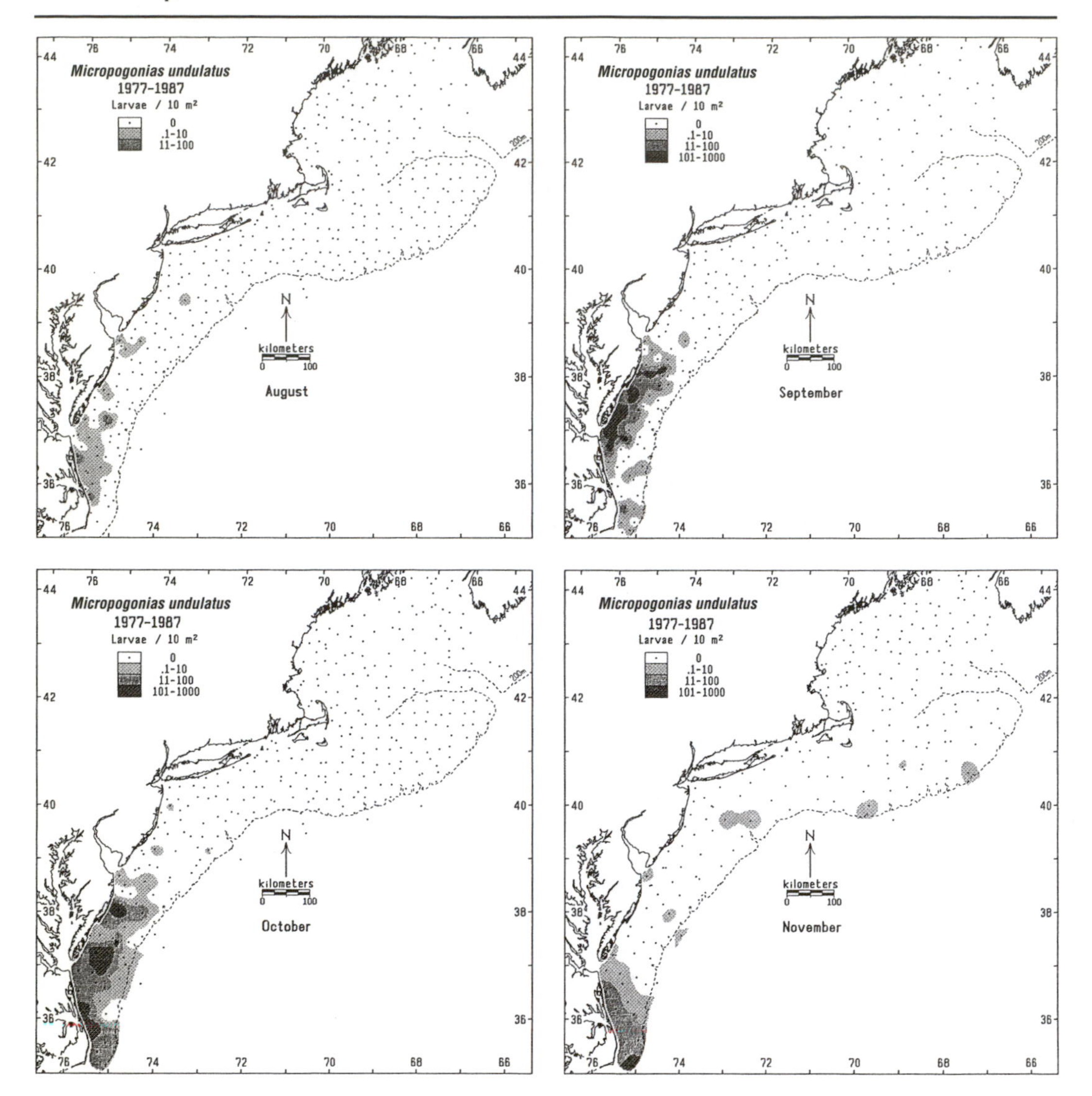

55.2 Monthly distributions of *Micropogonias undulatus* larvae during MARMAP surveys 1977–1987.

mouth is inferior, and several barbels are present at the tip of the lower jaw at sizes greater than about 30 mm. The eye diameter is less than the snout length. In comparably sized *Leiostomus xanthurus,* the eye diameter exceeds the snout length. Vert: 23–26; D: X, I, 27–30; A: II, 8–9; Pect: 17–18; Plv: I, 5.

THE FIRST YEAR

Larvae only rarely occur in shelf waters in the central part of the Middle Atlantic Bight (fig. 55.2). In some years, however, young stages are relatively common in this area and contribute significantly to the Delaware Bay fauna (Thomas 1971). During a 2-year study on the Beach Haven Ridge off Little Egg Inlet, pelagic larvae were rare in September 1991, somewhat more abundant in September and October of 1992. None was larger than 9.5 mm SL (K. W. Able and M. P. Fahay unpubl. observ.). At about 8 to 20 mm, pelagic young-of-the-year leave shelf waters and enter larger estuaries, eventually moving into nursery areas associated with low-salinity, tidal creeks. In collections inside Little Egg Inlet, this ingress occurred between September and December (fig. 55.3). These larvae averaged 11.4 mm SL, and 83% were in the postflexion stage (Witting 1995; D. A. Witting, K. W. Able, and M. P. Fahay in prep.). This ingress occurs as early

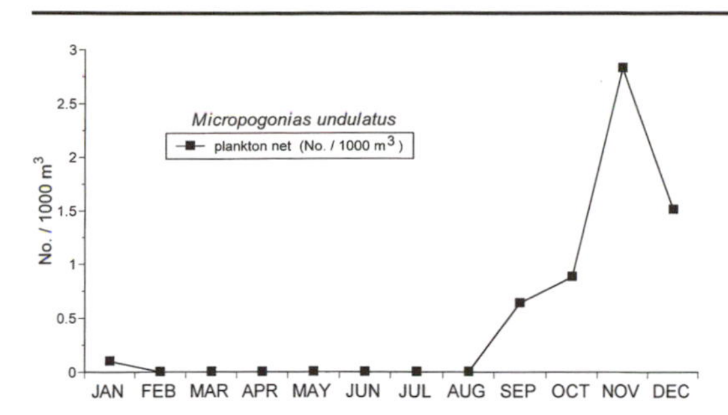

55.3 Monthly occurrences of *Micropogonias undulatus* larvae in the Great Bay–Little Egg Harbor study area 1989–1994 based on Little Sheepshead Creek Bridge 1-m plankton net collections 1989–1994 (n = 244).

as October in Delaware Bay nursery areas and continues into November at inlets in Virginia (Thomas 1971; Cowan and Birdsong 1985). Ingress reaches a strong peak in December in the Indian River Inlet, Delaware (at sizes between 8 and 46 mm) (Pacheco and Grant 1973).

Larvae gradually transform into the adult morphology with no sudden anatomical changes. Scales first appear along the lateral line of the caudal peduncle between 14 and 16 mm SL (Bridges 1971). Additional scale formation then proceeds anteriorly, until the full complement is complete at lengths between 31 and 38 mm SL. Early juveniles have the lower rays of the caudal fin elongate, a feature lost in adults. They also have relatively small eyes and a small, inferior mouth equipped with barbels. All these transient features indicate that these juveniles are relatively fast swimmers that feed in the lower water column (Chao and Musick 1977).

It is difficult to estimate growth rates based on New Jersey collections of young-of-the-year, because they are so rarely collected in this study area. Based on available evidence, it appears that very little growth occurs here, and young-of-the-year probably emigrate from the study area or do not survive winter cooling much after December (fig. 55.4). Other studies have reported a virtual lack of growth over the winter, with observations of young-of-the-year in the spring at sizes approximating those of the previous fall (e.g., Chao and Musick 1977). In Delaware Bay, they presumably reach lengths of 135 to 140 mm at age 1 (Thomas 1971).

Important habitats include coastal waters of the continental shelf for larvae and low-salinity habitats such as tributaries to major bay systems for the earliest settlement stages. The latter stages have been found to be more common in grass beds than in adjacent channel sites in the Chesapeake Bay (Olney and Boehlert 1988). In other studies, young-of-the-year were most commonly collected during the fall over a soft mud bottom at temperatures of 9.5 to 23.2 C and salinities of 24.0 to 33.7 ppt (Cowan and Birdsong 1985). Their first winter is spent in deeper portions of these habitats. A study of stomach contents of striped bass during winter in Chesapeake Bay indicates heavy predation on these overwintering young-of-the-year (Dovel 1968).

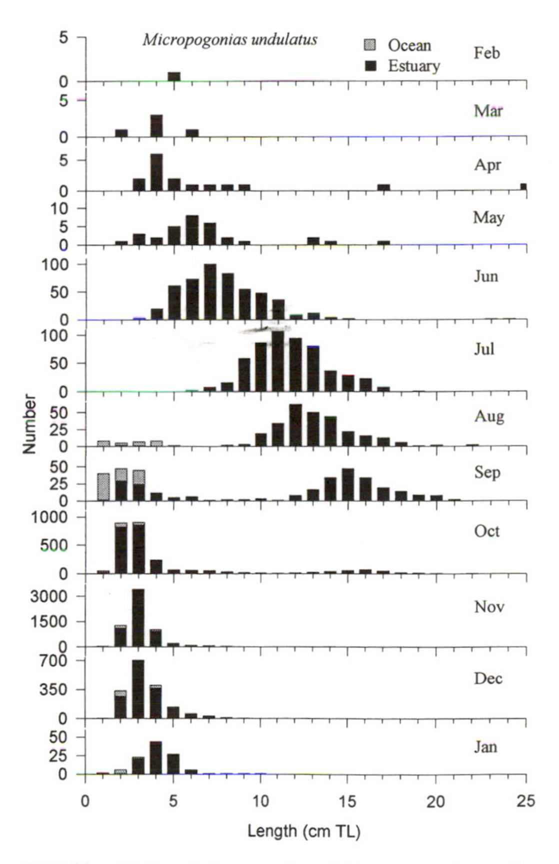

55.4 Monthly length frequencies of *Micropogonias undulatus* collected in estuarine and coastal ocean waters of the central Middle Atlantic Bight. Sources: MARMAP 1977–1987; Thomas 1971; Thomas et al. 1974; Public Service Electric & Gas Delaware Bay data courtesy of Environmental Consulting Services, Inc.; RUMFS: otter trawl (n = 13); 1-m beam trawl (n = 6); 1-m plankton net (n = 235); experimental trap (n = 1); LEO-15 Tucker trawl (n = 22); LEO-15 2-m beam trawl (n = 124); gear comparison (n = 6); seine (n = 1).

As seasonal cooling proceeds, young-of-the-year gradually occupy the deeper parts of nursery creeks where they are able to overwinter, at least in the Chesapeake Bay region. However, in unusually cold winters, mass mortalities of young-of-the-year have occurred in that bay (Chao and Musick 1977). In years when they have been abundant in Delaware and New Jersey estuaries, it is unknown whether they were able to tolerate ambient winter conditions. The most intensive sampling effort in our study area is that reported by Thomas (1971) in upper Delaware Bay. Despite a 3-year effort in that study, very few larvae or young-of-the-year were collected, most of these were taken during December, and none was taken after that month. Those taken in December were collected concurrently with collections made in the C&D Canal and upper Chesapeake Bay. The origin of these larvae is therefore in some question. As juveniles reach their first birthday in the fall, most leave Middle Atlantic Bight embayments with older fish and migrate offshore and south for their second winter. Descriptions of emigration of young from our study area are lacking.

CHAPTER 56

Pogonias cromis (Linnaeus) Black Drum

56.1 *Pogonias cromis* juvenile, 43.5 mm TL. Collected August 15, 1957, Absecon Island, Atlantic County, New Jersey. ANSP 122234. Illustrated by Nancy Arthur.

DISTRIBUTION

Pogonias cromis (fig. 56.1) is found from Argentina to Massachusetts (Hildebrand and Schroeder 1928) and occasionally to the Bay of Fundy (Bleakney 1963). Over this range, adults occur in a variety of habitats including sand, oyster reefs, clam shell deposits, and soft bottoms. Typically the adults are found in the high-salinity portion of estuaries, while the early life-history stages tolerate low salinities and even freshwaters. Historical records indicate adults were once much more common in northern New Jersey waters than at present. Despite intensive collections in Great Bay–Little Egg Harbor study area over the last decade, not a single young-of-the-year has been collected. In Middle Atlantic Bight estuaries, the eggs and larvae have been reported from the southern part of the bight to Delaware Bay, whereas only juveniles are recorded, sporadically, from there and north (table 4.2).

REPRODUCTION

Sexual maturity is reached at approximate sizes of 285 to 330 mm (Pearson 1929; Simmons and Breuer 1962). Spawning is variously reported to occur in the ocean near larger sounds and bays and in estuaries. Spawning peaks in the middle of May in lower Delaware Bay and adults are seldom captured there after the end of June (Wang and Kernehan 1979). Egg occurrence shows that spawning in Chesapeake Bay occurs in inshore areas (Joseph et al. 1964; Daniel and Graves 1994) and lasts from April to June (Richards 1973).

DESCRIPTION

Eggs are spherical (0.8–1.0 mm in diameter; Daniel and Graves 1994), with one to six oil globules (≤0.26 mm in diameter). Larvae are 1.9 to 3.0 mm TL at hatching. Larger larvae have a rounded caudal fin and stellate melanophores on the lateral surfaces above the base of the anal fin, below the base of the dorsal fin, and slightly posterior to the anus. Details of embryonic and larval development are available from Joseph et al. (1964) and Lippson and Moran (1974). Juveniles have a series of dark bars from the head to the base of the caudal fin (fig. 56.1). The second dorsal fin base is about twice the length of the anal fin base and the caudal fin is rounded. There are approximately 12 barbels on each side under the tip of the lower jaw. Vert: 24; D: X–XI, 21–23; A: II, 6–7; Plv: I, 5.

THE FIRST YEAR

Relatively little is known of events during the first year in the Middle Atlantic Bight. In Delaware Bay and Chesapeake Bay, the larvae and small juveniles move into low-salinity nursery areas after spawning occurs in the lower bays (Joseph et al. 1964; Thomas and Smith 1973). In Delaware Bay, some small individuals (5–10 mm) have been found as far up as the C&D Canal. The juveniles are 1 to 3 cm in June and increase rapidly in length until October when they range from 9 to 20 cm (fig. 56.2). Thus, the progression of modal lengths suggests an average growth of 1.2 mm per day.

Habitat use varies seasonally. Small juveniles were most abundant in the middle and upper parts of low-salinity (<6.0 ppt) marsh creeks, with little current and mud substrate in Delaware Bay (Thomas and Smith 1973) and Chesapeake Bay (Frisbie 1961). In late June and early July, the largest juveniles (30–50 mm) begin to move out of the creeks and can be found both inshore and in deeper waters of the lower Delaware River and upper Delaware Bay (Thomas 1971). By the end of August, there are few young in the Delaware River, and most are found in the lower bay by the end of October and through December (Abbe 1967; Daiber and Smith 1970). After October, the average size is smaller, suggesting that the larger individuals are the first to leave the bay (Thomas and Smith 1973). Overwintering locations are unknown.

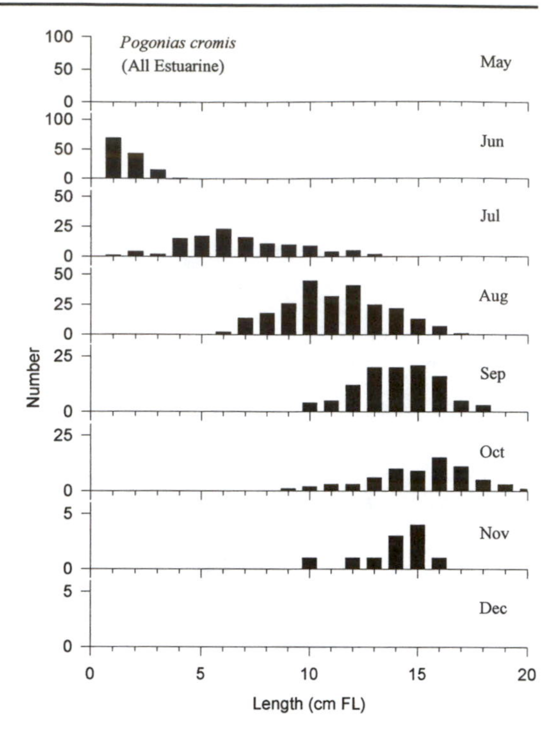

56.2 Monthly length frequencies of *Pogonias cromis* collected from the upper Delaware Bay during 1968–1969 (Thomas 1971).

CHAPTER 57

Chaetodon ocellatus Bloch
Spotfin Butterflyfish

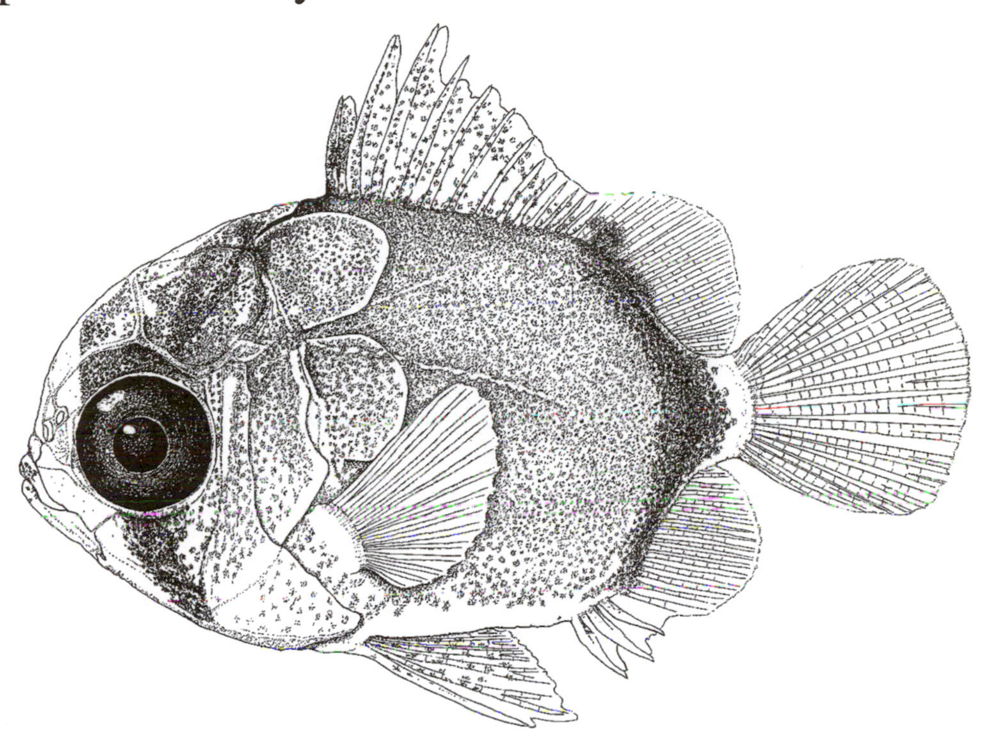

57.1 *Chaetodon ocellatus* "tholichthys" stage, 13.8 mm SL. Collected July 31, 1991, killitrap, RUMFS boat basin, Great Bay, New Jersey. ANSP 175238. Illustrated by Susan Kaiser.

DISTRIBUTION

Chaetodon ocellatus (fig. 57.1) is found from as far north as Nova Scotia to Tortugas and the Gulf of Mexico (Longley and Hildebrand 1941; Scott and Scott 1988), but all of the individuals collected north of Cape Hatteras are young-of-the-year (McBride and Able in press). These have been reported sporadically from Middle Atlantic Bight estuaries but primarily in those south of Long Island (table 4.2).

REPRODUCTION

Few data are available on reproduction, but presumably all spawning occurs south of Cape Hatteras.

DESCRIPTION

Eggs are pelagic, small (0.6–0.75 mm in diameter), and spherical (Moe 1976), but their characters are undescribed. Larvae are also undescribed. *Chaetodon* juveniles are distinctive and easily identified among Middle Atlantic Bight estuarine fishes. The "tholichthys" (early juvenile) is characterized by a bony enclosure over the head with earlike flaps extending to above the pectoral fin (Hubbs 1963; Burgess 1978). The body is disk-shaped, and the color is white, with a bright yellow fringe with prominent black bars (fig. 57.1). The most anterior of these bars crosses the head and passes through the eye. Note the high dorsal spines and the rounded caudal fin. This species differs from *C. capistratus* (fig. 57.2) in that it has a dark bar stretching from the dorsal to the anal fin and crossing the caudal peduncle, whereas *C. capistratus* has a dark smudge covering the caudal peduncle area. Vert: 24; D: XII–XIV, 18–21; A: III, 15–18; Pect: 16; Pelv: I, 5.

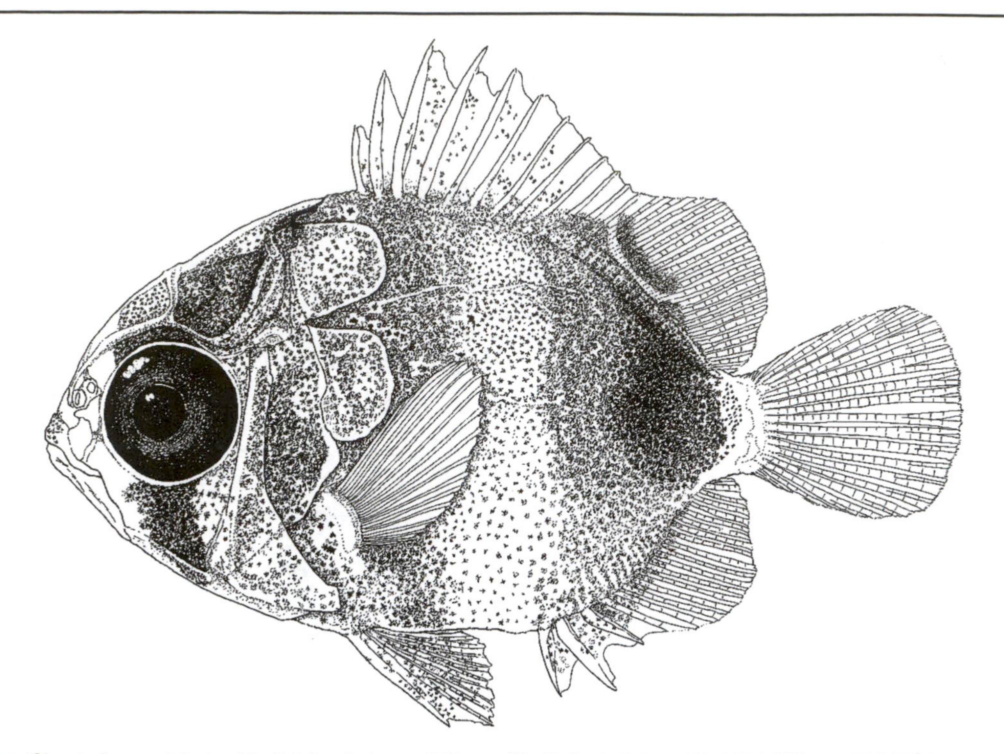

57.2 *Chaetodon capistratus* "tholichthys" stage, 13.7 mm SL. Collected June 29, 1991, killitrap, RUMFS boat basin, Great Bay, New Jersey. ANSP 175239. Illustrated by Susan Kaiser.

THE FIRST YEAR

Individuals collected in the water column are "tholichthys," typically 17 to 22 mm TL, when captured in southern New Jersey waters. The narrow size range of these fish is indicative of the size at settlement, because fish of this size range were also collected in traps on the bottom (McBride and Able in press). The head plate gradually becomes reduced and is absent by 25 to 27 mm TL (McBride and Able in press).

The young-of-the-year were consistently common in midsummer and early fall from early July through November over a period of several years (fig. 57.3). The modal size of fish collected in traps was smallest in July, but small fish (<30 mm TL) were present through October (fig. 57.4). Based on collections from a variety of trap types and mark-recapture experiments, the growth rate ranged from 0.09 to 0.21 mm per day (McBride and Able in press). By September, some young-of-the-year had reached lengths greater than 63 mm TL (fig. 57.4).

The habitats of young-of-the-year are variable. During the summer, the young-of-the-year occurred in intertidal and subtidal areas in a variety of structured and unstructured subtidal habitats, including around dock pilings, marsh peat reefs, and tidal creeks (Szedlmayer and Able 1996; McBride and Able in press). They show high fidelity to some sites, with small, marked individuals remaining in close proximity to the original release site; individuals were recaptured 1 to 39 days after release

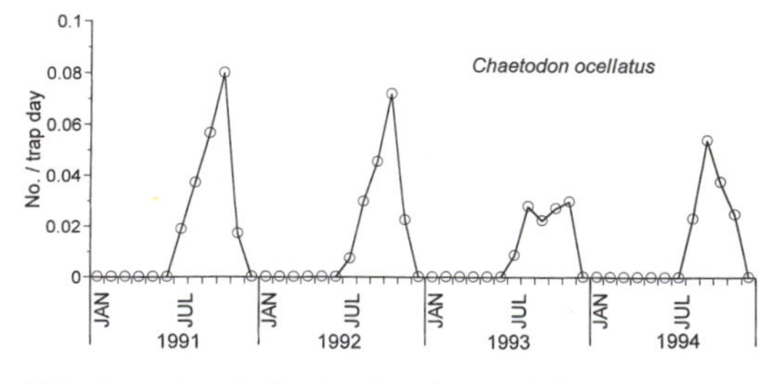

57.3 Annual variation in abundance of *Chaetodon ocellatus* juveniles in the Great Bay–Little Egg Harbor study area based on killitrap collections 1991–1994 ($n = 217$).

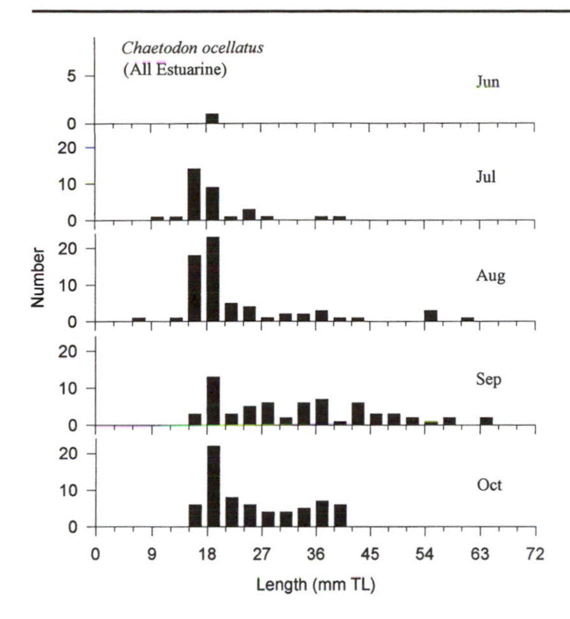

57.4 Monthly length frequencies of *Chaetodon ocellatus* collected in the Great Bay–Little Egg Harbor study area. Sources: RUMFS: beam trawl (n = 1); experimental trap (n = 58); LEO-15 2-m beam trawl (n = 1); gear comparison (n = 2); killitrap (n = 198); seine (n = 4); night-light (n = 23); weir (n = 4).

(McBride and Able in press). Laboratory and field observations indicate that they apparently remain in these habitats until they succumb to seasonally lowering temperatures. In the laboratory, young-of-the-year did not feed below 12 C, and no fish survived at temperatures below 10 C. In the field, individuals collected below 12 C were frequently disoriented and unable to remain upright (McBride and Able in press). Thus, individuals of this species that settle and stay in New Jersey waters cannot survive winter temperatures, do not migrate south for the winter, and are probably expatriates as first suggested by Gill (1904).

CHAPTER 58

Mugil cephalus Linnaeus
Striped Mullet

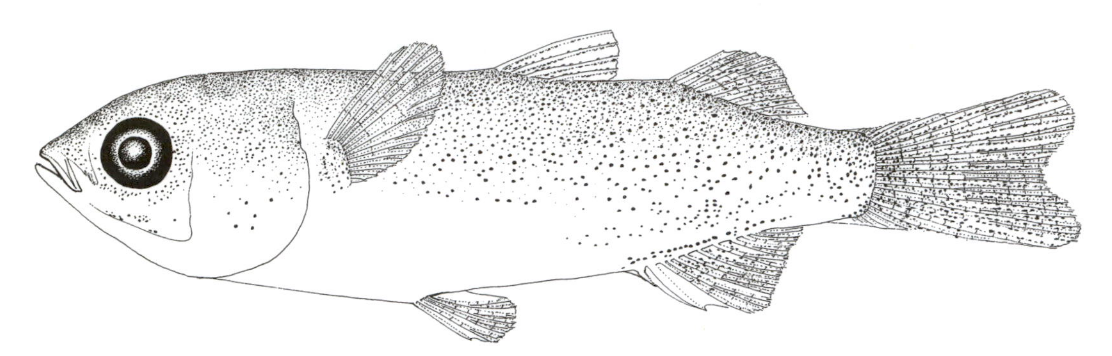

58.1 *Mugil cephalus* juvenile, 37.0 mm TL. Collected 1953, off Miami, Florida. ANSP 128648. Illustrated by Nancy Arthur.

DISTRIBUTION

Mugil cephalus (fig. 58.1) occurs worldwide in coastal waters and, by one estimate, is the most abundant and widespread inshore teleost (Odum 1970). In the western Atlantic, it is found between Maine and Brazil, and a single juvenile specimen has been recorded from Nova Scotia (Scott and Scott 1988). Other previous reports from Canadian waters were later found to refer to *M. curema* (Gilhen 1972). In the Middle Atlantic Bight, stages older than young-of-the-year are rarely collected (e.g., Hildebrand and Schroeder 1928), although a few reports of fishes exceeding 20 cm in the fall may refer to older year classes (Nichols and Breder 1927; Schaefer 1967). Juveniles have been recorded from most Middle Atlantic Bight estuaries (table 4.2), but eggs and larvae have not been reported from oceanic or estuarine waters in this region.

REPRODUCTION

Observations of spawning are few (e.g., Arnold and Thompson 1958), and establishing locations where reproduction occurs is based on larval occurrences. After a migration from freshwater (often in large schools), spawning occurs in South Atlantic Bight continental shelf waters over a range of depths extending from about the 36-m line into the Gulf Stream (Breder 1940; Gunter 1945; Broadhead 1953; Anderson 1958). The spawning season extends from October to February, with a peak in November and December (Anderson 1958) or January through April (Collins and Stender 1989) when continental shelf water temperatures are falling. Fish about 20 cm long (characterized as ripe) have been collected during the fall near New York (Nichols and Breder 1927), but this is the only report we have found suggesting any spawning activity in the central part of the Middle Atlantic Bight. These fish were deemed to be in spawning condition on the base of their robust morphology, not on gonad evaluation; thus, the evidence for spawning is nebulous.

DESCRIPTION

Eggs are spherical, transparent, and straw-colored, and have a smooth, unmarked chorion. The diameter averages 0.72 mm, and the single oil globule averages 0.28 mm. Hatching occurs about 48 hours after fertilization at larval sizes of about 2.4 mm TL. Larval eyes and mouth parts are undeveloped at hatching. The diagnostic count of two spines and nine rays is present in the anal fin at about 16 mm TL (Collins 1985b). After the development of adult complements of fin rays, mullets go through an early juvenile (querimana) stage while remaining in near-surface layers at sea. These pelagic juveniles are flat-sided and silvery, with a densely pigmented dorsum (see *M. curema,* fig. 59.1). Juveniles (fig. 58.1) have a rounded snout and small, terminal mouth. The two well-separated dorsal fins are both short-based, as is the anal fin. Scales do not cover the second dorsal and anal fins, however, as in *M. curema.* The total count of 11 anal fin elements (spines plus rays) separates juveniles of this species from those of *M. curema* (with a total of 12). Vert: 24; D: IV, I, 7–8; A: II, 9 (in larvae); III, 8 (in juveniles and adults); Pect: 14–18; Plv: I, 5.

THE FIRST YEAR

Small larvae have only been collected in shelf waters of the South Atlantic Bight (Anderson 1958), and these small stages (<6–10 mm SL) were most common over the outer half of the shelf (Powles 1981; Collins and Stender 1989). In quarterly sampling in the South Atlantic Bight, pelagic juveniles (<26.5 mm FL) were most common during the winter, with almost no occurrences in other seasons (Fahay 1975). The querimana, or pelagic-juvenile stage, ends at sizes between 35 and 45 mm, at which size a prominent adipose eyelid—which covers the eye with only a narrow slit over the pupil—forms, and the anal fin formula changes from two spines, nine rays to three spines, eight rays (Anderson 1958). The onset of scale formation is undescribed, but a figure of a fully scaled 25-mm querimana stage (Hildebrand and Schroeder 1928) indicates this development presumably occurs while pelagic juveniles are at sea. Scales are cycloid in young stages and become weakly ctenoid in adults (Hoese and Moore 1977).

In the South Atlantic Bight, a maximum size of 31 mm (mostly 20–25 mm) has been reported for pelagic juveniles occurring in offshore ocean waters (Anderson 1958; Fahay 1975; Collins and Stender 1989). A shoreward movement is then facilitated by favorable wind-driven drift (Powles 1981). Along the Georgia coast, young-of-the-year first arrive at barrier beaches in November and continue to occur there through April. They do not enter estuaries until January and then become more abundant through May and into June. They are never taken in abundance during summer and fall, however, a fact attributed to scattering of the young (Anderson 1958). In North Carolina, pelagic juveniles enter estuaries and sounds beginning in February, at sizes between 18 and 25 mm TL (Jacot 1920; Higgins 1928). Pelagic juveniles were most abundant near the Gulf Stream off the North Carolina coast during January in another study (Fahay 1975) and were presumably then subject to advection north into Middle Atlantic Bight waters. In sampling at Indian River Inlet, Delaware, young-of-the-year occurred December through May at sizes between 25 and 32 mm FL (Pacheco and Grant 1973). When juveniles reach 40 mm SL they are fully capable of osmoregulation and can then tolerate salinities from freshwater to full seawater (Nordlie et al. 1982). Young-of-the-year are reportedly most common during fall and winter in Delaware estuaries (Wang and Kernehan 1979).

Determining the growth rate of young-of-the-year in the central part of the Middle Atlantic Bight from available data is difficult. The continued presence of fishes 25 to 32 mm FL through the winter at Indian River Inlet (see above), combined with the progression of length modes later in the year (fig. 58.2), indicates growth may initially be retarded in young-of-the-year entering estuaries during the winter, but rapid growth then ensues through the first summer, and lengths of 9 to 19 cm are reached by fall. Size at age 1 in several Gulf of Mexico areas varies from 13 to 18 cm FL (Broadhead

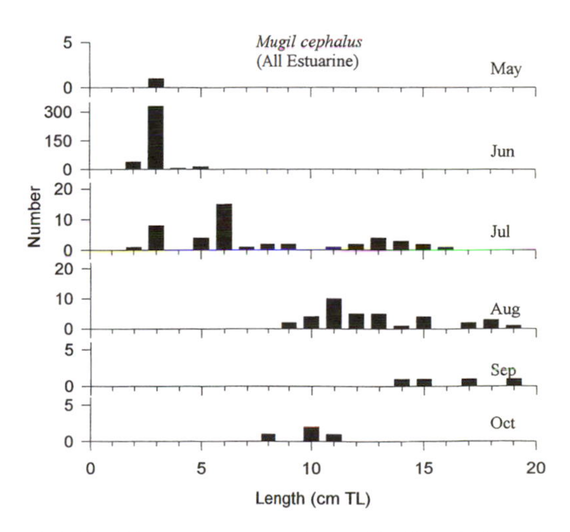

58.2 Monthly length frequencies of *Mugil cephalus* collected in estuarine waters of Great Bay and Delaware Bay. Sources: de Sylva et al. 1962; RUMFS: night-light ($n = 1$); weir ($n = 15$).

1958), and several growth curves derived from other areas in the South Atlantic Bight also suggest a size of about 15 cm as the year class enters its first fall (Anderson 1958). Lengths of 20 cm in these other studies equate with age-2 fish (Broadhead 1958). Assuming the modes shown in figure 58.2 represent a single year class, growth may be faster in the Middle Atlantic Bight, where these sizes are reached and exceeded by late summer. Alternately, perhaps the Middle Atlantic Bight is also visited by age-1 fish during the summer in some years, and the growth estimates represent more than a single year class. Most of these data were collected in the shore zone of Delaware Bay (de Sylva et al. 1962), and the occurrence of a smaller size class in October is curious. A similar size-group of *M. curema* appears in October (see *M. curema* chapter), and the possibility of misidentification exists. Clearly, a more thorough study of length-at-age and summer growth rates in both species of *Mugil* in the central part of the Middle Atlantic Bight is needed to resolve these seeming inconsistencies.

Habitat varies with life-history stage. Larvae and pelagic juveniles occur in the open ocean in near-surface layers (Powles 1981). Following the pelagic-juvenile stage, they occur along barrier beaches and then enter estuaries throughout their range. Water temperatures probably regulate the time that young-of-the-year are able to remain in estuaries (Sylvester et al. 1974). Young-of-the-year smaller than 50 mm SL have been reported from extremely shallow waters, including the surf zone and tide pools, despite temperatures approaching the upper lethal limit. Juveniles larger than 50 mm SL occur in somewhat deeper waters but may move to shallower areas during flood tides (Major 1978). Young are able to survive in marsh pools on Florida's gulf coast in temperatures as high as 34.5 C (Kilby 1949). Studies in Hawaii showed that young smaller than 50 mm SL prefer temperatures between 30.0 and 32.5 C, and larger juveniles 50–130 mm) prefer 19.5 to 20.0 C (Sylvester et al. 1974).

Juveniles of the two *Mugil* species appear to occur in different habitats during their first year. In the Great Bay–Little Egg Harbor study area, very few collections of this species were made in areas where *M. curema* was abundant, such as tidal creeks where salinities ranged from 23 to 33 ppt (Rountree and Able 1992a), suggesting the two species do not co-occur during their first summer. In Delaware Bay and tributaries, young-of-the-year may occur in freshwater (Smith 1971; Wang and Kernehan 1979) or salinities as low as 4 ppt. These are lower than salinities where *M. curema* occurs (de Sylva et al. 1962). This species occurred in a beach seine study of the tidal Delaware River near Philadelphia where *M. curema* did not (Weisberg et al. 1996). Conversely, *M. curema* was commonly collected in night-lighting in the high-salinity RUMFS boat basin in Great Bay, where *M. cephalus* was absent (Able et al. 1997).

Although large numbers of young-of-the-year overwinter in estuaries in the South Atlantic Bight (Nordlie et al. 1982), such is not the case in the Middle Atlantic Bight where most emigrate from estuaries in the fall. We have no records of occurrences after October in the Great Bay–Little Egg Harbor estuary, although a few have been recorded during winter months in Delaware Bay (de Sylva et al. 1962). The latter were small specimens (2–3 cm) and may have been early products of winter spawning in the South Atlantic Bight.

CHAPTER 59

Mugil curema Valenciennes
White Mullet

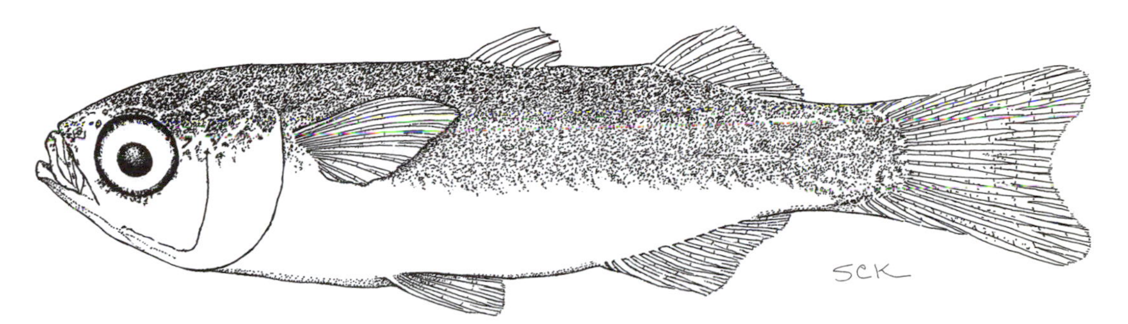

59.1 *Mugil curema* "querimana" stage juvenile, 24.2 mm SL. Collected June 11, 1986, dip net (night-light), RUMFS boat basin, Great Bay, New Jersey. ANSP 175240. Illustrated by Susan Kaiser.

DISTRIBUTION

All life-history stages of *Mugil curema* (fig. 59.1) are recorded from Canada to Uruguay (Collins 1985a). After the first year, however, this species is rarely, if ever, collected north of Florida (Anderson 1957). Juveniles have been reported from as far north as Nova Scotia (Scott and Scott 1988) and from many estuarine locations within the Middle Atlantic Bight (table 4.2), but eggs and larvae have not been reported from oceanic or estuarine waters in this region.

REPRODUCTION

There is no available evidence that spawning occurs in the Middle Atlantic Bight. The onset of spawning coincides with rising water temperatures in early spring in the South Atlantic Bight. Spawning occurs February through October, with a strong peak between April and June (Anderson 1957) or April and May (Collins and Stender 1989). An immense school of spawning fish was once observed off the coast of southern Florida, close to the axis of the Gulf Stream (Anderson 1957), and most observations of the smallest larvae have been across the South Atlantic Bight continental shelf, as far offshore as the Gulf Stream (Anderson 1957; Fahay 1975; Powles 1981; Collins and Stender 1989).

DESCRIPTION

The pelagic eggs have a finely etched chorion, very narrow perivitelline space, and a single oil globule 0.03 mm in diameter. Egg diameter averages 0.9 mm. Larvae hatch at 2.6 mm TL after 40 to 42 hours of incubation (Anderson 1957). Larval eyes and mouth parts are undeveloped at hatching. Total complements of fin rays are present in the dorsal and anal fins at 5.3 mm TL, and the diagnostic anal fin counts of two spines and 10 rays are present at 14.5 mm TL. After development of adult complements of fin rays, the young go through an early juvenile (querimana) stage while remaining in near-surface layers at sea. These pelagic juveniles are flat-sided and silvery, with a densely pigmented dorsum (fig. 59.1). Juveniles have a rounded snout and small terminal mouth. The two dorsal fins are well separated, and both are short-based, as is the anal fin. The total count of 12 anal fin elements (spines plus rays) separate juveniles of this species from those of *M. cephalus* (with a total of 11). Vert: 24; D: IV, I, 7–8; A: II, 10 (in larvae); III, 9 (juveniles and adults); Pect: 15–18; Plv: I, 5.

THE FIRST YEAR

In quarterly sampling in the South Atlantic Bight, pelagic juveniles were most common during spring (May), although small numbers continued to be present through the remaining seasons (Fahay 1975). Larvae and pelagic-juvenile stages are characterized by heavy dorsal pigmentation, a possible adaptation to high levels of solar radiation in surface layers of the ocean where they occur (Powles 1981). The spine/ray complement of the anal fin changes from two spines and 10 rays in larvae to three spines and 9 rays in early juveniles, at sizes of 30 to 40 mm TL (Anderson 1957). The onset of scale formation is undescribed, but at sizes larger than 6.5 cm, scales cover the second dorsal and anal fins.

In one study in the South Atlantic Bight, the smallest size class was distributed the farthest offshore, and sizes were inversely related to distance from shore (Powles 1981). This life-history pattern, involving spawning in relatively warm, highly saline water offshore, followed by a movement to cooler, less saline coastal water, was more strongly expressed in *M. cephalus* (during the winter) than in *M. curema* (during the spring). This study also found that circulation mechanisms in the South Atlantic Bight are more favorable for inshore transport during winter than during spring. Young-of-the-year of *M. curema* occur in South Atlantic Bight waters in two size modes, possibly deriving from reproduction in Caribbean waters and in South Atlantic Bight waters. The results of these differences are that, during the spring, the largest pelagic juveniles in the South Atlantic Bight are well-developed enough to be capable of actively swimming against the prevailing current and into estuaries there, but the smallest are subject to advection north via the Gulf Stream (Powles 1981).

Larvae and pelagic juveniles in the South Atlantic Bight are most common during April and May (Fahay 1975; Powles 1981; Collins and Stender 1989). Ingress into marshes in the South Atlantic Bight occurs in April and May and continues weakly through August (Anderson 1957). Ingress at Beaufort, North Carolina, begins in late April but peaks in May (Jacot 1920). They also first begin to occur in estuaries of the central part of the Middle Atlantic Bight in April, and small size classes continue to occur through June and into July. Results of night-lighting in Great Bay indicate they may enter the system in pulses (fig. 59.2). A report

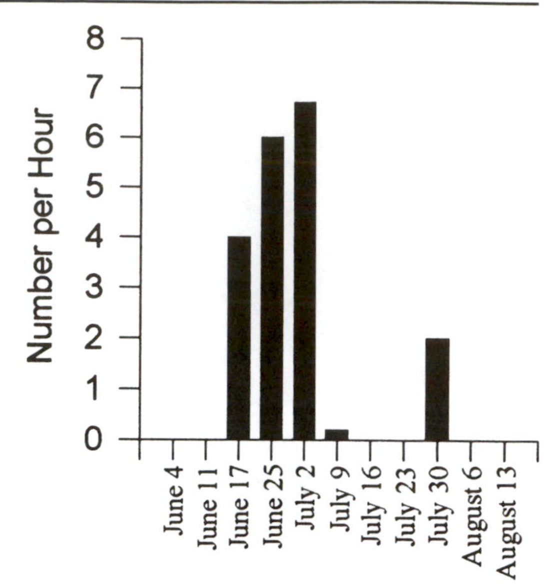

59.2 Weekly night-light collections of *Mugil curema* in the RUMFS boat basin, Great Bay, New Jersey.

of 41 specimens, 2 to 3 cm, collected during December from Delaware Bay (de Sylva et al. 1962) is inconsistent with other collections and is difficult to interpret.

Recent collections of larvae and pelagic juveniles (<25 mm) have been made during May in the Slope Sea off the Middle Atlantic Bight, where they were strongly associated with the Gulf Stream and waters transitional between the Gulf Stream and Slope Sea (J. A. Hare, M. P. Fahay, and R. K. Cowen in prep.). Most of these were collected in a neuston net, but smaller individuals were also taken in a plankton net, which sampled to a depth of 15 m. They occurred with an assemblage of other species, together categorized as "transients," members of which were spawned south of Cape Hatteras. Young-of-the-year of all members of this assemblage occupy Middle Atlantic Bight estuaries during the summer. Maximum size attained in offshore waters, and the size at which pelagic juveniles enter coastal habitats, is about 25 mm. Anderson (1957) took none larger than 25 mm in the ocean or when they began to enter a Georgia marsh. Maximum sizes taken were 26 mm (Fahay 1975) or 31 mm (Collins and Stender 1989) in the same oceanic area in other studies. Sizes at ingress in North Carolina are 20 to 21 mm, and this size continues to appear until September (Jacot 1920). In the Great Bay study area, entering pelagic juveniles are apparently large enough to avoid capture by a passive plankton net, because only two individuals (<21.2 mm SL) were collected in a 6-year study (Witting 1995; D. A. Witting, K. W. Able, and M. P. Fahay in prep.). Large numbers have been collected by night-lighting in this system, however (fig. 59.2; Able et al. 1997).

Our collections indicate growth resumes after ingress and in the Great Bay–Little Egg Harbor Estuary is most rapid between June and September when the observed size mode increases from 4 to 13 cm (fig 59.3). This rate equates with an increase of 1 mm per day, a value also estimated for Virginia young-of-the-year (Richards and Castagna 1970). In a laboratory experiment, larvae were 36 mm long after 36 days (Houde et al. 1976). Young-of-the-year reach 10 to 13 cm by the fall in our study area. This is similar to an observation from the South Atlantic Bight, where young-of-the-year reached 12 cm by October and November (Anderson 1957). This species is reported to reach 200 mm SL at age 1 (Alvarez-Lajonchere 1976). We cannot verify this because the species does not occur in the Middle Atlantic Bight after its first fall. However, there is a singular report of young-of-the-year ranging from 10 to 20 cm in November in the discharge canal of the Indian River Power Plant (Wang and Kernehan 1979).

Habitat use varies with life-history stage. Larvae and pelagic-juveniles occur in the open ocean and Gulf Stream, where they are primarily neustonic (Fahay 1975; Powles 1981). In the South Atlantic Bight, they occur over the entire shelf, from nearshore to the proximity of the Gulf Stream, but the smallest size classes have been collected near the 180-m contour (Powles 1981) or deeper (Collins and Stender 1989). Observations from Great Bay–Little Egg Harbor indicate that young-of-the-year spend the summer in higher-salinity portions of the estuary. In weir collections in high-salinity tidal creeks, *M. curema* (n = 1,450) vastly outnumbered *M. cephalus* (n = 15) (Rountree and Able 1992a). It has also been reported to be most abundant during summer and fall in higher-salinity portions of Delaware estuaries (Wang and Kernehan 1979). Young-of-the-year occur in salinities greater than 13 ppt in Delaware Bay (de Sylva et al. 1962) or 10 to 15 ppt at Woodland Beach, Delaware (Wang and Kernehan 1979), but none were found in low salinities or freshwater habitats in either of these studies.

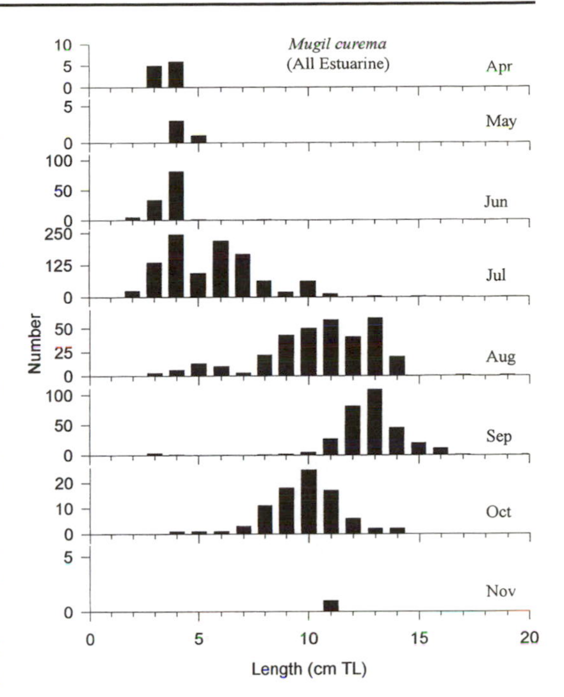

59.3 Monthly length frequencies of *Mugil curema* young-of-the-year collected from estuarine waters of Great Bay and Delaware Bay. Sources: de Sylva et al. 1962; RUMFS: beam trawl (n = 2); gear comparison (n = 4); seine (n = 11); night-light (n = 52); weir (n = 1,450).

Young-of-the-year leave estuaries in the central Middle Atlantic Bight in October. They also leave marshes in the Beaufort, North Carolina area in the fall (Jacot 1920), after which they migrate offshore (Anderson 1957). After this emigration, there is no indication that subsequent year classes return to the central part of the Middle Atlantic Bight.

Sphyraena borealis DeKay
Northern Sennet

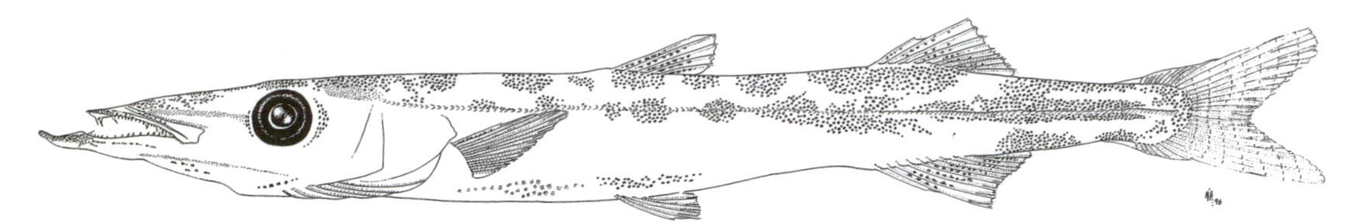

60.1 *Sphyraena borealis* juvenile, 56.0 mm TL. Collected June 27, 1973, Beasley's Point, New Jersey. ANSP 169550 (1 of 2). Illustrated by Nancy Arthur.

DISTRIBUTION

A pelagic schooling species, *Sphyraena borealis* (fig. 60.1) occurs in oceanic and estuarine waters from Nova Scotia to Florida and into the north-central Gulf of Mexico (Briggs 1958; Scott and Scott 1988). In the Middle Atlantic Bight, early life-history stages, represented only by juveniles, have been reported from most estuaries (table 4.2).

REPRODUCTION

Spawning may occur in the ocean because the only known collections of eggs occurred off Florida (Houde 1972). Larvae have been collected in Slope Sea waters north of Cape Hatteras during May. These were associated with a "frontal assemblage" of species, implying an origin in the Gulf Stream in the South Atlantic Bight and eventual flux into Middle Atlantic Bight continental shelf waters with increasing size (J. A. Hare, M. P. Fahay, and R. K. Cowen in prep.).

DESCRIPTION

The eggs are spherical (1.2 mm in diameter) with a single oil globule (0.2–0.3 mm in diameter) located at the anterior end of the yolk mass and a narrow perivitelline space. Larvae hatch at approximately 2.6 mm TL. The slender larvae have well-developed canine teeth and a fleshy knob at the tip of the lower jaw that develops by 5 mm. Fin rays are complete by 13.5 mm SL. Scale formation begins at 14.5 mm SL. Further details of embryonic and larval development are provided by Houde (1972). The juveniles are slender and cylindrical, and the snout is sharply pointed (fig. 60.1). The gill arches lack rakers. The two dorsal fins are widely separate, and the first has six spines. The second dorsal fin is opposite the anal fin, and both have short bases. The body pigment has characteristic square to rectangular blotches along the lateral and dorsal midlines. Scales on the posterior part of the lateral line are ridged, forming a keel-like structure along the caudal peduncle. Vert: 24; D: V–VI, I, 8–9; A: I–II, 8–9; Plv: I, 5.

THE FIRST YEAR

Larvae begin swimming 2 days after hatching in the laboratory and begin to feed by the third day. They average 5.5 mm SL at 7 days after hatching, 11 mm at 14 days, and about 13.5 mm at 21 days, although growth may have been unnaturally slow because of a lack of appropriate food late in these observations (Houde 1972). Larvae collected in Slope Sea waters north of Cape Hatteras during May 1993 ranged from 5.6 to 19.0 mm SL (mean = 8.0 mm SL) (M. P. Fahay unpubl. data). Juveniles first occur in Great Bay–Little Egg Harbor, New Jersey in June at 6 to 9 cm TL (fig. 60.2). By October, they range from 15 to 20 cm TL. Based on modal length-frequency progression, growth of young-of-the-year averaged approximately 1.2 mm per day up to September.

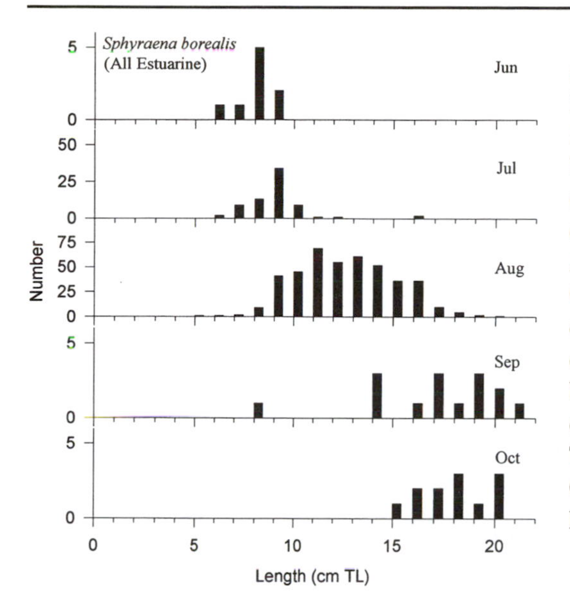

60.2 Monthly length frequencies of *Sphyraena borealis* young-of-the-year collected in the Great Bay–Little Egg Harbor study area. Sources: RUMFS: night-light (n = 5); weir (n = 341); pop net (n = 187).

The habitats, like much of the life history, are poorly known. Young-of-the-year have only been collected in estuarine waters in the Middle Atlantic Bight. They are presumed to be pelagic because they are seldom collected in nets fished on the bottom but can be common in nets that enclose the entire water column (Rountree and Able 1992a; Hagan and Able unpubl. data). In Great Bay–Little Egg Harbor, they have been collected during the day in subtidal marsh creeks (Rountree and Able 1993) and have been observed to move on to the marsh surface during flood tides (K. J. Smith pers. observ.). They are not collected in the estuary after October (fig. 60.2), presumably because they have migrated into the ocean for the winter.

CHAPTER 61

Tautoga onitis (Linnaeus)
Tautog

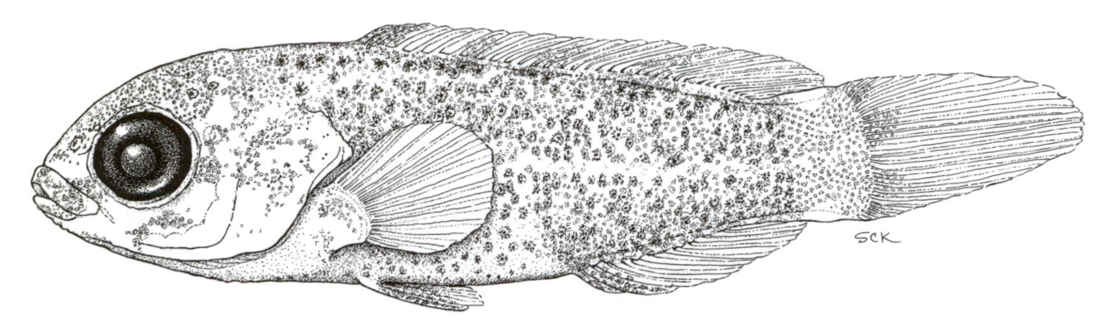

61.1 *Tautoga onitis* juvenile, 11.4 mm TL. Collected July 30, 1992, 1-m beam trawl, RUMFS boat basin, Great Bay, New Jersey. ANSP 175241. Illustrated by Susan Kaiser.

DISTRIBUTION

Tautoga onitis (fig. 61.1) occurs primarily on the inner continental shelf and the polyhaline portions of estuaries from Nova Scotia to South Carolina but is most abundant from Cape Cod to Delaware Bay (Bigelow and Schroeder 1953), where it occupies a variety of structured habitats from rocky reefs and mussel beds to eelgrass and the edges of deep channels. The populations may be highly localized (Cooper 1966), although seasonal inshore-offshore movements of large juveniles and adults are common (Olla et al. 1974). In the Middle Atlantic Bight, early life-history stages are reported from most estuaries, (table 4.2) but they are more abundant in the northern part of the bight.

REPRODUCTION

Spawning occurs on the inner portion of the continental shelf and in estuaries in May through June. It appears to follow a northward progression through the summer, beginning as early as April in the southern part of the bight and extending into the northern part by May. Peak spawning occurs in the central part of the bight in June and July, followed by a decline in August (Berrien and Sibunka in press). This timing is consistent with estimates derived from gonadosomatic indices off Maryland and Virginia (Eklund and Targett 1990) and the occurrence of larvae (fig. 61.2). In the Great Bay–Mullica River estuary, planktonic eggs were collected from April through August, with peak abundances in June and July (Sogard et al. 1992). The initial occurrence and peak abundance were earlier in the river than in the bay and adjacent inlet, suggesting that spawning began earlier in the upper part of the estuary and continued later in the summer in the lower estuary. Back-calculation of spawning dates from sagittal otoliths of juveniles collected in Great Bay–Little Egg Harbor found a mean date of 4 June and a range of 17 April–22 July (Sogard et al. 1992). This study and those in other locations in the Middle Atlantic Bight suggest that there is more spawning in estuaries and bays than on the inner continental shelf.

DESCRIPTION

The eggs are pelagic, spherical, and 0.97 to 1.00 mm in diameter. The perivitelline space is narrow, and oil globules are lacking. Larvae hatch at about 2.0 mm with unpigmented eyes and unformed mouth parts. Larger larvae are heavily pigmented over the anterior two-thirds of the body. The mouth is small throughout development. Most fins are well differentiated by about 10 mm, although the pelvic fins form later. In juveniles, the caudal peduncle is somewhat more lightly pigmented than the rest of the

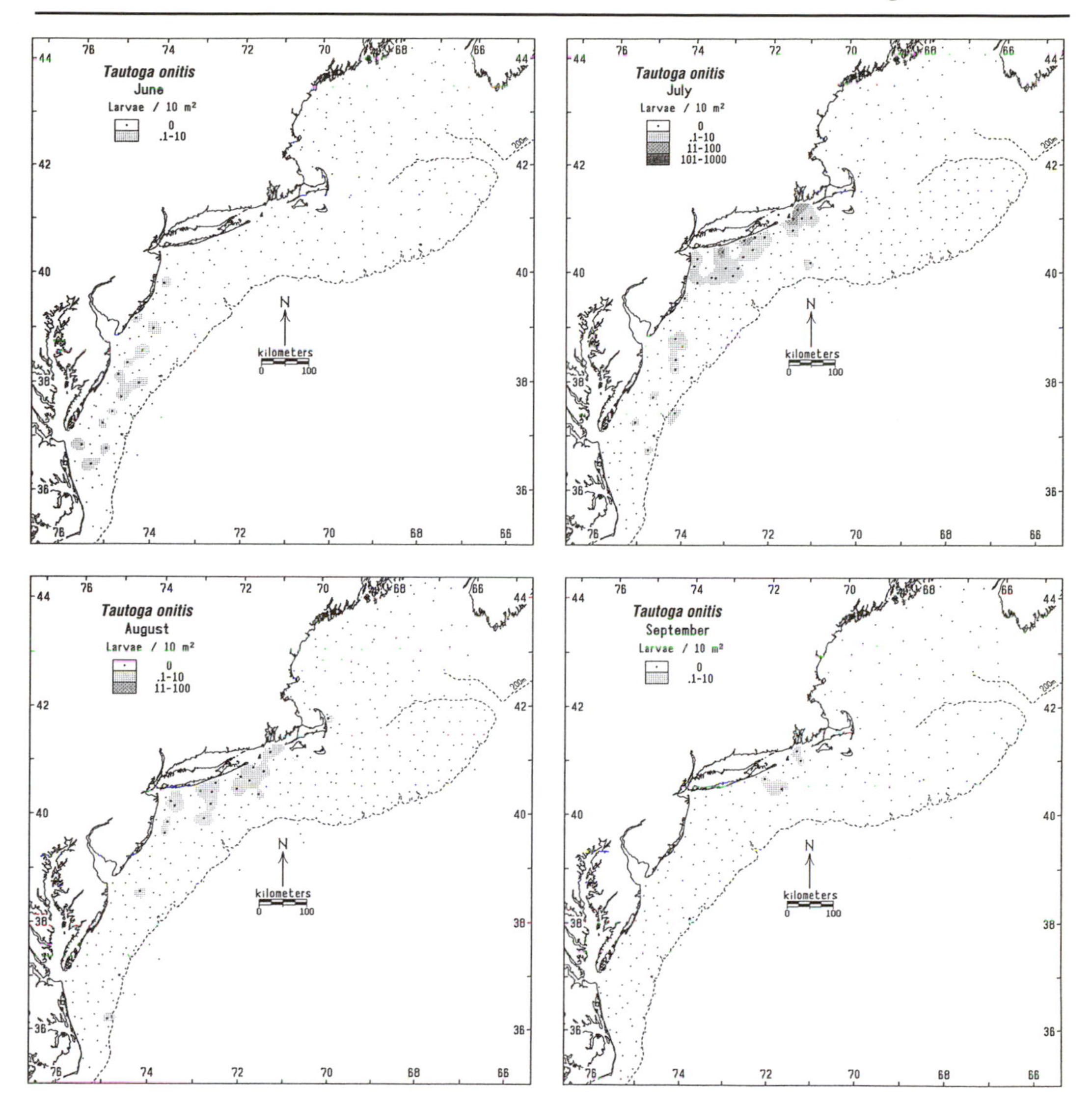

61.2 Monthly distributions of *Tautoga onitis* larvae during MARMAP surveys 1977–1987.

body (fig. 61.1). The mouth is small, the maxilla barely reaching the anterior edge of the eye. There is a single, continuous dorsal fin, with no separation between the spines and rays. The anterior dorsal fin rays lack the prominent spot found in *Tautogolabrus adspersus*. Vert: 34–35; D: XVI-XVII, 10–11; A: III, 7–8; Pect: 16; Plv: I, 5.

THE FIRST YEAR

Embryonic and larval development occur along the estuary-ocean gradient. In sampling along the Mullica River–Great Bay–Beach Haven Ridge corridor in southern New Jersey, the eggs were most abundant in Great Bay and the inner continental shelf and less so in the lower salinity portions of the estuary (Sogard et al. 1992). In the ocean, larvae were most abundant off New Jersey, Long Island, New York, and Rhode Island during the summer (fig. 61.2). The larvae have been collected infrequently in Great Bay ($n = 21$; Sogard et al. 1992; Witting 1995; D. A. Witting, K. W. Able, and M. P. Fahay in prep.), and the same can be said of the nearby Beach Haven Ridge area on the inner continental shelf (15 m depth). Elsewhere in the Middle Atlantic Bight, the larvae have been reported from large bays (Herman 1963; Croker

1965; Bourne and Govoni 1988) and barrier island estuaries (Allen et al. 1978; Monteleone 1992).

Daily increment formation in the sagittal otoliths has been validated, and they have a well-defined settlement mark. Thus, several aspects of the early life history can be interpreted (Sogard et al. 1992). Individuals examined from Great Bay–Little Egg Harbor estuary (n = 37) spent approximately 3 weeks in the plankton before settlement. This is in agreement with laboratory studies in which settlement occurred 17 days after hatching (Schroedinger and Epifanio 1997). Otoliths show that the smallest settled individuals collected in the field had been settled for 11 to 23 days (7.6–13.2 mm SL). Date of settlement for these individuals ranged from 6 May to 13 August, with a mean of 25 June. This corresponds well with peak abundance of early demersal stage individuals, which occurred in June through August (fig. 61.3, 61.4). The infrequent occurrence of early demersal-stage individuals in collections showed little evidence of successful settlement at Beach Haven Ridge. In Narragansett Bay the peak in young-of-the-year abundance occurred from July to August (Dorf and Powell 1997).

Abundance of postsettlement individuals collected in traps varied between years in the Great Bay–Little Egg Harbor estuary from 1991 to 1994 (fig. 61.5). Most were collected in the summers of 1992 and 1994, but they were consistently abundant for a long period of time in 1992. In 1994, a single large peak accounted for the highest catches. A hurricane caused a decline in young-of-the-year in Narragansett Bay immediately after the storm and during the following year (Dorf and Powell 1997).

Independent estimates of growth for young-of-the-year individuals during the first summer are similar. Analysis of modal length-frequency progressions and otolith ages indicate similar

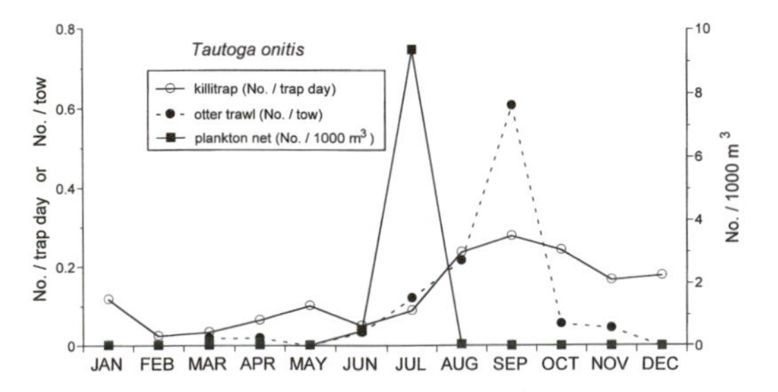

61.3 Monthly occurrences of *Tautoga onitis* young-of-the-year in the Great Bay–Little Egg Harbor study area based on 1-m plankton net collections 1989–1994 (n = 18), otter trawl collections 1988–1989 (n = 130), and killitrap collections 1991–1994 (n = 1,959).

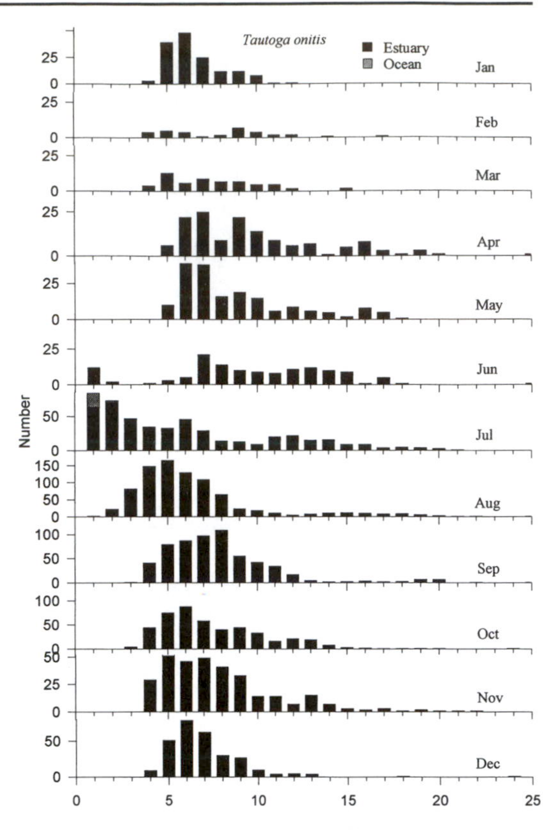

61.4 Monthly length frequencies of *Tautoga onitis* young-of-the-year collected in the Great Bay–Little Egg Harbor study area. Sources: RUMFS: otter trawl (n = 270); 1-m beam trawl (n = 131); 1-m plankton net (n = 142); experimental trap (n = 934); LEO-15 Tucker trawl (n = 22); LEO-15 2-m beam trawl (n = 21); gear comparison (n = 29); killitrap (n = 2,010); seine (n = 110); night-light (n = 6); weir (n = 45).

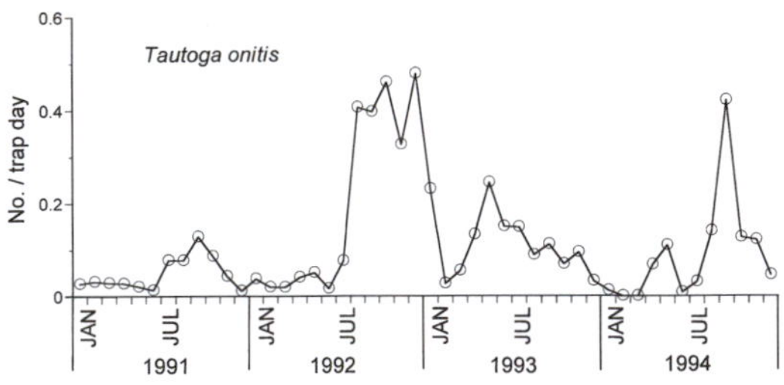

61.5 Annual variation in abundance of *Tautoga onitis* collected in Great Bay based on trapping 1991–1994 (n = 1,914).

estimates of growth for settled tautog (0.52 mm per day and 0.47 mm per day, respectively) (Sogard et al. 1992). In caging experiments, growth rates varied from −0.47 to +0.84 mm per day, but growth in vegetated habitats averaged 0.45 mm per day. Growth was usually fastest for the smallest fish, although it was strongly influenced by location and habitat (Sogard 1992; Able and Studholme 1994). Measurements of tagged fish were slightly lower with rates of less than 0.3 mm per day for free-swimming fishes (Hales and Able unpubl. data). Similar caging studies in the Hudson River, based on smaller individuals, found similar growth rates (Able and Studholme 1994). In Great Bay–Little Egg Harbor, young-of-the-year were 1 to 2 cm TL in June, and age-1 fish were approximately 4 to 12 cm (fig. 61.4) with most individuals 4 to 10 cm in October (fig. 61.4; Sogard et al. 1992). Comparison of these sizes with lengths of individuals older than age 1 the following June indicated only minor growth during the fall, winter, and spring. Age-1 fish reached a size of 110 to 170 mm SL by the end of their second summer, with a modal size in September of 155 mm SL. These size estimates of juveniles in New Jersey are larger than the mean lengths of individuals older than age 1 from Rhode Island (Cooper 1967; Dorf and Powell 1997) as well as those for Virginia (Hostetter and Munroe 1993).

A variety of studies have indicated that small juveniles prefer vegetated habitats. In southern New Jersey, they have been found abundantly in both sea lettuce (*Ulva lactuca*) and eelgrass (*Zostera marina*) (Nichols and Breder 1927; Able et al. 1989; Sogard and Able 1991) as well as in areas with shell and sponge (fig. 61.6; Szedlmayer and Able 1996). In association with sea lettuce, smaller juveniles (< 35 mm SL) are consistently a bright green, matching the color of the vegetation, while larger juveniles, which typically occur in unvegetated and deeper habitats have a dark, mottled coloration similar to that of adults (Nichols and Breder 1927; Sogard et al. 1992). Several other studies in the Middle Atlantic Bight have found higher abundances of juveniles in eelgrass (Briggs and O'Connor 1971; Orth and Heck 1980; Heck et al. 1989, Szedlmayer and Able 1996) and macroalgae (Dorf and Powell 1997). More re-

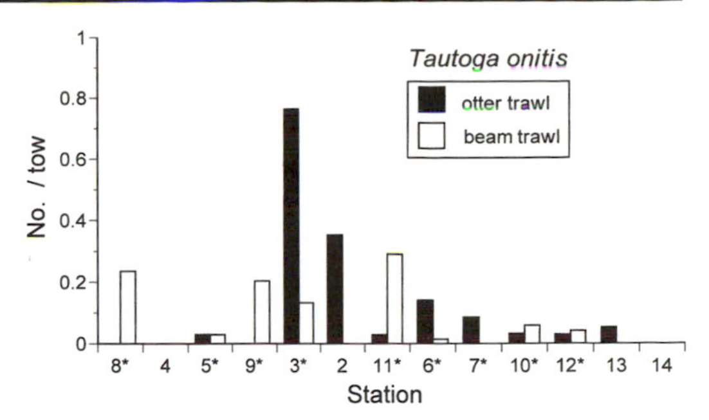

61.6 Distribution of *T. onitis* young-of-the-year by habitat type based on otter trawls (n = 144) and beam trawls (n = 68) deployed on regular stations 2–14 in Great Bay and Little Egg Harbor. Both otter trawl and beam trawl used on stations designated by asterisk; otherwise, only the otter trawl was used. See figure 2.3 for station locations; table 2.1 for specific habitat characteristics by station.

cent studies of heavily impacted and man-made habitats in the lower Hudson River estuary have found small juveniles associated with old pier pilings and interpier areas but not under large intact piers. Larger juveniles are typically found with other types of structured habitats such as rocks, jetties, and shipwrecks (Olla et al. 1974, 1979).

Based on tag-recapture studies of juveniles (25–190 mm TL) in Great Bay, there is strong site fidelity after settlement and during the first summer (L. S. Hales and K. W. Able unpubl. data). Of 1,148 individuals tagged at a single location, there were 278 recaptures (14%), with some individuals caught more than once (up to 13 times) over a 9-month period. Average distance moved was 19 m during all seasons. Of 48 individuals captured two or more times, 96% were recaptured within 5 m of a previous capture location.

Temperature has a profound affect on behavior of young-of-the-year based on laboratory observations (Hales and Able in press). As ambient seasonal temperatures declined, swimming frequency decreased sharply at 8 C, as did feeding at temperatures below 4 C. Burying occurred at low temperatures (2–7 C) and appeared to be a short-term response to sudden decreases in temperature. There was relatively little mortality at the lowest temperatures (down to 2 C). However, differences in fall and spring lengths from field collections suggested size-selective winter mortality on the smallest fish (Hales and Able in press).

CHAPTER 62

Tautogolabrus adspersus (Walbaum) Cunner

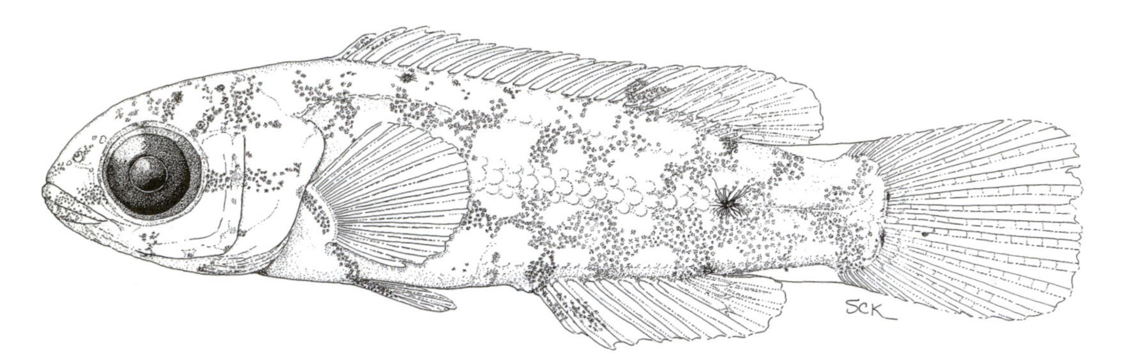

62.1 *Tautogolabrus adspersus* juvenile, 13.8 mm SL. Collected July 16, 1992, LEO-15 mooring, Beach Haven Ridge, New Jersey. ANSP 175242. Illustrated by Susan Kaiser.

DISTRIBUTION

Tautogolabrus adspersus (fig. 62.1) occurs from Newfoundland and the Gulf of St. Lawrence to the Chesapeake Bay area (Robins and Ray 1986). It is a marine species, only rarely penetrating low-salinity areas (Smith 1985) and is apparently more abundant around structures such as shipwrecks, pilings, rocky reefs, oyster beds, wharves, etc. In our study area, it occurs more commonly northeast of the New York Bight Apex. Thus, it is more common in Narragansett Bay than in the Hudson–Raritan Estuary and more common in the latter system than in Great Bay–Little Egg Harbor (Berg and Levinton 1985). It is apparently rare to uncommon in the Delaware Bay area, for only two specimens (adults) were captured in the surveys reported by de Sylva et al. (1962). In all estuaries where this species occur, they are less common during winter, when they either move offshore as temperatures decline, or simply become inactive, burrowing into the substrate and covering themselves with silt (Smith 1985). Sampling at an inner continental shelf site in the New York Bight Apex (Wilk et al. 1992) yielded large numbers of *T. adspersus* consisting of fishes age 1 and older between 10 and 20 cm. They were present nearly year-round, although they were rare or absent during June and July and were most abundant during the winter and early spring, suggesting that they occupy these shelf habitats during the coldest time of the year and coastal and estuarine habitats during summer. Early life-history stages have been reported from estuaries between Delaware Bay and Cape Cod (table 4.2), although all stages are rare in Delaware Bay (Stone et al. 1994), and the lack of egg occurrences in Nauset Marsh is most likely due to a lack of appropriate sampling (M. P. Fahay and K. W. Able unpubl. observ.).

REPRODUCTION

This species becomes sexually mature at about 5.4 to 6.0 cm SL, a size they usually reach at age 1+ in the Middle Atlantic Bight (Dew 1976). Spawning in this region occurs between May and October, based on the occurrences of eggs and larvae. Eggs were common constituents of MARMAP samples (1977–1987) over the inner continental shelf from May to November, with a peak in June and July (Berrien and Sibunka in press). Centers of abundance were between the New York Bight apex and Narragansett Bay; a separate center was over Georges Bank. There may be some spawning in estuarine waters near Little Egg and Hereford inlets, New Jersey, however, as there have been reports of numerous eggs and larvae in both places during the early summer (Milstein et al. 1977; Allen et al. 1978, respectively). Spawning begins earlier in waters north and east of the central part of the Middle Atlantic Bight. Thus, eggs appear as early as April in Massachusetts Bay (Collette and Hartel 1988).

DESCRIPTION

The pelagic eggs of this species are spherical and range from 0.84 to 0.92 mm in diameter. The perivitelline space is narrow, and they lack oil globules. The larvae hatch at a length of about 2.2 mm, with unpigmented eyes and a nonfunctional mouth. The snout-anus length is about 50% of the TL. A prominent melanophore occurs over the posterior end of the gut, and a pair of melanophores occurs midway between the anus and tail tip. Fin rays are completely formed by 8 to 10 mm. Further developmental details are in Kuntz and Radcliffe (1917) and Fahay (1983). Juveniles (fig. 62.1) can vary in color pattern, ranging from weakly to strongly mottled. In all patterns, however, a prominent melanophore occurs on the anterior part of the soft-ray portion of the dorsal fin, and there is no strong contrast between the caudal peduncle and the remainder of the body, as in *Tautoga onitis*. The dorsal fin is continuous (or with a very slight notch between the spines and rays). The mouth is small, with the maxilla barely reaching the anterior edge of eye. Vert: 36; D: XVIII, 9–10; A: III, 8–9; Pect: 16; Plv: I, 5.

THE FIRST YEAR

Larvae are more abundant in July and August over the continental shelf northeast of the drowned Hudson River valley than they are in the southern part of the Middle Atlantic Bight (Morse et al. 1987; Malchoff 1993) (fig. 62.2). Larvae occur continuously over Georges Bank from July through September. In plankton collections made at the Beach Haven Ridge near Little Egg Inlet, larvae were first collected during July, which is consistent with collections made during the MARMAP surveys between 1977 and 1987. They were only rarely collected, however, during another study in this area (Milstein et al. 1977) and then only in July. In a study of ichthyoplankton in Great Bay–Little Egg Harbor, larvae were collected during July and August during 4 years of the 6-year survey (Witting 1995; D. A. Witting, K. W. Able, and M. P. Fahay in prep.). Larvae are rare or absent in the lower Hudson River estuary (Dovel 1981).

In the Great Bay–Little Egg Harbor study area, larvae and juveniles appeared in plankton collections during July at sizes between 5.2 and 15.6 mm SL (Witting 1995; D. A. Witting, K. W. Able, and M. P. Fahay in prep.). Collections in a variety of small traps began in July and reached a peak in August (fig. 62.3). A large influx of larvae (all <10 mm) appeared in Long Island waters (mostly along the ocean shore) during June (Perlmutter 1939). Small larvae then decreased in abundance, whereas juveniles up to 34 mm continued to occur in July. As is the case with other labrids, *T. adspersus* settles at a relatively wide range of sizes (8–14 mm) (Tupper and Boutilier 1995a). It tends to utilize a variety of habitats but suffers extreme postsettlement mortality in less structurally complex habitats (Levin 1991, 1993; Tupper and Boutilier 1995a). In the few weeks following settlement, young-of-the-year are typically associated with rocky bottom, pilings, and debris, or seagrass or macroalgae beds. In localities where it is locally abundant, it may also settle in very high densities (>6 individuals/m^2) and is known to exhibit density-dependent growth and survival (Tupper and Boutilier 1995a).

Growth during the first year is slow. In collections near Little Egg Inlet the July ingress of larvae and early-benthic-stage juveniles is clearly depicted (fig. 62.4). The slow rate of growth (about 0.4 mm/day) is apparent, and the mode at the end of summer is only about 5 cm. Published growth rates from the Scotian Shelf and Gulf of Maine are similar at about 0.5 mm per day (Bigelow and Schroeder 1953; Tupper 1994; Tupper and Boutilier 1995a). There is little or no growth through the winter, and age-1 fish are apparently only 5 to 12 cm TL, with a mode between 6 and 8 cm TL. In the region of Hereford Inlet, New Jersey, young (20–28 mm) first occur in seine hauls in August and only reach 40 to 60 mm before disappearing in early winter (Allen et al. 1978). The observed length frequencies the following

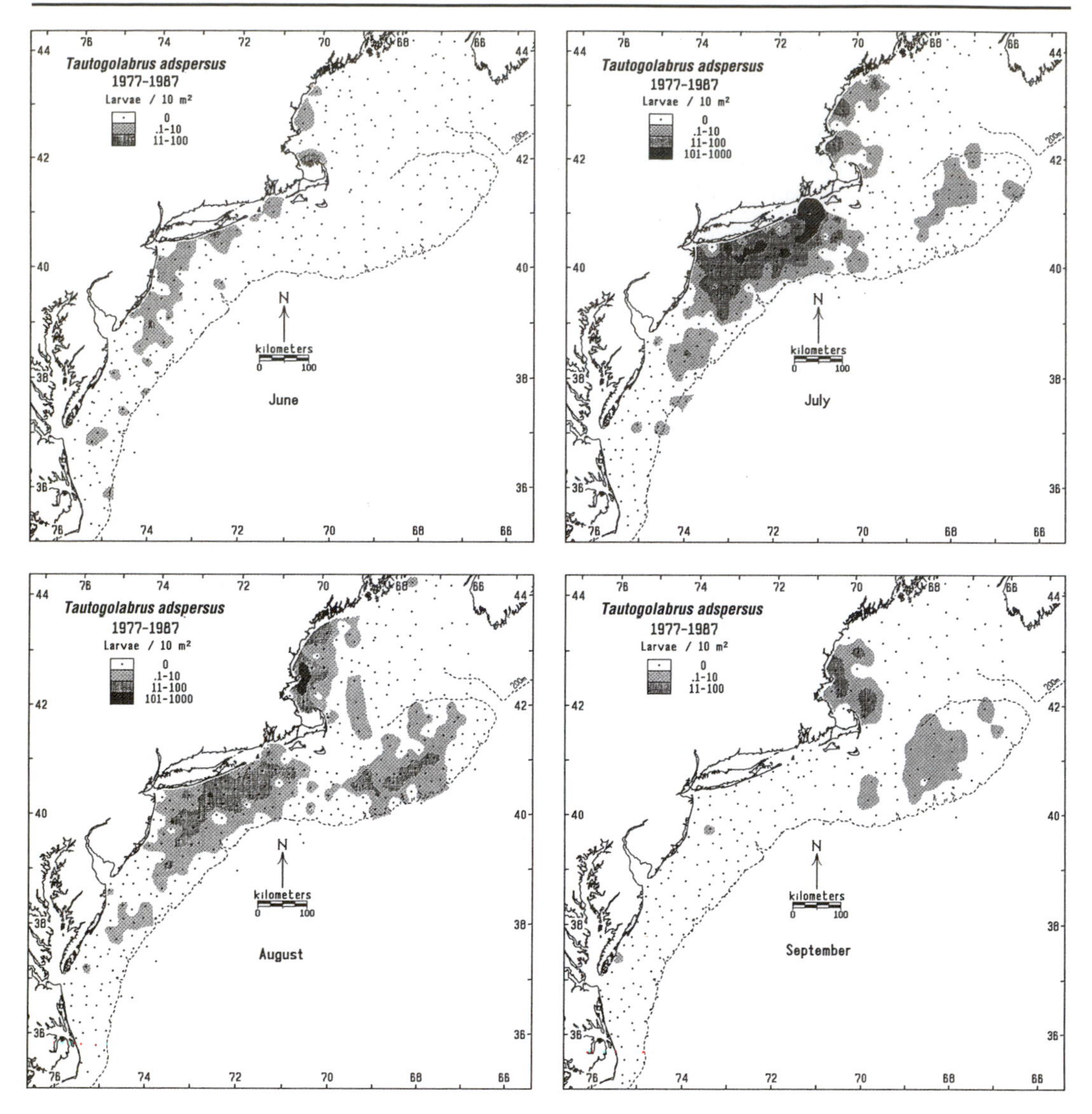

62.2 Monthly distributions of *Tautogolabrus adspersus* larvae during MARMAP surveys, 1977–1987.

spring may be affected by mortality during the winter, because it has been shown that this mortality affects smaller size classes more than larger (L. S. Hales and K. W. Able in press).

Mark/recapture studies indicate a high degree of habitat fidelity in this species during the summer (L. S. Hales and K. W. Able unpubl. observ.; Tupper and Boutilier 1995a). In the Hudson River, all age classes (including young-of-the-year) are more abundant in pile field habitats than in interpier or underpier habitats (Able and Studholme 1994). They have also been reported to take shelter in eelgrass beds during

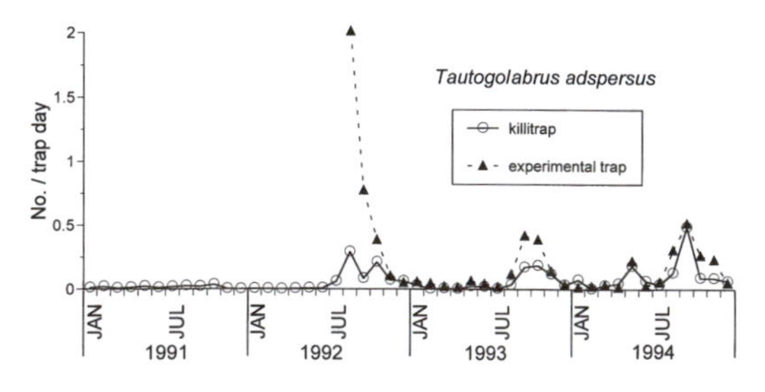

62.3 Annual variation in abundance of *Tautogolabrus adspersus* young-of-the-year in the Great Bay–Little Egg Harbor study area based on killitrap collections 1991–1994 (n = 764) and experimental trap collections 1992–1994 (n = 488).

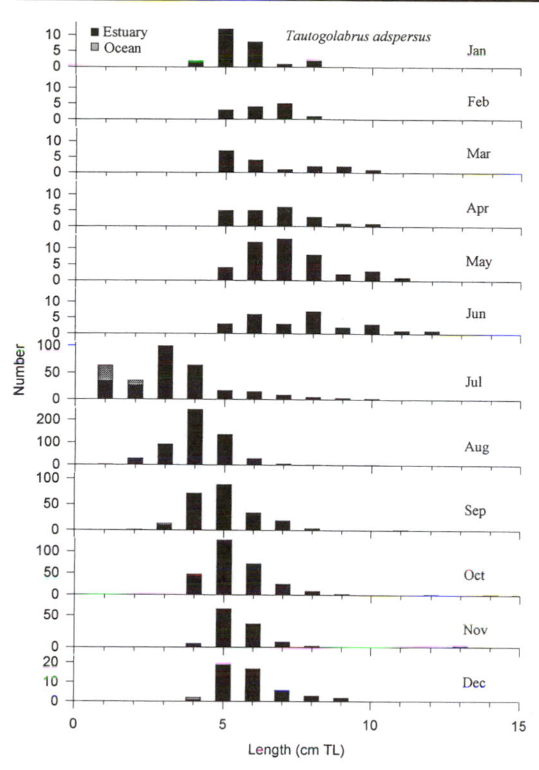

62.4 Monthly length frequencies of *Tautogolabrus adspersus* young-of-the-year collected from estuarine and adjacent coastal ocean waters of Long Island and Great Bay. Sources: Perlmutter 1939; RUMFS: otter trawl (*n* = 7); 1-m beam trawl (*n* = 24); 1-m plankton net (*n* = 29); LEO-15 Tucker trawl (*n* = 32); LEO-15 2-m beam trawl (*n* = 11); gear comparison (*n* = 8); killitrap (*n* = 647); seine (*n* = 4); night-light (*n* = 1).

winter (Nichols and Breder 1927). They are only rarely collected in Sandy Hook Bay and then only in October and November (Berg and Levinton 1985). Traps are effective collecting devices, presumably because the young of this species occur in structured habitats. Some transforming juveniles may settle on the continental shelf and remain uncollected. Divers have observed large numbers of young-of-the-year around submerged wrecks on the continental shelf off New Jersey (D. Witting pers. observ.), and these may also provide critical habitat during the first year.

Judging from trap collections, *T. adspersus* occurs in the Great Bay–Little Egg Harbor study area throughout the year, but a decline during December and January (fig. 62.3) suggests many either become inactive or migrate away from estuaries during the winter. Several marked fish were recaptured at intervals through the winter in one Great Bay study (L. S. Hales and K. W. Able unpubl. observ.). In Cape Cod and Gulf of Maine areas, *T. adspersus* young-of-the-year enter metabolic torpor at temperatures below 5 C and then usually estivate under rocks near their settlement sites, at least during their first winter (Dew 1976; Curran 1992; Tupper 1994). Numbers remain low through the spring until the following year's progeny begin to be collected (fig. 62.4).

CHAPTER 63

Pholis gunnellus (Linnaeus) Rock Gunnel

63.1 *Pholis gunnellus* juvenile, 56 mm TL. Collected August 11, 1994, Suction Device, Submersible *Delta,* dive no. 3309, 30-m depth, east of Little Egg Inlet, New Jersey. ANSP 175243. Illustrated by Nancy Arthur.

DISTRIBUTION

A boreal species, *Pholis gunnellus* (fig. 63.1) is found on both sides of the Atlantic Ocean. In the western Atlantic, it is distributed from Labrador to Delaware Bay, but it is rare south of southern New England. It frequently occurs in tide pools and other intertidal habitats, where it is usually associated with structures (stones, crevices, etc.), under which it hides (Collette 1986; Scott and Scott 1988). It inhabits the north shore of Long Island but avoids the generally warmer bays along the south shore (Perlmutter 1939). It has also been collected at considerable distances from shore, as at 183 m depth on Georges Bank (Schroeder 1933; Bigelow and Schroeder 1953). The distribution of early life-history stages in Middle Atlantic Bight estuaries is usually not reported (table 4.2).

REPRODUCTION

Spawning presumably takes place during winter after adults vacate intertidal areas in the fall. This conclusion is based on observations in New Hampshire where: (1) females in November contained eggs about 1.0 mm in diameter; (2) they disappeared during the winter; and (3) those returning to inshore waters in March were in a spent condition (Sawyer 1967). Egg masses have been found in Long Island Sound and Peconic Bays (Gudger 1927). Nest sites are guarded by both parents and have been described from shallow waters where eggs have been found deposited in empty oyster shells (Nichols and Breder 1927) or as deep as 22 m (Smith 1887).

DESCRIPTION

Eggs are adhesive and occur in masses deposited on a variety of bottom types. Egg diameters average 1.4 mm (Breder and Rosen 1966) or range from 1.7 to 2.2 mm (Wheeler 1969) and have a single oil globule. Larvae are elongate and the snout-anus length is slightly more than 50% the SL. Characteristic pigment includes a row of melanophores along the ventral body margin from the anus to the caudal fin base, and a "stitching" pattern along the venter between the gill covers and the anus. See Barsukov (1959) and Fahay (1983) for further details of early development, including methods to separate them from very similar stichaeid and anarhichadid larvae. Juveniles (fig. 63.1) have long dorsal and anal fins, greatly reduced pelvic fins, and wide, fan-shaped pectoral fins. The ornate pigment pattern is unlike that of either larvae or adults and consists of a series of large spots along the dorsal fin base and another series along the lateral line. Vert: 86–89; D: 73–86 (all spines); A: II, 37–44; Pect: 10–14; Plv: I, 1.

THE FIRST YEAR

Hatching occurs after 42 to 70 days incubation at 6 C (Breder and Rosen 1966) at a larval size of about 9.0 mm (Nichols and Breder 1927). Larvae are common constituents of the ich-

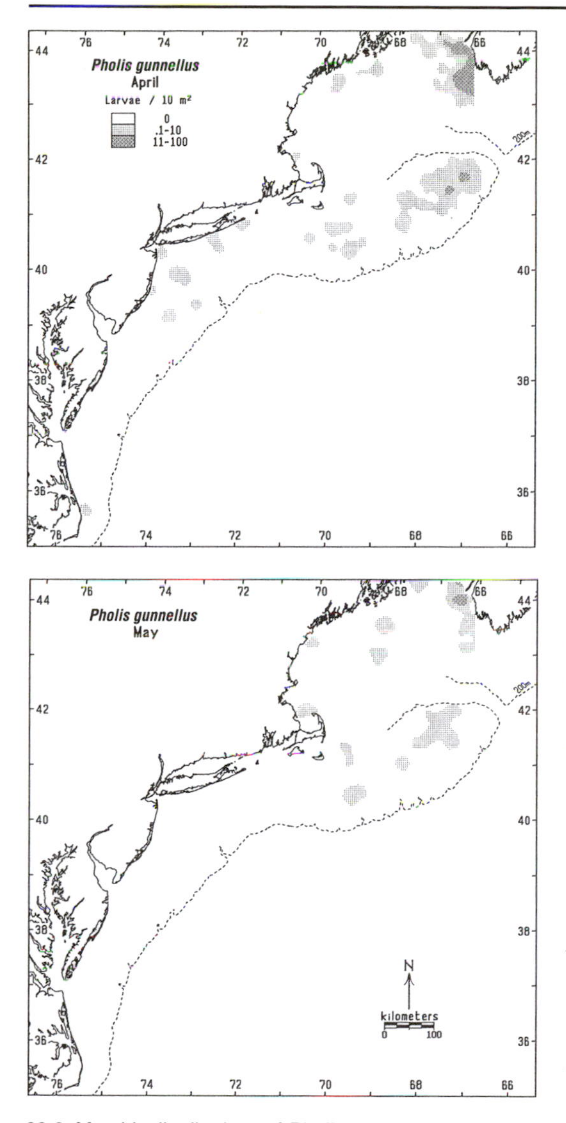

63.2 Monthly distributions of *Pholis gunnellus* larvae during MARMAP surveys 1977–1987. Average distributions during April and May.

thyoplankton communities over the continental shelf and in certain large New England embayments, such as Beverly-Salem Harbor, Massachusetts (Elliott et al. 1979). Larvae collected during the MARMAP surveys (1977–1987) occurred throughout the Gulf of Maine, Georges Bank, and northern part of the Middle Atlantic Bight, between January and June, with the southernmost occurrences off New Jersey in April and May (fig. 63.2). Larvae have also been collected in Newark Bay during January, March and April (S. J. Wilk pers. comm.), but juveniles are rarely collected, presumably because of their cryptic behavior. Although rare, juveniles collected in Long Island Sound in July and early August in the Perlmutter study (1939) were reportedly young-of-the-year (46–103 mm).

Pelagic larvae apparently reach a maximum size of 30 to 35 mm SL before descending to the bottom, judging from collections made during the 11-year MARMAP study and during an ichthyoplankton study inside Little Egg Inlet (fig. 63.3), but details of settlement are unknown. The early juvenile (52 mm SL) illustrated for this chapter (fig. 63.1) was collected with a suction device from a submersible operating on the bottom at 30-m depth, just off Little Egg Inlet. Presumably, settlement occurs at sizes greater than 35 mm SL, or between 30 and 40 mm as suggested by Bigelow and Schroeder (1953). No information suggests that all young-of-the-year actively enter bays and estuaries, although sizes of those collected in the Little Sheepshead Creek plankton study (Witting 1995; D. A. Witting, K. W. Able, and M. P. Fahay in prep.) were consistently larger

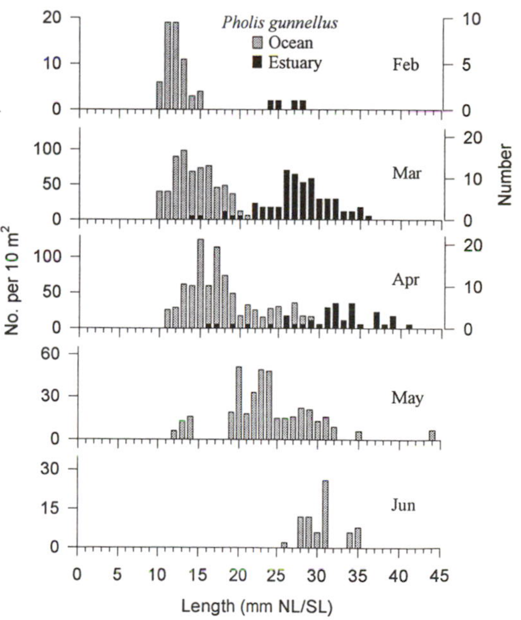

63.3 Monthly length frequencies of *Pholis gunnellus* larvae collected over the continental shelf during MARMAP study (1977–1987) (shown as number per 10 m² of sea surface, left axis) and in the Great Bay–Little Egg Harbor study area (1991–1994) (shown as number collected per length interval, right axis). Sources: MARMAP (bongo net); RUMFS: 1-m plankton net (n = 132).

than those collected during the same months farther offshore during the MARMAP study (fig. 63.3) suggesting an inshore movement with growth.

Size at age 1 has been calculated as 68 mm in the Bay of Fundy and 73 mm in New Hampshire (Sawyer 1967). The difficulty in sampling for this species prevents an appraisal of habitats used by young-of-the-year in bays and estuaries in the central part of the Middle Atlantic Bight.

CHAPTER 64

Astroscopus guttatus Abbott
Northern Stargazer

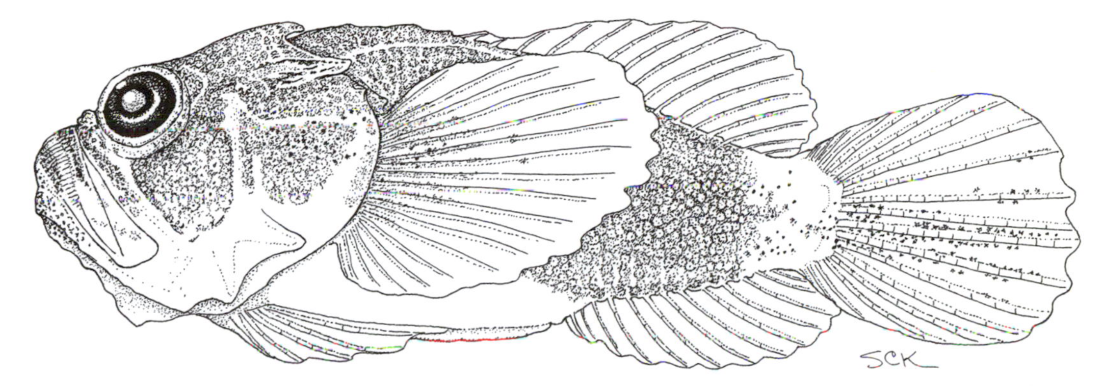

64.1 *Astroscopus guttatus* juvenile, 17.0 mm SL. Collected October 3, 1991, dip net (night-light), RUMFS boat basin, Great Bay, New Jersey. ANSP 175244. Illustrated by Susan Kaiser.

DISTRIBUTION

Astroscopus guttatus (fig. 64.1) is one of a few identified endemic species in the Middle Atlantic Bight. It is uncommon throughout its range from New York to North Carolina. It buries itself in the substrate with the eyes and top of head exposed and is known from rivers (Smith 1985), bays, coastal areas, and the ocean to a depth of 200 m (Robins and Ray 1986). Juveniles have been reported from Hudson River–Raritan Bay south to Chesapeake Bay (table 4.2), but larvae have only been reported from Great Bay and Chesapeake Bay.

REPRODUCTION

Details of spawning are not well known. Based on collections of larvae (identified as "Uranoscopidae"), reproduction occurs in Middle Atlantic Bight continental shelf waters between June and October, with most activity in the midshelf region off the Delmarva Peninsula and Chesapeake Bay mouth (fig. 64.2). Spawning reportedly occurs in lower Chesapeake Bay during May and June (Murdy et al. 1997).

DESCRIPTION

Eggs are undescribed. Larvae have a large mouth and very deep body anterior to the anus. Pigment is dense between the head and anus. Early juveniles (fig. 64.1) have a large head and robust body. Knoblike protuberances are present on top of the head, and blunt spines project posteriorly from the opercle. Young-of-the-

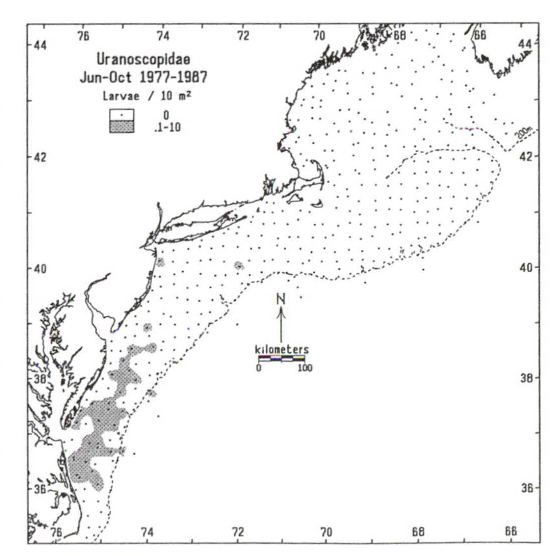

64.2 Occurrences of larval uranoscopids between June and October (inclusive) during MARMAP surveys 1977–1987. These larvae were not identified to the species level, but based on the distribution of adults, larvae in this region are probably *Astroscopus guttatus*.

year have their eyes situated near the top of the head with small electric organs behind them. Characteristic pigment in juveniles includes a pair of moustache marks at the tip of the lower jaw. Vert: 25; D: IV–V, 13–15; A: I, 12; Pect: 19–21; Plv: I, 5.

THE FIRST YEAR

Electric organs begin to develop at lengths greater than 12 mm (Dahlgren 1927) and complete their development and merge to form a single organ at 33 to 45 mm (White 1918). The migration of the eyes to a dorsal position on the head is completed by 25 mm (Pearson 1941).

Pelagic larvae (up to 11.5 mm SL) were collected from the Beach Haven Ridge study site on the inner continental shelf between July and September of 2 years (K. W. Able, M. P. Fahay, D. A. Witting, R. S. McBride, and L. S. Hales in prep.). Lengths of planktonic larvae collected by bongo net during the MARMAP surveys of the Middle Atlantic Bight continental shelf compared to lengths of larvae collected in the Tucker trawl at Beach Haven Ridge adjacent to Little Egg Inlet suggest that larger larvae occur in coastal areas (fig. 64.3). The differences in length distributions may be attributed to the different gears used, but the smallest larvae in midshelf waters suggest an inshore movement with growth. Plankton-net sampling near Little Egg Inlet collected larvae and early juveniles during the same time period as continental shelf collections (June–October) (Witting 1995; D. A. Witting, K. W. Able, and M. P. Fahay in prep.) and at the same

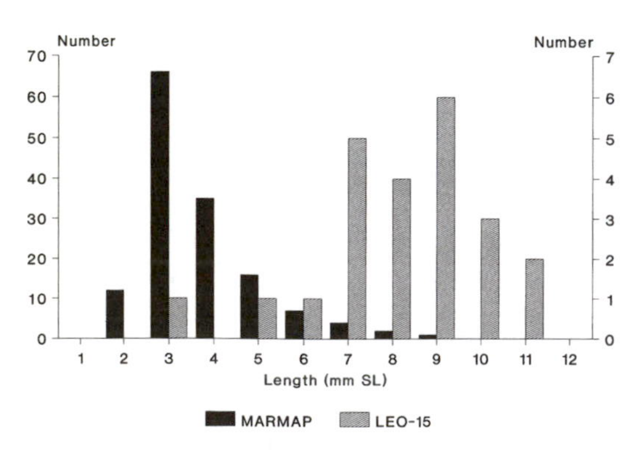

64.3 Comparison of lengths in *Astroscopus guttatus* larvae and juveniles collected from the continental shelf (MARMAP) and the inner continental shelf (LEO-15 study area).

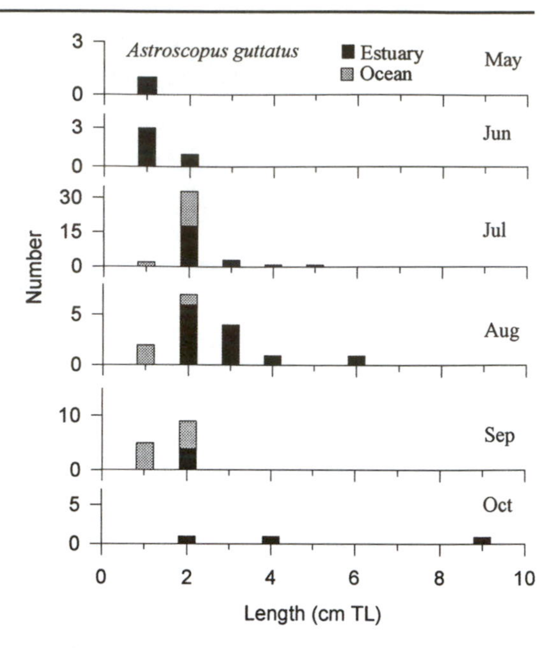

64.4 Monthly length frequencies of *Astroscopus guttatus* collected in the Great Bay–Beach Haven Ridge study area. Sources: RUMFS: 1-m plankton net (n = 17); LEO-15 Tucker trawl (n = 23); LEO-15 2-m beam trawl (n = 7); seine (n = 4); night-light (n = 38); gear comparison (n = 1).

sizes as those from the Beach Haven Ridge site. It is not known whether young-of-the-year only occupy estuarine waters or whether they settle to the bottom on the inner continental shelf as well.

Larvae (5–15 mm SL) collected from the water column under night-lights have been observed in laboratory conditions to settle to a sandy bottom. These individuals were able to bury themselves in the substrate in 1 to 2 seconds, using a combination of buccal pumping that liquefied the coarse sand below them, digging with the pectoral fins, and a peculiar swimming motion that redistributed the sand above them (D. Witting pers. comm.). Further indications that larvae settle to the bottom at these lengths are provided by bottom trawl collections at the Beach Haven Ridge (LEO-15) study site where the smallest specimens were about 13 mm SL (fig. 64.3).

Because of the scarcity of collections of this species, we can only make limited estimates of growth rate. On the basis of 40 young-of-the-year specimens collected near Great Bay–Little Egg Inlet (fig. 64.4), *Astroscopus guttatus* appears to reach 6 cm by its first fall. It is a lie-in-wait predator that buries in sandy substrate with only the eyes and tips of jaws protruding. Although it is not often collected in trawl tows, fish kills associated with low dissolved-oxygen levels often result in mass mortalities of this species in the Raritan Bay area (Fahay pers. observ.). There is no information on seasonal limits of estuarine occupation or mechanisms associated with emigration from the estuary by this species, nor are details of overwintering habitat known.

Hypsoblennius hentz (Lesueur) Feather Blenny

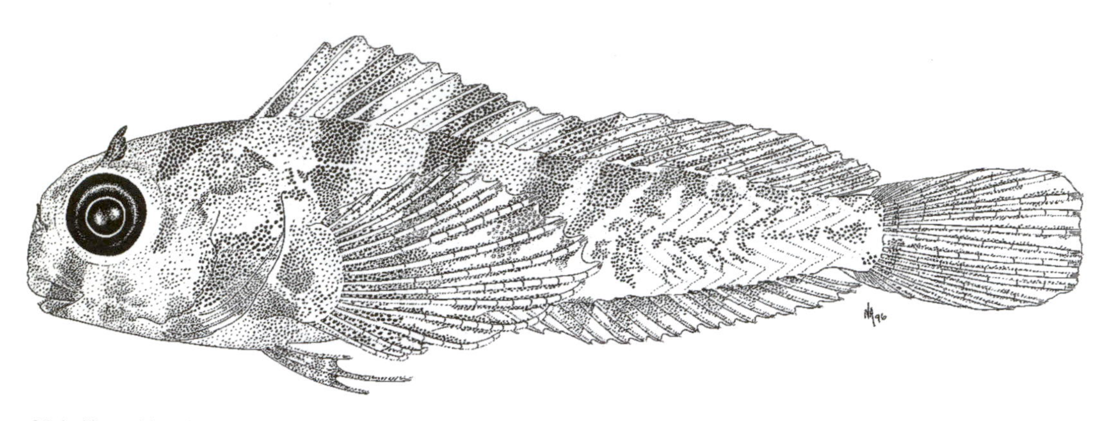

65.1 *Hypsoblennius hentz* early juvenile, 14.8 mm TL. Collected June 22, 1942, Bird Shoal, Beaufort, North Carolina. ANSP 74120. Illustrated by Nancy Arthur.

DISTRIBUTION

Hypsoblennius hentz (fig. 65.1) is a resident of inner continental shelf and higher-salinity estuarine waters from New Jersey and occasionally Nova Scotia to Florida and into the Gulf of Mexico (Robins and Ray 1986). It is abundant in Chesapeake Bay (Musick 1972). Relatively little is known of the habitat for this species, but it is reported at higher salinities where it is found in oyster reefs and shallow flats during the summer and in deeper channels and holes during the winter (Musick 1972). The early life-history stages occur in the southernmost estuaries of the Middle Atlantic Bight as far north as New Jersey (table 4.2) although they are not abundant there (Milstein and Thomas 1976a).

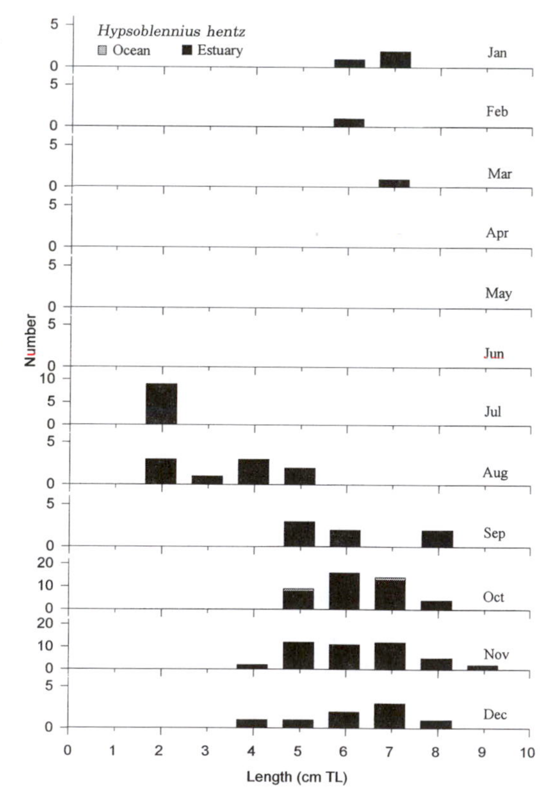

65.2 Monthly length frequencies of *Hypsoblennius hentz* young-of-the-year collected in the Great Bay–Little Egg Harbor study area. Sources: RUMFS: otter trawl (n = 4); 1-m beam trawl (n = 1); experimental trap (n = 32); LEO-15 2-m beam trawl (n = 10); gear comparison (n = 2); killitrap (n = 77).

REPRODUCTION

The occurrence of small demersal juveniles in July and August indicates that spawning occurs during the summer in the Great Bay–Little Egg Harbor study area (fig. 65.2). Spawning is also reported from May through September in Chesapeake Bay (Hildebrand and Schroeder 1928; Hildebrand and Cable 1938).

DESCRIPTION

The eggs are demersal, oblong, and flattened along one side at the point of attachment to the substrate, with the major axis 0.72 to 0.80 mm in diameter and the minor axis 0.64 to 0.68 mm. A variable number of oil globules is concentrated near the blastoderm. The chorion is sculptured. At hatching, the larvae are 2.6 to 2.8 mm long and are robust anteriorly with a thin tapering body. Pigment includes a heavy accumulation over the gut and a line of spots along the ventral edge of the tail (Hildebrand and Cable 1938). Details of larval development are provided in Hildebrand and Cable (1938). Juveniles have the laterally flattened body and prominent feathery appendages over the eye typical of this species (fig. 65.1). Vert: 32–34; D: XII, 13–15; A: II, 16–7; Pect: 13–15; Pelv: 3.

THE FIRST YEAR

The eggs are deposited on a hard substrate, such as oyster shell reefs, and they are guarded by the male (Hildebrand and Cable 1938). Time from development to hatching takes 11 days at 26 C. The larvae were moderately abundant in lower Chesapeake Bay from May through October (Olney 1983; Olney and Boehlert 1988). Demersal young-of-the-year first appear in Great Bay estuarine collections in July and August, but they are most abundant in October and November (figs. 65.2, 65.3). In Great Bay, they grow at an approximate rate of 0.5 mm per day during the summer and fall. By November, they have reached 4 to 9 cm (fig. 65.2). These sparse data suggest that the population in New Jersey is made up primarily of a single year class.

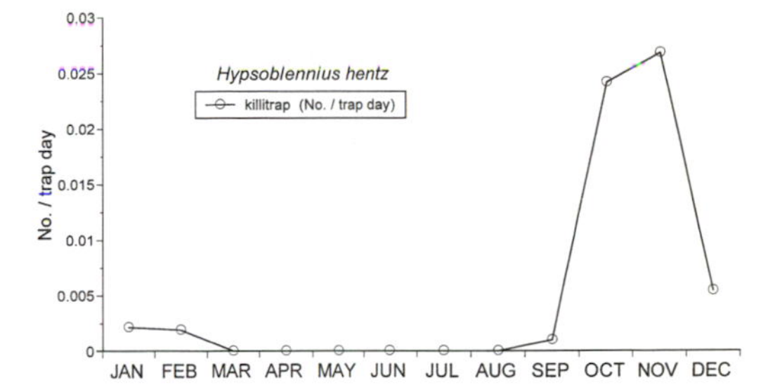

65.3 Monthly occurrences of *Hypsoblennius hentz* young-of-the-year in the Great Bay study area based on trapping 1991–1994 ($n = 77$).

CHAPTER 66

Ammodytes americanus DeKay
American Sand Lance

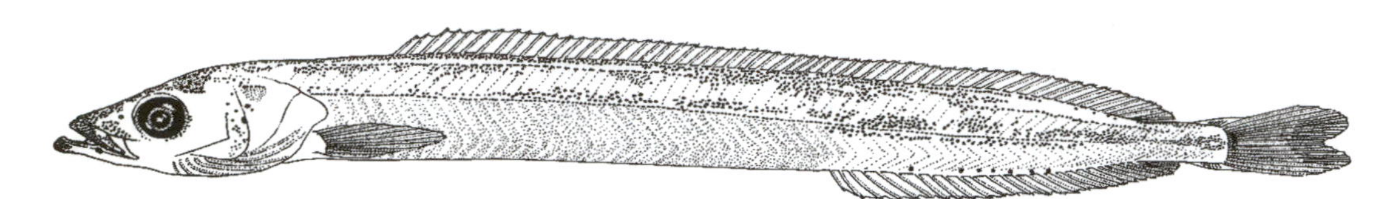

66.1 *Ammodytes americanus* juvenile, 45.5 mm TL. Collected June 6, 1995, seine, Tucker Island, Little Egg Inlet, New Jersey. ANSP 175245. Illustrated by Nancy Arthur.

DISTRIBUTION

Many aspects of the morphology and ecology of *Ammodytes* spp. along the East Coast of the United States are potentially confounded by the taxonomic problems in differentiating between *A. americanus* (fig. 66.1) and *A. dubius* (Nizinski et al. 1990). *Ammodytes dubius* is distinguished by more oblique folds of skin (plicae) on the sides (124–147 vs. 106–126 in *A. americanus*) and by differences in other meristic characters (see Nizinski et al. 1990). *Ammodytes americanus* is found from very shallow water of the inner continental shelf and estuaries from Labrador to Delaware Bay (Nizinski et al. 1990). Specimens we collected from Nauset Marsh were included in the study by Nizinski et al. (1990) and verified as *A. americanus*. We have also identified this species in our collections from Great Bay–Little Egg Harbor, but museum collections from Beach Haven Ridge, just outside Great Bay, included *A. dubius*. As a result, we conclude that while most estuarine collections of *Ammodytes* are *A. americanus*, special care should be taken with identification, especially with collections from the coastal ocean. Early life-history stages have been reported from almost all estuaries in the Middle Atlantic Bight (table 4.2), but some of these records may be confounded by identification problems.

REPRODUCTION

The timing of the maturation of field-collected individuals in the laboratory indicates that spawning occurs in the winter (Smigielski et al. 1984). Spawning may occur 1 to 2 months before hatching at low winter temperatures (see below). Small larvae occurred in the vicinity of Great Bay–Little Egg Harbor during March, April, and May (fig. 66.2).

DESCRIPTION

The eggs are demersal and adhesive and range from 0.9 to 1.0 mm in diameter with a narrow perivitelline space and a single oil globule of 0.28 to 0.38 mm (Smigielski et al. 1984). The larvae hatch at 5.7 to 6.3 mm and have elongate bodies with wide dorsal and ventral finfolds and pigmented eyes. Details of embryonic and larval development, including standardized developmental stages are available from Smigielski et al. (1984). The juveniles have a slender, elongate body and a deeply forked caudal fin, which is separated by a gap from the dorsal and anal fins (fig. 66.1). Pelvic fins are lacking. The snout is sharply pointed, and the lower jaw protrudes. The single, spineless dorsal fin is about twice as long as the single anal fin. Vert: 62–70; D: 52–61; A: 26–33; Pect: 11–15; Plv: none.

THE FIRST YEAR

Many details of embryonic development are available from laboratory studies (Smigielski et al. 1984). Time to hatching varied in duration with temperature and ranged from 61 days at 2 C to 25 days at 10 C. Yolk sac absorption by the larvae was complete at 6.3 to 6.8 mm. Transformation to juvenile morphology occurred at 29 mm and 131 days after hatching at 4 C and 102 days at 7 C. Individuals held at 2 C still had not transformed after 149 days. These same laboratory-raised individuals began schooling at 25 to 30 mm at 90 days after hatching. They began burying in a sand substrate at 35 to 40 mm at 133 days after hatching.

In the Great Bay–Little Egg Harbor estuary, larvae are a consistent component of the ichthyoplankton in early spring in every year from 1989 to 1994 (fig. 66.2; Witting 1995; D. A. Witting, K. W. Able, and M. P. Fahay in prep.). They first appear in coastal ocean and estuarine waters in January (fig. 66.3). Subsequently, small larvae have been collected only in the estuary from February to April. After the larvae reach a peak in abundance in plankton samples in late spring, they become available to seines towed in shallow water, especially in May and June (fig. 66.4). The young-of-the-year become less available to our collecting gears in the estuary during summer but can become more abundant in the fall (fig. 66.3).

Growth in the laboratory varied with temperature (2–10 C) and ranged from approximately 0.4 to 1.0 mm per day (Smigielski et al. 1984). The young-of-the-year appear as a single cohort from January through May. This same year class appears to be represented by larger individuals (8–15 cm TL) during October through February, and we interpret this as the approximate size at age 1.

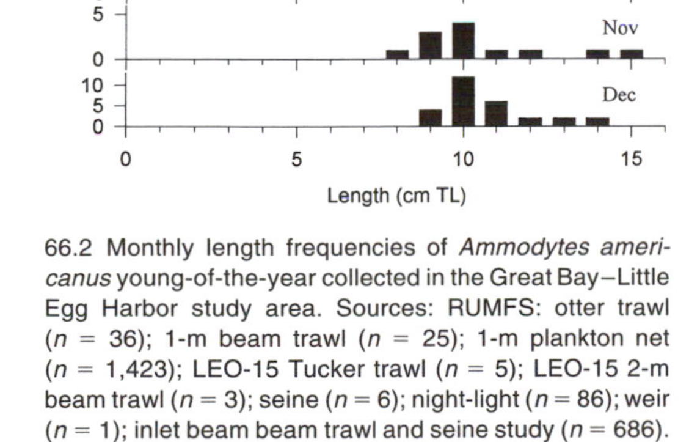

66.2 Monthly length frequencies of *Ammodytes americanus* young-of-the-year collected in the Great Bay–Little Egg Harbor study area. Sources: RUMFS: otter trawl (n = 36); 1-m beam trawl (n = 25); 1-m plankton net (n = 1,423); LEO-15 Tucker trawl (n = 5); LEO-15 2-m beam trawl (n = 3); seine (n = 6); night-light (n = 86); weir (n = 1); inlet beam beam trawl and seine study (n = 686).

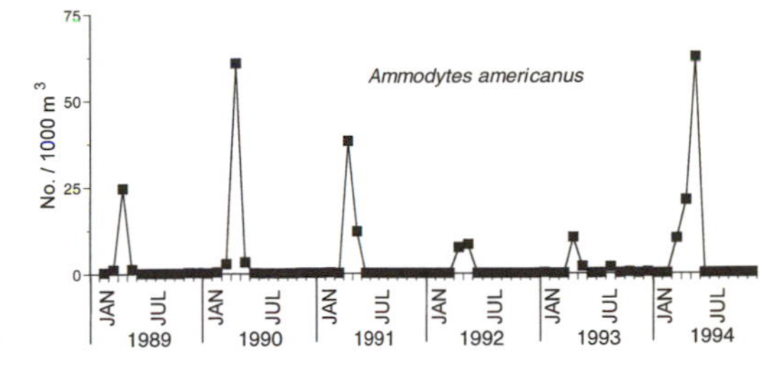

66.3 Annual variation in abundance of *Ammodytes americanus* larvae in the Great Bay–Little Egg Harbor study area based on Little Sheepshead Creek Bridge 1-m plankton net collections 1989–1994 (n = 2,352).

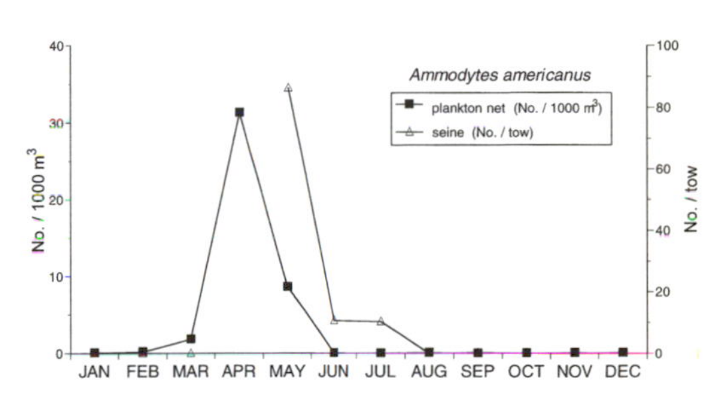

66.4 Monthly occurrences of *Ammodytes americanus* larvae and young-of-the-year in the Great Bay–Little Egg Harbor study area based on 1-m plankton net collections 1989–1994 (n = 2,352), and inlet beam trawl and seine study 1994–1995 (n = 1,322).

The habitat of young-of-the-year is poorly known. This species is pelagic much of the time but is capable of diving into sandy substrates very quickly, as we have observed on numerous occasions at Nauset Marsh. In Little Egg Inlet, young-of-the-year can be found over sandy substrates (D. Haroski and K. W. Able unpubl. data).

CHAPTER 67

Gobionellus boleosoma (Jordan and Gilbert) Darter Goby

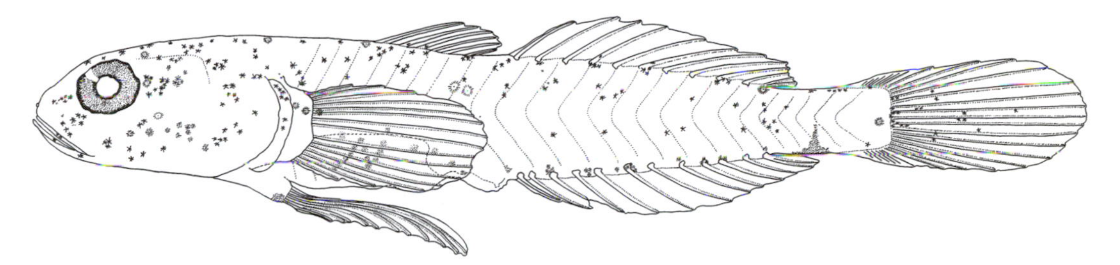

67.1 *Gobionellus boleosoma,* 9.1 mm TL. After D. M. Wyanski and T. E. Targett, in prep.

DISTRIBUTION

Gobionellus boleosoma (fig. 67.1) was reported to occur from North Carolina to Brazil (Böhlke and Chaplin 1968), but it has also been found, rarely, in Chesapeake Bay (Massmann 1957), at Indian River Inlet (de Sylva et al. 1962), and in southern New Jersey (McDermott 1971; Milstein and Thomas 1976a). It occurs in a variety of shallow estuarine habitats over muddy bottoms (Hildebrand and Cable 1938). The early life history stages have only rarely been collected in Middle Atlantic Bight estuaries (table 4.2).

REPRODUCTION

This species matures at approximately 25 to 30 mm (Hildebrand and Cable 1938) and probably spawns at age 1 (see below). The occurrence of the smallest larvae in August in Great Bay–Little Egg Harbor estuary indicates that reproduction occurs in the summer. Spawning may be prolonged, because small individuals continue to occur through November (fig. 67.2).

DESCRIPTION

The eggs are demersal and irregularly shaped, with a diameter of about 0.3 mm and a tuft of fibrous strands. Hatching occurs at about 1.2 mm. The snout-anus length in larvae is about 50% of the SL, and the body is elongate. The head is broad, the eyes are large, and the mouth is vertically oriented. Characteristic pigment includes melanophores on the dorsal and posterior surfaces of the air bladder, and a prominent melanophore on the venter near the end of the anal fin. All fins (except first dorsal spines) are complete at about 10 mm. Further details of development are in Kuntz (1916) and Hildebrand and Cable (1938) and summarized in Fahay (1983). Juveniles and adults have a diagnostic, small subcutaneous spot on the nape (fig. 67.1). Vert: 26; D: VI, 11; A: I, 10–12; Pect: 16.

THE FIRST YEAR

Although adults are rare, the larvae are among the common species collected planktonically in Great Bay–Little Egg Harbor (Witting 1995; D. A. Witting, K. W. Able, and M. P. Fahay in prep.). They first occur in August, but small larvae (<10 mm) have been collected as late as November (fig. 67.2). Growth, based on monthly modal length frequencies, averages approximately 0.2 mm per day into October when the modal size is 20 mm and the maximum is 40 mm. There is no growth during the winter, although sample sizes are small. Given that the maximum size is approximately 55 mm (Hildebrand and Cable 1938), they probably spawn at the end of the first year.

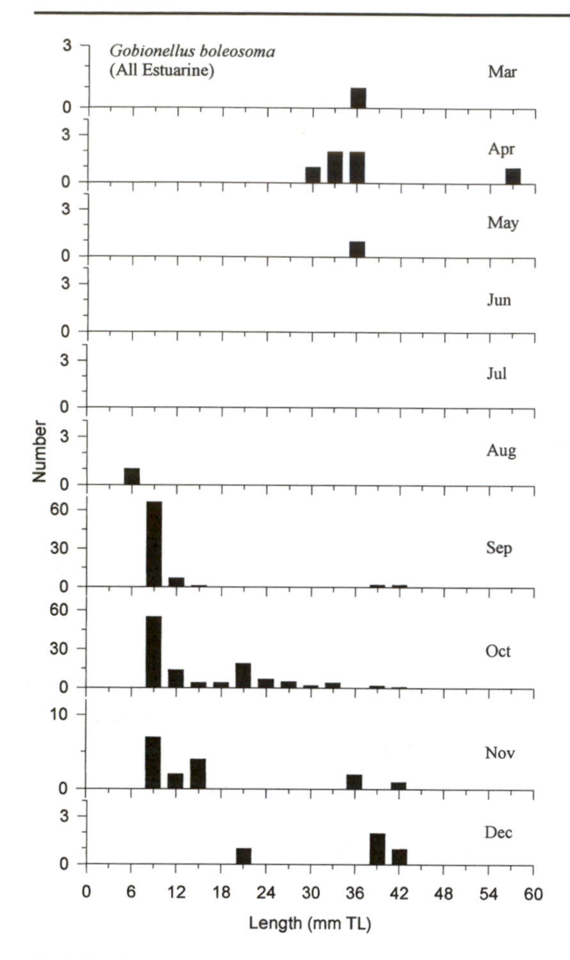

67.2 Monthly length frequencies of *Gobionellus boleosoma* young-of-the-year collected in estuarine waters in the Great Bay–Little Egg Harbor study area. Sources: RUMFS: beam trawl ($n = 36$); 1-m plankton net ($n = 162$); gear comparison ($n = 6$); killitrap ($n = 11$); seine ($n = 1$); weir ($n = 8$).

Little is known about habitat use, in part because *Gobionellus boleosoma* is never abundant in collections. In North Carolina, larvae have been collected from May to November with the peak in July and August (Hildebrand and Cable 1938). Most of these were taken in the estuary, but a few were collected offshore, suggesting that the species may spawn in coastal waters as well. Larvae are reportedly more abundant in plankton collections near the bottom than at the surface (Hildebrand and Cable 1938). The juveniles have been collected in the subtidal portions of a polyhaline marsh creek in Great Bay–Little Egg Harbor.

CHAPTER 68

Gobiosoma bosc (Lacepède) Naked Goby

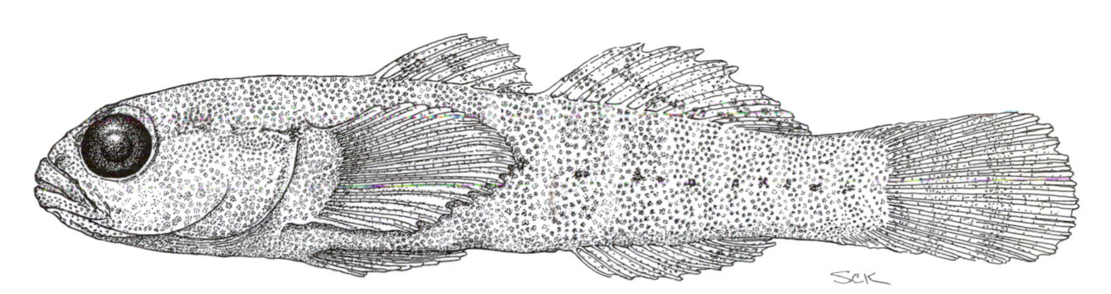

68.1 *Gobiosoma bosc* juvenile, 15.0 mm SL. Collected August 3, 1990, dip net, RUMFS boat basin, Great Bay, New Jersey. ANSP 175246. Illustrated by Susan Kaiser.

DISTRIBUTION

Gobiosoma bosc (fig. 68.1) occurs in estuaries from Connecticut to Campeche, Mexico (Pearcy and Richards 1962; Dawson 1966). It typically occurs in structured habitats, including oyster reefs, eelgrass, and marsh creeks (Musick 1972; Orth and Heck 1980; Sogard and Able 1991; Able et al. 1996b). It can be found far up estuaries (Wang and Kernehan 1979) and has been collected at salinities as low as 8 ppt (de Sylva et al. 1962). In the Middle Atlantic Bight, early life-history stages are common in estuaries south of New Jersey and sporadically in estuaries to the north (table 4.2).

REPRODUCTION

This species reaches sexual maturity at age 1 between 14.5 and 23 mm (Hildebrand and Cable 1938; Dawson 1966) (see below). Spawning occurs from April or May through August or September in Middle Atlantic Bight estuaries (Lippson and Moran 1974; Nero 1976; Wang and Kernehan 1979; Shenker et al. 1983). In New Jersey, the occurrence of small larvae and juveniles (fig. 68.2) indicates that spawning peaks in the spring and early summer but may extend through the summer.

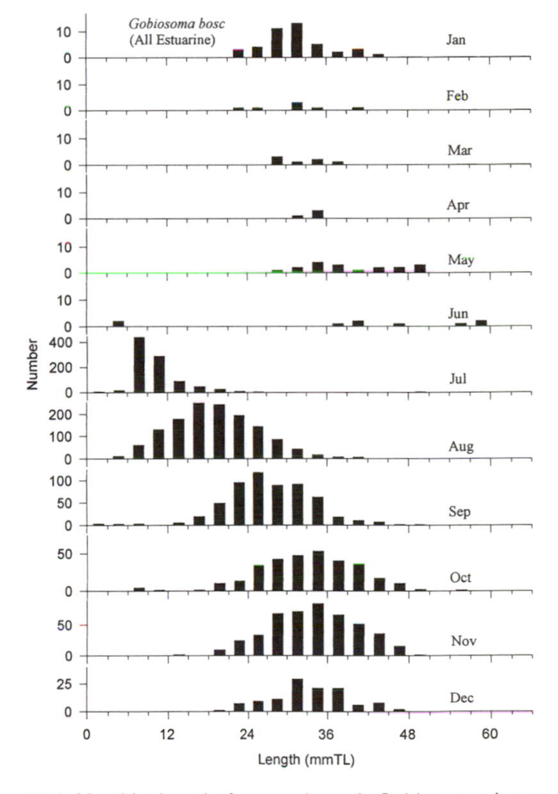

68.2 Monthly length frequencies of *Gobiosoma bosc* young-of-the-year collected in estuarine waters in the Great Bay–Little Egg Harbor study area. Sources: RUMFS: beam trawl (n = 456); 1-m plankton net (n = 504); LEO-15 2-m beam trawl (n = 42); experimental trap (n = 2,517); killitrap (n = 308).

DESCRIPTION

The eggs are demersal, elliptical (long axis 1.2–1.4 mm), and are attached to the substrate by fibrous strands. Larvae hatch at lengths of 2.0 to 2.6 mm. Pigment spots cover the dorsal aspect of the air bladder. All fins are completely formed by 10 to 15 mm (Hildebrand and Cable 1938). Juveniles have a rounded snout, a somewhat tubular body, and pelvic fins united by a membrane (fig. 68.1). Both juveniles and adults lack scales. The adults have two dorsal fins, separated by a very narrow gap, with seven spines in the first fin and 12 to 13 rays in the second fin. The tips of the pelvic fin rays fall well short of the anal fin origin. There are no prominent scales at the base of the caudal fin as in *G. ginsburgi.* Vert: 27; D: VI–VIII, 12–14; A: 10–12; Pect: 16–19; Plv: I, 5.

THE FIRST YEAR

Eggs are deposited and guarded in nests, typically in oyster shells. The larvae hatch in about 4 to 5 days (Kuntz 1916; Nelson 1928). In Great Bay–Little Egg Harbor estuary, the planktonic larvae are usually more abundant in July but can be found through November (fig. 68.3). In several Middle Atlantic Bight estuaries, the larvae move to low-salinity nursery areas soon after hatching (Massmann et al. 1963; Wang and Kernehan 1979.) This up-estuary movement may occur at rates of 1 km per day and is presumably the result of tidal stream transport (Shenker et al. 1983). Prior to settlement, individuals begin schooling near the bottom at sizes of 6 to 10 mm SL. Settled individuals have completed development of the pelvic fins (Breitburg 1989, 1991). The overlap between the largest planktonic individuals and the smallest demersal individuals in southern New Jersey and Chesapeake Bay indicates that settlement occurs over 8 to 13 mm TL (fig. 68.4) (Breitburg 1991). In Chesapeake Bay, settlement occurs during the day and night, and the transforming individuals aggregate in low-flow areas on the down-current sides of rocks. This active response influenced settlement sites (Breitburg 1991; Breitburg et al. 1995). In Great Bay–Little Egg Harbor, settled individuals become increasingly abundant after the larval peak in July, and reach greatest numbers in November (fig. 68.5). Their abundance appeared quite variable between years, such that in 1993 they had greater abundance over a larger period of time (fig. 68.3).

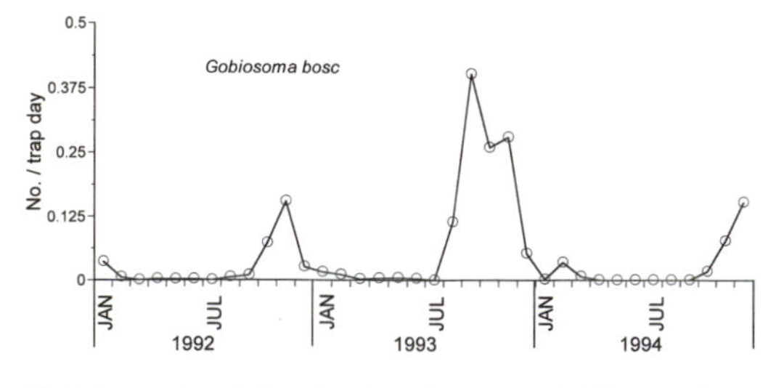

68.3 Annual variation in abundance of *Gobiosoma bosc* young-of-the-year collected in the Great Bay–Little Egg Harbor study area based on killitrap collections 1992–1994 (n = 336).

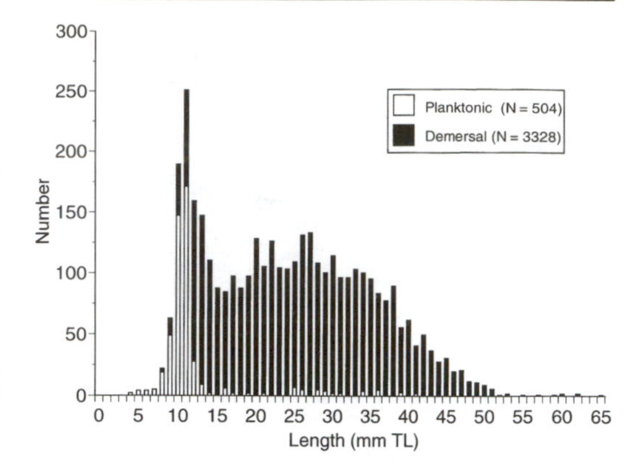

68.4 Composite length frequencies of planktonic and demersal stages of *Gobiosoma bosc* collected in the Great Bay-Little Egg Harbor estuary.

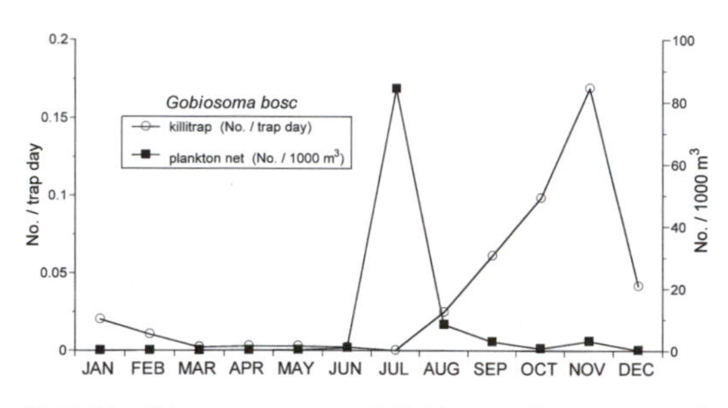

68.5 Monthly occurrences of *Gobiosoma bosc* young-of-the-year collected in the Great Bay–Little Egg Harbor study area based on killitrap collections 1992–1994 (n = 339) and 1-m plankton net collections 1989–1994 (n = 1,082).

During this period, modal changes in length frequency show that growth averages approximately 0.25 mm per day. The young-of-the-year reach approximately 21 to 48 mm TL by

November (fig. 68.2). Little growth is evident during the winter. By the following May and June, approximately 12 months after hatching, the larger young-of-the-year are the same size (27–60 mm TL) as reproductively mature adults (Nero 1976). The recently hatched young-of-the-year are beginning to appear at this time as well (fig. 68.2); thus, this species is capable of spawning the first year, as observed for a Virginia population (Nero 1976). There is little evidence that they survive their second summer because larger individuals are absent from the length-frequency data (fig. 68.2).

The habitat of this species is exclusively estuarine. Intensive sampling on the adjacent inner continental shelf on Beach Haven Ridge has failed to collect this species, although *G. ginsburgi* has been collected frequently. A recent examination of the distribution relative to habitat type in the polyhaline portions of Great Bay–Little Egg Harbor estuary found that young-of-the-year collected with beam trawls were common across a variety of substrate types (fig. 68.6). They occurred in sandy to very muddy substrate, where the dominant physical structure ranged from vegetation (eelgrass, sea lettuce; Sogard and Able 1991) to other biogenic forms such as amphipod tubes and shell. The young-of-the-year also occurred in inter- and subtidal marsh creeks (Rountree and Able 1992a; Able et al. 1996b).

Elsewhere in the Middle Atlantic Bight in the spring, summer, and fall, *G. bosc* has

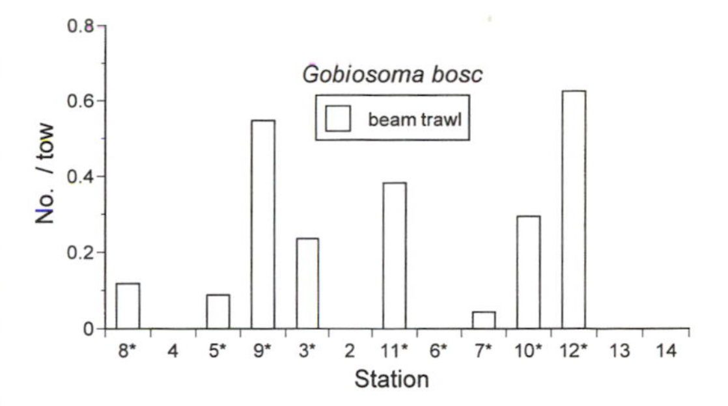

68.6 Distribution of *Gobiosoma bosc* young-of-the-year by habitat type based on beam trawls (n = 159) deployed on regular stations 2–14 in Great Bay and Little Egg Harbor. Both otter trawl and beam trawl used on stations designated by asterisk; otherwise, only the otter trawl was used. See figure 2.3 for station locations; table 2.1 for specific habitat characteristics by station.

been collected in areas with man-made structures such as piers and pile fields (Able and Studholme 1994) and eelgrass beds (Orth and Heck 1980), but it appears to be most abundant in oyster reefs (Nelson 1928; Dovel 1971; Dahlberg and Conyers 1973; Breitburg 1991). A number of studies have identified dissolved oxygen as a habitat variable that negatively influences the behavior of larvae (Breitburg et al. 1994) and their survival when exposed to predators (Breitburg 1992; Breitburg et al. 1994). For recently hatched larvae (2.4–2.6 mm), the critical dissolved-oxygen level was 0.5 m per liter (Saksena and Joseph 1972). During the winter, the species may move to channels and channel edges (Musick 1972).

CHAPTER 69

Gobiosoma ginsburgi Hildebrand and Schroeder Seaboard Goby

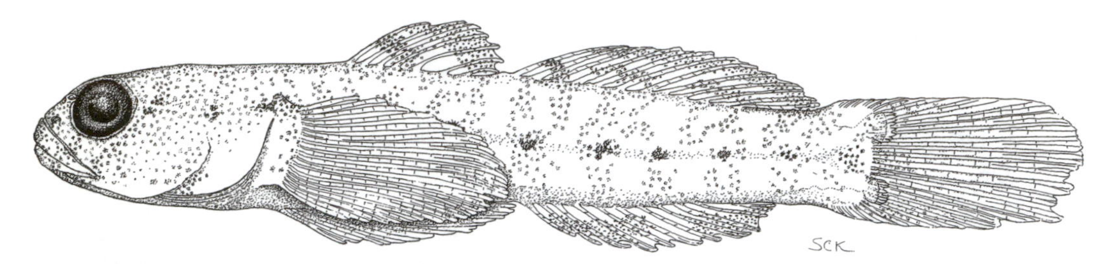

69.1 *Gobiosoma ginsburgi* juvenile, 14.8 mm SL. Collected August 1991, bottom grab sample, Delaware Bay. ANSP 175247. Illustrated by Susan Kaiser.

DISTRIBUTION

Gobiosoma ginsburgi (fig. 69.1) is distributed in the high-salinity portions of estuaries and coastal ocean waters from Massachusetts to Georgia (Ginsburg 1933; Dawson 1966; Lux and Nichy 1971). In the Middle Atlantic Bight, early life-history stages have been reported most consistently from New Jersey south (table 4.2) although they occur sporadically as far north as Narragansett Bay and Massachusetts (Collette and Hartel 1988).

REPRODUCTION

This species presumably spawns at age 1 (see below). In New Jersey, the occurrence of gravid females (19.0–38.7 mm SL) and larvae (E. Duval and K. W. Able unpubl. data) indicates that spawning occurs in estuaries from approximately June through September. This period overlaps with the presumed periods of reproduction in Rhode Island (Munroe and Lotspeich 1979) and Long Island (Greeley 1938) but it occurs slightly later than the May spawning in Chesapeake Bay (Hildebrand and Schroeder 1928). In Georgia, eggs were deposited in empty oyster valves in subtidal areas, where the males guarded and aggressively defended them until hatching (Dahlberg and Conyers 1973).

DESCRIPTION

The eggs are undescribed, and the larvae are not well described. Pigment includes melanophores over the dorsal aspect of the air bladder and an elongate melanophore on the venter midway between the anus and caudal fin base (Massmann et al. 1963). The smallest demersal individuals (11–15 mm SL) have complete fin ray counts and pigmentation similar to that of the adults (fig. 69.1). The basiocaudal ctenoid scales, which are diagnostic for this species, are also present at this size. The pelvic fin tips reach the anal fin origin. Vert: 27; D: VIII, 11–13; A: 10–12; Pect: 17–20; Plv: I, 5.

THE FIRST YEAR

The eggs have been collected only from subtidal oyster valves in Georgia (Dahlberg and Conyers 1973), but given their presumed preference for mollusk shells, it would not be surprising to find adults laying eggs in these habitats in the Middle Atlantic Bight. The larvae have been collected at similar times from estuaries throughout the range including June through August in Rhode Island (Munroe and Lotspeich 1979), July through October in Delaware (de Sylva et al. 1962), and June through September or December in Chesapeake Bay (Dovel 1971; Olney 1983). In the Great Bay–Beach Haven Ridge area in southern New Jersey, planktonic larvae (3–9 mm SL) were collected from the estuary and inner continental shelf in July (fig. 69.2) through September or October. Recently settled juveniles (9–12 mm SL) were most abundant in July through Octo-

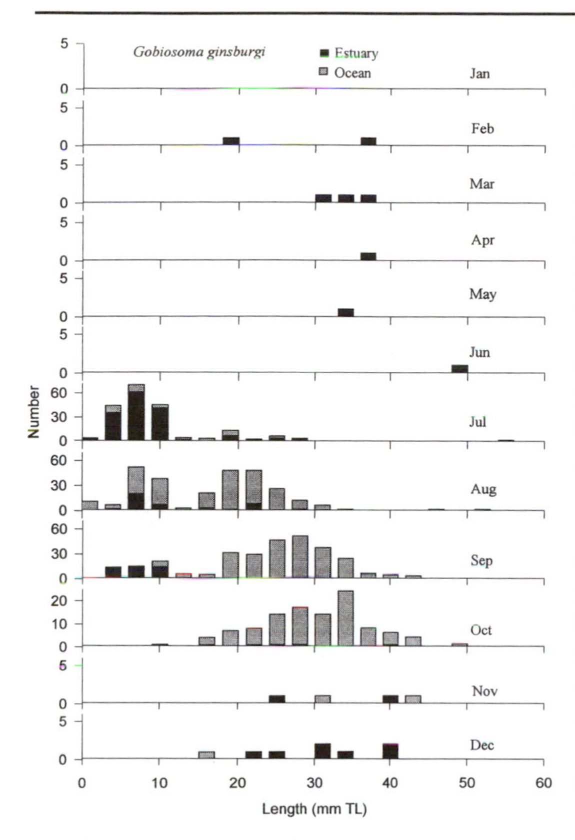

69.2 Monthly length frequencies of *Gobiosoma ginsburgi* collected in estuarine and coastal ocean waters of the Great Bay–Little Egg Harbor study area. Sources: RUMFS: beam trawl ($n = 29$); 1-m plankton net ($n = 235$); experimental trap ($n = 4$); LEO-15 Tucker trawl ($n = 102$); LEO-15 2-m beam trawl ($n = 521$).

ber, both in the estuary and on an adjacent portion of the inner continental shelf (E. Duval and K. W. Able, unpubl. data).

Based on modal length-frequency progression and examination of sagittal otoliths, most individuals that occur in New Jersey are young-of-the-year (E. Duval and K. W. Able unpubl. data). In the summer, planktonic and demersal young-of-the-year range from 6 to 24 mm SL (fig. 69.2). By October, they are 9 to 39 mm SL. Little growth occurs during the winter so that by April they range from 21 to 39 mm SL. These individuals overlap in size with gravid females collected in the spring and summer; thus, this species apparently spawns at age 1. There is little evidence in their lengths that adults survive through their second summer (fig. 69.2).

In New Jersey, the habitats range from the estuary to the inner continental shelf. Planktonic larvae, juveniles, and adults are reported from mid- to high-salinity (15 to 33 ppt) estuarine areas and the adjacent inner continental shelf in depths from 0.5 m (estuary) to 21 m (inner continental shelf) (E. Duval and K. W. Able unpubl. data). Habitats in Great Bay ranged from mud to sand substrate with shell, worm tubes, and hydroids over the substrate. Elsewhere *Gobiosoma ginsburgi* has been reported from depths to 10 m off Georgia (Dahlberg and Conyers 1973) and to 45 m elsewhere (Dawson 1966). In Chesapeake Bay, it has been collected at salinities between 15 and 31 ppt at oyster reefs and deeper flats during spring through summer and in higher-salinity channels in winter (Musick 1972). Its occurrence in mollusk shells is reported throughout its range. It is not clear whether seasonal migrations occur.

CHAPTER 70

Peprilus triacanthus (Peck) Butterfish

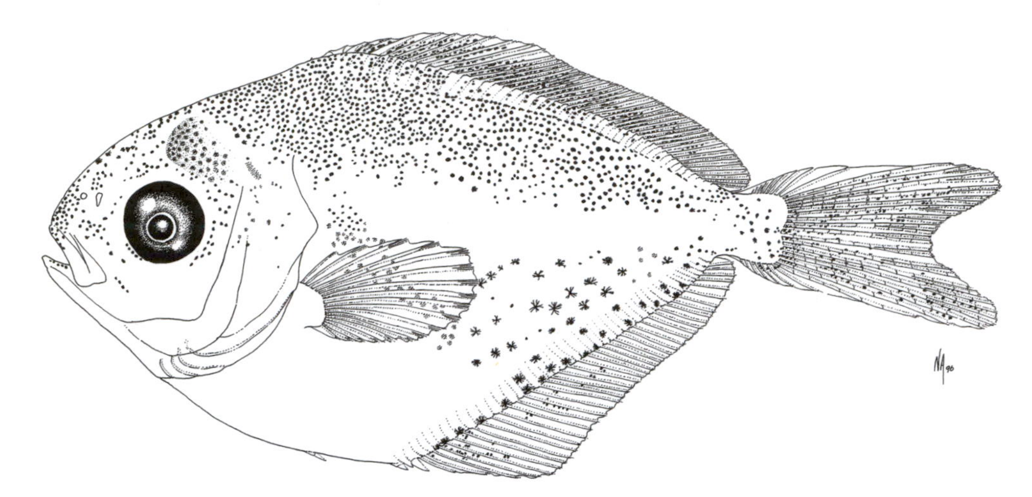

70.1 *Peprilus triacanthus* juvenile, 24.0 mm TL. Collected August 5, 1974, ca. 3.7 km off Holgate Peninsula, Long Beach Island, New Jersey. ANSP 175248. (CBM-74-151). Illustrated by Nancy Arthur.

DISTRIBUTION

A pelagic species, *Peprilus triacanthus* occurs from Nova Scotia to South Carolina (Bigelow and Schroeder 1953) and in deeper offshore water as far south as Florida (Nichols and Breder 1927). The population undergoes seasonal migrations, triggered by temperature changes, whereby they are found in deep water in the southern part of their range during winter, and summer movements are northward and toward shore (Fritz 1965; Horn 1970a; Waring 1975). Eggs, larvae, and juveniles occur in most estuaries in the northern part of the Middle Atlantic Bight (table 4.2, Lux and Wheeler 1992), although they have not been reported from Nauset Marsh or the Connecticut River. In the southern part of the bight, eggs have been reported only from the Chesapeake Bay and Delaware Bay, but larvae and juveniles are found more commonly.

REPRODUCTION

Spawning occurs during spring and summer in the Middle Atlantic Bight, and as early as February (or year-round, with a peak in spring) in the South Atlantic Bight (Fahay 1975). In one New York Bight study, gonadosomatic indices were highest during May and June (Wilk et al. 1990), and a peak in June has also been reported for the Middle Atlantic Bight as a whole (Horn 1970a). Eggs have been found during the spring along the edge of the continental shelf between Cape Hatteras and Georges Bank (Berrien and Sibunka in press). As water temperatures on the shelf increase, eggs are found increasingly closer to the coast, in a south to north progression. Eggs reach a peak in July in the central part of the Middle Atlantic Bight (Berrien and Sibunka in press). Sampling of larvae during MARMAP surveys (1977–1987) also indicates a northward progression of spawning activity in the Middle Atlantic Bight (fig. 70.2). Other studies have described spawning between May and October (Kawahara 1978; Colton et al. 1979; Morse et al. 1987). Although most summer spawning occurs in continental shelf waters of the Middle Atlantic Bight, reports of eggs and larvae (table 4.2) indicate reproductive activity also occurs in coastal and estuarine waters.

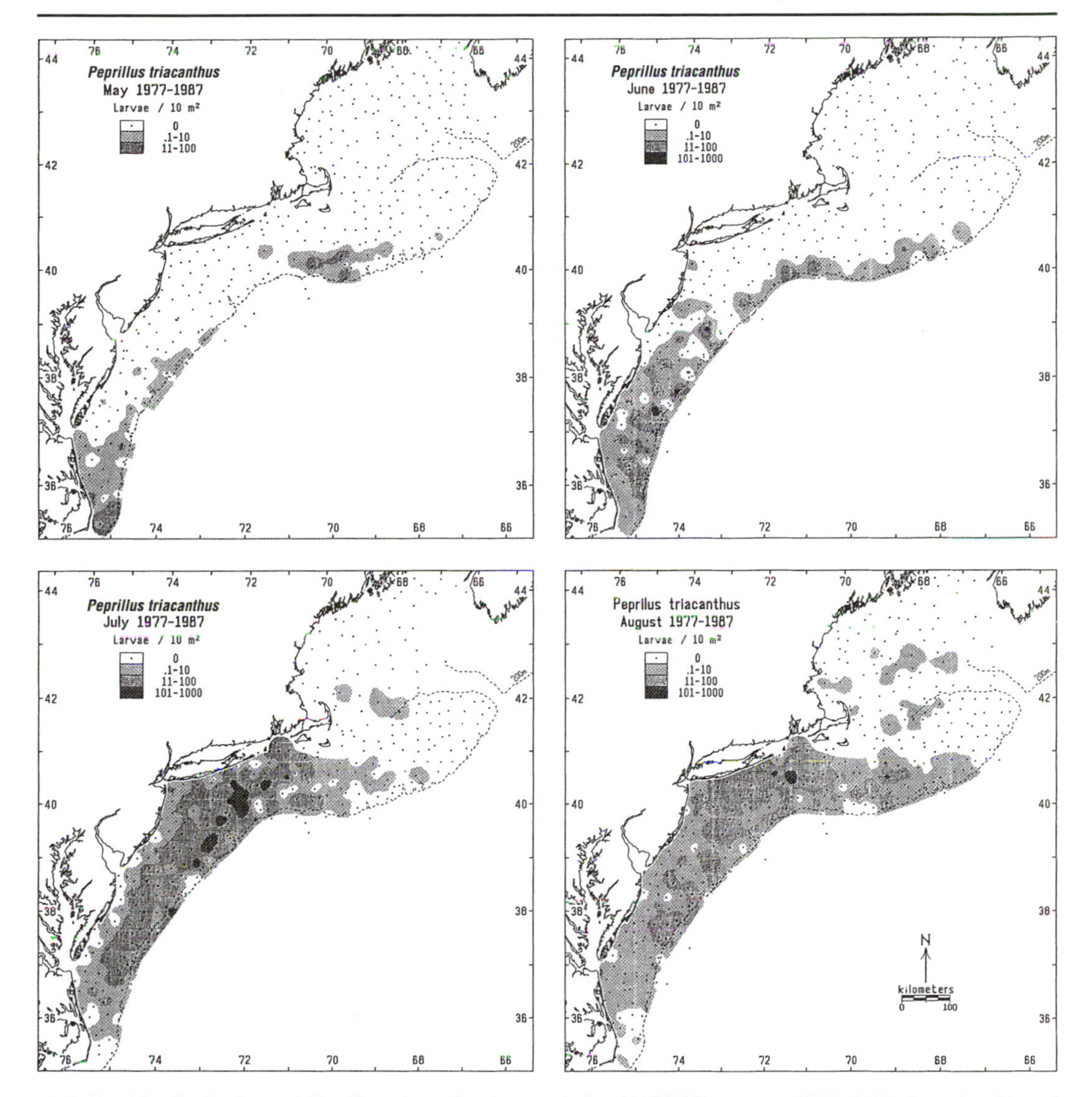

70.2 Monthly distributions of *Peprilus triacanthus* larvae during MARMAP surveys, 1977–1987. Low densities of larvae near Cape Hatteras in March–April and farther north in September–October not shown.

DESCRIPTION

Eggs are pelagic and spherical, with a diameter ranging from 0.68 to 0.83 mm. The (usually) single oil globule ranges from 0.17 to 0.21 mm, and the perivitelline space is narrow. Larvae hatch at about 1.7 mm in an undeveloped condition. The body deepens early in development, and the snout-anus length is about 50% TL. Fin rays begin to form at 5.0 mm and are complete by 12 to 18 mm. Pigment is light but includes a line of melanophores along the anal fin base and a pattern resembling stitching along the venter of the gut. See Horn (1970a), Ditty and Truesdale (1983), and Fahay (1983) for further details of larval development. Juveniles are recognized by their deep, laterally flattened body and lack of pelvic fins (fig. 70.1). The dorsal and anal fin bases are long and composed of a few spines and many rays. The mouth is small and terminal. The color is silvery, perhaps with a few blotches along the upper sides. Vert: 30–33; D: II–IV, 40–48; A: II–III, 37–44; Pect: 17–22; Plv: none.

THE FIRST YEAR

A few larvae occur in the Cape Hatteras area during March and April, and they increase in abundance in the southern part (and near the continental shelf edge in the northern part) of the Middle Atlantic Bight during May and June. Larvae reach peak abundance in the central part of the bight during July and August (fig. 70.2), and reduced numbers linger into September and October. Pelagic juveniles (11.3–52.1 mm FL) occur year-round in the South Atlantic Bight, and spawning there may be continuous, with a peak in spring (Fahay 1975). A recent otolith study has discovered a bimodality in birthdates for young-of-the-year collected in the Middle Atlantic Bight (Rotunno 1992), with an early mode the result of spawning as early as February. During May, larvae and pelagic juveniles have also been collected in the Slope Sea (J. A. Hare, M. P. Fahay, and R. K. Cowen in prep.). These Slope Sea larvae were part of a transient assemblage comprising species spawning in the South Atlantic Bight, whose larvae were found associated with water masses with properties transitional between the Gulf Stream and Slope Sea. Therefore, these collections may be the result of advection from spawning in southern waters and provide some evidence explaining the origin of the early mode reported by Rotunno (1992).

Larvae gradually assume juvenile characters and there is no sudden morphological transformation. During early summer, young-of-the-year move from continental shelf waters to bays and other protected inshore nursery areas (Horn 1970b) where the smallest stages are neustonic (Lux and Wheeler 1992). Larger juveniles may dominate collections in some systems, but, because they are a pelagic schooling species (at times found near bottom), they are often undersampled, as in Narragansett Bay, for example (Oviatt and Nixon 1973). Other evidence suggests they congregate near bottom during the day, and disperse upward at night (Waring 1975). Several authors have reported aggregations of young-of-the-year under the bells of coelenterates (Mansueti 1963; Horn 1970b; Scott and Scott 1988). This relationship is presumably not an obligate one, because juveniles have also frequently been observed swimming freely near the surface (Bigelow and Schroeder 1953).

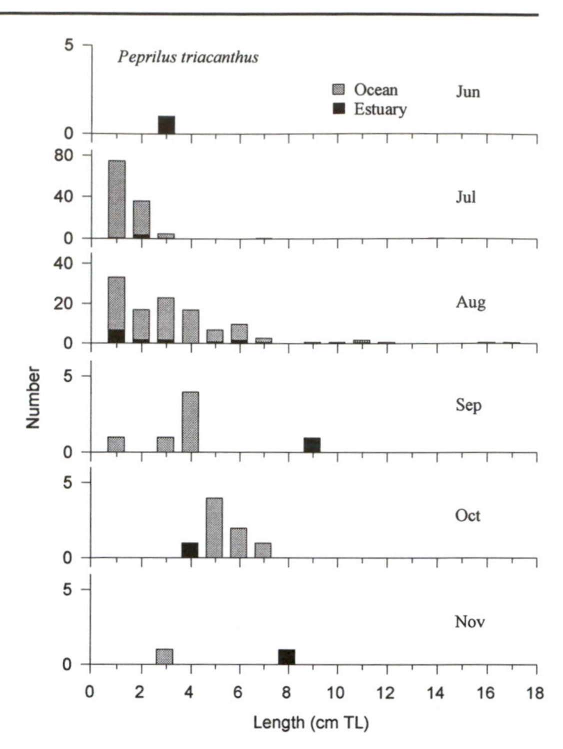

70.3 Monthly length frequencies of *Peprilus triacanthus* young-of-the-year collected in the Great Bay–Beach Haven Ridge study area 1988–1994. Sources: RUMFS: otter trawl ($n = 23$); 1-m beam trawl ($n = 3$); 1-m plankton net ($n = 13$); LEO-15 Tucker trawl ($n = 150$); LEO-15 2-m beam trawl ($n = 62$); gear comparison ($n = 2$); night-light ($n = 1$); killitrap ($n = 1$); weir ($n = 3$).

Our estimates of growth during the first year are based on 258 specimens collected from Great Bay–Little Egg Harbor and the nearby Beach Haven Ridge (fig. 70.3). The smallest size classes (1 cm TL) occur from July through September, and by November, the few remaining individuals range from 3 to 8 cm TL. Larval growth rates of 0.23 mm per day for fish up to 30 mm have been reported, based on analyses of otolith increments (Rotunno 1992). Young-of-the-year collected in the New York Bight displayed a modal length of 2 cm FL during July and grew to a mode of 9 cm FL in November (Wilk et al. 1975). Growth then appeared to slow over the first winter. The mode in March for these fish was 11 cm; by June (1 year after the peak in spawning), it had reached 14 cm. Other estimates of size at age 1 range from 9 to 13 cm FL (DuPaul and McEachran 1973; Waring 1975; Kawahara 1978).

Peprilus triacanthus does not survive temperatures below 10 C (Colton 1972), and, in response to seasonal cooling, all age groups migrate toward the edge of the continental shelf in the fall and spend the winter near the 183-m contour where temperatures are warmer (Pentilla and Dery 1988). They then disperse over the shelf during April and May. South of Delaware Bay, where winter cooling is not as severe, this offshore–inshore migration is not as pronounced (Waring and Murawski 1982).

CHAPTER 71

Scophthalmus aquosus (Mitchill) Windowpane

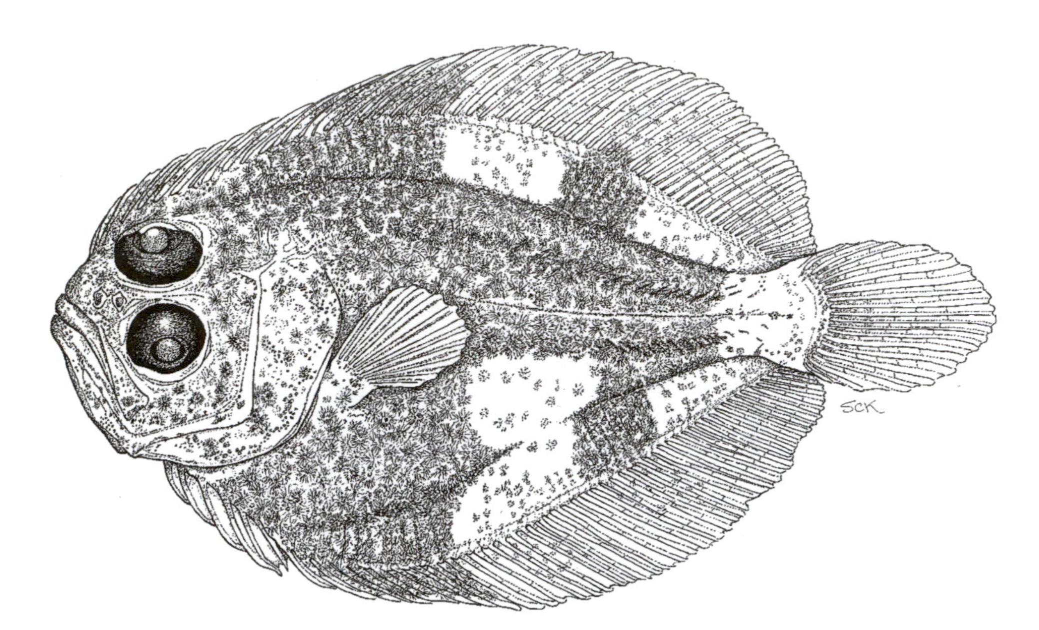

71.1 *Scophthalmus aquosus* juvenile, 12.5 mm SL. Collected May 28, 1991, 1-m plankton net, Little Sheepshead Creek, Great Bay, New Jersey. ANSP 175249. Illustrated by Susan Kaiser.

DISTRIBUTION

Scophthalmus aquosus (fig. 71.1) is distributed from the Gulf of St. Lawrence to Florida (Scott and Scott 1988) but is most abundant from Georges Bank to Chesapeake Bay (Morse and Able 1995). It is found in shallow, sandy to sand/silt or mud substrates in inner continental shelf waters (usually < 56 m; Wenner and Sedberry 1989; Thorpe 1991) and estuaries. In the Middle Atlantic Bight, most early life-history stages have been reported from all the estuaries for which we have data (table 4.2).

REPRODUCTION

Spawning on the continental shelf occurs from February to November, as shown by gonadosomatic indices (Wilk et al. 1990; Morse and Able 1995) and distribution and abundance of eggs (Berrien and Sibunka in press). Spawning begins in February or March in inner shelf waters, reaches a peak in the bight in May, and extends onto Georges Bank during the summer. Spawning returns to the southern portion of the bight in the fall. In the central part of the bight, there is a split spawning season with peaks in the spring and fall (Morse and Able 1995; Berrien and Sibunka in press). In the vicinity of Little Egg Inlet, this split spawning season is evident with eggs in the spring from the inner shelf into the estuary (fig. 71.2). It occurs again in the fall, although very few eggs are in the estuary at this time. Some spawning may also occur in the high-salinity portions of other estuaries in the Middle Atlantic Bight, including Great South Bay (Monteleone 1992), Sandy Hook Bay (Croker 1965), inside Hereford Inlet (Allen et al. 1978), and other locations in the bight (see Morse and Able 1995). This species is reported to spawn in the evening or at night (Ferraro 1980).

DESCRIPTION

The eggs are buoyant, spherical and variable in diameter (0.9–1.4 mm) with a single oil glob-

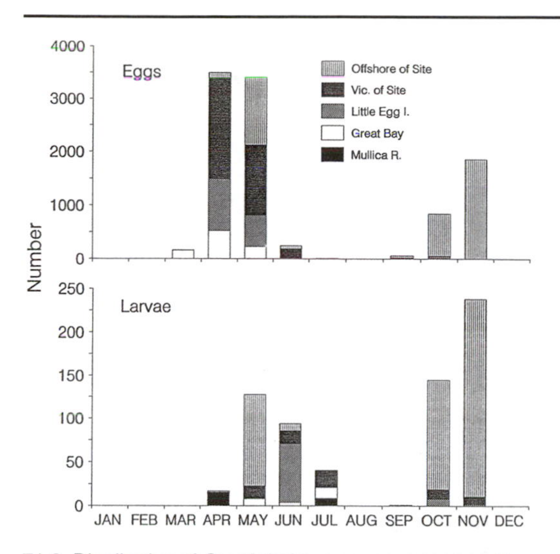

71.2 Distribution of *Scophthalmus aquosus* eggs and larvae along the Mullica River–Great Bay–Little Egg Inlet–Beach Haven Ridge (Sites 1 & 2) transect. After Morse and Able 1995.

ule (0.2–0.3 mm in diameter) (Wheatland 1956). The larvae hatch at approximately 2.0 mm. The body is darkly pigmented from the head to approximately one-third the distance from the anus to the tip of the tail. As development proceeds, the body becomes deeper and more laterally compressed. Fin ray formation is complete at 11.5 mm. Details of development are provided by Moore (1947). In juveniles (fig. 71.1), the body is oval and wider (60%–70% of SL) than in other left-eyed flounders. The first 10 to 12 rays of the dorsal fin are separate, often somewhat elongate, and branched at the tips. The body and fins are heavily pigmented in larger young-of-the-year, whereas smaller individuals are characterized by broad, alternating swaths of dark and light bands. The mouth is very large, extending to the eye or beyond. The lateral line is arched over the pectoral fin. Vert: 34–36; D: 63–73; A: 46–56; Pect: 11; Plv: 6.

THE FIRST YEAR

In Great Bay–Little Egg Harbor, the eggs are found throughout the polyhaline portions of the estuary in the spring but are more concentrated on the continental shelf during the fall spawning (fig. 71.2; Morse and Able 1995). The larvae hatch in 8 days at 10.6 to 13.3 C (Bigelow and Schroeder 1953). The larval distribution mirrors that of the eggs both in space and time; that is, they are found throughout the polyhaline portion of the estuary in the spring but primarily on the shelf in the fall (fig. 71.2; Morse and Able 1995). On the continental shelf of the Middle Atlantic Bight the smallest larvae are distributed over the middle to inner shelf and onto the shallow portions of Georges Bank (fig. 71.3). With increasing size, larvae maintain a similar distribution pattern, but by 11 to 20 mm they are infrequent in the Middle Atlantic Bight but still very abundant on Georges Bank.

Some aspects of metamorphosis and settlement are known. Eye migration during metamorphosis occurs between 6.5 and 13.0 mm (Colton and Marak 1969). Larvae are reported to settle to the bottom by 10 mm (Bigelow and Schroeder 1953), but some, especially those on Georges Bank, remain planktonic up to 20 mm (Morse and Able 1995). Based on collections from southern New Jersey, it appears that settlement occurs both in the estuary and on the shelf for spring-spawned individuals and primarily on the shelf for fall-spawned individuals (Morse and Able 1995).

The growth patterns of young-of-the-year in southern New Jersey vary with the timing of spawning. The spring-spawned fish grow quickly (fig. 71.4). Based on modal length-frequency progression, they achieve sizes of approximately 11 to 19 cm by September, approximately 4 months after spawning (Morse and Able 1995). By the following spring, they are difficult to separate by length alone, but most members of this cohort appear to be larger than 16 cm (fig. 71.4). The fall-spawned fish are most obvious in December when they are 4 to 7 cm. Perhaps because they are exposed to winter temperatures soon after settlement, they do not grow during the winter and reach only 4 to 8 cm by March (fig. 71.4; Morse and Able 1995). The fall-spawned fish increase in length through the summer growing season and reach approximately 18 to 26 cm by October when they are approximately age 1. Other studies in the Middle Atlantic Bight indicate much slower rates (Moore 1947) or are in close agreement (Grosslein and Azarovitz 1982; Thorpe 1991) to the rates we have estimated (Morse and Able 1995).

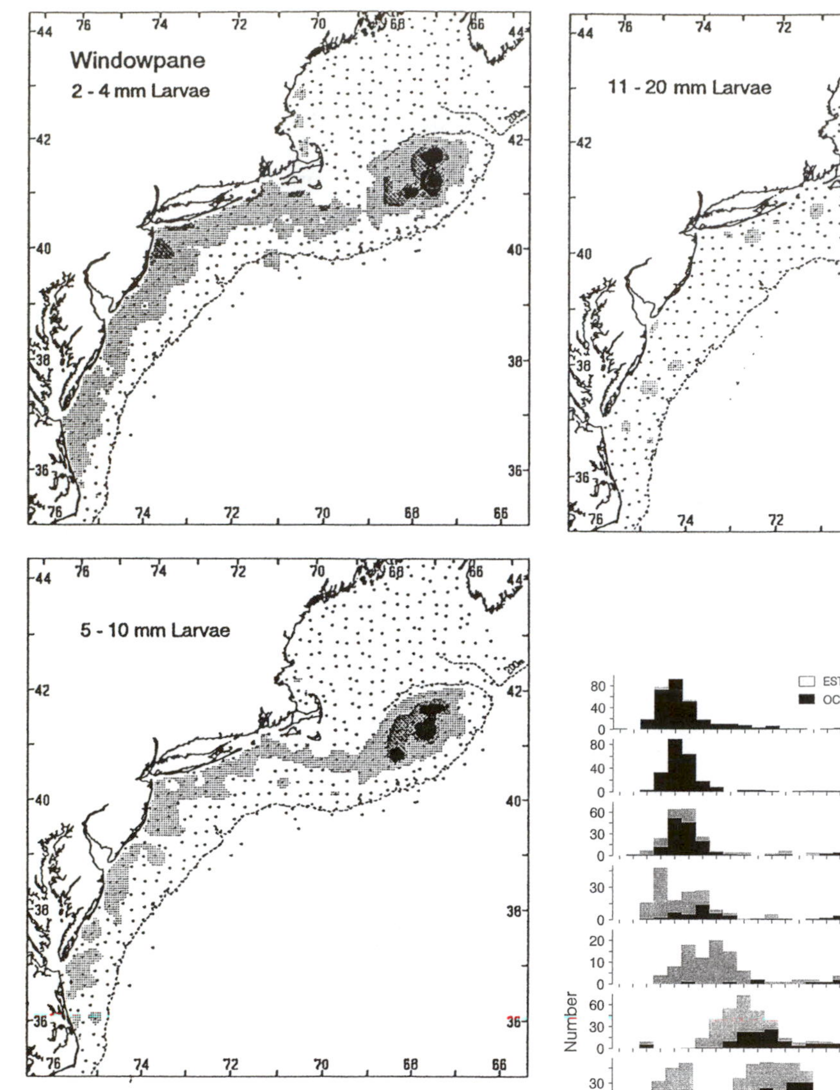

71.3 Distribution and abundance of three size classes of *Scophthalmus aquosus* larvae collected on MARMAP surveys 1977–1987. After Morse and Able 1995.

Based on intensive collections in the Great Bay–Little Egg Inlet–Beach Haven Ridge corridor, it appears that both the inner shelf and adjacent estuaries serve as nurseries for the spring-spawned young-of-the-year (M. J. Neuman and K. W. Able unpubl. data). The inner shelf appears to serve as the primary nursery area for the fall-spawned young-of-the-year (fig. 71.4). Although the adults are commonly collected over sand bottoms (Bigelow and Schroeder 1953; de Sylva et al. 1962), the habi-

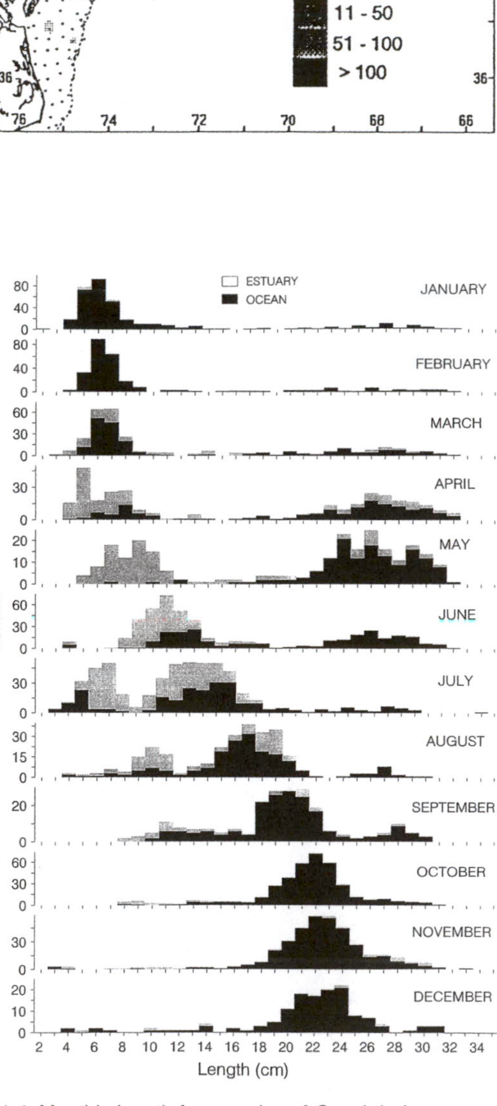

71.4 Monthly length frequencies of *Scophthalmus aquosus* young-of-the-year collected from the Great Bay–Beach Haven Sand Ridge study area. After Morse and Able 1995.

tats of the spring- and fall-spawned juveniles are not well defined. In extensive collections in estuarine shallows, juveniles were never collected in intertidal areas, but they occurred frequently along subtidal shores (Able et al. 1996) and in a variety of deeper (<1–8 m) habitats (Szedlmayer and Able 1996). It appears that both young-of-the-year cohorts move out of the estuary and offshore for the winter, and at least some of the fall-spawned individuals are found in the estuary the following summer (fig. 71.4).

In a laboratory study, early demersal (8–18 mm SL) and larger juveniles (32–89 mm SL) preferred sand over mud substrate (Neuman and Able in press). During these observations early demersal individuals buried less often and exhibited larval pigmentation more than juveniles.

CHAPTER 72

Etropus microstomus (Gill) Smallmouth Flounder

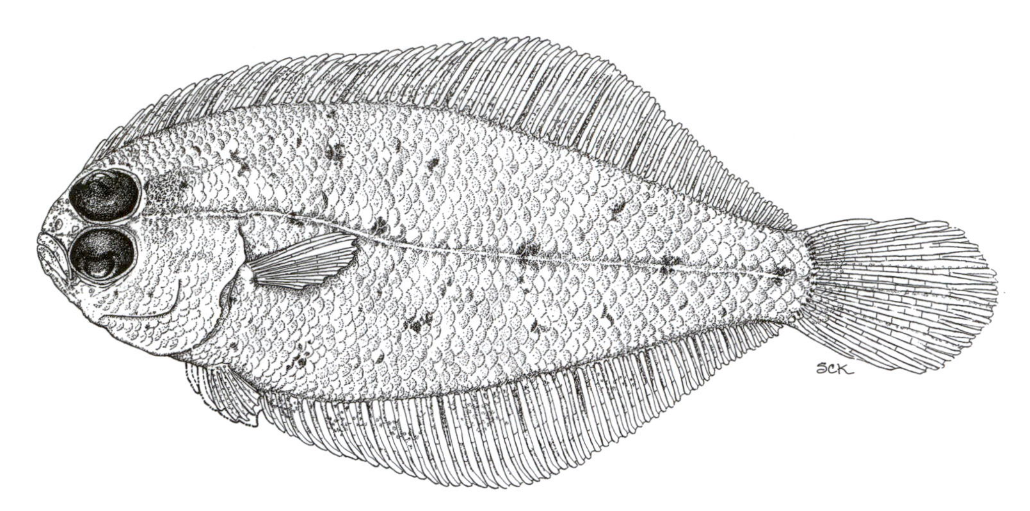

72.1 *Etropus microstomus* juvenile, 19.9 mm SL. Collected September 25, 1990, seine, Great Bay, New Jersey. ANSP 175250. Illustrated by Susan Kaiser.

DISTRIBUTION

Etropus microstomus (fig. 72.1) is most abundant from New England to North Carolina (Parr 1931) but apparently strays as far south as Florida (Leslie and Stewart 1986). In the Middle Atlantic Bight, the demersal juveniles and adults are found on the inner portion of the continental shelf (Richardson and Joseph 1973). The early life-history stages have been collected in the polyhaline portions of adjacent estuaries from North Carolina to Long Island and less consistently in the northern portion of the bight (table 4.2).

REPRODUCTION

Because of the occurrence of small larvae, Richardson and Joseph (1973) and Smith et al. (1975) reported spawning in the ocean in the Middle Atlantic Bight. Spawning begins in the spring off North Carolina, proceeds northward during the summer, and retreats southward in the fall (Smith et al. 1975). Based on more extensive collections over the continental shelf from 1977 to 1987, the smallest larvae (<3.9 mm) occurred between May and October and were most abundant in July and August (table 72.1). The presence of eggs in Great South Bay on Long Island (Monteleone 1992) and Hereford Inlet in southern New Jersey (Keirans 1977) indicates that spawning also occurs in estuaries.

DESCRIPTION

The eggs are undescribed. Larvae are characterized by several prominent melanophores along the dorsal and ventral body edges, as well as internally along the notochord. Early larvae (2.5–8.0 mm) have spines along the edge of the preopercle. Dorsal, anal, and caudal fin rays are complete by 8 mm SL. Further details of larval development (and comparisons with similar larvae) are in Richardson and Joseph (1973) and Tucker (1982). Juveniles (and adults) of this left-eyed flatfish are characterized by a small mouth, with maxilla barely reaching the anterior margin of the eye (fig. 72.1). In larger juveniles (and adults), each scale on the left side of the body has a smaller accessory scale at its base. Vert: 31–33; D: 67–84; A: 50–63; Pect: 8–12; Plv: 5.

THE FIRST YEAR

Larvae are distributed from Cape Hatteras to Massachusetts (Morse et al. 1987) but are most abundant from New Jersey and south, where they occur over the entire continental shelf

Table 72.1. Abundance (number/10 m²) of *Etropus microstomus* Larvae from MARMAP Survey Data by Length and Month, 1977–1987

Length (mm NL)	Jan	Feb	Mar	Apr	May	Jun	Jul	Aug	Sep	Oct	Nov	Dec
1.0–1.9	—	—	—	—	—	—	—	34	1	—	—	—
2.0–2.9	—	—	—	—	5	190	303	929	133	41	—	—
3.0–3.9	3	—	5	—	105	252	1629	3339	762	203	—	—
4.0–4.9	—	—	3	—	66	326	1835	5019	1475	191	14	—
5.0–5.9	5	—	—	—	20	485	2161	4238	1716	196	18	—
6.0–6.9	10	—	—	—	13	272	1225	2372	1496	96	9	—
7.0–7.9	—	—	—	—	3	30	754	1281	1173	77	30	—
8.0–8.9	—	—	—	3	8	20	433	864	889	50	20	—
9.0–9.9	—	—	—	—	5	9	301	401	871	57	27	—
10.0–10.9	6	—	—	—	—	15	174	245	518	57	28	—
11.0–11.9	—	—	—	—	2	3	125	144	550	31	13	3
12.0–12.9	—	—	—	6	—	—	47	78	215	27	4	—
13.0–13.9	—	—	—	—	—	3	22	22	132	16	5	—
14.0–14.9	—	—	—	—	—	—	—	15	58	11	4	—
15.0–15.9	—	—	—	—	—	—	3	33	37	22	2	—
16.0–16.9	—	—	—	—	—	—	6	28	50	11	—	—
17.0–17.9	—	—	—	—	—	—	10	3	17	33	—	—
18.0–18.9	—	—	—	—	—	—	—	3	5	5	5	—
19.0–19.9	—	—	—	—	—	—	—	—	5	—	—	—

NOTE: Sampling area includes Middle Atlantic Bight. Occurrences based on 870 total stations.

(fig. 72.2). Results of three studies agree that larvae occur most frequently in depths less than 40 m (Richardson and Joseph 1973; Smith et al. 1975; this study, fig. 72.3). In an intensive study of the diel vertical distribution of small larvae (average 3–5 mm), the larvae occurred in deeper (15–30 m) tows during the day and closer to the surface (0–4 m) at night (Kendall and Naplin 1981). In recent collections in Great Bay–Little Egg Harbor, they were most abundant from 1991 to 1993 and less abundant in other years (fig. 72.6). They occur in the plankton at sizes of up to 2 cm (table 72.1). In the vicinity of Beach Haven Ridge, larval duration, based on examination of daily otolith increments, was up to 23 days at sizes less than or equal to 14.8 mm SL (L. S. Hales and K. W. Able unpubl. data).

Metamorphosis, as indicated by eye migration, occurs over sizes of 10 to 12 mm (Richardson and Joseph 1973) or at 8 to 15 mm SL at Beach Haven Ridge (L. S. Hales and K. W. Able unpubl. data). Settlement in New Jersey occurs from July through October in the estuary and the ocean, based on the presence of the smallest juveniles (<2 cm TL, fig. 72.5). More detailed studies at Beach Haven Ridge found the smallest early demersal individuals ranging from 11.8 to 22.0 mm (L. S. Hales and K. W. Able unpubl. data). Demersal juveniles occur in Great Bay–Little Egg Harbor during the summer and fall but are most abundant in September and October, and their numbers decline during the early winter (fig. 72.6).

Studies of the growth of demersal juveniles and adults (R. Bush and K. W. Able unpubl. data) indicate that estuarine and inner continental shelf collections from New Jersey were dominated by young-of-the-year (12–90 mm TL), which made up 73% of all collections. Age-1 fish ranged from 45 to 120 mm TL and comprised 26% of the collections. These estimates were based on comparisons of length frequency and sagittal otolith increments, which

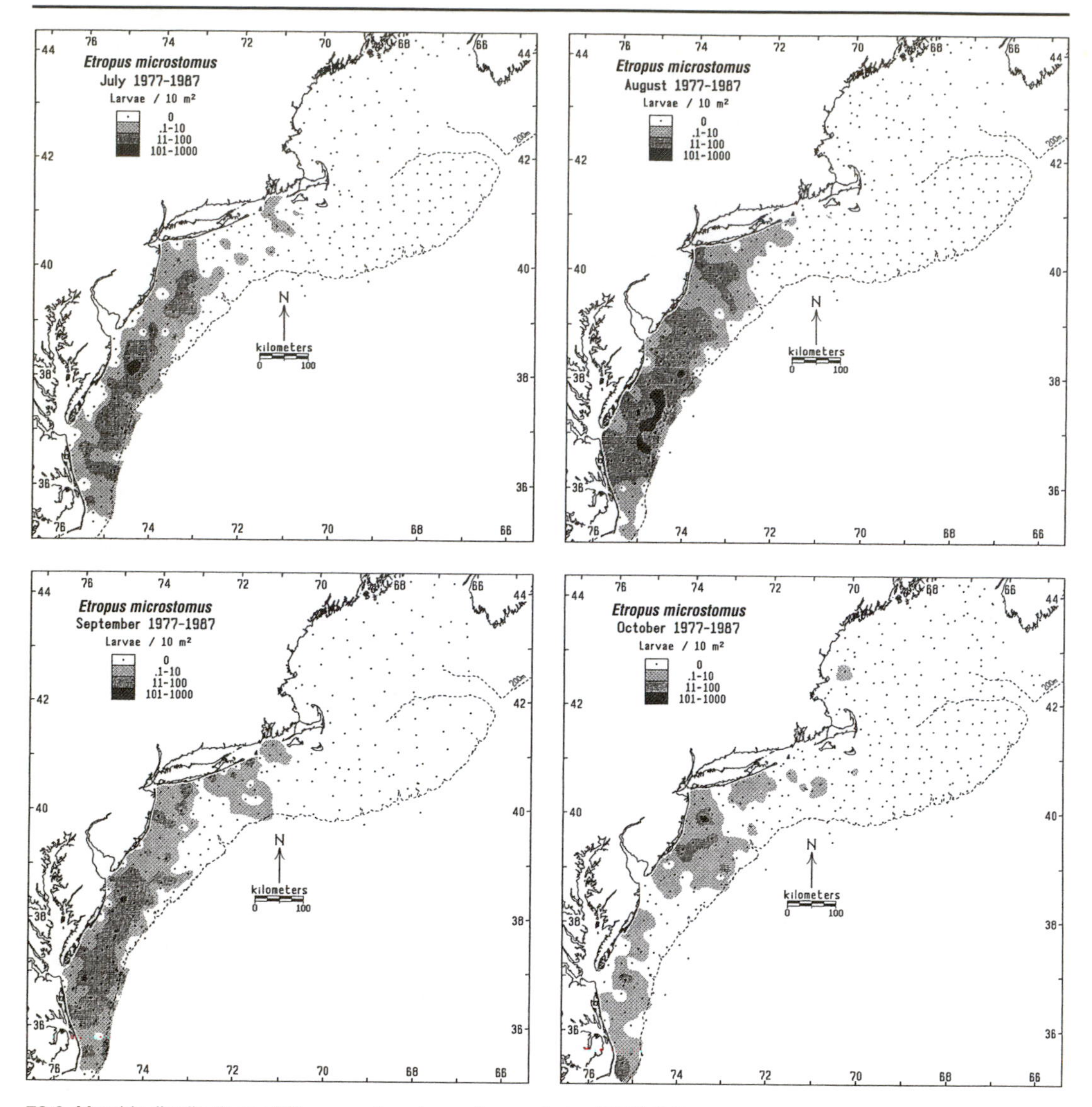

72.2 Monthly distributions of *Etropus microstomus* larvae during MARMAP surveys, 1977–1987.

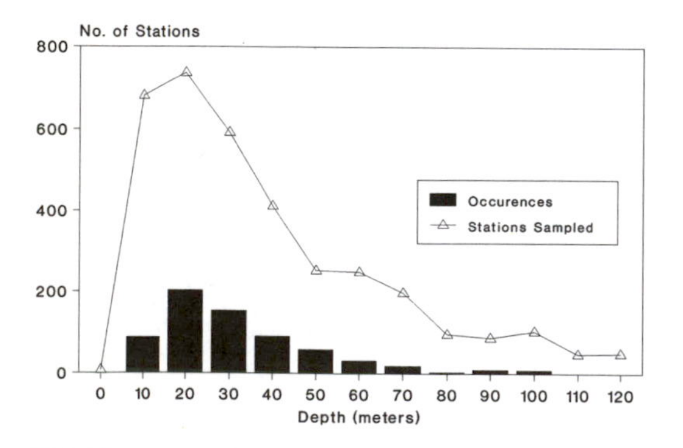

72.3 Frequency of occurrence of *Etropus microstomus* larvae by water depth based on collections in the Middle Atlantic Bight between 36 N and 40 N during MARMAP surveys 1977–1987.

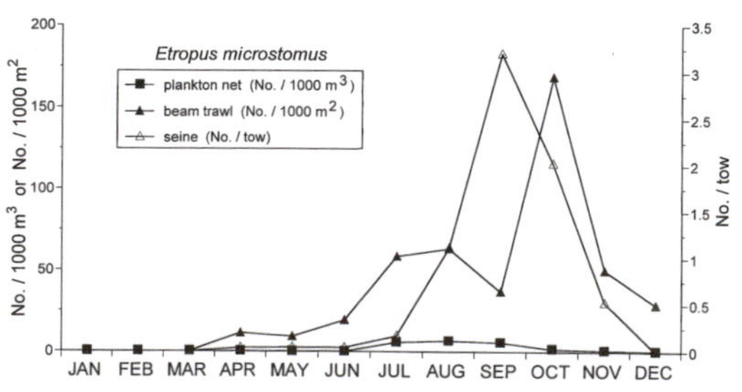

72.4 Annual variation in abundance of *Etropus microstomus* larvae in Little Sheepshead Creek Bridge 1-m plankton net collections 1989–1994 ($n = 724$).

were in close agreement. The monthly progression of lengths indicate that, after settlement, growth in the ocean and the estuary is relatively slow so that the young-of-the-year are approximately 2 to 10 cm TL with a mode at 3 by November (fig. 72.5). There is little evidence of growth through the winter. This lack of growth at low temperatures has been confirmed in laboratory experiments (Hales and Able in press). Based on the occurrence of larger individuals after April, growth resumes in the spring.

Young-of-the-year were found across an array of habitats in Great Bay–Little Egg Harbor estuary, but they did not occur in shallow eelgrass or in areas with high silt or clay content. Both of these habitat types were the farthest from the inlet (fig. 72.7; Szedlmayer and Able 1996). They also did not occur in subtidal creeks (Rountree and Able 1992a) but were collected along shallow subtidal shorelines (Able et al. 1996b). At a site on the inner continental shelf adjacent to Great Bay the juveniles and adults were one of the most abundant species collected by beam trawl (K. W. Able and M. P. Fahay unpubl. data). They were only infrequently collected during the winter. This might be explained by movement into deeper water or avoidance of the collecting gear by burial in the substrate, which is known to occur at low temperatures in the laboratory (Hales and Able in press) and is inferred from seasonal collections near Cape Hatteras (Leslie and Stewart 1986). There is evidence of size-selective mortality for young-of-the-year at low winter temperatures (Hales and Able in press).

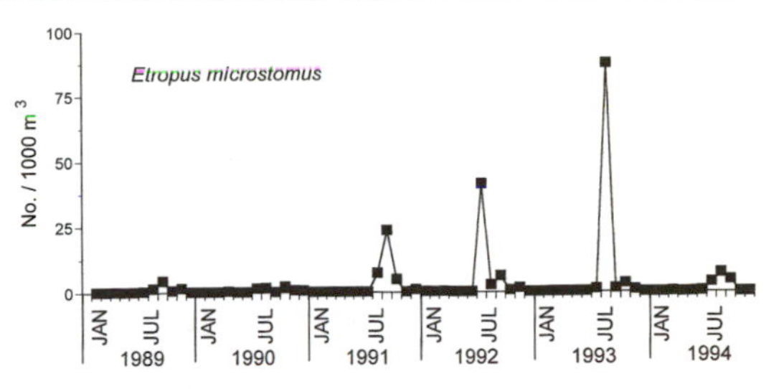

72.6 Monthly occurrences of *Etropus microstomus* larvae and young-of-the-year in the Great Bay-Beach Haven Sand Ridge study area based on 1-m plankton net collections 1989–1994 ($n = 724$), seine collections 1990–1991 ($n = 798$), and LEO-15 2-m beam trawl collections 1991–1992 ($n = 342$).

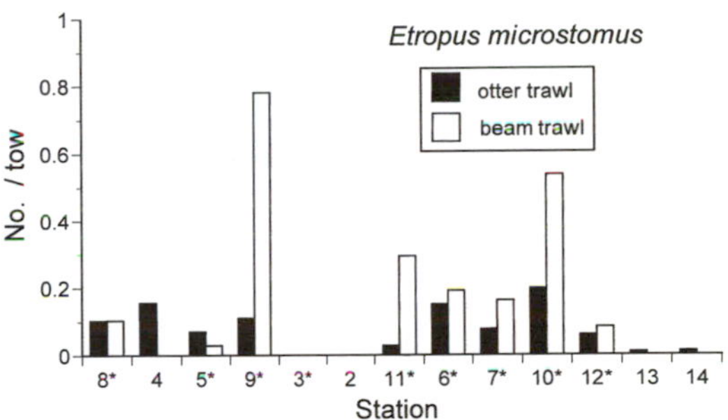

72.7 Distribution of *Etropus microstomus* young-of-the-year by habitat type based on otter trawls ($n = 94$) and beam trawls ($n = 146$) deployed on regular stations 2–14 in Great Bay and Little Egg Harbor. Both otter trawl and beam trawl used on stations designated by asterisk; otherwise, only the otter trawl was used. See figure 2.3 for station locations; table 2.1 for specific habitat characteristics by station.

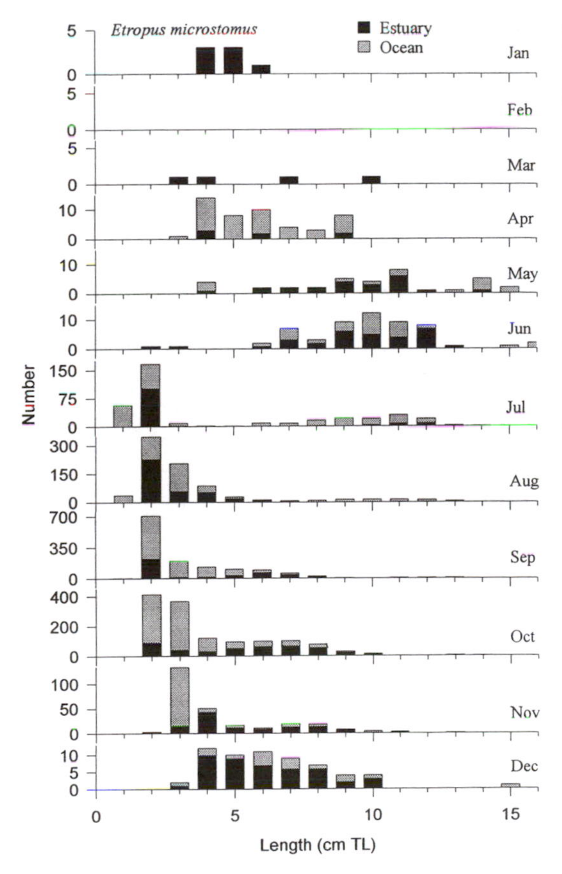

72.5 Monthly length frequencies of *Etropus microstomus* young-of-the-year and older stages collected in estuarine and coastal ocean waters of the Great Bay–Little Egg Harbor study area. Sources: RUMFS: otter trawl (n = 137); 1-m beam trawl (n = 212); 1-m plankton net (n = 610); experimental trap (n = 33); LEO-15 Tucker trawl (n = 306); LEO-15 2-m beam trawl (n = 2,431); gear comparison (n = 200); killitrap (n = 29); seine (n = 358); night-light (n = 3).

CHAPTER 73

Paralichthys dentatus Linnaeus
Summer Flounder

73.1 *Paralichthys dentatus* larva (18.0 mm SL) and juvenile (27.0 mm SL). Collected March 15, 1990 and January 7, 1990, respectively, 1-m plankton net, Little Sheepshead Creek, Great Bay, New Jersey. ANSP 175251 and ANSP 175252. Illustrated by Susan Kaiser.

DISTRIBUTION

Paralichthys dentatus (fig. 73.1) occurs from Nova Scotia, Canada, to the eastern coast of Florida (Gutherz 1967; Gilbert 1986; Scott and Scott 1988). Although there has been considerable discussion concerning the occurrence of separate stocks or populations, the issue is unresolved (see Able and Kaiser 1994). The species is most abundant on the continental shelf from Cape Cod, Massachusetts to North Carolina (Able and Kaiser 1994), but it is plentiful in estuaries as well, where it occupies a variety of habitats over sand, mud, and vegetated substrate including marsh creeks. It has pronounced seasonal migrations, with movements into estuaries or coastal areas in the spring, estuarine residence through the summer, and movements out of estuaries and nearshore areas in late summer and fall. It overwinters at the edge of the continental shelf.

In the Middle Atlantic Bight, the larvae and juveniles are reported from most estuaries (table 4.2). However, some of these records may be questionable. Except for a report of nine larvae collected in the Hudson River Estuary (Dovel 1981) and a single report cited in Berg and Levinton (1985), which may pertain to a single larva, we have not found published evidence of larvae occurring in the Hudson/Raritan system, nor have we been able to find a reliable record of eggs ever occurring there, despite the appraisal in Stone et al. (1994).

REPRODUCTION

Spawning occurs on the continental shelf in the Middle Atlantic Bight from September through January, with the peak in October and November, based on examination of gonads and the occurrence of eggs (Morse 1981; Able et al. 1990; Wilk et al. 1990; Able and Kaiser 1994; Berrien and Sibunka in press). The report of spawning in Narragansett Bay beginning in late July (Herman 1963) is inconsistent with the other available data sources and is questionable. The pattern of larval distribution suggests that spawning occurs as the adults are migrating offshore because the eggs are initially found nearshore and then spread to the entire continental shelf but, by December and January, are only found at the edge of the continental shelf. Egg distributions indicate that some limited spawning occurs in the southern part of the bight in the spring (Berrien and Sibunka in press).

DESCRIPTION

The eggs are pelagic and range from 0.9 to 1.1 mm in diameter with an oil globule of 0.18 to 0.31 mm (Smith and Fahay 1970). At hatching, the larvae are small (2.4–2.8 mm) and have unpigmented eyes and undeveloped mouth parts. Fin rays begin development at about 6 mm; flexion occurs at about 8 to 10 mm. All of the larval stages are heavily pigmented and can be distinguished from congeners by vertebral counts and the presence of melanophores along the edges of the dorsal and anal fins and internal melanophores along the notochord (Fahay 1983; Powell and Henley 1995). Recently transformed juveniles are characterized by the typical flatfish shape with eyes on the left side of the head and a large mouth (fig. 73.1, top). They eventually develop a series of three prominent ocelli on the body just forward of the caudal peduncle (fig. 73.1, bottom). Vert: 41–42; D: 92–98; A: 67–74; Pect: 18–19; Plv: 6–7.

THE FIRST YEAR

Eggs hatch in 2 to 9 days depending on temperature (Smith and Fahay 1970; Smigielski 1975; Johns et al. 1981). Larval survival depends on the length of delay for first feeding after the mouth becomes functional (Bisbal and Bengston 1995). The point of no return for nonfeeding larvae, that is, the time at which feeding is no longer possible and mortality is certain, is 6 to 7 days after hatching. Under natural conditions, the pelagic larvae develop in continental shelf waters at sizes from 2 mm SL to approximately 13 mm SL (Able et al. 1990). Most larvae occur from October through January over the entire Middle Atlantic Bight shelf and Georges Bank (fig. 73.2). Peak larval abundance appears to occur in November and December.

This species is one of the few flatfishes for which we have some detailed information on metamorphosis and settlement. Larvae of 13 mm or larger have been collected in November, December, and March through May on the continental shelf and in adjacent estuaries (Olney 1983; Olney and Boehlert 1988). Larvae at these larger sizes are initiating eye migration (fig. 73.3, stages F−, F, G). Low winter temperatures may influence stage duration at least during metamorphosis (Keefe and Able 1993). Individuals in the laboratory averaged 24.5 days (range 20–32 days) to complete metamorphosis (from Stage F− to Stage I) at ambient spring temperatures of 16.6 C, but partial metamorphosis (from Stage H− to Stage I) at colder average temperatures (6.6 C), required as much as 92.9 days (range 67–99 days).

Movement of larvae, at similar and slightly larger sizes (8–15 mm) and at later stages (fig. 73.3, stage G through H+), into Great Bay–Little Egg Harbor occurs during the fall and winter (fig. 73.4), but the exact timing and duration appears to vary between years (fig. 73.5). During the winters of 1989–1990, 1990–1991, and 1992–1993, most larvae occurred in January through March, whereas few, if any, occurred in the fall. In the winter of 1991–1992, however, they were clearly more abundant during October through January. Prior

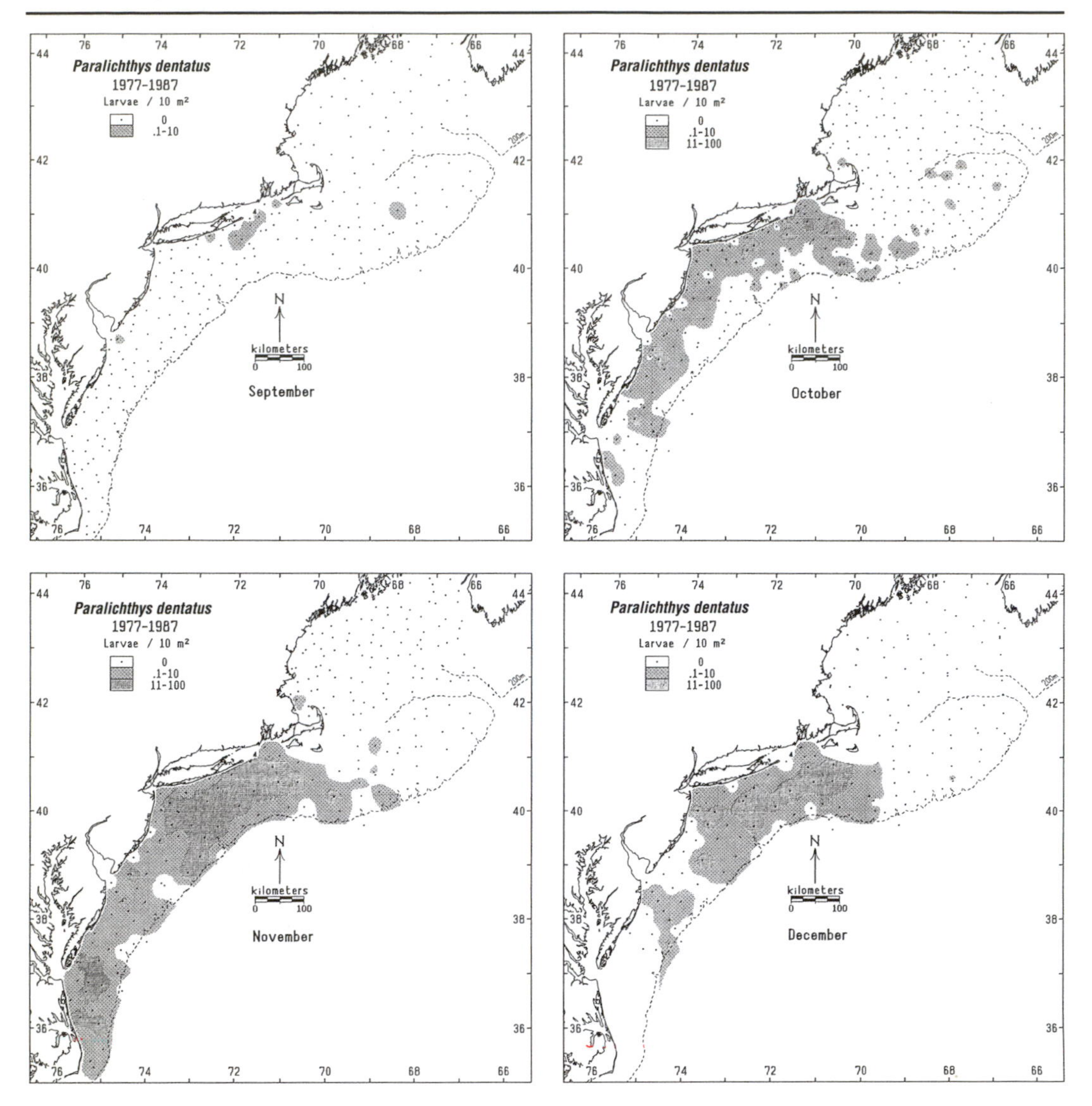

73.2 Monthly distributions of *Paralichthys dentatus* larvae collected during MARMAP surveys 1977–1987.

surveys (1962–1972) in nearby Manasquan Inlet, New Jersey found all larvae during October through December despite extensive effort in the midwinter and spring (Able et al. 1990).

Movement into the estuary may involve intermittent settling to take advantage of tidal stream transport before permanent settlement. When transforming larvae collected from Great Bay were held in aquaria in the absence of tidal currents, they swam up into the water column more often at night (Keefe and Able 1994). They were capable of periodically resting on the substrate on their right sides even before completing eye migration. Some transforming larvae were capable of partial burial at early stages (G to H−) but were not able to bury completely until late in eye migration (H+). Laboratory experiments (Witting and Able 1993, 1995) show that stage I individuals (11–16 mm SL) may be preyed on by the sevenspine bay shrimp (*Crangon septemspinosa*). The morphological transition from larvae to juveniles lasts beyond settlement (Keefe and Able 1993) to approximately 27 mm SL when scale formation is complete, the lateral line is formed, and adult pigmentation is apparent (Keefe and Able 1993).

Although planktonic larvae are common in New Jersey estuaries, small juveniles (<80 mm

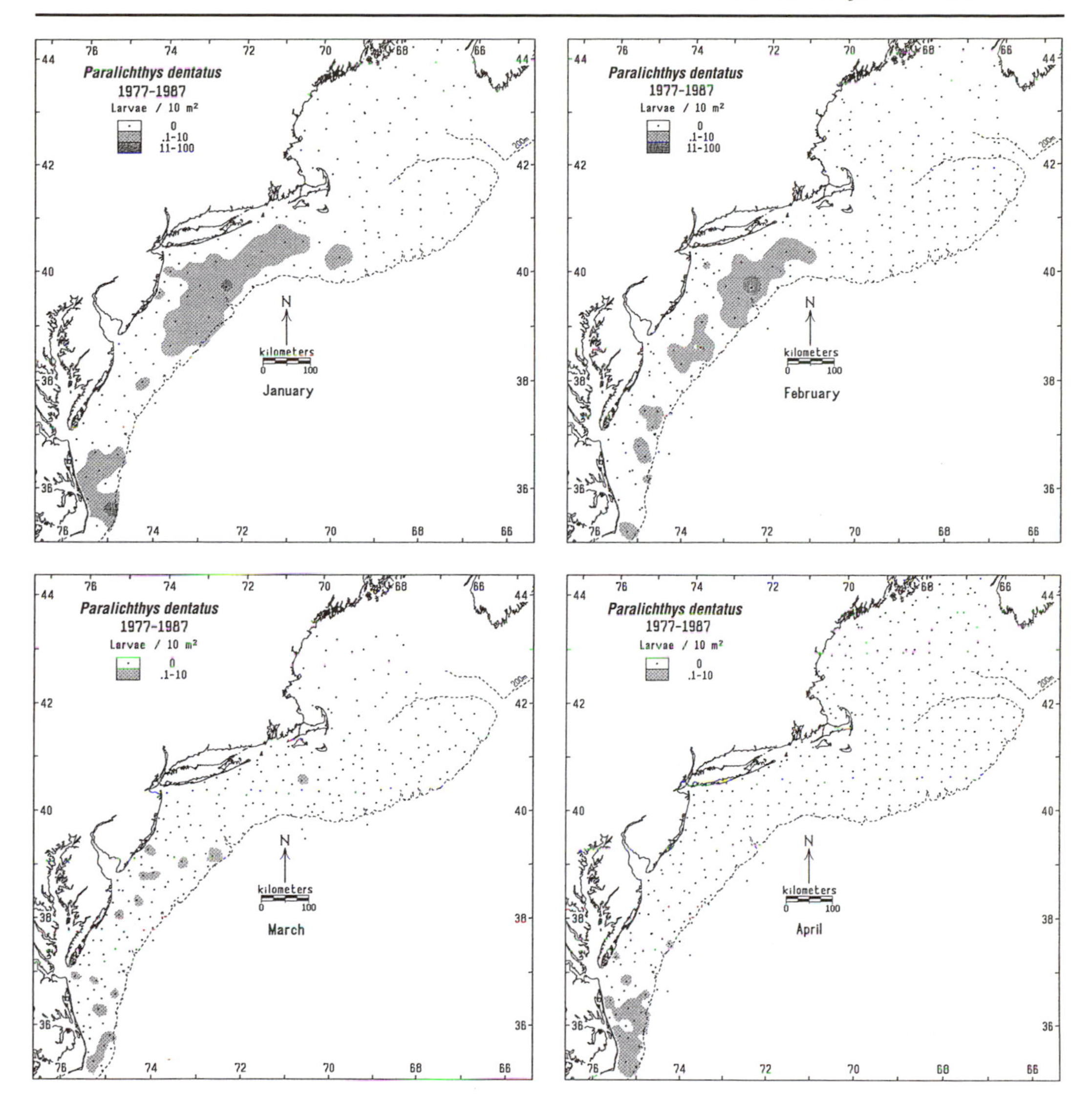

73.2 Monthly distributions of *Paralichthys dentatus* larvae collected during MARMAP surveys 1977–1987.

TL) are not abundant in sampling to date (fig. 73.6; Able et al. 1990; Szedlmayer et al. 1992; Keefe and Able 1993). They have been collected frequently on Virginia's eastern shore (Wyanski 1988) and more commonly in South Atlantic Bight estuaries (see Able and Kaiser 1994). Low winter temperatures may have significant effects on early demersal individuals that enter the estuary in the winter. Small juveniles confined in natural habitats in New Jersey during the winter grew a negligible amount (−0.6 and 0.01 mm per day), and there was little change in development stage (M. Keefe and K. Able unpubl. data). Transforming larvae and juveniles exposed to temperatures below 2 to 3 C suffer significant mortality, at least in the laboratory (Malloy and Targett 1991; Szedlmayer et al. 1992). The impact of low temperatures may vary between years, depending on the severity of the winter. In some years, the larvae enter much later (fig. 73.5) and thus may encounter rising temperatures during the spring warming and more favorable conditions for survival and growth.

During the spring (May–June), growth rates of caged small juveniles (12–41 mm) ranged from approximately 0.5 to more than 1.0 mm per day for small juveniles (Keefe and Able

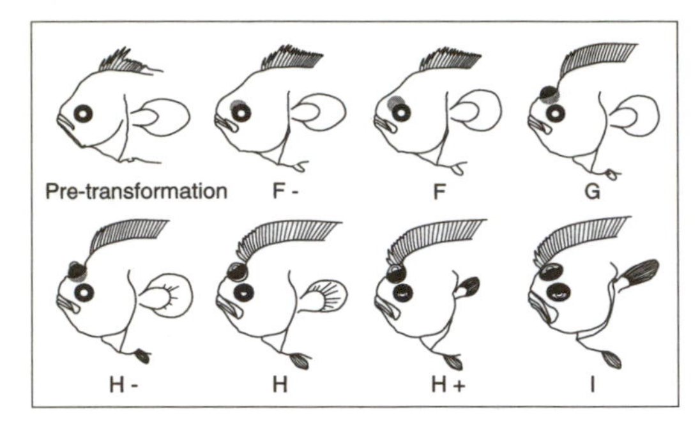

73.3 Classification of metamorphic stages for *Paralichthys dentatus* based on degree of eye migration (Keefe and Able 1993). The right and left eyes are bilateral and symmetrical in premetamorphs. At the first stage of metamorphosis, F−, the eyes are bilateral but asymmetrical with the right eye just dorsal to the left eye. By Stage G, the right eye has reached the dorsal midline and is visible from the left side of the fish. Stage H− differs from Stage G in that the cornea of the eye is visible from the left side of the fish. At Stage H, the right eye has migrated halfway and is on the midline at the dorsal edge of the head. By Stage H+, the right eye has reached the left surface but has not yet reached its final resting place. At Stage I, the eye is set in the socket and the dorsal canal has closed.

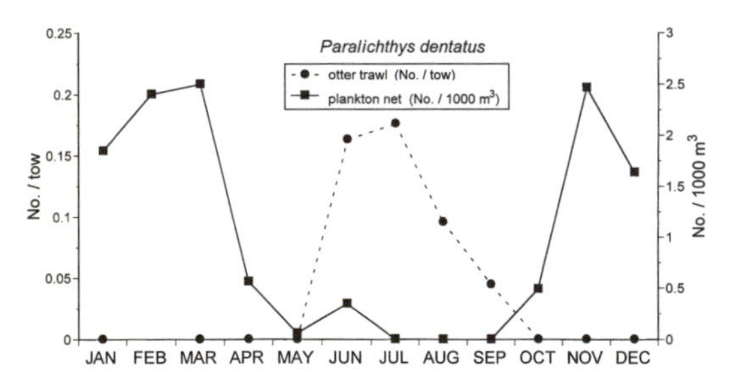

73.4 Monthly occurrences of *Paralichthys dentatus* young-of-the-year in the Great Bay–Little Egg Harbor study area based on 1-m plankton net collections 1989–1994 (n = 671), and otter trawl collections 1988–1989 (n = 55).

1992). At these growth rates, the increments in the sagittal otoliths are deposited daily (Szedlmayer and Able 1992) and thus may be useful in determining more detailed growth rates. Larger juveniles (>80 mm TL) are commonly collected during the summer and early fall (fig. 73.6). In Great Bay–Little Egg Harbor, juveniles grow very quickly, averaging 1.5 to 1.9 mm per day (Rountree and Able 1992b; Szedlmayer et al. 1992) so that by September they range from 200 to 300 mm TL.

In New Jersey estuaries, the transforming larvae and juveniles can be found across a variety of high-salinity, subtidal habitats including subtidal marsh creeks, coves, large bays, and inlets in both vegetated and unvegetated habitats (fig. 73.7; Able et al. 1989; Rountree and Able 1992b; Szedlmayer et al. 1992). The temperature and salinity ranges where most life-history stages are encountered are wide because of their extensive movement patterns (Able and Kaiser 1994). In Great Bay–Little Egg Harbor, the recently settled juveniles (<27 SL) have been found in many habitats including salt marsh creeks, shallow coves, and shallow portions of bays. In North Carolina, they are consistently found on tidal flats (Burke 1995) or sand substrate or a transition from fine sand to silt and clay (Powell and Schwartz 1977). During the summer, many of the larger juveniles (>80 mm TL) are found in marsh creeks where they undergo diel movements. Ultrasonically tagged individuals (210–254 mm TL) followed a regular pattern of movements in a 1-km long subtidal creek in Great Bay–Little Egg Harbor (Szedlmayer and Able 1993). These individuals spent most of the time at the mouth of the creek during the July–September study period.

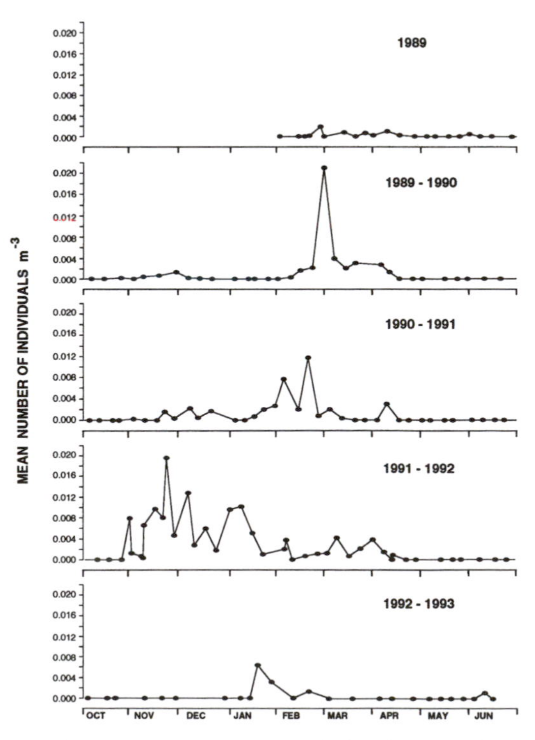

73.5 Annual variation in seasonal abundance of *Paralichthys dentatus* metamorphosing larvae in New Jersey waters (Able and Kaiser 1994).

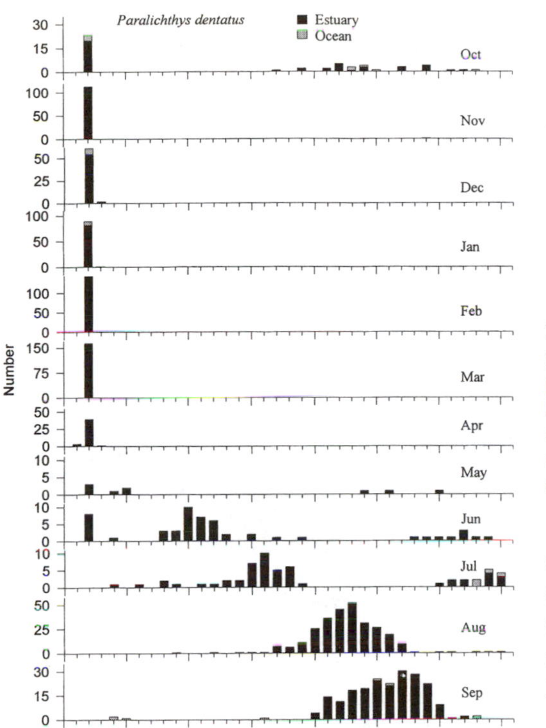

73.6 Monthly length frequencies of *Paralichthys dentatus* young-of-the-year collected in estuarine and coastal ocean waters of the Great Bay–Little Egg Harbor study area. Sources: RUMFS: otter trawl (*n* = 124); 1-m beam trawl (*n* = 13); 1-m plankton net (*n* = 613); experimental trap (*n* = 3); LEO-15 Tucker trawl (*n* = 8); LEO-15 2-m beam trawl (*n* = 33); gear comparison (*n* = 28); killitrap (*n* = 3); seine (*n* = 11); night-light (*n* = 16); weir (*n* = 443).

Movements up the creek typically occurred on night high tides, followed by a return down the creek on the following ebb tide. Movement up the creek appeared to be associated with periods of feeding on marsh creek fishes and

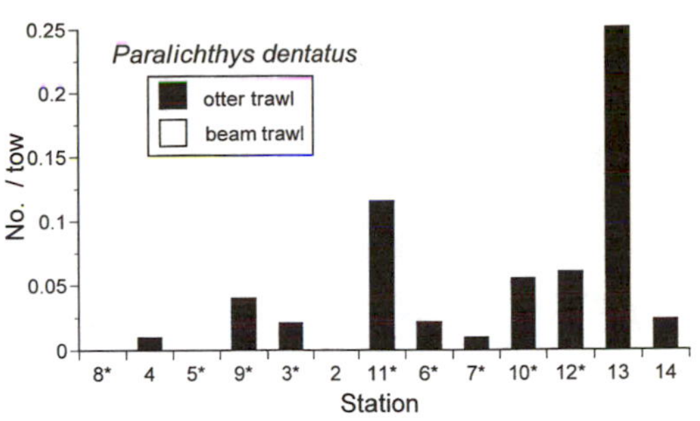

73.7 Distribution of *Paralichthys dentatus* young-of-the-year by habitat type based on otter trawls (*n* = 63) deployed on regular stations 2–14 in Great Bay–Little Egg Harbor. Both otter trawl and beam trawl used on stations designated by asterisk; otherwise, only the otter trawl was used. See figure 2.3 for station locations; table 2.1 for specific habitat characteristics by station.

crustaceans (Rountree and Able 1992b). Movements down the creek may have been caused by the need to avoid low dissolved-oxygen conditions in the upper portion of the creek on night low tides (Szedlmayer and Able 1993).

Larger young-of-the-year first occur offshore during the late summer or early fall (Able et al. 1990; Rountree and Able 1992b; Szedlmayer et al. 1992), and the movement out of the estuaries continues until almost all of them have left by November or December (fig. 73.6). In the fall, they are clearly most abundant on the inner continental shelf throughout the Middle Atlantic Bight (Able and Kaiser 1994). During the winter, this same year class was found farther offshore, some to the edge of the shelf. By spring, some individuals had moved back to coastal areas, but others could still be found at the edge of the shelf. Beginning about this time, they reenter estuaries for the spring and summer (Able et al. 1989).

CHAPTER 74

Pseudopleuronectes americanus Walbaum
Winter Flounder

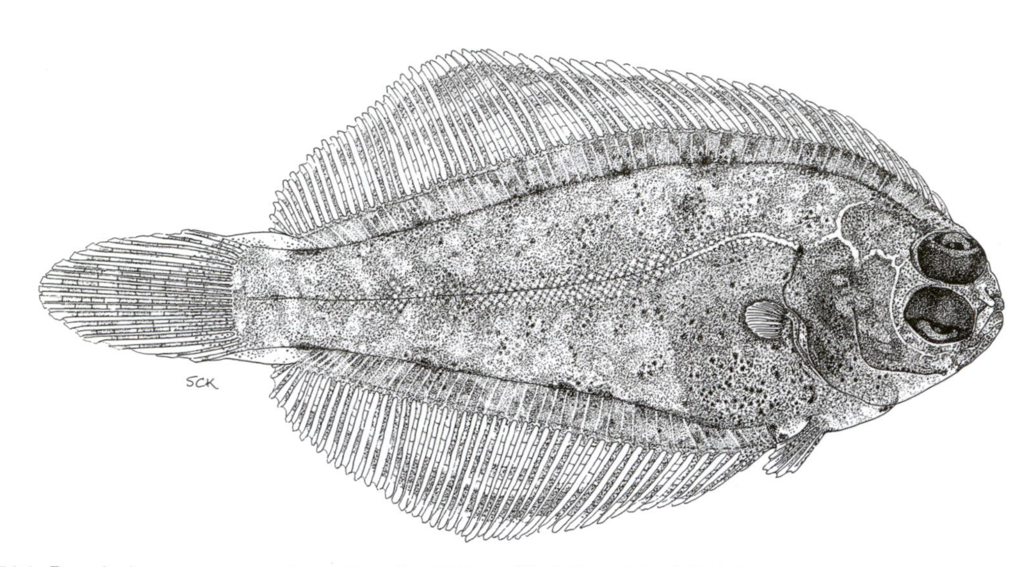

74.1 *Pseudopleuronectes americanus* juvenile, 14.0 mm SL. Collected April 1991, beam trawl, small cove at Holgate, Great Bay, New Jersey. ANSP 175253. Illustrated by Susan Kaiser.

DISTRIBUTION

Pseudopleuronectes americanus (fig. 74.1) occurs in estuarine and continental shelf habitats from Labrador to Georgia (Scott and Scott 1988). Over this range, there are numerous distinct populations, including one on Georges Bank where it was once considered a separate species (Pierce and Howe 1977). The adults in southern New England are presumed to move offshore during the summer as estuarine temperatures warm and then back into estuaries in the fall to spend the winter (Bigelow and Schroeder 1953). Some components of the populations in the New York Bight are known to spend the winter offshore (Phelan 1992; Able and Hagan 1995). Early life-history stages occur in estuaries throughout the Middle Atlantic Bight (table 4.2), but they are more abundant in the northern portion (Jeffries and Johnson 1974; Howe et al. 1976). This is also evident along the coast of New Jersey, where they are quite abundant in Raritan Bay but become increasingly less so in the southern part of the state (Scarlett 1991).

REPRODUCTION

This species spawns demersal eggs (Pearcy 1962b) in estuaries in the winter to early spring (Bigelow and Schroeder 1953; Crawford and Carey 1985); the exact timing is probably temperature-dependent and thus varies with latitude. For example, in Massachusetts, spawning peaks in February and March (Bigelow and Schroeder 1953; Howe et al. 1976), while it extends into April in Connecticut (Pearcy 1962a). In New Jersey, spawning occurs primarily from January to March (Scarlett and Allen 1992). In a Rhode Island salt pond, spawning sites appeared spatially distinct (Crawford and Carey 1985). Spawning also occurs in the offshore waters of Georges Bank (Bigelow and Schroeder 1953; Smith et al. 1975). Tatham et al. (1974) suggested that spawning may occur in the ocean near Atlantic City, but these observations and those by Smith et al. (1975) suggest that larvae are carried out of estuaries on ebb tides as has been reported by Pearcy (1962a).

DESCRIPTION

The demersal eggs are 0.7 to 0.9 mm in diameter and lack an oil globule until late in development (Bigelow and Schroeder 1953). The larvae are 2.4 to 3.5 mm at hatching and have a distinctive wide vertical band of pigment approximately halfway between the anus and the tip of the tail. Notochord flexion occurs by 6.6 mm. Fin rays are complete by 13 mm. Details of the development can be found in Sullivan (1915), Lippson and Moran (1974), and Laroche (1981). Juveniles of this right-eyed flounder (fig. 74.1) can be distinguished from those of the only other right-eyed flounder in estuaries in the Middle Atlantic Bight, *Trinectes maculatus,* by a more pointed snout and by the presence of a pectoral fin. Vert: 34–40; D: 60–76; A: 44–58; Pect: 10–11.

THE FIRST YEAR

The duration of incubation varies with temperature (Williams 1975). Hatching has been reported to occur at 11 to 63 days at 8.0 to −1.8 C and in 23 to 40 days at 3.5 to 0 C. In the same study, survival was lower and more variable at temperatures above 10 C. In a series of comprehensive studies of a population in the Mystic River estuary in Connecticut (Pearcy 1962a), it was determined that the larvae were most common from March to June, where initially they were found in the upper estuary and subsequently, with development, became more abundant in the lower estuary. The larvae were typically more abundant near the bottom, and based on laboratory observations, were occasionally demersal. Even with this vertical distribution pattern, in a vertically stratified system, it was estimated that 3% per day were swept from the estuary and lost to the population. However, larvae in a Rhode Island salt pond were retained, apparently as a result of the hydrodynamics of the system (Crawford and Carey 1985). In the Mystic River estuary study, other sources of natural mortality included predation on the larvae by the medusae of *Sarsia tubulosa.* Together these resulted in an estimated loss of 20% per day for small larvae and 4% per day for larger larvae.

In the Great Bay–Little Egg Harbor estuarine system planktonic larvae were a consistent component of the larval fish fauna (fig. 74.2) and were abundant from mid-March to June (fig. 74.3), but the timing of peak density varied from year to year (Witting 1995; Witting and Able in review). The peaks occurred as early as 10 April and as late as 14 May and appeared to vary with water temperature during the period of spawning and egg development. In general, late peaks in larval density occurred when temperatures were primarily below 5 C from January to April. Early peaks occurred when temperatures were predominantly at or above 5 C during this period.

Settlement occurs at sizes of 9 to 13 mm SL (Pearcy 1962a; Witting 1995; Witting and Able in review). Over this size range, eye migration is completed, and juvenile pigmentation appears. Laboratory observations clearly show that winter flounder express higher variation in age than size at metamorphosis (Chambers and Leggett 1987) and that variation in age at metamorphosis is greatly influenced by temperature (Laurence 1975). The timing of settlement in

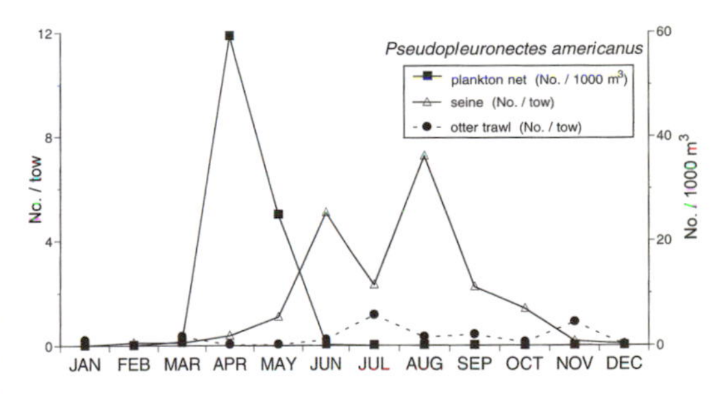

74.2 Monthly occurrences of *Pseudopleuronectes americanus* young-of-the-year in the Great Bay study area based on 1-m plankton net collections 1989–1994 (n = 3,750), otter trawl collections 1988–1989 (n = 329), and seine collections 1990–1991 (n = 218).

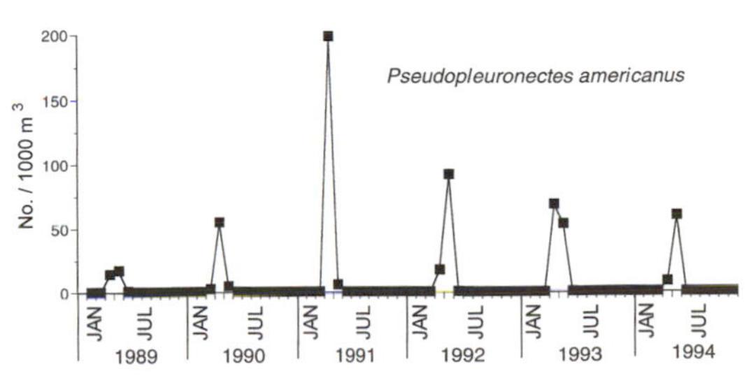

74.3 Annual variation in abundance of *Pseudopleuronectes americanus* young-of-the-year in Little Sheepshead Creek Bridge 1-m plankton net collections 1989–1994 (n = 3,750).

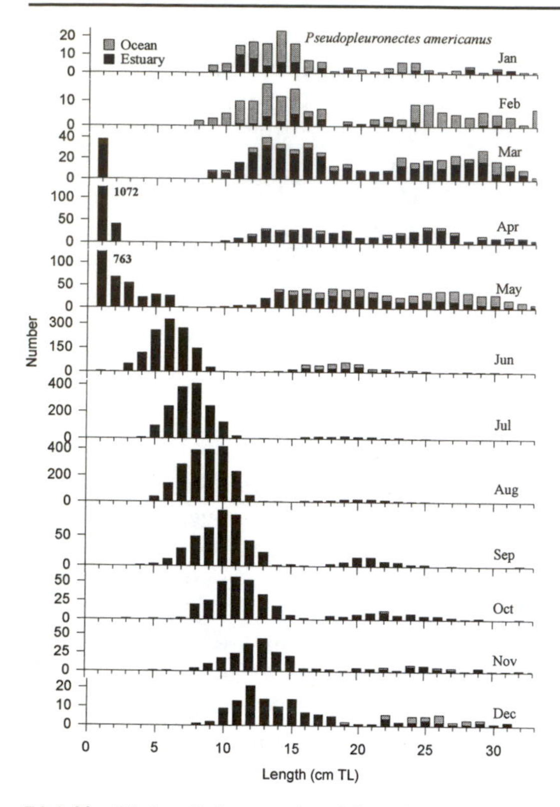

74.4 Monthly length frequencies of *Pseudopleuronectes americanus* young-of-the-year and older stages collected in the Great Bay–Little Egg Harbor study area. Sources: RUMFS: otter trawl (n = 753); 1-m beam trawl (n = 200); 1-m plankton net (n = 1,909); experimental trap (n = 216); LEO-15 Tucker trawl (n = 31); LEO-15 2-m beam trawl (n = 3); seine (n = 1,288); gear comparison (n = 284); killitrap (n = 335); night-light (n = 40); weir (n = 522).

the Great Bay–Little Egg Harbor system appears to be correlated with and follow the peaks in larval density (Witting 1995; Witting and Able in review). In this estuary, settlement appears to be localized. Over several years, very high densities of recently settled individuals occurred in a small cove just inside Little Egg Inlet (fig. 74.5). In a single year, the average density of early demersal individuals reached a peak of 4.1/m^2. Subsequently the abundance declined as some individuals were preyed on by benthic crustaceans (Witting 1995; Witting and Able in review) and others dispersed to other habitats within the estuary. In the Mystic River estuary (Pearcy 1962a), average density of settled juveniles was in excess of 1/m^2 in the spring.

This species is one of the few in which estimates of growth and mortality are available. Growth in the Mystic River estuary during the summer and fall averaged 0.28 to 0.35 mm per day in Connecticut (Pearcy 1962a) and was relatively fast with most individuals attaining 60 to 80 mm by early winter. Average monthly mortality during the first year was 31%. Total mortality during the larval and juvenile stages was approximately 99.98 to 99.99%, demonstrating the importance of events during the first year. Growth in the Great Bay–Little Egg Harbor system, subsequent to settlement, ranged from 0.23 to 0.47 mm per day among eight different habitats, but the highest growth rates were in the settlement cove and nearby habitats (Witting 1995). Growth rates in experimental cages at a variety of habitats ranged from negative values to 1.3 mm per day (Sogard 1992). Growth in a Rhode Island estuary had daily rates similar to the above (Mulkana 1966).

Sampling of populations in the Great Bay system indicated that growth of young-of-the-year occurs over the spring through early winter (December) with little evidence of growth during the middle of the winter based on modal progression of monthly length frequencies (fig. 74.4). The young-of-the-year are clearly evident in May when they range up to 6 cm; by October, they are 7 to 16 cm. These same-size individuals occur through the following winter and, by May (at approximately 12 months), this year class is approximately 11 to 23 cm TL. At this time, the difference in the length-frequency distribution between the young-of-the-year and age 1 individuals becomes quite

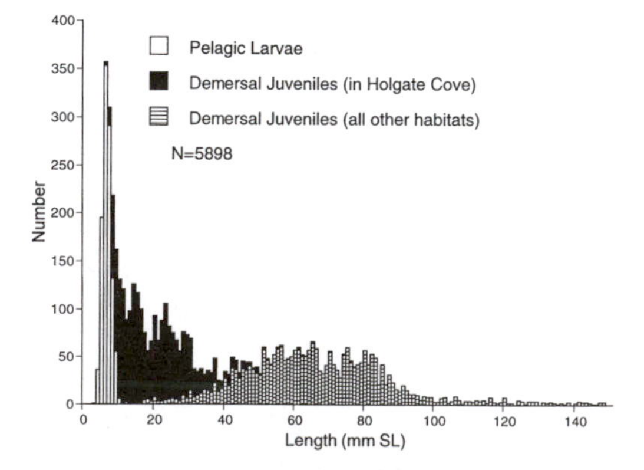

74.5 Composite length frequencies of planktonic and demersal stages of *Pseudopleuronectes americanus* young-of-the-year collected in the Great Bay–Little Egg Harbor estuary (Witting 1995).

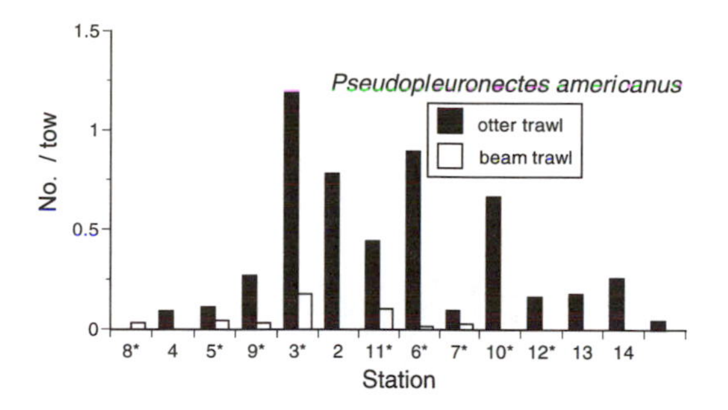

74.6 Distribution of *Pseudopleuronectes americanus* young-of-the-year by habitat type based on otter trawls ($n = 517$) and beam trawls ($n = 27$) deployed on regular stations 2–14 in Great Bay and Little Egg Harbor. Both otter trawl and beam trawl used on stations designated by asterisk; otherwise, only the otter trawl was used. See figure 2.3 for station locations; table 2.1 for habitat characteristics by station.

evident (fig. 74.4). This estimate of size at age 1 overlaps that based on examination of the otoliths (Haas and Recksiek 1995).

In Great Bay–Little Egg Harbor Estuary, juveniles (>25 mm) occur across a variety of habitat types, regardless of sediment and structure. However, most are found in shallower depths (1–3 m) over sandy substrates (<20% silt/clay) (fig. 74.6). In very shallow habitats (<1 m), young-of-the-year were most common over unvegetated substrates (Sogard and Able 1991), but this and other sampling also found them associated with macroalgae (Able et al. 1989). In the same system, young-of-the-year were also collected in the subtidal portions of polyhaline marsh creeks (Rountree and Able 1992a). Although there does not appear to be a distinct habitat preference for young-of-the-year (>25 mm) during the first summer, the above observations suggest that loss of estuarine habitat may be as large a contributor to mortality as fishing (Boreman et al. 1993).

Dissolved-oxygen levels may influence habitat quality as well (Bejda et al. 1992). Laboratory studies of the effect of low dissolved oxygen on growth indicated that growth was significantly reduced under low dissolved-oxygen (2.2 mg/liter) conditions but was intermediate for individuals held at fluctuating levels (2.5–6.5 mg/liter). Under fluctuating levels, 60% died when nighttime concentrations fell below 1.4 mg per liter for several hours.

The habitat of young-of-the-year is influenced by their seasonal behavior. The young-of-the-year spend most of the first year in the estuary. They are most abundant in the spring as larvae and subsequently as settled juveniles during the summer (figs. 74.2, 74.4). From spring through December, all young-of-the-year were collected in the estuary, but by January a large proportion of the young-of-the-year individuals captured were in the ocean as were larger and older fish (fig. 74.4). This occurrence in the ocean in the winter differs from the expected pattern in which they are presumed to overwinter in the estuary (Bigelow and Schroeder 1953), but large numbers have also been reported from the shelf in the New York Bight apex (Phelan 1992) and outside the Great Bay Estuary (Able and Hagan 1995). Alternatively, at least larger individuals are reported to bury in the mud as deep as 12 to 15 cm during the winter (Fletcher 1977), and this may account for the reduced numbers collected in the estuary during the winter (fig. 74.4). These contradictions in seasonal use of estuaries and the ocean require further study.

CHAPTER 75

Trinectes maculatus (Bloch and Schneider) Hogchoker

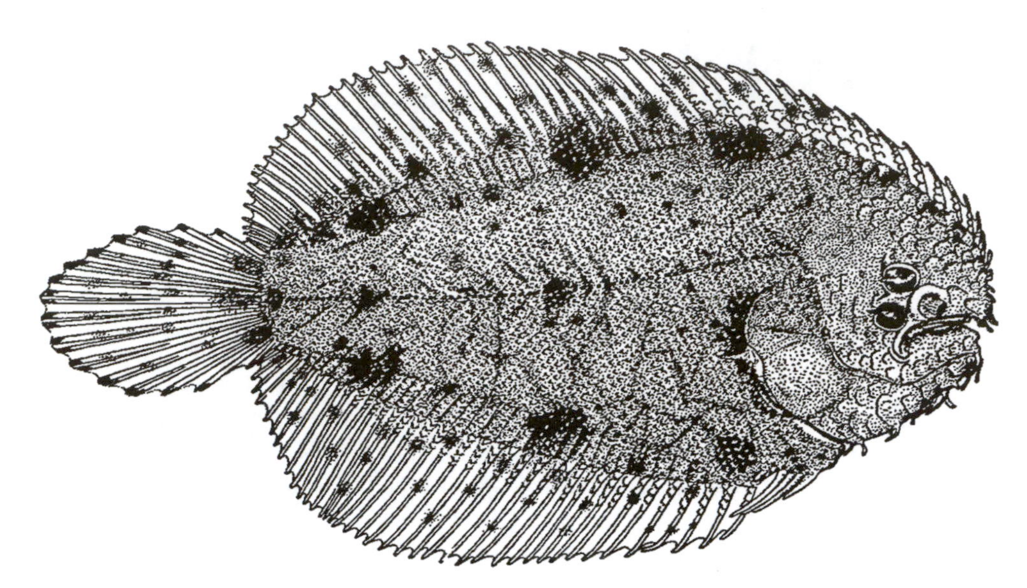

75.1 *Trinectes maculatus* juvenile, 18.0 mm TL. After Hildebrand and Cable 1938.

DISTRIBUTION

Trinectes maculatus (fig. 75.1) occurs in estuarine and marine waters from Massachusetts to Florida and through the Gulf of Mexico to the coast of Venezuela (Bigelow and Schroeder 1953). Individual estuaries may have resident populations in which the adults undergo seasonal movements (Dovel et al. 1969). Early life-history stages occur in most estuaries in the Middle Atlantic Bight (table 4.2) but are most abundant from Chesapeake Bay (Musick 1972) to the south.

REPRODUCTION

Egg collection shows that spawning occurs throughout lower Chesapeake Bay (Olney 1983). In the Patuxent River, a tributary of Chesapeake Bay, spawning occurs from May through September, with a peak when water temperatures are greater than 25 C (Dovel et al. 1969). Eggs have been found as far as 9.7 km off the mouth of Chesapeake Bay (Hildebrand and Cable 1938). In intensive estuarine collections in the Patuxent River, the eggs occurred from 0 to 24 ppt but were concentrated in 10 to 16 ppt (Dovel et al. 1969). Spawning in the Delaware Bay occurs in the upper bay, and the larvae are found into the Delaware River (Wang and Kernehan 1979).

DESCRIPTION

The eggs are slightly buoyant, spherical, or slightly oblong. Size varies with salinity (0.66 to 0.92 mm in diameter), with the smallest eggs found in the highest salinities (Hildebrand and Cable 1938; Dovel et al. 1969). They have numerous (15–35) small oil globules and a narrow perivitelline space. Hatching occurs at 1.7 to 2.0 mm, and the larvae have a short body with wide fin folds and a large oval yolk sac with numerous oil globules. Areas of dark pigment occur on the dorsal and anal finfolds, and these are retained as the fins develop. A pronounced hump is evident on the head early in development. Additional details of development are available from Hildebrand and Cable (1938). The early demersal individuals have a well-rounded snout, and the pectoral fins are absent or vestigial (fig. 75.1). The color pattern on the right side of the body consists of a variety of bars, blotches, or spots, while the left side is a dirty white. Vert: 28–29; D: 51–55; A: 41–42; Pect: none; Plv: 5.

THE FIRST YEAR

The eggs and larvae were dominant components of the Chesapeake Bay ichthyoplankton in the summer, especially during August when they were distributed throughout the bay (Olney 1983; Olney and Boehlert 1988). The most intensive studies in the Middle Atlantic Bight have been conducted in the Patuxent River, Maryland, and these (Dovel et al. 1969) provide the basis for the following interpretation. After spawning in the lower estuary, the eggs develop quickly (e.g., the eggs hatch in 26 to 36 hrs at 23.3 to 24.5 C; Hildebrand and Cable 1938), and the recently hatched larvae (1.6–6.0 mm) are most abundant in the same area. During this period, the pectoral fins degenerate and are only visible as small flaps by approximately 6.0 mm. The left eye migrates across the top of the head to the adult position on the right side at 6.0 to 10.0 mm (fig. 75.1).

In the Patuxent River and elsewhere, the size at settlement is unknown, largely because collections of 11 to 21 mm individuals are infrequent. Scale formation occurs before 18 mm. Growth of young-of-the-year is relatively slow. Examination of length-frequency modes in the Patuxent River (Dovel et al. 1969) suggests that young-of-the-year are approximately 10 to 50 mm by November, do not grow over the winter, and are approximately 20 to 60 mm by the following spring when they are age 1. Schwartz (1964a) reported individuals of 27 to 34 mm from February through May in Isle of Wight Bay and Assawoman Bay. Growth estimates based on scales from fish collected in the Patuxent River indicate an average back-calculated length of 40 mm at age 1 and 66 mm at age 2 (Mansueti and Pauly 1956). Thus, this resident form is among the slowest-growing flatfishes in estuaries in the Middle Atlantic Bight.

Surprisingly, little is known of the habitats of this abundant form. Larger larvae and recently transformed juveniles are reported to migrate rapidly upriver to low-salinity nursery areas where the smallest juveniles are typically collected. In Delaware Bay, juveniles were found over mud substrate in creeks during the late summer and early fall (Smith 1971). In Chesapeake Bay, the young are found on shallow mud flats during the summer (Musick 1972). The upstream limit of the young-of-the-year nursery area extends into freshwater (Hildebrand and Cable 1938; Massmann 1954). The young are reported to overwinter in the upper parts of estuaries, and they and the adults migrate to higher-salinity waters in the spring.

CHAPTER 76

Sphoeroides maculatus (Bloch and Schneider) Northern Puffer

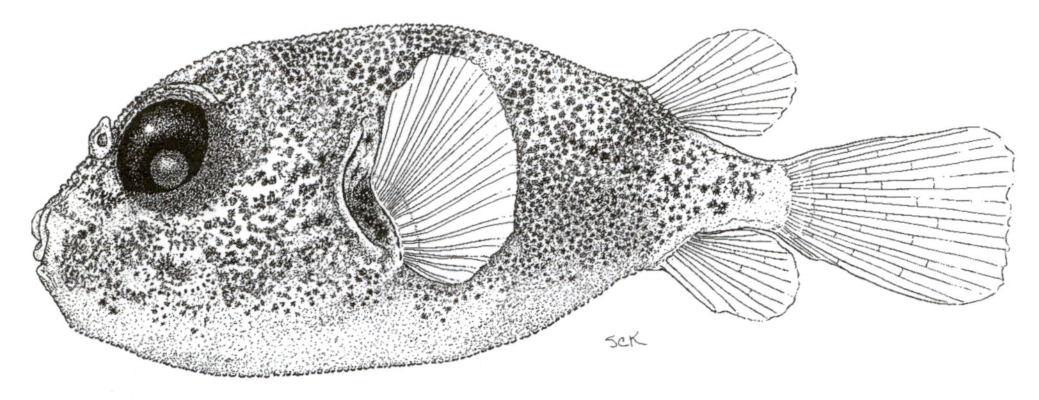

76.1 *Sphoeroides maculatus* juvenile, 9.7 mm SL. Pigmentation based on examination of two uncatalogued specimens. Collected June 4, 1991, night-light, dip net, RUMFS boat basin, Great Bay, New Jersey. ANSP 175254. Illustrated by Susan Kaiser.

DISTRIBUTION

Sphoeroides maculatus (fig. 76.1) occurs on the Atlantic coast of North America from Newfoundland to northern Florida (Shipp and Yerger 1969). It is abundant in the Middle Atlantic Bight where it makes annual coastal migrations, northward and inshore in spring, and southward and offshore in fall (Laroche and Davis 1973; Shipp 1974; Bigelow and Schroeder 1953). In our study area, *S. maculatus* occurs in near-coastal waters, bays, and estuaries over a variety of substrates (Nichols and Breder 1927; de Sylva et al. 1962) and is often found around piers and other structural habitats (Schwartz 1964a). It was the most abundant species collected during a 2-year (1962–1963) survey of the surf zone of Long Island (Schaefer 1967), but it was not present (in any life-history stage) in a 6-year survey of the nearby Hudson River estuary (Kahnle and Hattala 1988). It was once much more common than it is today off the coast of New Jersey. During collections made from 1929 to 1933, it ranked fifth in frequency of occurrence and ninth by numerical abundance, but failed to rank among the top 20 in comparable collections in the 1970s (Thomas and Milstein 1974). It was reported to be one of the most abundant sport fishes in Great Bay in June 1969 (Hamer 1972) but declined drastically through the 1970s (Murawski and Festa 1979). Larvae and juveniles have been reported from most estuaries in the Middle Atlantic Bight, but eggs have only been reported from Barnegat Bay (table 4.2).

REPRODUCTION

Spawning occurs in the ocean close to shore and in estuaries between May and August (Perlmutter 1939; Bigelow and Schroeder 1953; Laroche and Davis 1973). Observations of a captive pair indicate that eggs are partially buried in a circular depression in the substrate (Breder and Clark 1947). Age-1 fish as small as 88 mm are capable of spawning (Laroche and Davis 1973). Peak spawning in the New York area occurs in June (Perlmutter 1939) and in the Chesapeake Bay area in June and July (Laroche and Davis 1973). Spawning continues through summer and into fall (Sibunka and Pacheco 1981), and young-of-the-year as small as 15 mm FL have been collected as late as November in the Indian River Inlet area (Pacheco and Grant 1973). In the Great Bay–Little Egg Harbor study area, the presence of small juveniles (<2 cm) indicates that spawning is continuous from June to October.

DESCRIPTION

The demersal eggs are spherical, transparent, and equipped with a smooth, adhesive covering. They adhere to substrates, to submerged objects, and to each other. Diameter of the egg ranges from 0.85 to 0.91 mm, but the adhesive covering makes it appear larger. Many colorless oil globules (average diameter 0.34 mm) are massed in a cluster (Welsh and Breder 1922). Larvae hatch at about 2.4 mm (Welsh and Breder 1922). Pigmentation includes many patches of red, orange, yellow, and black chromatophores, which are most dense along the ventral half of the anterior two-thirds of the body. At 7.0 mm, larvae resemble adults and are capable of inflating their bodies (Welsh and Breder 1922). Juveniles are recognizable by their inflatable body (fig. 76.1). The fins are small and inconspicuous, the eyes are located near the top of the head, and the mouth is small and beaklike. Color patterns vary with age, but may consist of dark bars and blotches against a yellow or white background. Prickles covering the skin begin to form when juveniles are about 10 mm (Shipp and Yerger 1969). Dark black "pepper" spots appear on the body in fishes between 40 and 100 mm (Shipp and Yerger 1969). Pelvic fins are lacking. Vert: 19; D: 8; A: 7; Pect: 15–16; Plv: none.

THE FIRST YEAR

Larvae are about 2.4 mm at hatching after about 3.5 days incubation at 19.4 C (Welsh and Breder 1922). Larvae were rarely collected over the Middle Atlantic Bight continental shelf during 11 years of MARMAP sampling (1977–1987), although larger, pelagic-juveniles (identified only to the family level) occurred infrequently, particularly in the southern part of the bight (fig. 76.2). Larvae are also rarely collected in major bays and estuaries throughout our study area despite intensive sampling (see Tables 3.1, 3.2, 3.3). Nevertheless, in certain embayments and during some years, larvae have been ranked among the most abundant. For example, larvae were the fifth most abundant species collected in Great South Bay from May to September (Monteleone 1992). Conversely, larvae have been collected in the Patuxent River but not the Hudson River in comparable sampling (Dovel 1981). In Long Island Sound, intensive plankton sampling failed to find more than two larvae in July, yet juveniles were abundantly collected in September and October (Richards 1959). Larvae are also rare in Narragansett Bay (Bourne and Govoni 1988). In Great Bay–Little Egg Harbor, young-of-the-year are most abundant during July (fig. 76.3). Unlike the adults, which are solitary, juveniles are usually found in large schools (Fish 1954).

Growth is rapid through the first summer and fall. In the Great Bay–Little Egg Harbor

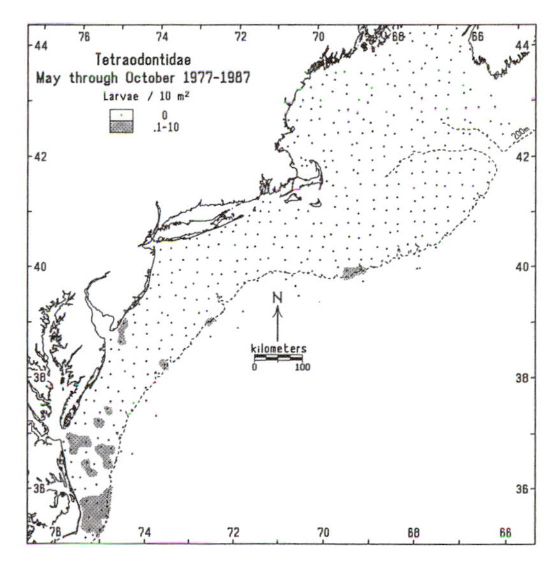

76.2 Occurrences of Tetraodontidae larvae collected during MARMAP surveys 1977–1987. May through October combined. These larvae were not identified to species level.

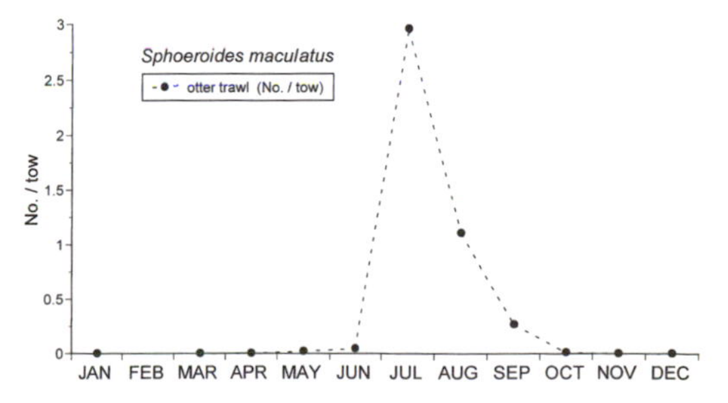

76.3 Monthly occurrences of *Sphoeroides maculatus* young-of-the-year in the Great Bay–Little Egg Harbor study area based on otter trawl collections 1988–1989 ($n = 541$).

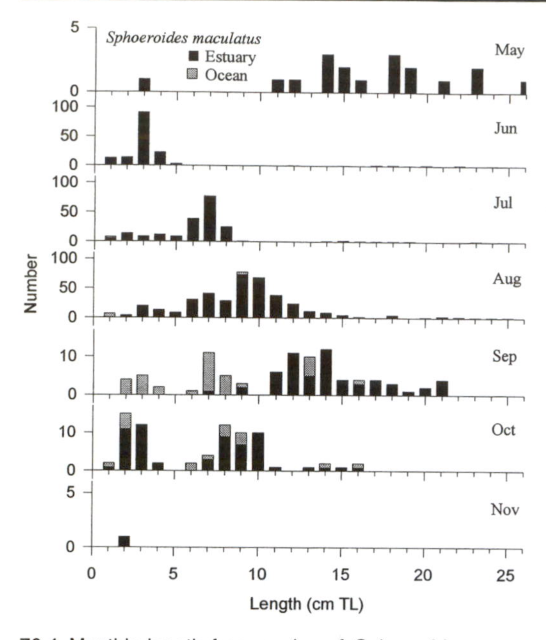

76.4 Monthly length frequencies of *Sphoeroides maculatus* collected in estuarine and coastal ocean waters of the Great Bay–Little Egg Harbor study area. Sources: RUMFS: otter trawl (n = 452); 1-m beam trawl (n = 19); 1-m plankton net (n = 23); LEO-15 Tucker trawl (n = 8); LEO-15 2-m beam trawl (n = 57); gear comparison (n = 341); killitrap (n = 5); seine (n = 29); experimental trap (n = 1); night-light (n = 5); weir (n = 8).

study area, young-of-the-year first appear in May and June when they are 3 cm or smaller (fig. 76.4). This cohort can be followed through September when the size mode is about 14 cm. Larger individuals apparently first begin to leave the system in October, and they can be detected returning the following spring at sizes between 11 and 23 cm. Incursions of cold water, caused by upwelling, have been observed to slow growth rates in Great Bay (S. Weiss unpubl. observ.). In this study during 1988, juvenile growth was 0.47 mm per day between July and August (when temperatures were cold), but 1.54 mm per day between August and September when temperatures returned to normal. In a Barnegat Bay study, the growth rates of young-of-the-year have been calculated as 1.11 mm per day over 45 days and 0.93 mm per day over a 60-day period (Marcellus 1972). In the Chesapeake Bay, lengths of 158 to 187 mm TL have been attained by November, and sizes of 203 mm have been reported for fishes reaching age 1 (Laroche and Davis 1973).

Most habitat information is restricted to juveniles because, although larvae have been collected only rarely (see above), juveniles are common constituents of several studies undertaken in the Middle Atlantic Bight. In a study of habitat associations of estuarine fishes in Great Bay–Little Egg Harbor, *S. maculatus* was the sixth most abundant fish species collected (Szedlmayer and Able 1996). Young-of-the-year occurred most abundantly in a variety of depths (0.6–3.7 m), substrate types (3.8%–54.4% silt), and structure (no structure to complex eelgrass beds) (fig. 76.5) but were not found in intertidal habitats. Young-of-the-year have also been collected in subtidal marsh creeks in Great Bay (Rountree and Able 1992a). Larvae apparently inhabit shallow, sandy beaches of estuaries in Long Island Sound (Wheatland 1956). One report describes juveniles as semidemersal, inhabiting the intertidal zone over smooth bottoms (Merriman 1947). Salinity may influence the distribution of young-of-the-year. Most collections of larvae and young in the Delaware Bay area have been made in higher salinities (Wang and Kernehan 1979), although some have been made in salinities as low as 13 ppt (de Sylva et al. 1962). Juveniles have been found in salinities up to 32.2 ppt on the eastern shore of Virginia (Richards and Castagna 1970). A Chesapeake Bay study found larvae and juveniles in salinities between 12 and 21 ppt and temperatures between 16 and 26 C (Dovel 1971).

Several studies have indicated that temperature exerts a strong influence on the distribution

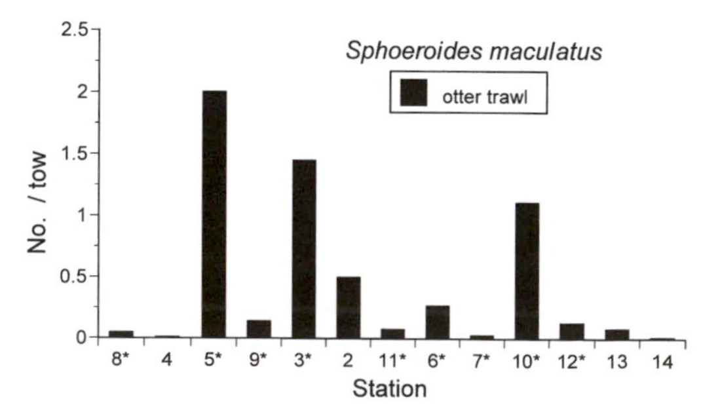

76.5 Distribution of *Sphoeroides maculatus* young-of-the-year by habitat type based on otter trawls (n = 563) deployed on regular stations 2–14 in Great Bay and Little Egg Harbor. Both otter trawl and beam trawl used on stations designated by asterisk; otherwise, only the otter trawl was used. See figure 2.3 for station locations; table 2.1 for specific habitat characteristics by station.

and abundance of this species, and this is particularly true for young-of-the-year. In one Virginia study, juveniles (13–90 mm) occurred only at temperatures between 20.4 and 29.4 C, whereas adults occurred between 10 and 25.9 C (Richards and Castagna 1970). In Barnegat Bay, young-of-the-year appeared as early as April when temperatures exceeded 15 C, but they did not appear until May in years when temperatures were colder (Vouglitois 1983).

In and near Great Bay–Little Egg Harbor, young-of-the-year become less abundant in estuarine habitats during September and October when some begin to appear in ocean collections (Figs. 76.3, 76.4). In the surf zone of Long Island, adults are present throughout the summer and are joined by young-of-the-year (about 100 mm TL) in August and September (Schaefer 1967). All year classes are rare in October and absent in November in these collections. Young-of-the-year and adults then migrate to deeper continental shelf waters where they spend the winter in a quiescent state on the bottom (Bigelow and Schroeder 1953). Temperatures may be critical during the winter. Wicklund (1970) reported a massive kill involving this species on May 10, 1969 from Sandy Hook to Manasquan Inlet, presumably because an upwelling event subjected the fish to a sudden drop in temperature.

CHAPTER 77

The First Year in the Life of Estuarine Fishes

We have attempted to identify patterns in reproduction, development, growth, habitat, and spatial and temporal use of estuaries based on the foregoing 70 species accounts. Because our immediate goal is to determine broad patterns, and the data for individual variation are usually unavailable, we have based the following summary on mean values. The variance around these means, and the processes that affect it, should be the focus for future efforts, such as that of Chambers (1997) for eggs and larvae of marine fishes.

The clearest difference is that between residents (species that spend their entire life spans in estuaries) and transients (species that spend only a portion of their lives there). In instances where this distinction may be equivocal, we relied on the first year in the life of these fishes in order to make the allocation. As a result, it appears that most Middle Atlantic Bight estuarine species for which we have good information (tables 77.1, 77.2) are transients (60%), fewer are residents (28%), fewer still are strays or infrequent visitors (6%), and a similar number are not classified (6%). For a number of rarely collected species ($n = 50$), we have little early life-history information. These are listed with collection details in table 77.3.

SPATIAL AND TEMPORAL PATTERNS IN ESTUARINE USE

The spatial and temporal patterns we have observed in estuarine use by fishes during their first year of life vary among and within resident and transient species and cohorts, as well. The following are representative and recurring patterns for fishes in the central part of the Middle Atlantic Bight (table 77.4). Evidence for assignment to a group is in the individual species accounts.

Group I includes several species that are facultative users of estuaries because their nurseries are in estuaries or the inner continental shelf. They typically spawn during summer, and larvae and early juveniles occur in both estuaries and the coastal ocean for the remainder of the summer before they emigrate from these nurseries into deeper oceanic waters in the fall. The most completely documented example is *Centropristis striata,* but this group also includes such species as the spring-spawned cohort of *Scophthalmus aquosus.*

Group II includes species whose adults migrate into estuaries to spawn in spring or summer. Their progeny remain in the estuaries before emigrating to the ocean in the fall. The best documentation of this pattern of obligatory spawning in estuaries is for *Menidia menidia* and *Syngnathus fuscus,* but the group also includes *Hippocampus erectus, Strongylura marina,* and *Sphoeroides maculatus. Gasterosteus aculeatus* follows a similar pattern but enters the estuary in the winter and spawns in early spring. The juveniles then emigrate in early summer.

Group III includes anadromous species, such as *Morone saxatilis* and *Alosa* spp. In this group, adults migrate through estuaries in order to spawn in freshwaters. Young-of-the-year use freshwaters as well as saline portions of estuaries as nurseries, and most emigrate before winter, although those of *Morone saxatilis* remain in the estuary during their first winter.

Groups IV to VI spawn exclusively in the ocean, but the location, timing and manner in which young-of-the-year use estuaries vary. Most species in Group IV (e.g., *Urophycis regia, U. tenuis, Pollachius virens*) spawn in the Middle Atlantic Bight during the spring, but at least some of their offspring enter estuaries, stay only a short time, and then leave to spend the remainder of their first year on the continental shelf. In *Leiostomus xanthurus* and *Paralichthys dentatus,* the young-of-the-year enter estuaries in early spring (or earlier) but remain through the fall. In Group V, spawning occurs in the Middle Atlantic Bight during summer and fall, and transforming larvae settle and spend their first winter on the continental shelf. They then enter estuaries for the first time the following spring, remain through the summer, and then emigrate in the fall. Examples include *Ophidion marginatum, Prionotus carolinus,* and the fall-spawned cohort of *Scophthalmus aquosus.* This pattern is slightly different for *Urophycis chuss,* which enters estuaries in the winter or spring after an initial stage involving a commensal association with sea scallops on the continental shelf, but they do not stay through the summer.

Species in Group VI spawn outside the Middle Atlantic Bight, yet their progeny regularly occur in the Gulf Stream, then ultimately use estuaries in the Middle Atlantic Bight as nurseries. Certain of these (*Mugil curema, Caranx hippos, Sphyraena borealis,* and the spring-spawned cohort of *Pomatomus saltatrix*) spawn in the South Atlantic Bight. Others (*Anguilla rostrata* and *Conger oceanicus*) reproduce in the Sargasso Sea. Larvae of all these species eventually arrive (either by passive transport, active swimming, or a combination of both) in waters of the Middle Atlantic Bight, from which they migrate into estuarine nurseries.

In Group VII, larvae result from distant spawning, usually in the South Atlantic Bight. They arrive in Middle Atlantic Bight estuaries in the summer, but do not successfully emigrate back to the South Atlantic Bight, and succumb to falling temperatures during fall and winter. Examples include *Chaetodon ocellatus* and *Monacanthus hispidus* (Moss 1973), but the evidence is less clear regarding the fate of several other regularly occurring transients (table 77.4).

Resident estuarine fishes in the central part of the Middle Atlantic Bight exhibit less diverse patterns of estuarine use than transients, because their entire life histories are limited to the estuary. Despite this spatial constraint, there are differences. Group VIII is the largest group and is represented by primarily shallow-water species that spawn in the summer and develop in the immediate vicinity of their spawning sites. These include *Cyprinodon variegatus, Menidia beryllina,* and *Apeltes quadracus.* Group IX comprises a few species that are spawned in the winter and spring (*Pseudopleuronectes americanus, Ammodytes americanus,* and *Myoxocephalus aenaeus*). Group X differs from the above groups because they undergo a spawning migration within the estuary. One of these, *Trinectes maculatus,* appears to be semicatadromous because it leaves fresh or low-salinity portions of the estuary and moves downstream into higher-salinity areas to spawn pelagic eggs. Others in this group are semianadromous because they leave the saltier portions of the estuary and move into freshwater to spawn. These include *Morone americana, Microgadus tomcod,* and *Osmerus mordax.*

Group XI includes two labrid species (*Tautoga onitis, Tautogolabrus adspersus*) that are difficult to classify because it appears that some components of their populations live and spawn in estuaries, whereas others live and spawn on the adjacent continental shelf. We are also unable to place *Micropogonias undulatus* in any other group, because spawning in the Middle Atlantic Bight reaches a peak in the fall, and larvae then enter estuaries where they remain in deepwater habitats through the winter, a pattern unlike any other.

PATTERNS IN REPRODUCTION

Fishes that use estuaries in the central part of the Middle Atlantic Bight spawn in a variety of locations, some quite distant from the estuaries where the early stages spend much of their first year. Most spawning occurs in the Middle Atlantic Bight in freshwaters, estuaries, or on the continental shelf (fig. 77.1). However, a number of these fishes are derived from spawning in the South Atlantic Bight (*Mugil curema, Caranx hippos*), and some may come from a combination of South and Middle Atlantic Bight spawning (*Pomatomus saltatrix, Peprilus triacanthus*). This pattern is best demonstrated by *Pomatomus saltatrix,* in which individuals spawning in both bights use Middle Atlantic Bight estuaries during the first year (Hare and Cowen 1993). The same may be true for *Brevoortia tyrannus,* based on the timing of spawning, larval ingress, and back-calculation of hatching dates (Warlen et al. in press). The most distant spawning site for species using estuarine nurseries in the Middle Atlantic Bight

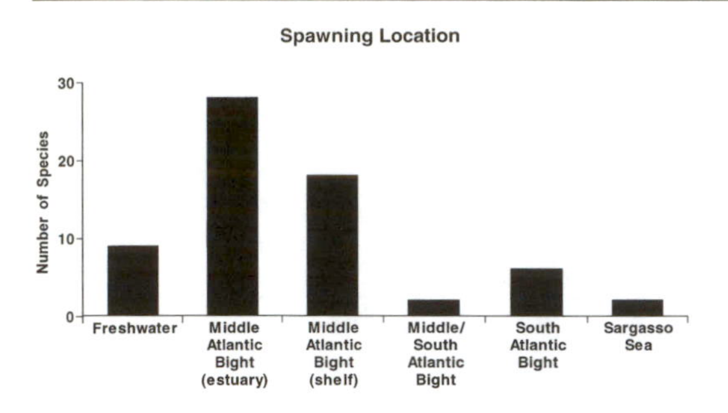

77.1 Spawning locations for selected estuarine species that use the central part of the Middle Atlantic Bight as nurseries.

is the Sargasso Sea (*Anguilla rostrata, Conger oceanicus*).

Most residents in Middle Atlantic Bight estuaries spawn during the spring and summer, whereas transients are spawned during every season (table 77.2). The timing of spawning may also vary within a species. Some species may spawn in a distinctly bimodal temporal pattern. Examples from the Middle Atlantic Bight include *Urophycis regia* and *Scophthalmus aquosus.* Others may spawn in different areas, and the progeny produce a bimodal size pattern. The clearest example is *Pomatomus saltatrix,* which spawns in the South Atlantic and Middle Atlantic bights, and the resulting cohorts retain distinct bimodal size distributions.

PATTERNS IN DEVELOPMENT AND GROWTH

Most residents have demersal eggs (79%), whereas the majority of transients have pelagic eggs (73%) (fig. 77.2). Demersal eggs may be typical of most resident estuarine fish assemblages (Pearcy and Richards 1962; Able 1978; Haedrich 1983; Dando 1984) because this characteristic increases retention in the nursery. We have information on larval stage duration for only 19% of the species we treat, and these range from fewer than 20 to 60 days, with an exceptional few that are larvae for more than 100 days (table 77.1). Settlement size ranges from less than 10 to 70 mm (plus *Conger oceanicus* at 100 mm), but these data are also lacking for 65% of our species (table 77.1).

Settlement warrants further study because important morphological, physiological, and behavioral transitions occur while fishes are undergoing habitat transitions (Moser 1981; Balon 1984; Chambers et al. 1988; Youson 1988; Levin 1991; Kaufman et al. 1992). The small size of recently settled fishes may make them especially susceptible to predators, and several reviews and species-specific studies have clearly identified high-mortality rates during these early juvenile periods (Gulland 1965; Cushing 1974; Sissenwine 1984; Sissenwine et al. 1984; Smith 1985; Houde 1987; Elliot 1989; Doherty 1991; Beverton and Iles 1992; Tupper and Boutilier 1995a,b; Cushing 1996). The period of metamorphosis and settlement may be especially important to estuarine fishes because they also encounter osmoregulatory challenges during these transitions. The coincidence between metamorphosis and settlement is evident in fishes in the Middle Atlantic Bight (see previous species accounts) as well as estuarine species in the South Atlantic Bight of the United States (Hoss and Thayer 1993), the Gulf of Mexico (Yanez-Arancibia 1985), Spain (Arias and Drake 1990), South Africa (Day 1981; Beckley 1984, 1985, 1986), and Australia (West and King 1996). It does not apply to all species, however. All of the *Urophycis* spp. in the Middle Atlantic Bight have a pelagic-juvenile stage, and they do not settle to the bottom until they are older and larger than most other species.

This transition is still the most poorly known in the life-history of fishes (Hempel 1965; Houde 1987; de Lafontaine et al. 1992). It

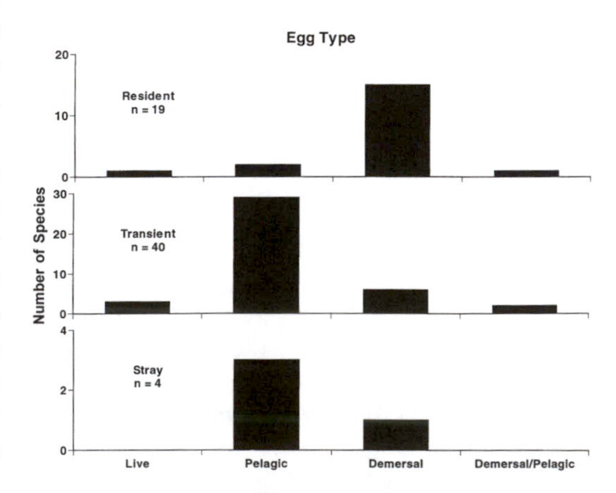

77.2 Egg types in resident, transient and stray components of the estuarine fish fauna in the central part of the Middle Atlantic Bight. Based on species listed in Table 77.2.

is longer in duration than typically perceived, at least in the morphological sense, because it does not end until the major sensory and organ systems are complete. For many species, a convenient morphological end point is completion of scale formation when the lateral line system is also complete (Fuiman and Higgs 1977). This end point has been used by others (Copp and Kovac 1996) and can be applied to species with "direct" or "indirect" development (Balon 1990).

Many of our growth estimates are based on modal length-frequency progressions, which provide reasonable estimates when compared with other aging techniques (e.g., *Urophycis tenuis, Centropristis striata, Tautoga onitis, Paralichthys dentatus* species accounts). These estimates range from near zero to 1.7 mm per day (fig. 77.3). Rates observed in some transients (typically greater than for residents) are among the fastest reported for the young-of-the-year of any fish species, and this is consistent with the prediction that migratory species will have faster growth rates (Roff 1988). Slower growth rates in resident species in a Massachusetts estuary were attributed to their using more predictable, but lower-quality food sources (Werme 1981; Teal 1986). Growth may appear to be slow because many residents are small as adults (e.g., fundulids, gasterosteids, soleids, etc.), and if we express growth as a function of adult size, these estimates would be more similar to those of transient species. During the winter, growth rates are consistently low or nil for both residents and transients (fig. 77.3). The slower estimated growth rates

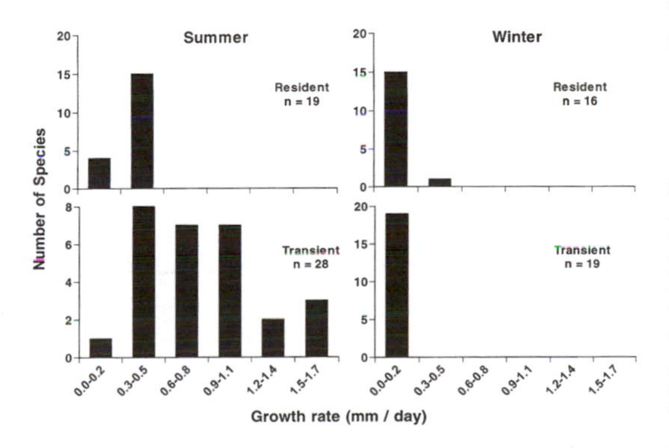

77.3 Growth rates in resident and transient fishes in the summer and winter in the central part of the Middle Atlantic Bight. Based on species listed in Table 77.2.

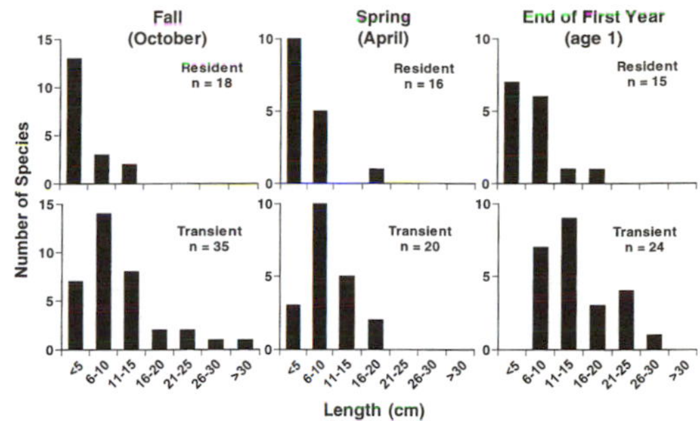

77.4 Size attained by the fall, spring, and end of first year for resident and transient fishes in the central part of the Middle Atlantic Bight. Based on species listed in Table 77.1.

for many residents is reflected in the smaller sizes they reach during the fall, following spring and at the end of the first year (fig. 77.4). In species that use both estuaries and coastal ocean habitats during the first year, growth rates can be similar in both habitats, as in *Centropristis striata* (Able et al. 1995a), or higher in the estuary, as in *Urophycis regia* (Barans 1972).

Some evidence shows that growth is fastest immediately following settlement. Growth rates from caged *Pseudopleuronectes americanus* in different estuarine habitats were fastest in the smallest sizes (<50 mm TL), and this pattern was similar for uncaged individuals (Sogard 1991, 1992; Able et al. 1996b). Other studies of recently settled fishes have also determined that fast growth rates are essential for survival (Campana et al. 1989; Tupper and Boutilier 1995a,b; Campana 1996), and this is consistent with the hypothesis that growth rate and stage duration are important determinants of subsequent survival and year-class strength (Houde 1987).

MORTALITY

One of the most important shortcomings in our knowledge of the first year in the life of estuarine fishes is an estimate of the rate of mortality. It has been suggested (Gunter 1961) and often repeated that there may be relaxed predation pressure in estuaries relative to the ocean, and this may account, in part, for the large number of juvenile fishes surviving the first year in estuaries (Joseph 1972; Whitfield and Blaber 1978; Blaber 1980; Blaber and Blaber 1980).

Unfortunately, there are few estimates of mortality rates for pre- or postsettlement estuarine fishes in general and, in particular, for species in the Middle Atlantic Bight. Even where estimates have been made, the influence of confounding factors (i.e., gear avoidance, inaccessible habitats, etc.) make it difficult to determine the factors influencing survival during the first year. This makes it very difficult to evaluate survival among different early life-history stages or cohorts and the role of growth, habitat, etc., and their influence on recruitment to the adult stage. One problem is that the mobility of the juveniles of many estuarine forms makes it difficult to separate emigration from actual mortality when calculating loss rates (Herke 1977; Weinstein 1983; Sogard 1989). The exceptions may be for some pelagic species such as *Anchoa mitchilli* for which gear avoidance is less of an issue (Houde 1987) or for individual demersal fish, which can be tagged and followed with acoustic techniques (Szedlmayer and Able 1993). Based on intensive sampling in the Mystic River estuary in Connecticut, estimates of mortality for young-of-the-year *Pseudopleuronectes americanus* averaged 31% per month (Pearcy 1962a). Day (1981) suggested, as a result of the closing of the mouth of a South African estuary, that the estimates of 78% to 80% during the period of estuarine residence and 30% to 31% as monthly estimates may be useful guides for future research.

Another source of mortality during the first year of temperate estuarine fishes is that which occurs during the winter, either as a result of low temperatures or a loss of energy reserves during the period of little or no feeding (Conover and Present 1990; Hales and Able in press; Hare and Cowen 1996; M. C. Curran and K. W. Able unpubl. data). Often this mortality is size selective, with the smaller young-of-the-year suffering the highest mortality. The size selectivity may be especially important because those individuals that grow slowly due to a variety of possible influences (i.e., poor food resources, threat of predation, and reduced access to food, disease, parasitism, etc.) may ultimately suffer higher mortality rates during the winter. Besides those species already indicated in the literature as being susceptible to overwinter mortality (*Centropristis striata, Menidia menidia, Morone saxatilis*), we would add other species (*Hippocampus erectus, Syngnathus fuscus, Menidia beryllina, Hypsoblennius hentz, Gobiosoma bosc, Gobiosoma ginsburgi*) because of the absence of small individuals in early spring relative to the previous fall (see species accounts). We suggest this differential is due to size-selective mortality in the absence of growth during the winter. These suggestions should be tested with further observations or experiments.

HABITATS

Our knowledge of estuarine habitat use, especially by young-of-the-year fishes, is incomplete. Much of the current level of understanding is based on nonquantitative capture techniques that can only provide a snapshot of actual habitat use. Additionally, most of the habitat use data are based solely on daytime collecting, and where comparisons between day and night collections have been made, they provide a very different view of habitat use (e.g., Rountree and Able 1993; Sogard and Able 1994). Fishes are also highly mobile, and their capture may indicate only a fraction of the habitats they use. Such dynamic use of fluctuating estuarine habitats (Deegan and Day 1986) might require a broader, landscape approach to the study of habitats (Ray 1991; Hoss and Thayer 1993; Ray 1997).

Several general characteristics of estuaries may make them better habitats for young-of-the-year fishes than open coasts. Estuaries may offer more protection by reducing wave surge and providing structurally complex shelters such as *Zostera* beds and *Spartina* marshes (Day 1981; Day et al. 1989). The usually turbid waters of estuaries might provide for concealment from predators (Livingston 1975), or the response to turbidity may favor either the predator or the prey (Minello et al. 1987). Alternatively, estuarine predators may have adapted to turbid conditions or rely on other sensory modalities. The general and usually extensive shallowness of estuaries may also result in an increase in the area of potential refuge (Nixon 1980; Kneib 1984; Ruiz et al. 1993).

There do not appear to be many habitat-specific use patterns for fishes in the Middle Atlantic Bight, but it is just as possible that they do exist and we simply lack the in situ observations necessary to detect them in typically turbid estuaries. Possible habitat associations in Middle Atlantic Bight estuaries include the following. A number of residents (several fundulids, *Menidia beryllina, Cyprinodon variegatus*) and transients (*Menidia menidia, Gasterosteus aculeatus*) use the marsh surface for spawning. Of these, *Fundulus heteroclitus heteroclitus* shows the greatest degree of adaptation to this habitat as it consistently deposits its eggs in the intertidal zone, either at the base of blades of *Spartina alterniflora* or in the valves of *Geukensia demissa.* It also shows a high degree of site fidelity, with a home range that is approximately 18 m in the summer (Lotrich 1975) but somewhat larger in the fall and winter (Fritz et al. 1975). Coves near inlets in Great Bay–Little Egg Harbor annually have very high abundances of recently settled *Pseudopleuronectes americanus* (Witting 1995; Witting and Able in review; Curran and Able unpubl. data), although the importance of this habitat type has not been verified for other estuaries. At least some young-of-the-year *Paralichthys dentatus* show continuous use of salt marsh creeks during the summer (Rountree and Able 1992b; Szedlmayer and Able 1993). Other species are consistently found in eelgrass, *Zostera marina,* including *Apeltes quadracus,* which builds its nests there (Reisman 1963), and *Syngnathus fuscus, Hippocampus erectus, Opsanus tau,* and *Tautoga onitis,* which use these habitats during the summer. This preference for eelgrass is also demonstrated by *Bairdiella chrysoura,* but in years when it is abundant it can also be found in many other habitats. A few species are consistently collected over sand substrates in the lower portion of the estuary, including *Ammodytes americanus* (D. Haroski and K. W. Able unpubl. data) and *Scophthalmus aquosus* (M. Neuman and K. W. Able unpubl. data). Other species have specific habitat preferences such as *Gobiosoma bosc* in association with oyster reefs (Breitburg 1991) and *Centropristis striata* in these and other hard substrates (Able et al. 1995a).

The lack of obvious strong habitat associations within estuaries for most juveniles during their first year may be due, in part, to their general mobility (Herke 1977; Sogard 1989; Sogard and Able 1994). Because of this mobility, it is often difficult to determine residence time in discrete habitats. One of the few examples is based on studies of *Leiostomus xanthurus* in Chesapeake Bay salt marsh creeks. In this instance, the marked population was resident for up to 181 days with an average duration of 91 days (Weinstein et al. 1984). Werme (1981) reported extensive tidal movements by many components of a salt marsh creek fauna in Massachusetts. Similar tidal movements have also been observed in salt marshes in New Jersey (Rountree and Able 1992a), Virginia (McIvor and Odum 1988), and North Carolina (Hettler 1989). Extensive movements, over hundreds of meters in marsh creeks, have been recorded in a single tidal stage for ultrasonically tagged *Paralichthys dentatus* young-of-the-year in Great Bay (Szedlmayer and Able 1993). Other estuarine studies have identified diel influences on the mobility of juvenile estuarine fishes (Rountree and Able 1993; Sogard and Able 1994).

Despite the general mobility of estuarine young-of-the-year, several species have demonstrated a degree of site fidelity, on a scale of meters, based on tag/recapture studies. This has been shown for *Centropristis striata* (Able and Hales 1997), *Tautoga onitis* and *Tautogolabrus adspersus* (L. S. Hales and K. W. Able unpubl. observ.), *Chaetodon ocellatus* (McBride and Able in press), and *Pseudopleuronectes americanus* (Saucerman and Deegan 1991).

Several caging experiments have been designed to determine habitat quality as measured by its effect on growth of young-of-the-year of demersal fishes in the Middle Atlantic Bight. Studies comparing vegetated to unvegetated substrates indicated that the growth response is species-specific, with *Tautoga onitis* and *Pseudopleuronectes americanus* clearly benefiting from vegetated sites, although this varies somewhat with location (Sogard 1992). Another species in this study (*Gobiosoma bosc*) was most abundant in *Z. marina* beds but it did not grow well there, indicating a possible trade-off between slower growth and enhanced protection from predation. A similar approach has been used to evaluate man-made habitats such as piers and pile fields in the Hudson River (Able et al. 1995b) with the result that there was decreased abundance and growth under large piers and greater abundance and growth in open areas and in pile fields.

CONCLUSIONS

The blurring of the seaward boundary of the estuary (Smith 1966; McHugh 1967; Haedrich 1983; Day et al. 1989) brings into question the concept of estuarine-dependent species, formalized by McHugh (1967) but based on earlier observations (Gunter 1950, 1967) and more recent analyses (Pacheco 1973; Chambers 1992). These clearly indicate that many species of fishes use estuaries as nurseries. It is important to understand that many species are facultative estuarine users, however, and their patterns of use may differ within their geographic range and may be influenced by changes over geologic time scales (Walford 1966; Schubel and Hirschberg 1978; McDowall 1985). Often the ability to make the allocation of obligate estuarine use is hampered by lack of comparative, quantitative data from ocean habitats. For example, *Centropristis striata* was originally considered estuarine dependent (McHugh 1967). Recent data have demonstrated that this species, during the first summer, has similar densities and growth rates in an estuary and on the inner continental shelf; thus, both areas serve as nurseries (Able et al. 1995a). In another example, transforming larvae of *Ophidion marginatum* settle on the continental shelf but apparently migrate into estuaries the following spring (Fahay 1992). Some species appear to have resident populations in the estuary, and others on the continental shelf (e.g., *Tautogolabrus adspersus*). The degree to which these examples are exceptional or typical has yet to be determined and must await further quantitative comparisons with appropriate sampling techniques in the dynamic interface between estuarine and oceanic waters. The easiest distinctions are for those species in which at least one stage of the life history is an obligate inhabitant of the estuary.

In South Africa, extensive surveys in coastal ocean and estuarine waters have helped to define the degree of estuarine dependence for a variety of species (Beckley 1984, 1985; Wallace et al. 1984; Whitfield 1994). In Australia, studies have demonstrated that estuarine-dependent species use inshore oceanic waters as alternative nurseries and that the degree of estuarine dependence may vary with location (Lenanton 1982; Lenanton and Hodgkin 1985; Lenanton and Potter 1987). Finally, on the West Coast of the United States, the use of estuaries by *Paralichthys californicus* varies interannually (Kramer 1991).

Previously posed questions about the importance of estuarine habitats to year-class success in fishes remain unanswered and pertinent. (1) Are there obligate estuarine stages in fishes? (2) Are there significant differences in growth and mortality in and out of estuaries? (3) What particular and measurable benefits does the estuary provide (Haedrich 1983)? (4) Are there differences in the productivity, growth, and mortality for species in individual habitats along the estuarine coenocline? (5) If so, what are the causes of these differences? (6) Do these differences take the form of inequalities in the carrying capacity of the areas or in differences in other factors that may affect growth and survival rates and the type of species utilizing the area (Weinstein 1985)? More extensive study of these and other questions is required if we are to understand, protect, or enhance our estuarine-based fisheries, faunas, and their associated habitats.

Table 77.1. Summary of Selected Morphological Characteristics during the First Year for Dominant Estuarine Fishes in the Central Part of the Middle Atlantic Bight

					Size range TL at				
Species	Egg size (mm)	Hatching size (mm)	Larval duration (yr, mo, d)	Settlement/ transformation size (mm)	Onset of scale formation (mm)	Juvenile pigmentation (mm)	Fall migration or end of summer (Oct.) (cm)	End of winter (Apr.) (cm)	End of first year (12 mo) (cm)
Carcharhinidae									
Mustelus canis	—	28–39	—	—	—	—	55–70	?	?
Anguillidae									
Anguilla rostrata	0.6–1.2*	?	<1 yr	50–60	?	≈60	10–15	?	?
Congridae									
Conger oceanicus	?	?	3–6 mo	100	?	≈60	10–15?	15–20	20?
Clupeidae									
Alosa aestivalis	0.8–1.1	3.1–5.0	25–35 d	20	25–29	25–40	5–10	5–10	10–15
A. mediocris	0.9–1.6	5.2–6.5	?	≈18?	?	?	?	?	15–20
A. pseudoharengus	0.8–1.3	2.5–5.0	?	20	25–29	30–40	9–11	10–12	10–12
A. sapidissima	2.5–3.8	5.7–10.0	21–28 d	≈27	34	?	7–12	10–15	10–15
Brevoortia tyrannus	1.3–1.9	2.4–4.5	?	30	?	?	?	?	?
Clupea harengus	1.0–1.4	4.0–10.0	5–7 mo	30–42	?	45	?	?	?
Engraulidae									
Anchoa hepsetus	1.2–1.7 long axis	3.6–4.0	?	?	45	?	?	?	?
A. mitchilli	0.8–1.1 long axis	2.0	?	?	?	?	2–7	5–10	6–10
Osmeridae									
Osmerus mordax	≈1.0	5.5–6.0	?	?	36+	?	?	?	?
Synodontidae									
Synodus foetens	?	2.5	?	30	?	30–50	?	?	?
Gadidae									
Microgadus tomcod	1.4–1.7	5.0–7.0	?	12–23	?	?	10–15	15–22	15–22
Pollachius virens	1.0–1.2	3.0–4.0	?	50	?	≈25	?	?	?

Continued

Table 77.1. *Continued*

Species	Egg size (mm)	Hatching size (mm)	Larval duration (yr, mo, d)	Settlement/ transforma-tion size (mm)	Size range TL at				
					Onset of scale formation (mm)	Juvenile pigmen-tation (mm)	Fall migration or end of summer (Oct.) (cm)	End of winter (Apr.) (cm)	End of first year (12 mo) (cm)
Phycidae									
Urophycis chuss	0.6–1.0	<2.0	≈60 d?	23–49	?	?	10	10	15–17
U. regia	0.6–1.0	<2.0	?	25–30	?	35–50	4–5	5–6	15–20
U. tenuis	0.7–0.8	<2.0	50–60 d	50–80	?	67–76	20–25	?	25–30
Ophidiidae									
Ophidion marginatum	Oblong 0.8–1.0 × 0.9–1.1	≈2.0	?	≈22	?	?	2–4	2–4	12–18
Batrachoididae									
Opsanus tau	5.0–5.5	≈7	—	—	—	?	2–9	2–9	2–9
Belonidae									
Strongylura marina	3.0	14.0	?	?	?	?	19–40	?	?
Cyprinodontidae									
Cyprinodon variegatus	1.2–1.4	4.2–5.2	?	—	?	?	2–6	3–6	3–6
Fundulidae									
Fundulus heteroclitus	1.7–2.0	4.8–5.5	?	—	12	≈20	4–6	4–7	4–8
F. luciae	1.7–2.2	5.3–6.0	?	—	10–11	11	3–5	3–5	3–5
F. majalis	2.0–3.0	7.0	?	—	?	14	3–7	3–7	6–10
Lucania parva	1.0–1.3	4.0–5.5	?	—	?	?	3–4	3–4	4
Poeciliidae									
Gambusia holbrooki	—	7–10	—	—	7	9–10	2–3	3	3–4
Atherinidae									
Membras martinica	0.7–0.8	2.1	?	?	<14	<30	?	?	?
Menidia beryllina	0.8–0.9	3.5–4.0	?	?	?	?	4–5	5–6	5–6
M. menidia	1.0–1.5	3.5–4.0	?	?	16–24	?	5–10	6–11	8–12

Gasterosteidae									
Apeltes quadracus	1.3–1.6	4.2–4.5	?	?	?	?	3–4	3–5	3–5
Gasterosteus aculeatus	1.5–1.9	4.2	?	?	?	?	?	?	?
Syngnathidae									
Hippocampus erectus	—	?	?	?	?	?	4–9	4–9	7–10
Syngnathus fuscus	—	10–12 at birth	?	40?	?	?	6–20	8–20	12–19
Triglidae									
Prionotus evolans	1.0–1.2	2.8	18–19 d	7–12	>10	?	3–6	5?	9
P. carolinus	0.9–1.5	3.0	18–19 d	7–12	?	?	2–3	4–5	8
Cottidae									
Myoxocephalus aenaeus	1.5–1.7	5.4	55 d	9–10?	?	?	7	7	6–7
Percichthyidae									
Morone americana	0.7–1.0	≈2.6	≈42 d	20–30	16–35	?	≈5?	≈7?	7–9
M. saxatilis	1.3–4.6	2.0–3.7	?	25–36	?	≈40?	10	10	10
Serranidae									
Centropristis striata	0.9–1.0	2.1	?	10–16	?	?	2–10	5–12	8–17
Pomatomidae									
Pomatomus saltatrix	1.0	2.0–2.4	60 d	40–100	?	?	15–20	?	≈23?
Carangidae									
Caranx hippos	0.7–0.9	1.7	?	?	?	?	4–19	?	?
Lutjanidae									
Lutjanus griseus	0.7–0.8	?	?	10–15	?	?	?	?	?
Sparidae									
Stenotomus chrysops	0.8–1.0	≈2.0	?	15–30	?	?	10	12–13	13
Sciaenidae									
Bairdiella chrysoura	0.7–0.8	1.8	?	?	?	?	3–15	?	?
Cynoscion regalis	0.8–0.9	1.5–1.7	?	?	14	?	5–15	?	?

Continued

Table 77.1. *Continued*

Species	Egg size (mm)	Hatching size (mm)	Larval duration (yr, mo, d)	Settlement/ transforma- tion size (mm)	Size range TL at				
					Onset of scale formation (mm)	Juvenile pigmen- tation (mm)	Fall migration or end of summer (Oct.) (cm)	End of winter (Apr.) (cm)	End of first year (12 mo) (cm)
Leiostomus xanthurus	0.7–0.9	1.6–1.7	82 d	9–20	23–25	?	10–11	?	?
Menticirrhus saxatilis	0.8–0.9	2.0–2.5	?	?	?	?	1–23?	?	20–30?
Micropogonias undulatus	?	?	?	8–20	14–16	?	?	?	13–14
Pogonias cromis	0.8–1.0	1.9–3.0	?	?	?	?	9–20	?	?
Chaetodontidae									
Chaetodon ocellatus	?	?	?	>22	?	?	1–4	?	?
Mugilidae									
Mugil cephalus	0.6–0.9?[†]	2.2–3.6	20–24 d	10–44	11	8–15[‡]	10	?	?
M. curema	0.9	1.6–1.8	?	7–40	7	?	10	?	?
Sphyraenidae									
Sphyraena borealis	1.2	2.6	?	?	14.5	>14.5	15–21	?	?
Labridae									
Tautoga onitis	0.9–1.0	2.2	11–23 d	8–13	?	?	3–15	4–12	4–15
Tautogolabrus adspersus	0.8–0.9	?	?	≈10	?	?	5	5	5–12
Pholidae									
Pholis gunnellus	1.7–2.2	9.0	?	30–40	?	40–50	?	?	6–7
Uranoscopidae									
Astroscopus guttatus	?	?	?	23?	?	?	?	?	?
Blenniidae									
Hypsoblennius hentz	0.7–0.8	2.6–2.8	?	?	?	?	2–6	?	?
Ammodytidae									
Ammodytes americanus	0.9–1.0	5.7–6.3	?	?	?	?	8–12	?	?

Gobiidae									
Gobionellus boleosoma	0.3	1.2	?	?	?	?	1–3	3–4	?
Gobiosoma bosc	1.4–2.0	2.0	?	8–13	—	?	2–4	2–4	3–6
G. ginsburgi	?	?	?	9–12	?	?	1–4	?	?
Stromateidae									
Peprilus triacanthus	0.7–0.8	0.7	?	—	?	?	9	11	14
Scophthalmidae									
Scophthalmus aquosus	0.9–1.4	2.0	?	10–20	?	?	11–19 (spring) 4–7 (fall)	>16 (spring) 4–8 (fall)	? (spring) 18–26 (fall)
Paralichthyidae									
Etropus microstomus	?	?	?	<20	?	?	2–10	3–9	6–12
Paralichthys dentatus	0.9–1.1	2.4–2.8	?	11–16	<27	<27	20–30	?	20–30
Pleuronectidae									
Pseudopleuronectes americanus	0.7–0.9	3.0–3.5	?	9–13	?	?	8–15	9–20	9–20
Soleidae									
Trinectes maculatus	0.6–0.9	1.7–2.0	?	?	<18	?	?	?	?
Tetraodontidae									
Sphoeroides maculatus	0.8–0.9	≈2.4	?	?	≈10[§]	?	2–16	11–16?	?

NOTE: Size at settlement refers to demersal forms that are pelagic as larvae. For pelagic species, this is size at transformation. ? = unknown; — = not applicable.

*Ovarian eggs.

†No data from Western Atlantic.

‡Special prejuvenile stage included.

§In the form of "dermal prickles."

Table 77.2. Summary of Selected Ecological Characteristics and Life-History Characteristics during the First Year of Estuarine Fishes in the Central Part of the Middle Atlantic Bight

Species	General pattern of utilization*	Spawning time†	Spawning location‡	Egg type	Season in estuary†	Growth rate (mm/day)		Habitat	
						Summer	Winter	Summer	Winter
Carcharhinidae									
Mustelus canis	Transient	Sp	Estuary/MAB	Live	Sp–Su	1.9	?	Estuary	Ocean
Anguillidae									
Anguilla rostrata	Transient	Sp	SS	?	All	?	?	Estuary	Estuary
Congridae									
Conger oceanicus	Transient	Su–Wi	SS	?	?	?	?	Estuary	?
Clupeidae									
Alosa aestivalis	Transient	Sp	FW	Pelagic	Sp–Fa	0.4	0	Estuary	Ocean
A. mediocris	Transient	Sp	FW	Demersal/Pelagic	?	?	?	?	?
A. pseudoharengus	Transient	Sp	FW	Pelagic	All	0.6	0	Estuary	Ocean
A. sapidissima	Transient	Sp	FW	Demersal/Pelagic	Sp–Fa	0.5	0	FW/Estuary	Ocean
Brevoortia tyrannus	Transient	Fa, Sp	MAB/SAB	Pelagic	Sp–Su	1.0	?	Estuary	Ocean
Clupea harengus	Transient	Sp	MAB	Demersal	Sp	?	?	?	?
Engraulidae									
Anchoa hepsetus	Transient	Su	MAB	Pelagic	Su–Fa	?	?	Estuary/Ocean	Estuary/Ocean
A. mitchilli	Transient	Su	Estuary/MAB	Pelagic	Sp–Fa	0.2	?	Estuary	Ocean
Osmeridae									
Osmerus mordax	Resident	Sp	FW	Demersal	Sp–Su	<0.5	?	Brackish	Estuary
Synodontidae									
Synodus foetens	Transient	?	SAB	?	Su–Fa	?	?	?	Ocean
Gadidae									
Microgadus tomcod	Resident	Wi	FW	Demersal	Sp–Su	<0.5?	?	Estuary/FW	FW
Pollachius virens	Transient	Fa–Wi	MAB	Pelagic	Sp–Su	0.8	?	Estuary	Ocean

Phycidae									
Urophycis chuss	Stray	Su	MAB	Pelagic	Sp	1.0+	0?	Ocean	Ocean
U. regia	Transient	Su–Fa, Sp	MAB	Pelagic	Sp	1.0+	0?	Ocean	Ocean
U. tenuis	Transient	Sp	MAB (Slope Sea)	Pelagic	Sp	1.0	0	Ocean	Ocean
Ophidiidae									
Ophidion marginatum	Transient	Su–Fa	MAB	Pelagic	Sp–Su	1.0	1.0	Estuary/Ocean	Ocean
Batrachoididae									
Opsanus tau	Resident	Sp–Su	Estuary	Demersal	All	0.3	0	Estuary	Estuary
Belonidae									
Strongylura marina	Transient	Sp	Estuary	Demersal	Su	1.5	?	Estuary	?
Cyprinodontidae									
Cyprinodon variegatus	Resident	Sp–Su	Estuary	Demersal	All	0.3	0	Marsh	Estuary
Fundulidae									
Fundulus heteroclitus	Resident	Sp–Su	Estuary	Demersal	All	0.3	0	Marsh	Estuary
F. luciae	Resident	Sp–Su	Estuary	Demersal	All	0.2	0	Marsh	Estuary
F. majalis	Resident	Sp–Su	Estuary	Demersal	All	0.5	0	Creeks/Shores	Estuary
Lucania parva	Resident	Sp–Su	Estuary	Demersal	All	0.3	0	Marsh	Estuary
Poeciliidae									
Gambusia holbrooki	Resident	Su	FW	Live	All	0.1	0	FW/Estuary	FW/Estuary
Atherinidae									
Membras martinica	?	?	?	Demersal	?	?	?	?	?
Menidia beryllina	Resident	Sp–Su	Estuary	Demersal	All	0.3	0	Marsh	Estuary
M. menidia	Transient	Sp–Su	Estuary	Demersal	Sp–Fa	0.6	0	Estuary	Ocean
Gasterosteidae									
Apeltes quadracus	Resident	Sp	Estuary	Demersal	All	0.1	0	Eelgrass	Estuary
Gasterosteus aculeatus	Transient	Sp	Estuary	Demersal	Wi–Sp	?	?	Marsh	Ocean
Syngnathidae									
Hippocampus erectus	Transient	Sp–Su	Estuary/MAB	Live	Sp–Fa	?	0	Estuary	Ocean
Syngnathus fuscus	Transient	Su	Estuary	Live	Sp–Fa	?	0	Estuary	Ocean

Continued

Table 77.2. *Continued*

Species	General pattern of utilization*	Spawning time†	Spawning location‡	Egg type	Season in estuary†	Growth rate (mm/day)		Habitat	
						Summer	Winter	Summer	Winter
Triglidae									
Prionotus carolinus	Transient	Su–Fa	MAB (Estuary?)	Pelagic	Sp–Su	0.5	0	Estuary/Ocean	Ocean
P. evolans	Transient	Su–Fa	MAB (Estuary?)	Pelagic	Su–Wi	0.5	0	Estuary/Ocean	Ocean
Cottidae									
Myoxocephalus aenaeus	Resident	Wi	Estuary/MAB	Demersal	All	0.3	0	Estuary/Ocean?	Estuary/Ocean?
Percichthyidae									
Morone americana	Resident	Sp	FW	Demersal/Pelagic	All	0.3	0	Estuary/FW	Estuary
M. saxatilis	Transient	Sp	FW	Pelagic	All	0.4	0	Estuary/FW	Estuary
Serranidae									
Centropristis striata	Transient	Sp–Fa	MAB	Pelagic	Sp–Fa	0.4	0	Estuary/Ocean	Ocean
Pomatomidae									
Pomatomus saltatrix	Transient	Sp, Su	SAB/MAB	Pelagic	Sp–Su	1.0	?	Estuary	Ocean§
Carangidae									
Caranx hippos	Transient	?	SAB	Pelagic	Su	1.3	?	Estuary	?
Lutjanidae									
Lutjanus griseus	Stray	Su	SAB	Pelagic	Su–Fa	?	?	?	?
Sparidae									
Stenotomus chrysops	?	Sp–Su	Estuary‖	Pelagic	Sp–Su	0.8	0?	Estuary	Ocean
Sciaenidae									
Bairdiella chrysoura	Transient	Su	?	Pelagic	Su–Fa	0.8	?	Estuary	?
Cynoscion regalis	Transient	Sp–Su	Estuary/MAB	Pelagic	Sp–Fa	1.0+	0?	Estuary	Ocean
Leiostomus xanthurus	Transient	Wi	MAB**	Pelagic	Sp–Fa	0.7	?	Estuary	Ocean
Menticirrhus saxatilis	Transient	Su	MAB	Pelagic	Su–Fa	?	?	Ocean/Estuary	Ocean
Micropogonias undulatus	Transient	Su–Fa	MAB**	Pelagic	All	0.7	0	Estuary	Estuary
Pogonias cromis	Transient	Su	MAB	Pelagic	Su–Fa	1.2	?	Estuary	Ocean

Chaetodontidae									
Chaetodon ocellatus	Stray	?	SAB	Pelagic	Su–Fa	0.1–0.2	?	Estuary	?
Mugilidae									
Mugil cephalus	Transient	Wi	SAB	Pelagic	Sp–Su	?	?	Estuary/FW	Ocean
M. curema	Transient	Sp	SAB	Pelagic	Sp–Su	?	?	Estuary	Ocean
Sphyraenidae									
Sphyraena borealis	Transient	Sp	SAB	Pelagic	Su–Fa	0.8	?	Estuary	?
Labridae									
Tautoga onitis	Transient	Sp–Fa	Estuary/MAB	Pelagic	All	0.5	0	Estuary	Estuary
Tautogolabrus adspersus	Resident	Sp–Fa	MAB	Pelagic	All?	0.4	0	Estuary	Estuary/Ocean
Pholidae									
Pholis gunnellus	Stray	Wi	Estuary/MAB	Demersal	Sp–Su–Fa	?	?	Estuary	Ocean
Uranoscopidae									
Astroscopus guttatus	?	Su	Estuary/MAB	?	?	?	?	Estuary/Ocean	?
Blenniidae									
Hypsoblennius hentz	Stray	Su	Estuary	Demersal	Su–Wi	0.4	0	Estuary	Estuary
Ammodytidae									
Ammodytes americanus	Resident	Wi	?	Demersal	Fa–Sp	0.3	0	Estuary	Estuary
Gobiidae									
Gobionellus boleosoma	Resident?	Su	Estuary	Demersal	All	?	0	Estuary	Estuary
Gobiosoma bosc	Resident	Sp–Su	Estuary	Demersal	All	0.2	0	Estuary	Estuary
G. ginsburgi	Resident	Su	Estuary	Demersal	All	0.2	?	Estuary/Ocean	?
Stromateidae									
Peprilus triacanthus	Transient	Sp, Su	Estuary/MAB	Pelagic	Su	0.4?	?	Estuary/Ocean	Ocean
Scophthalmidae									
Scophthalmus aquosus	Transient	Sp, Fa	Estuary/ MAB	Pelagic	Sp–Fa	1.1 (Sp) ? (Fa)	? (Sp) 0 (Fa)	Estuary/Ocean	Ocean

Continued

Table 77.2. *Continued*

Species	General pattern of utilization*	Spawning time[†]	Spawning location[‡]	Egg type	Season in estuary[†]	Growth rate (mm/day)		Habitat	
						Summer	Winter	Summer	Winter
Paralichthyidae									
Etropus microstomus	Transient	Sp–Fa	MAB	Pelagic	Sp–Fa	?	0	Estuary/Ocean	Ocean
Paralichthys dentatus	Transient	Fa–Wi	MAB	Pelagic	All	1.5–1.9	0	Estuary	Estuary
Pleuronectidae									
Pseudopleuronectes americanus	Resident	Wi	Estuary/MAB	Demersal	All	0.5	0.3	Estuary	Estuary/Ocean?
Soleidae									
Trinectes maculatus	Resident	Sp–Fa	Estuary	Pelagic	All	0.3	?	Estuary	Estuary
Tetraodontidae									
Sphoeroides maculatus	Transient	Sp–Su	Estuary	Demersal	Su–Fa	1.0+	0	Estuary	Ocean

*Resident, Stray, Transient (nursery, seasonal resident).

[†] Spring, summer, fall, winter—Sp = March–May, Su = June–August, Fa = September–November, Wi = December–February.

[‡] Estuarine, freshwater, Middle Atlantic Bight continental shelf, South Atlantic Bight continental shelf, Sargasso Sea.

[§] Unresolved for young-of-the-year.

[‖] Large bays and inner continental shelf.

**Southern only.

Table 77.3. Summary of Collection Details for Rare Species in Great Bay–Little Egg Harbor Estuary

Species	No. collected	Years	Primary collecting gear	Length range	Seasonal occurrence
Rajidae					
Raja eglanteria	16	1988–91	Otter trawl	4–43 cm BW	Sp–Su
Elopidae					
Elops saurus	1	1991	Plankton net	3 cm SL	Fa
Albulidae					
Albula vulpes	2	1992, 1994	Plankton net/dip net	33.5 SL–60.0 mm TL??	Su–Fa
Muraenidae					
Gymnothorax sp.	2	1989–90	Misc.	7–9 cm TL	Su
Ophichthidae					
Myrophis punctatus	32	1990–94	Plankton net	5–24 cm TL	All
Ophichthus gomesi	9	1989, 1991–92	Plankton net/dip net	7–9 cm TL	Su–Fa
Clupeidae					
Sardinella aurita	66	1988–90	Weir/seine	4–8 cm TL	Su
Engraulidae					
Engraulis eurystole	125	1989–93	Plankton net	1–5 cm TL	Su–Fa
Gadidae					
Gadus morhua	1	1992	Plankton net	1 cm TL	Sp
Phycidae					
Enchelyopus cimbrius	99	1989–94	Plankton net	1–6 cm TL	Sp–Fa
Merlucciidae					
Merluccius bilinearis	14	1988–90	Otter trawl	8–11 cm TL	Fa–Wi
Ophidiidae					
Ophidion welshi	1	1989	Plankton net	3 cm SL	Fa
Hemiramphidae					
Hyporhamphus meeki	2	1987, 1990	Killitrap/dip net	1–4 cm TL	Sp–Su
Gasterosteidae					
Gasterosteus wheatlandi	1	1992	Dip net	3 cm TL	Sp
Pungitius pungitius	2	1979, 1982	Dip net	5 cm TL	Sp
Fistulariidae					
Fistularia tabacaria	3	1990	Seine	161 mm SL–420 mm TL?	Fa
Serranidae					
Epinephelus niveatus	1	1991	Gear comp	10 cm TL	Su
E. striatus	1	1992	Killitraps	8 cm TL	Fa
Mycteroperca microlepis	84	1991, 1993	Killitraps	5–26 cm TL	Su–Fa
Apogonidae					
Phaeoptyx pigmentaria	1	1992	Trap	4 cm FL	Fa
Carangidae					
Caranx crysos	2	1992	Dip net	4 cm TL	Su
Decapterus macarellus	1	1989	Dip net	10 cm TL	Su
Selene vomer	10	1988–90	Otter trawl	28 mm SL–117 mm FL??	Su–Fa
Seriola zonata	8	1986, 1992	Dip net	5–12 cm FL	Su
Trachinotus falcatus	17	1989–91	Weir/seine	2–5 cm SL	Su–Fa
T. carolinus	3	1996	Seine	27–32 mm TL	Su

Continued

Table 77.3. *Continued*

Species	No. collected	Years	Primary collecting gear	Length range	Seasonal occurrence
Gerreidae	85	1989–90, 1993–94	Weir/seine	1–6 cm SL	Su–Fa
Haemulidae					
Orthopristis chrysoptera	17	1996–97	Trap	5–15 cm FL	Su
Sparidae					
Lagodon rhomboides	6	1991	Gear comp	32 mm TL–100 mm FL??	Su
Chaetodontidae					
Chaetodon capistratus	28	1990–91, 1993	Killitrap	2–5 cm TL	Su–Fa
Sphyraenidae					
Sphyraena barracuda	1	1989	Plankton net	3 cm SL	Su
Stichaeidae					
Lumpenus lumpretaeformis	1	1990	Plankton net	2 cm SL	Su
L. maculatus	2	1989	Plankton net	2–3 cm SL	Su
Ulvaria subbifurcata	25	1989–90, 1992	Plankton net	1–2 cm SL	Sp
Blenniidae					
Chasmodes bosquianus	10	1991–92	Killitrap	1–2 cm	Su–Fa
Hypleurochilus geminatus	2	1992	Killitrap/dip net	6 cm TL	Su–Fa
Scombridae					
Scomber japonicus	1	1990	Dip net	9 cm TL	Su
S. scombrus	2	1990	Dip net	2–3 cm TL	Su
Scomberomorus maculatus	8	1989–91	Gear comp	31 mm SL–53 mm TL??	Su
Stromateidae					
Peprilus alepidotus	1	1988	Otter trawl	2 cm FL	Su
Bothidae					
Bothus sp.	3	1990, 1992–93	Plankton net	20 mm SL–23 mm TL??	Su–Fa
Paralichthyidae					
Citharichthys spilopterus	2	1993–94	Plankton net	2 cm SL	Fa
Etropus crossotus	2	1990–91	Plankton net	1–7 cm SL	Fa
Cynoglossidae					
Symphurus sp.	156	1990–93	Plankton net	1–6 cm SL	Fa
S. plagiusa	2	1990, 1992	Seine/beam trawl	5–14 cm TL	Fa
Monacanthidae					
Aluterus schoepfi	1	1989	Plankton net	8 cm TL	Su
A. scriptus	6	1989, 1994	Otter trawl	6–33 cmTL	Su–Fa
Monocanthus hispidus	26	1988–89, 1991–93	Otter trawl	3–15 cm TL	Su–Fa
Ostraciidae					
Lactophrys sp.	2	1990–91	Misc.	8 mm SL–20 mm TL??	Su
Tetraodontidae					
Chilomycterus schoepfi	94	1988–90, 1993	Otter trawl	2–28 cm TL	Sp, Su

Table 77.4. Transient and Resident Estuarine Fishes in the Central Portion of the Middle Atlantic Bight Grouped According to Temporal Patterns of Estuarine Use and Spawning Site

Transients
Group I: Facultative Estuarine Users *Centropristis striata, Scophthalmus aquosus* I, *Prionotus evolans, Menticirrhus saxatilis?, Anchoa hepsetus?, Astroscopus guttatus?, Ammodytes americanus?, Gobiosoma ginsburgi, Etropus microstomus, Pomatomus saltatrix* II, *Peprilus triacanthus* II, *Brevoortia tyrannus* II, *Gobionellus boleosoma?, Cynoscion regalis?*
Group II: Seasonal Residents *Menidia menidia, Syngnathus fuscus, Hippocampus erectus, Strongylura marina, Sphoeroides maculatus, Gasterosteus aculeatus, Mustelus canis, Anchoa mitchilli, Bairdiella chrysoura, Pogonias cromis, Cynoscion regalis?*
Group III: Anadromous *Morone saxatilis, Alosa aestivalis, A. mediocris, A. pseudoharengus, A. sapidissima*
Group IV: Early Users *Urophycis regia* I and II, *U. tenuis, Pollachius virens, Clupea harengus, Leiostomus xanthurus, Paralichthys dentatus*
Group V: Delayed Users *Ophidion marginatum, Prionotus carolinus, Scophthalmus aquosus* II, *Urophycis chuss*
Group VI: Distant Spawners *Mugil cephalus, M. curema, Caranx hippos, Pomatomus saltatrix* I, *Anguilla rostrata, Conger oceanicus, Brevoortia tyrannus* I, *Sphyraena borealis, Peprilus triacanthus* I
Group VII: Expatriates *Chaetodon ocellatus, C. capistratus, Monacanthus hispidus, Sardinella aurita?, Synodus foetens?, Lutjanus griseus?, Mycteroperca microlepis?, Chilomycterus schoepfi?*
Residents
Group VIII: Summer Spawners *Apeltes quadracus, Cyprinodon variegatus, Fundulus heteroclitus, F. luciae, F. majalis, Lucania parva, Menidia beryllina, Gambusia holbrooki, Gobiosoma bosc, Opsanus tau, Gobionellus boleosoma?*
Group IX: Winter Spawners *Myoxocephalus aenaeus, Pseudopleuronectes americanus, Ammodytes americanus*
Group X: Migrating Spawners *Microgadus tomcod, Osmerus mordax, Morone americana, Trinectes maculatus*
Unclassified
Group XI *Tautogolabrus adspersus, Tautoga onitis, Micropogonias undulatus, Pholis gunnellus*

NOTE: Roman numerals for some species refer to different cohorts. I indicates the products of an early spawning; II indicates those of a later spawning.

Identification of Juvenile Stages of Estuarine Fishes

The larvae of fishes that use estuaries of the Middle Atlantic Bight are well-described, and sources are available to help in identifying these stages (Fritzsche 1978; Hardy 1978a, b; Johnson 1978; Jones et al. 1978; Martin and Drewry 1978; Fahay 1983; Moser et al. 1984). Keys and guides are also available for identifying adult fishes in the Middle Atlantic Bight (e.g., Flescher 1980; Robins and Ray 1986). The intervening juvenile stages are usually not addressed in these treatments, however, and can sometimes present identification problems to field workers and collectors. These juvenile stages have adult complements of fin rays, gill rakers, and vertebrae, yet they often lack scales or are characterized by unique pigment patterns and body shapes that can make identification difficult. A description of morphological characters, readily observable in the field, is critical in the present study because our focus is on the first year in the life history of these species, when these transient characters are expressed.

The following dichotomous (or trichotomous) key presents two or three choices on a step-by-step basis, leading to an identification. Certain characters are size-specific, such that a specimen smaller or larger than those we have examined may exhibit slightly different values. We have attempted to minimize these instances, however, and focused on characters that apply to a range of sizes. In some cases, for completeness, we have included species in the key that are not covered in detail in the species accounts, in order to account for rarely occurring, or primarily oceanic, species. After arriving at a potential identification in the key, readers should refer to the appropriate chapter for a more thorough description of applicable juvenile characters. Figures are original except for the following: 5a–Hildebrand and Cable 1938; 12a–Collette 1966; 12b, 14c, 46c–Lippson and Moran 1974; 19b, 25a, 58a, 59a–Fowler 1945; 22a–Cooper 1978; 24a–Hildebrand 1963; 26a–Kuntz and Radcliffe 1917; 28a, 28b, 29a–Mansueti and Hardy 1967; 29b–Mansueti 1962; 35a–Fish 1932; 43a–Pearson 1941; 46a–Hardy 1978b; 50b–D. Wyanski and T. Targett, in prep.; 59b–Berry 1959; 64a–Bigelow and Schroeder 1953; 66b–Kuntz 1914a, b; 67b–Pearson 1929; 68b–Hildebrand and Cable 1934; 69a–Mansueti 1958; 69b–Mansueti 1964.

KEY TO JUVENILES OF ESTUARINE FISHES IN THE CENTRAL PART OF THE MIDDLE ATLANTIC BIGHT

1a. Body flattened; both eyes on one side of head 2
1b. Body not flattened; eyes on opposite sides of head 6

2a. Eyes on left side of head 3
2b. Eyes on right side of head 5

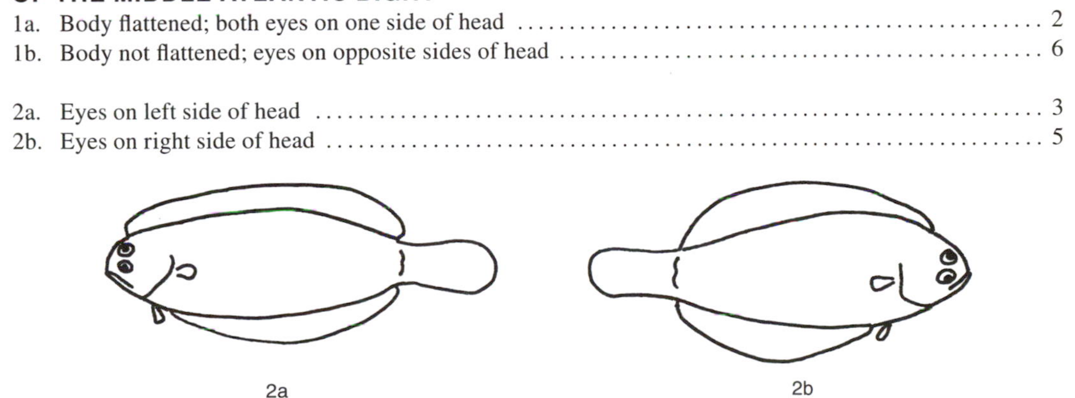

3a. Mouth small, barely reaching eye; lateral line straight .... *Etropus microstomus*
3b. Mouth large, reaching to eye or beyond; lateral line arched over pectoral fin 4

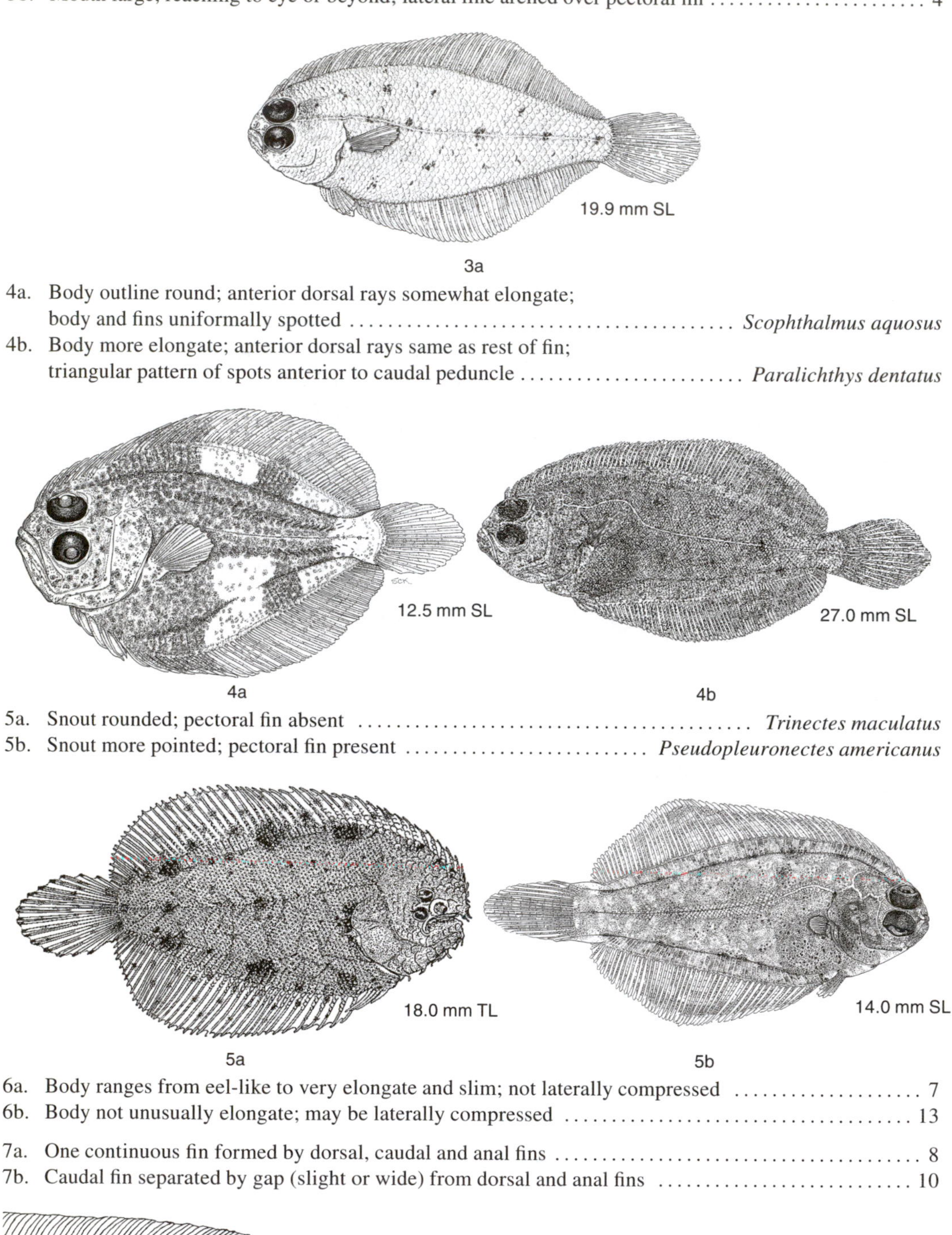

3a

4a. Body outline round; anterior dorsal rays somewhat elongate; body and fins uniformally spotted .... *Scophthalmus aquosus*
4b. Body more elongate; anterior dorsal rays same as rest of fin; triangular pattern of spots anterior to caudal peduncle .... *Paralichthys dentatus*

4a 4b

5a. Snout rounded; pectoral fin absent .... *Trinectes maculatus*
5b. Snout more pointed; pectoral fin present .... *Pseudopleuronectes americanus*

5a 5b

6a. Body ranges from eel-like to very elongate and slim; not laterally compressed 7
6b. Body not unusually elongate; may be laterally compressed 13

7a. One continuous fin formed by dorsal, caudal and anal fins 8
7b. Caudal fin separated by gap (slight or wide) from dorsal and anal fins 10

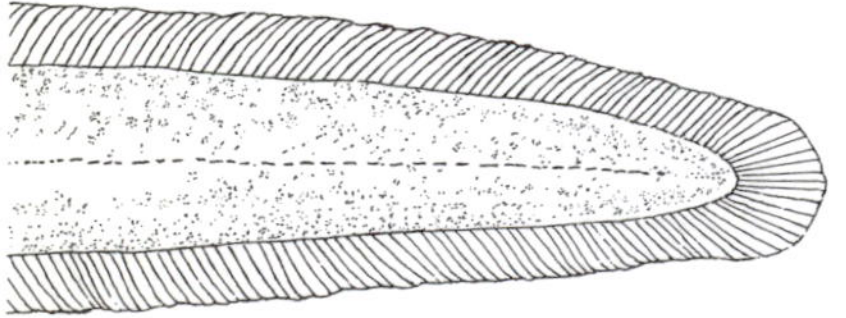

7a 7b

8a. Barbel-like pelvic fin rays between chest and chin; plain color pattern . . . . *Ophidion marginatum*
8b. A tiny pelvic fin with one spine and one ray; color pattern includes distinct ocelli along base of dorsal fin and bar through eye . . . . *Pholis gunnellus*
8c. No pelvic fins; dull, drab color pattern, uniformally brown along dorsum 9

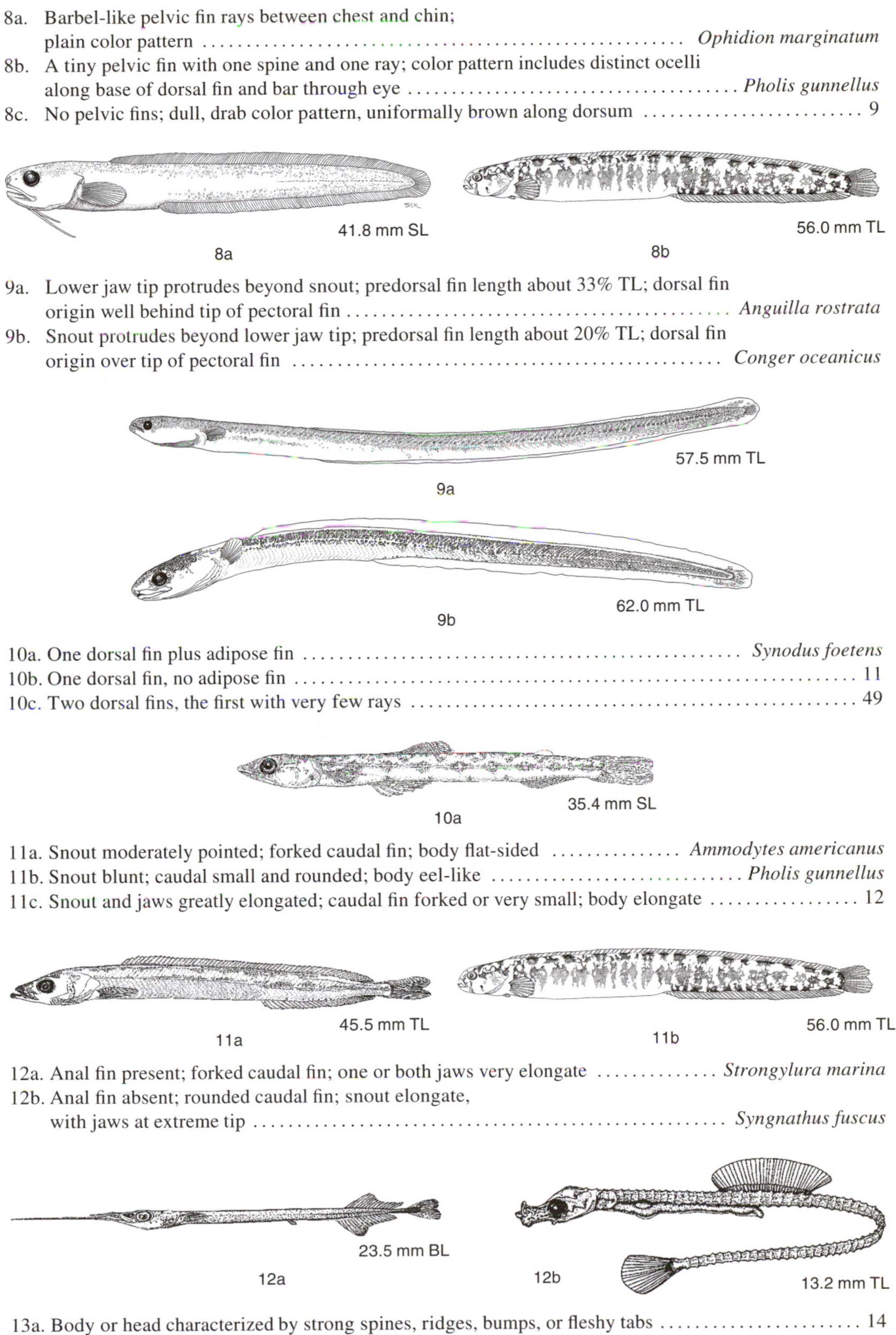

41.8 mm SL
8a

56.0 mm TL
8b

9a. Lower jaw tip protrudes beyond snout; predorsal fin length about 33% TL; dorsal fin origin well behind tip of pectoral fin . . . . *Anguilla rostrata*
9b. Snout protrudes beyond lower jaw tip; predorsal fin length about 20% TL; dorsal fin origin over tip of pectoral fin . . . . *Conger oceanicus*

57.5 mm TL
9a

62.0 mm TL
9b

10a. One dorsal fin plus adipose fin . . . . *Synodus foetens*
10b. One dorsal fin, no adipose fin 11
10c. Two dorsal fins, the first with very few rays 49

35.4 mm SL
10a

11a. Snout moderately pointed; forked caudal fin; body flat-sided . . . . *Ammodytes americanus*
11b. Snout blunt; caudal small and rounded; body eel-like . . . . *Pholis gunnellus*
11c. Snout and jaws greatly elongated; caudal fin forked or very small; body elongate 12

45.5 mm TL
11a

56.0 mm TL
11b

12a. Anal fin present; forked caudal fin; one or both jaws very elongate . . . . *Strongylura marina*
12b. Anal fin absent; rounded caudal fin; snout elongate, with jaws at extreme tip . . . . *Syngnathus fuscus*

23.5 mm BL
12a

12b
13.2 mm TL

13a. Body or head characterized by strong spines, ridges, bumps, or fleshy tabs 14
13b. Body and head not as above; usual or spherical fish-shape without ornamentation 19

14a. Dorsal spines few, isolated, prominent; very thin caudal peduncle; sides of body may have overlapping bony plates 15
14b. Dorsal spines connected by membrane, forming one fin; wider caudal peduncle; bony plates lacking on body 16
14c. Dorsal spines lacking; caudal fin lacking; prehensile tail tip; unique body shape, unmistakable . . . . *Hippocampus erectus*

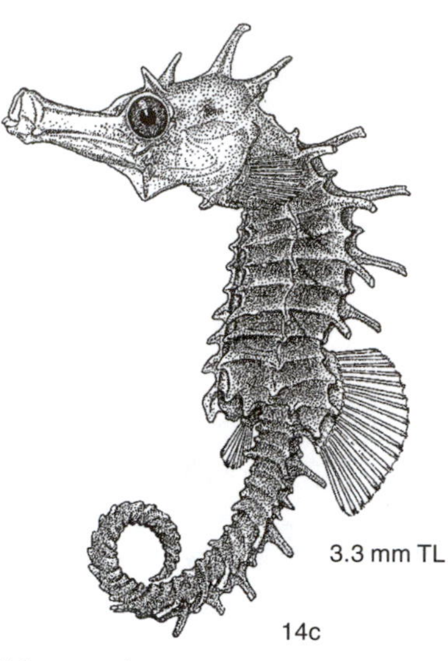

14c

15a. Dorsal spines three, last two widely spaced . . . . *Gasterosteus aculeatus*
15b. Dorsal spines four, last two widely spaced . . . . *Apeltes quadracus*

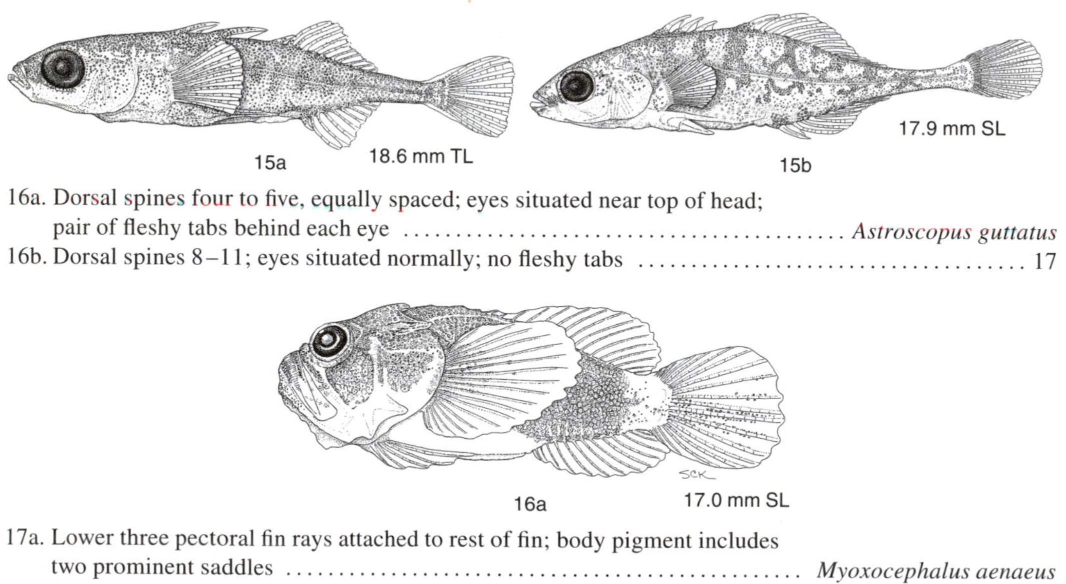

15a 15b

16a. Dorsal spines four to five, equally spaced; eyes situated near top of head; pair of fleshy tabs behind each eye . . . . *Astroscopus guttatus*
16b. Dorsal spines 8–11; eyes situated normally; no fleshy tabs 17

16a

17a. Lower three pectoral fin rays attached to rest of fin; body pigment includes two prominent saddles . . . . *Myoxocephalus aenaeus*
17b. Lower three pectoral fin rays separate, feelerlike 18

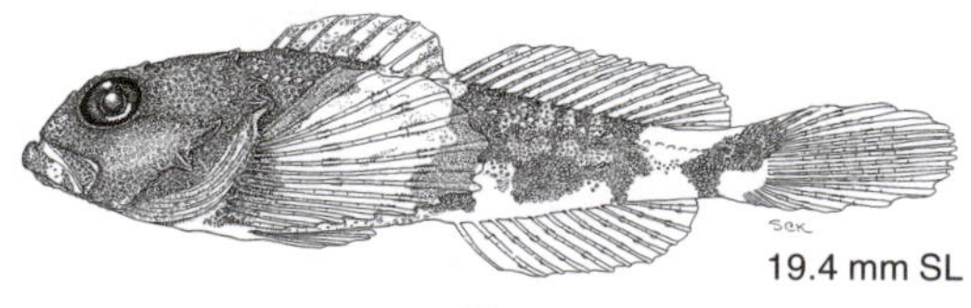

17a

18a. Body pigment includes two prominent saddles; pectoral fin reaches to anal fin origin .... *Prionotus carolinus*
18b. Body pigment dark overall with unpigmented caudal peduncle; pectoral fin reaches well past anal fin origin .... *Prionotus evolans*

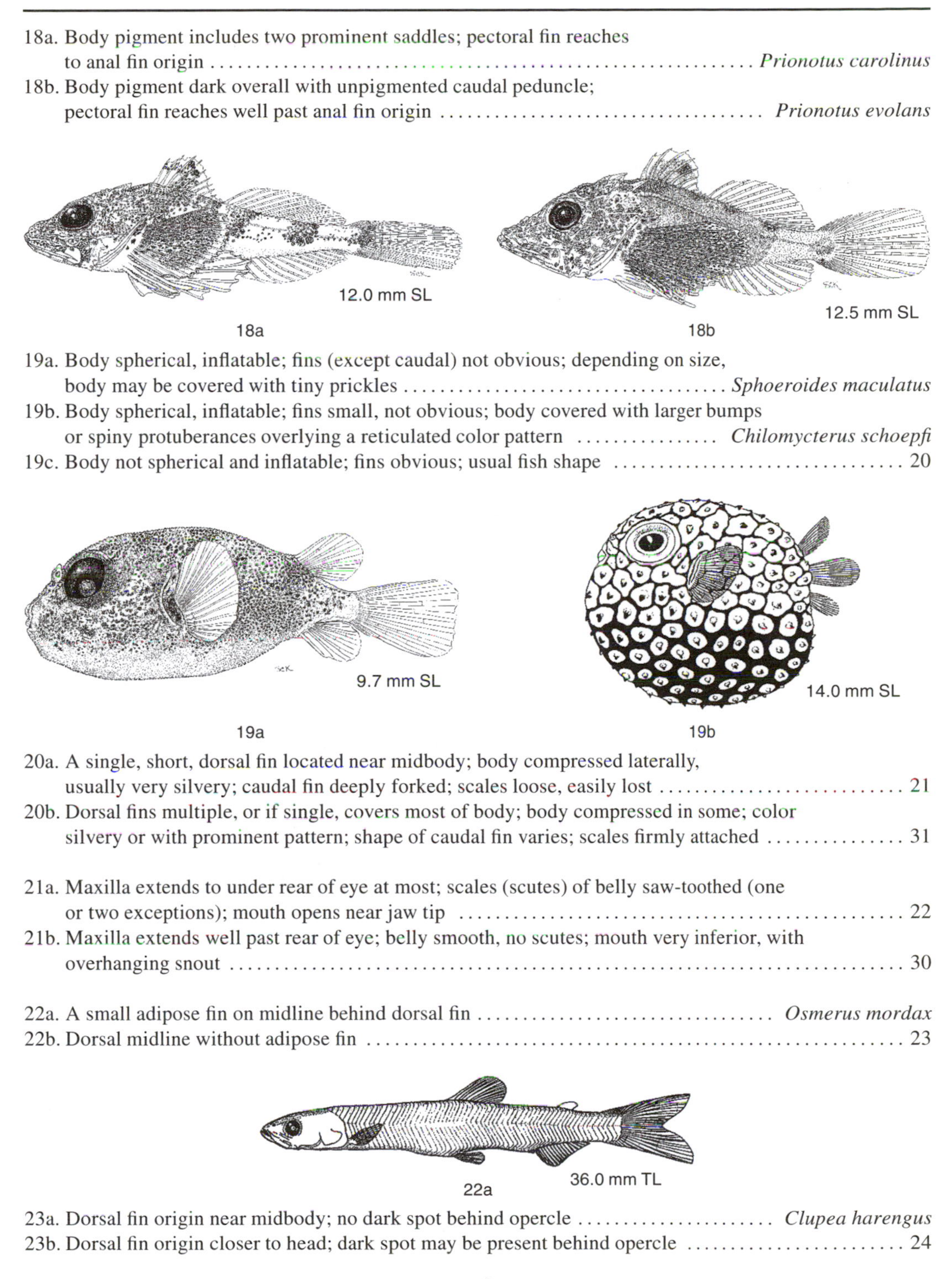

12.0 mm SL

18a

12.5 mm SL

18b

19a. Body spherical, inflatable; fins (except caudal) not obvious; depending on size, body may be covered with tiny prickles .... *Sphoeroides maculatus*
19b. Body spherical, inflatable; fins small, not obvious; body covered with larger bumps or spiny protuberances overlying a reticulated color pattern .... *Chilomycterus schoepfi*
19c. Body not spherical and inflatable; fins obvious; usual fish shape 20

9.7 mm SL

19a

14.0 mm SL

19b

20a. A single, short, dorsal fin located near midbody; body compressed laterally, usually very silvery; caudal fin deeply forked; scales loose, easily lost 21
20b. Dorsal fins multiple, or if single, covers most of body; body compressed in some; color silvery or with prominent pattern; shape of caudal fin varies; scales firmly attached 31

21a. Maxilla extends to under rear of eye at most; scales (scutes) of belly saw-toothed (one or two exceptions); mouth opens near jaw tip 22
21b. Maxilla extends well past rear of eye; belly smooth, no scutes; mouth very inferior, with overhanging snout 30

22a. A small adipose fin on midline behind dorsal fin .... *Osmerus mordax*
22b. Dorsal midline without adipose fin 23

36.0 mm TL

22a

23a. Dorsal fin origin near midbody; no dark spot behind opercle .... *Clupea harengus*
23b. Dorsal fin origin closer to head; dark spot may be present behind opercle 24

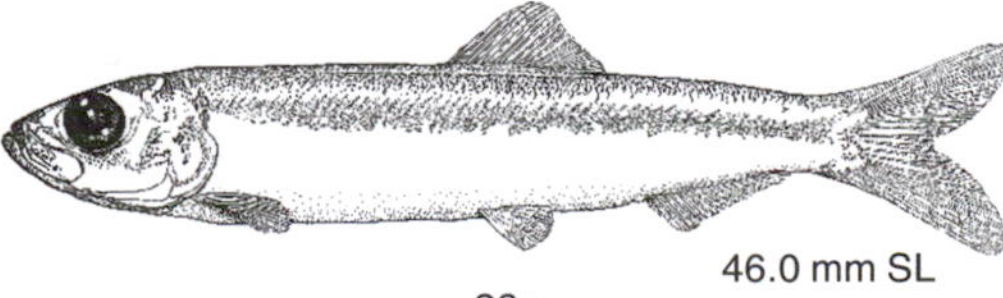

46.0 mm SL

23a

24a. Belly smooth, without sawtooth scutes; coastal species only rarely collected in estuaries; not treated herein (round herring) .... *Etrumeus teres*
24b. Belly with scutes 25

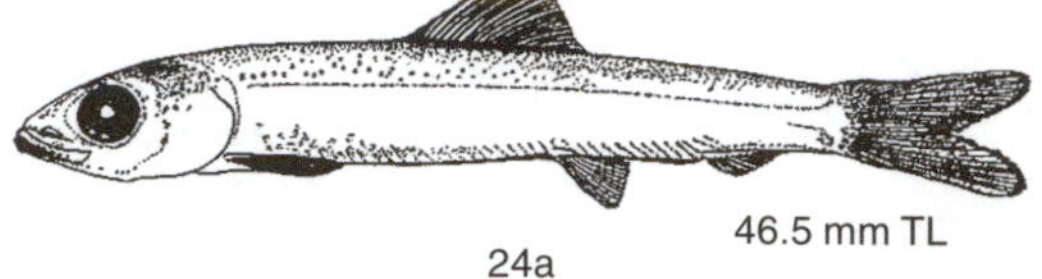

24a

25a. Last ray of dorsal fin elongate and filamentous or just beginning to elongate; snout blunt, overhanging; dark spot behind opercle; last anal fin ray not longer than others .... *Dorosoma cepedianum*
25b. Last dorsal fin ray not elongate; snout not overhanging mouth; no dark spot behind opercle; last anal fin ray longer than other anal rays .... *Sardinella aurita*
25c. Last dorsal fin ray not elongate; snout not overhanging mouth; dark spot may be present behind opercle; last anal fin ray not longer than others 26

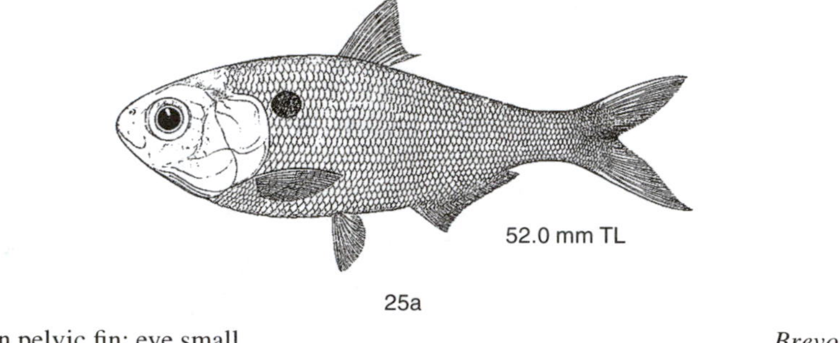

25a

26a. Seven rays in pelvic fin; eye small .... *Brevoortia tyrannus*
26b. Nine rays in pelvic fin; eye larger 27

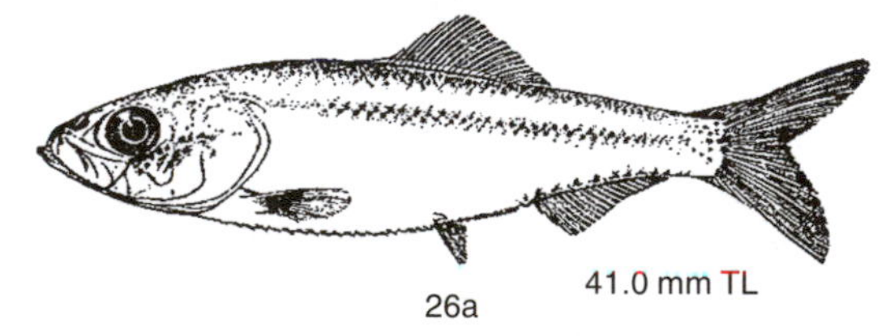

26a

27a. Bony plate on side of head under eye wider than deep; usually a single dark spot behind opercle 28
27b. Bony plate on side of head under eye deeper than wide; dark spot behind opercle usually followed by series of smaller spots 29

28a. Eye diameter greater than snout length; body deep; interior lining of gut pale with dusky spots; gill rakers on lower limb of first arch usually 39–41; 12–16 dorsal fin rays .... *Alosa pseudoharengus*
28b. Eye diameter less than snout length; body slimmer; interior lining of gut black or dark gray; gill rakers on lower limb of first arch usually 44–50; 16–18 dorsal fin rays .... *Alosa aestivalis*

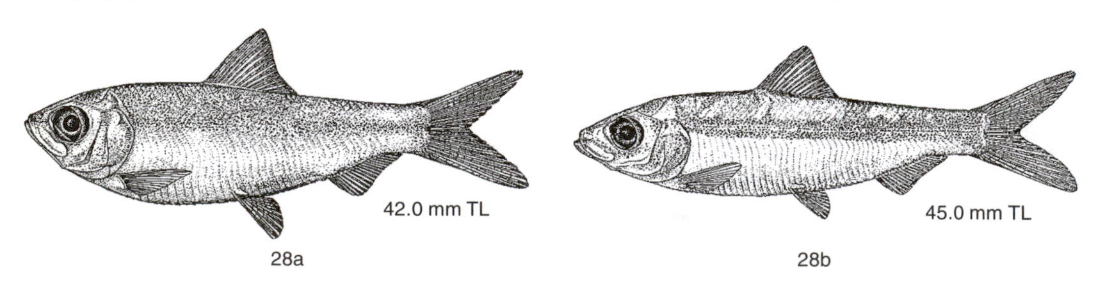

28a 28b

29a. Dark spot behind opercle usually followed by cluster of smaller spots; deep bodied; interior lining of gut pale or silvery; snout extends farther anteriorly than lower jaw tip; gillrakers on lower limb of first arch usually 59–76 .... *Alosa sapidissima*

29b. Dark spot behind opercle usually followed by row of smaller spots; body slimmer; interior lining of gut black or dark gray; lower jaw tip extends farther anteriorly than snout tip; gill rakers on lower limb of first arch usually 18–23 .... *Alosa mediocris*

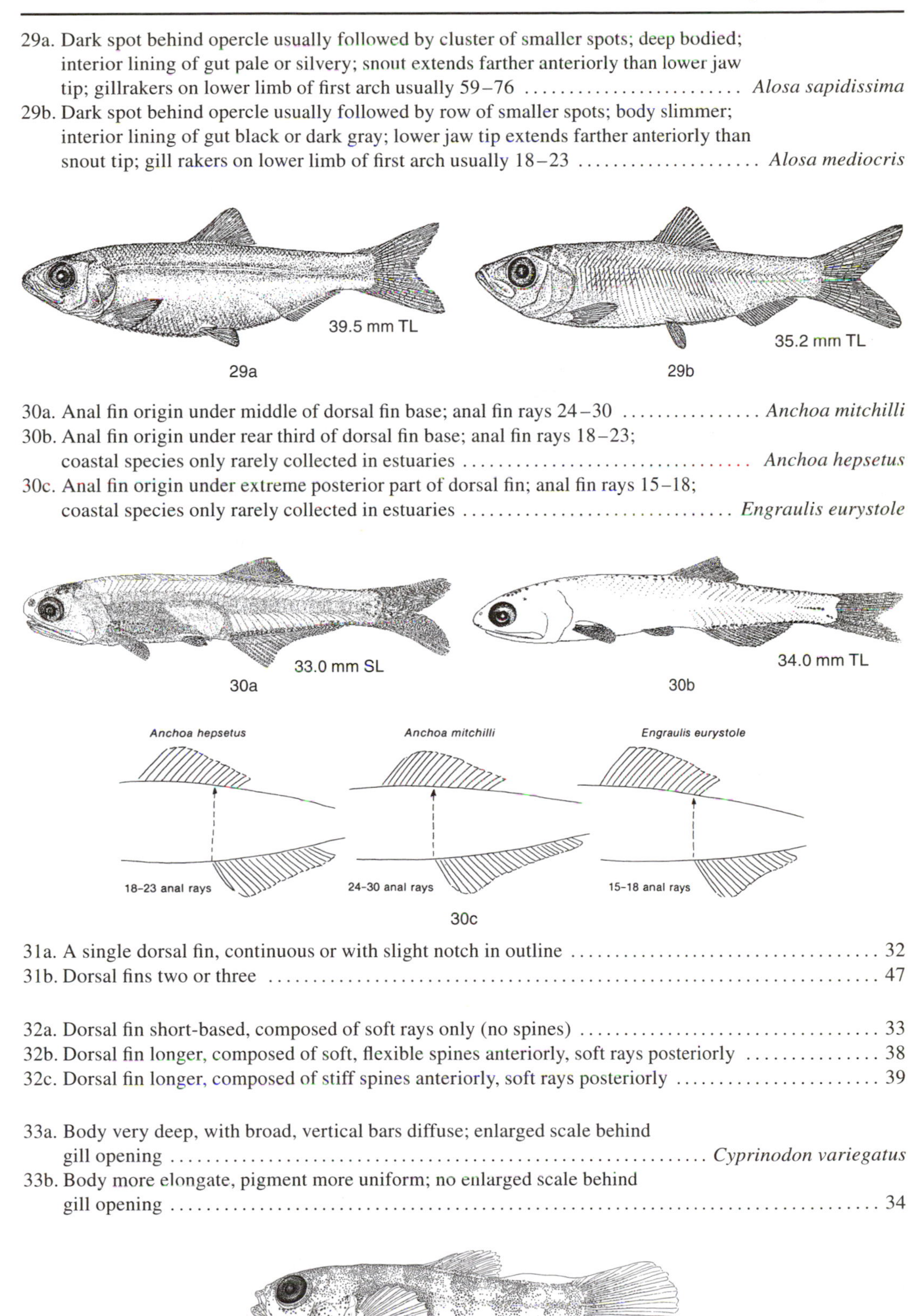

30a. Anal fin origin under middle of dorsal fin base; anal fin rays 24–30 .... *Anchoa mitchilli*

30b. Anal fin origin under rear third of dorsal fin base; anal fin rays 18–23; coastal species only rarely collected in estuaries .... *Anchoa hepsetus*

30c. Anal fin origin under extreme posterior part of dorsal fin; anal fin rays 15–18; coastal species only rarely collected in estuaries .... *Engraulis eurystole*

31a. A single dorsal fin, continuous or with slight notch in outline 32

31b. Dorsal fins two or three 47

32a. Dorsal fin short-based, composed of soft rays only (no spines) 33

32b. Dorsal fin longer, composed of soft, flexible spines anteriorly, soft rays posteriorly 38

32c. Dorsal fin longer, composed of stiff spines anteriorly, soft rays posteriorly 39

33a. Body very deep, with broad, vertical bars diffuse; enlarged scale behind gill opening .... *Cyprinodon variegatus*

33b. Body more elongate, pigment more uniform; no enlarged scale behind gill opening 34

34a. Teeth in single row; longitudinal scale rows < 30 .... *Lucania parva*
34b. Teeth in multiple rows; longitudinal scale rows > 30 35

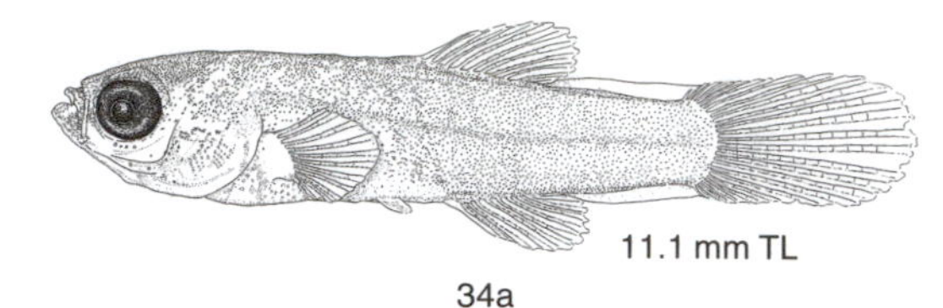

34a

35a. Longitudinal scale rows > 40 .... *Fundulus diaphanus*
35b. Longitudinal scale rows 32–39 36

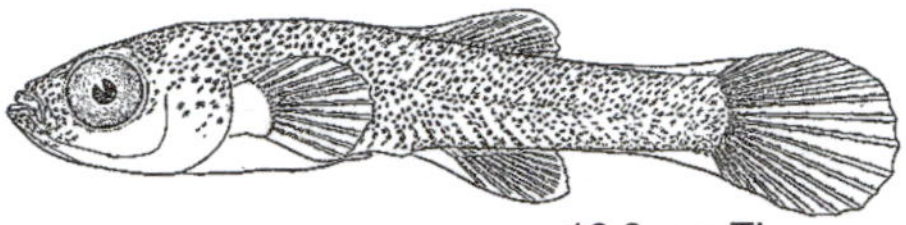

12.3 mm TL

35a

36a. Dorsal fin rays 8–9 (usually 8); anal fin base longer than dorsal fin base; gill opening restricted, its upper end near upper end of pectoral fin base .... *Fundulus luciae*
36b. Dorsal fin rays 10–15; anal fin base shorter than dorsal fin base; gill opening not restricted, its upper end well above pectoral fin base 37

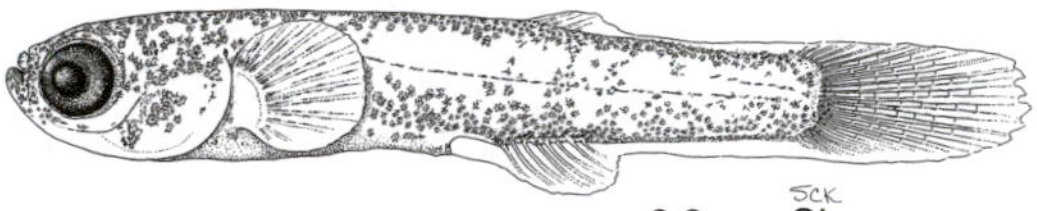

6.9 mm SL

36a

37a. Snout long and pointed (snout length about twice eye diameter); dorsal fin rays 13–15 .... *Fundulus majalis*
37b. Snout short and rounded (snout length about equal to eye diameter); dorsal fin rays 10–12 .... *Fundulus heteroclitus*

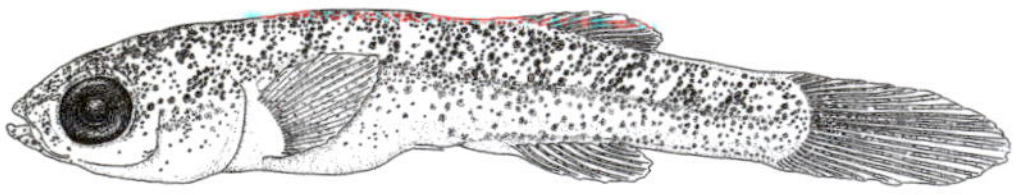

14.5 mm TL

37a

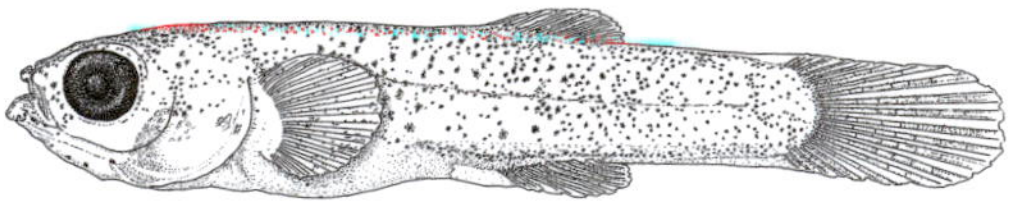

11.0 mm TL

37b

38a. Area between eyes flat; pectoral fin rays usually 12; total dorsal fin elements usually 29–30; no prominent featherlike structures above eyes .... *Chasmodes bosquianus*
38b. Area between eyes concave; pectoral fin rays usually 14; total dorsal fin elements usually 25–27; feathery structures may be above eyes .... *Hypsoblennius hentz*
38c. Area between eyes convex; pectoral fin rays 19–21; total dorsal fin elements two or three feeble spines plus 24–28 rays; no structures above eyes in juveniles .... *Opsanus tau*

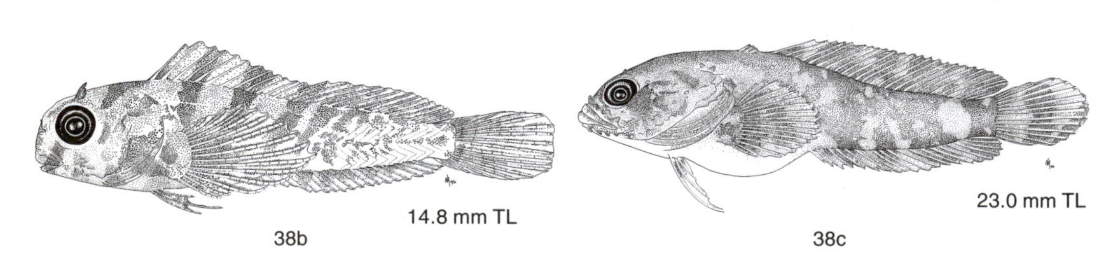

38b

38c

39a. Strong color pattern, usually including some yellow; a bony enclosure surrounding head; body laterally compressed, deep (depth subequal to SL) 40
39b. Color pattern silvery, vague, or including dark stripe laterally; no bony enclosure around head; body laterally compressed or rounder in cross section 41

40a. Body with two dark bars, one crossing caudal peduncle, the other starting at dorsal fin origin and crossing through eye to ventral ouline . . . . *Chaetodon ocellatus*
40b. Body with dark smudge covering caudal peduncle area, as well as a bar extending from dorsal fin origin through eye . . . . *Chaetodon capistratus*

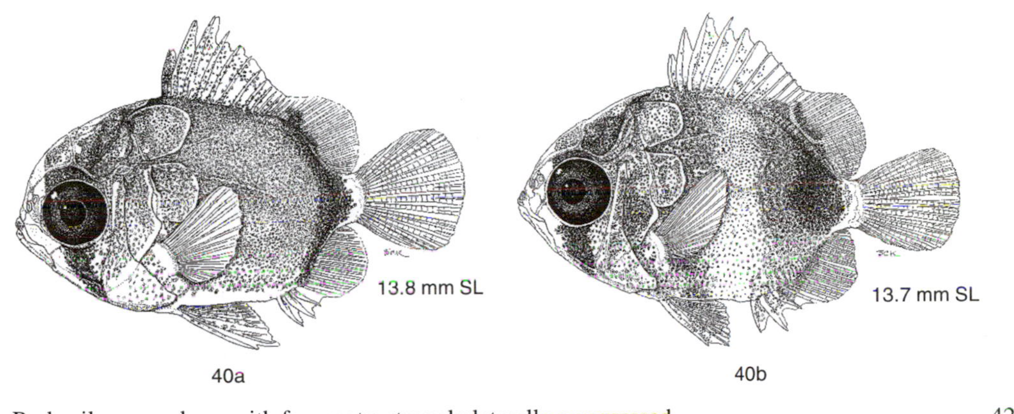

40a 40b

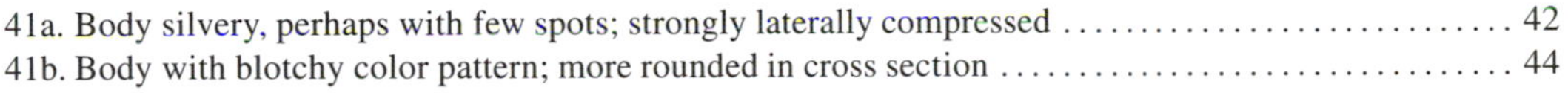
41a. Body silvery, perhaps with few spots; strongly laterally compressed 42
41b. Body with blotchy color pattern; more rounded in cross section 44

42a. Rounded snout; dorsal and anal fin bases and outlines almost equal, anterior portions of both elevated; dorsal fin spines feeble; pelvic fin absent 43
42b. Pointed snout; anal fin base shorter than dorsal fin base; dorsal spines sharp, strong; pelvic fin present . . . . *Stenotomus chrysops*

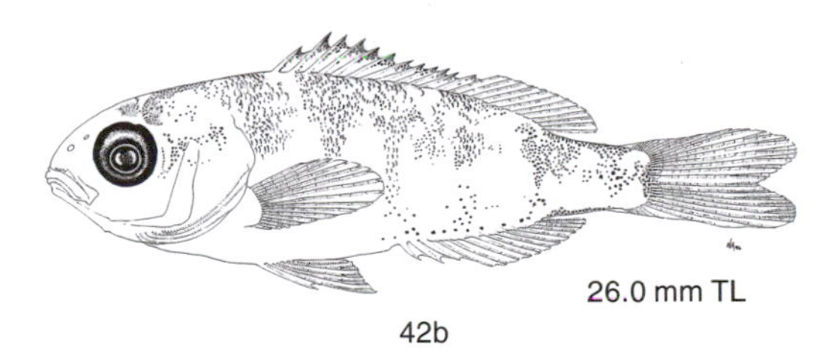

42b

43a. Anterior portion of dorsal and anal fins strongly elevated; body deep .... *Peprilus alepidotus*
43b. Anterior portions of dorsal and anal fins only slightly elevated; body not as deep .... *Peprilus triacanthus*

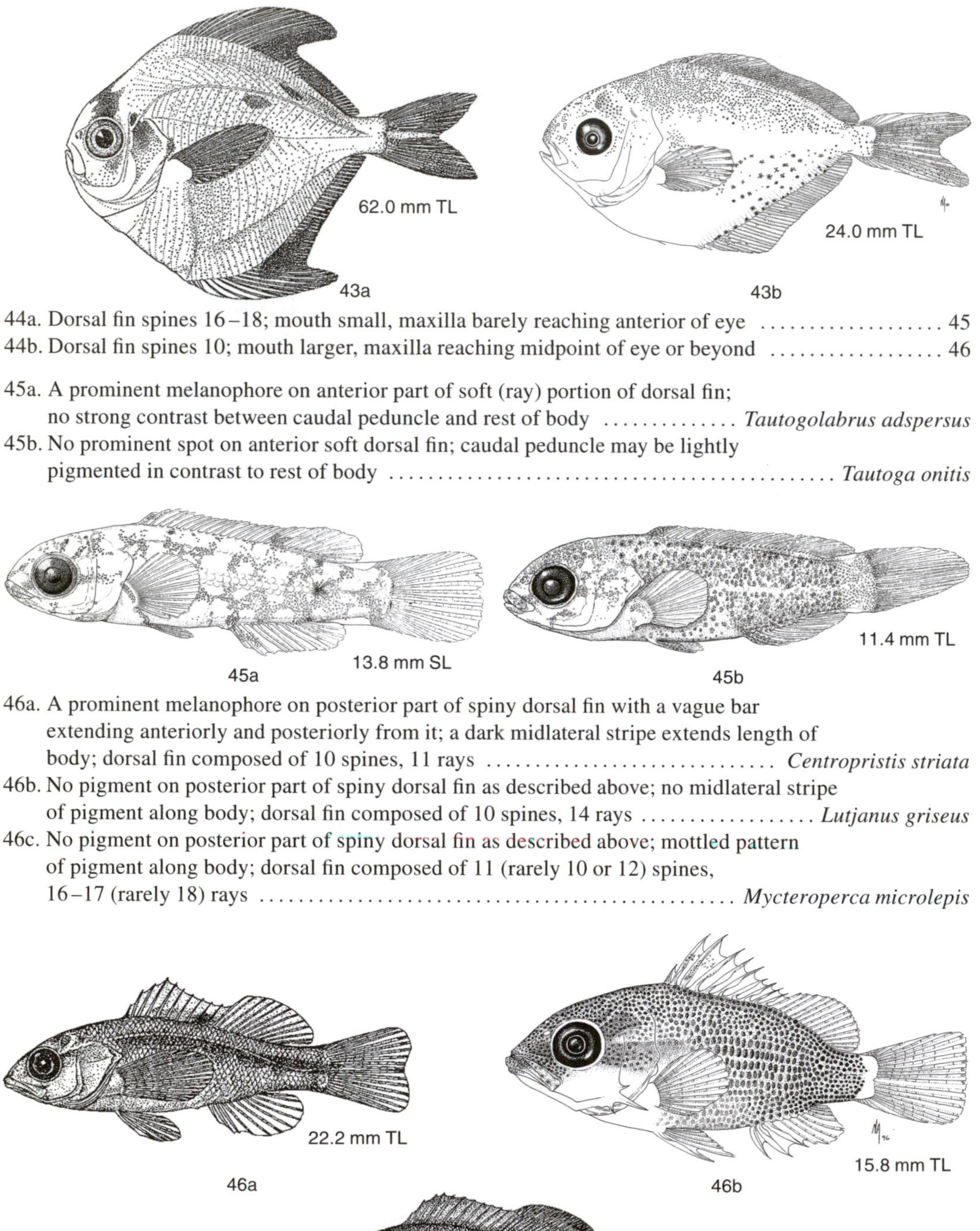

44a. Dorsal fin spines 16–18; mouth small, maxilla barely reaching anterior of eye 45
44b. Dorsal fin spines 10; mouth larger, maxilla reaching midpoint of eye or beyond 46

45a. A prominent melanophore on anterior part of soft (ray) portion of dorsal fin; no strong contrast between caudal peduncle and rest of body .... *Tautogolabrus adspersus*
45b. No prominent spot on anterior soft dorsal fin; caudal peduncle may be lightly pigmented in contrast to rest of body .... *Tautoga onitis*

46a. A prominent melanophore on posterior part of spiny dorsal fin with a vague bar extending anteriorly and posteriorly from it; a dark midlateral stripe extends length of body; dorsal fin composed of 10 spines, 11 rays .... *Centropristis striata*
46b. No pigment on posterior part of spiny dorsal fin as described above; no midlateral stripe of pigment along body; dorsal fin composed of 10 spines, 14 rays .... *Lutjanus griseus*
46c. No pigment on posterior part of spiny dorsal fin as described above; mottled pattern of pigment along body; dorsal fin composed of 11 (rarely 10 or 12) spines, 16–17 (rarely 18) rays .... *Mycteroperca microlepis*

47a. Three distinct and separate dorsal fins; two anal fins 48
47b. Two distinct dorsal fins, separated by gap (may be short) at their base; one anal fin 49
47c. Two distinct dorsal fins, touching at the base; one anal fin 55

48a. Prominent barbel on tip of lower jaw; snout tip extends farther forward than lower jaw; one ray of pelvic fin elongate, filamentous; caudal fin rounded; blotchy color pattern . . . . *Microgadus tomcod*
48b. Lower jaw barbel tiny or lacking; lower jaw extends beyond snout tip; pelvic fin nearly normal in outline with no greatly elongated rays; caudal fin forked; color pattern uniform . . . . *Pollachius virens*

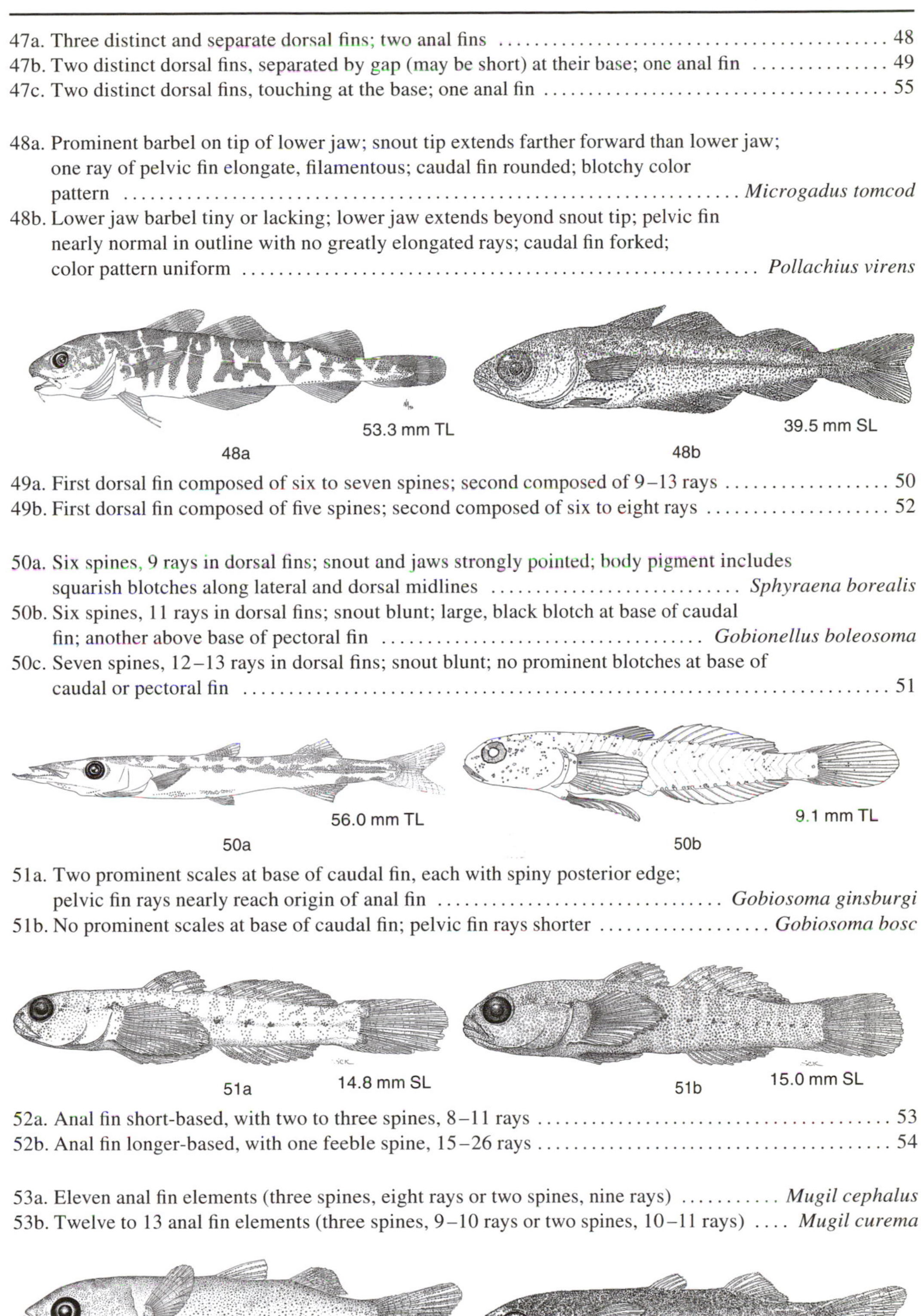

53.3 mm TL

48a

39.5 mm SL

48b

49a. First dorsal fin composed of six to seven spines; second composed of 9–13 rays 50
49b. First dorsal fin composed of five spines; second composed of six to eight rays 52

50a. Six spines, 9 rays in dorsal fins; snout and jaws strongly pointed; body pigment includes squarish blotches along lateral and dorsal midlines . . . . *Sphyraena borealis*
50b. Six spines, 11 rays in dorsal fins; snout blunt; large, black blotch at base of caudal fin; another above base of pectoral fin . . . . *Gobionellus boleosoma*
50c. Seven spines, 12–13 rays in dorsal fins; snout blunt; no prominent blotches at base of caudal or pectoral fin 51

56.0 mm TL

50a

9.1 mm TL

50b

51a. Two prominent scales at base of caudal fin, each with spiny posterior edge; pelvic fin rays nearly reach origin of anal fin . . . . *Gobiosoma ginsburgi*
51b. No prominent scales at base of caudal fin; pelvic fin rays shorter . . . . *Gobiosoma bosc*

51a 14.8 mm SL

51b 15.0 mm SL

52a. Anal fin short-based, with two to three spines, 8–11 rays 53
52b. Anal fin longer-based, with one feeble spine, 15–26 rays 54

53a. Eleven anal fin elements (three spines, eight rays or two spines, nine rays) . . . . *Mugil cephalus*
53b. Twelve to 13 anal fin elements (three spines, 9–10 rays or two spines, 10–11 rays) . . . . *Mugil curema*

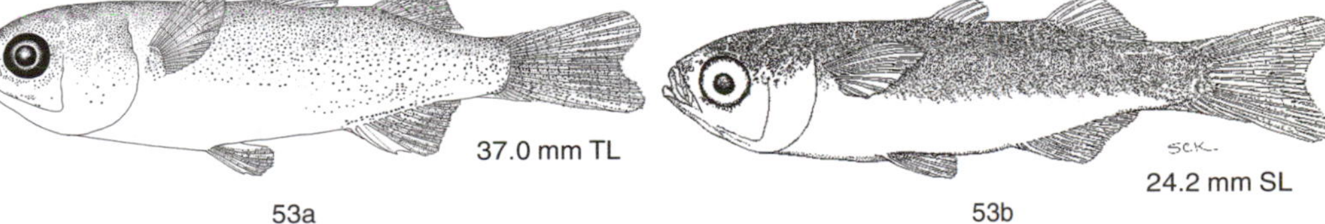

37.0 mm TL

53a

24.2 mm SL

53b

54a. Fifteen to 21 anal fin rays; dorsal pigment restricted to a few, widely spaced spots along dorsal midline, upper flanks clear .... *Membras martinica*
54b. Twenty-three to 26 anal fin rays; dorsal pigment scattered along (and restricted to) both sides of dorsal midline .... *Menidia menidia*
54c. Sixteen to 19 anal fin rays; dorsal pigment scattered along dorsal midline and also extends down onto upper sides of body .... *Menidia beryllina*

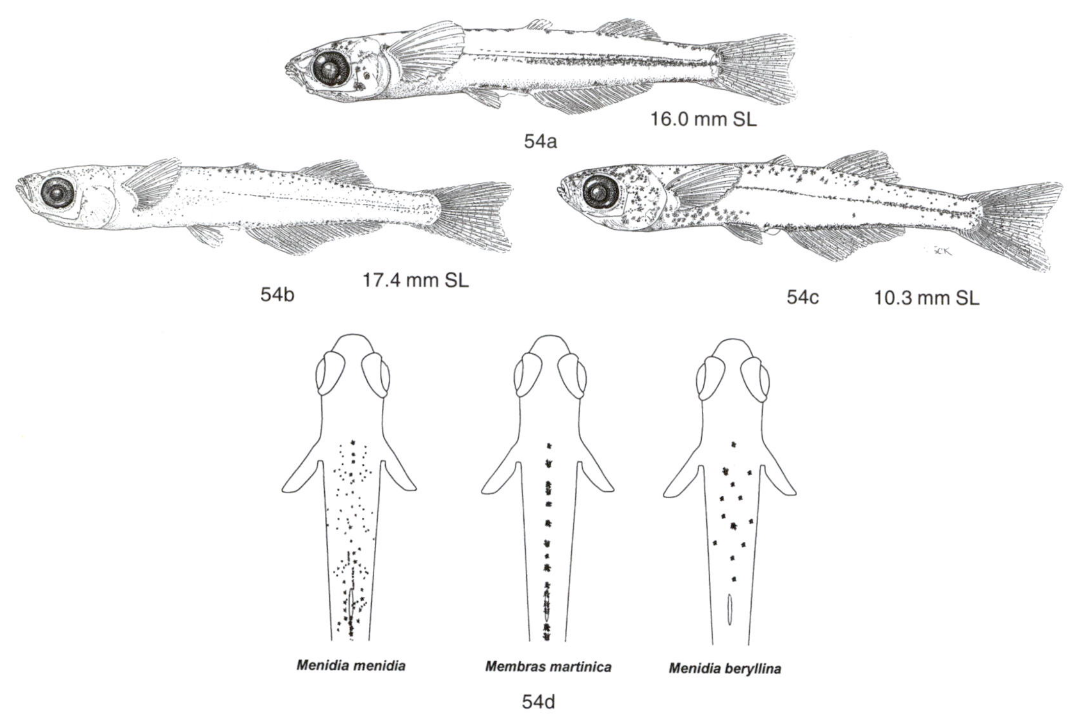

55a. Fins composed of soft rays only, no sharp spines; pelvic fin rays long, filamentous 56
55b. Fins composed of both sharp spines and soft rays; pelvic fin rays not filamentous 57

56a. Pigment on first dorsal fin includes a dark, circular spot surrounded by white margin; 43–51 rays in second dorsal fin; 30–32 rays in caudal fin; three gill rakers on upper limb of first gill arch .... *Urophycis regia*
56b. Pigment on first dorsal fin dusky; 53–64 rays in second dorsal fin; 28–34 rays in caudal fin; three gill rakers on upper limb of first gill arch .... *Urophycis chuss*
56c. Pigment on first dorsal fin dusky; 50–58 rays in second dorsal fin; 33–39 rays in caudal fin; two gill rakers on upper limb of first gill arch .... *Urophycis tenuis*

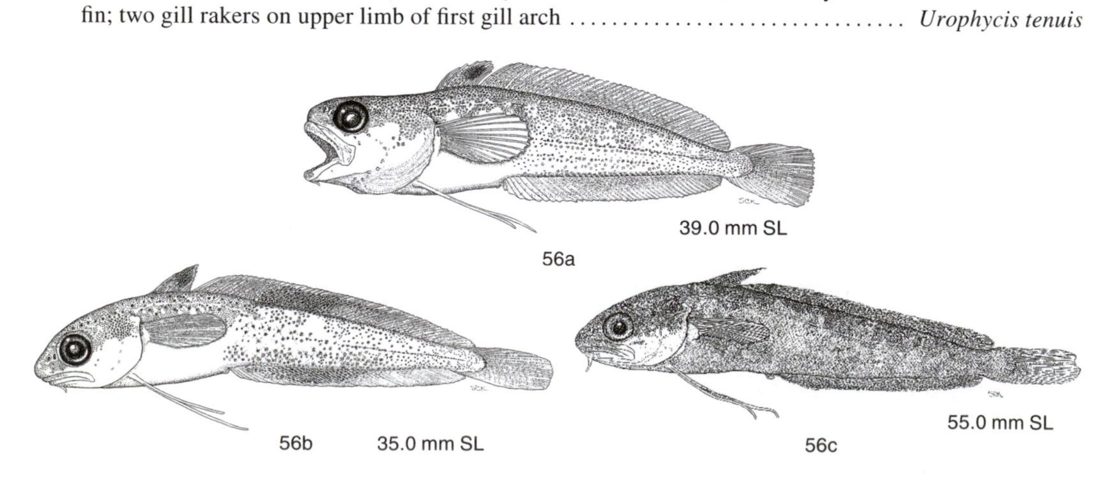

57a. Second dorsal fin about same length as anal fin 58
57b. Second dorsal fin about twice the length of anal fin 60
57c. Second dorsal fin slightly longer than anal fin 69

58a. Dorsal fin with nine spines, 10 rays; anal fin with three spines, seven rays; fin spines weak; mouth greatly protrusible .... *Eucinostomus* sp. (mojarras)
58b. Dorsal fin with nine spines, 24–25 rays; anal fin with three spines, 26–28 rays; fin spines strong; mouth not greatly protrusible .... *Pomatomus saltatrix*
58c. Dorsal fin with nine spines, 16–18 rays; anal fin with three spines, 16–18 rays 59

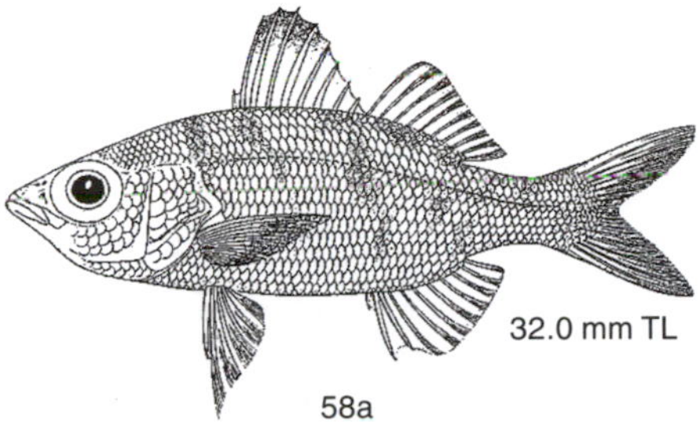

32.0 mm TL

58a

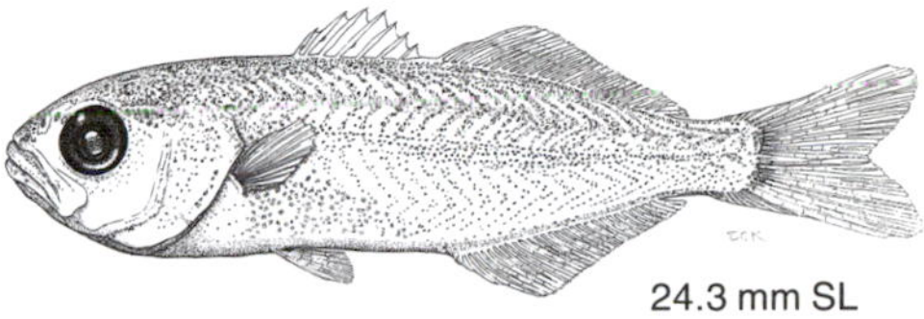

24.3 mm SL

58b

59a. Body silvery; head profile extremely steep; body very compressed; anterior dorsal and anal fin rays elongated; juveniles <50 mm TL have greatly elongated filaments on second and third dorsal spines .... *Selene vomer*
59b. Body with pronounced bars; head profile less steep; body moderately compressed; anterior dorsal and anal rays not greatly elongate; dorsal spines as high as rays .... *Caranx hippos*

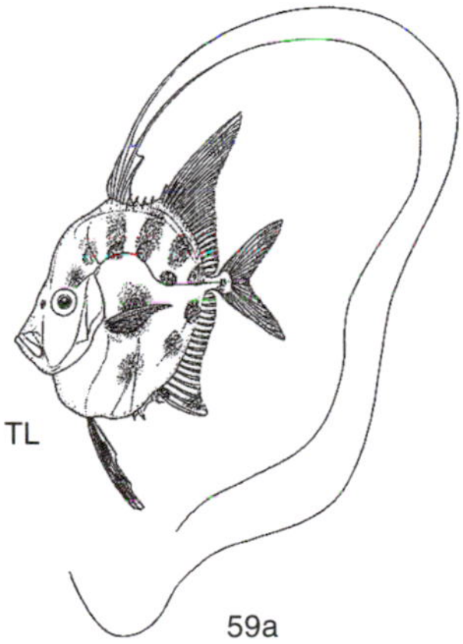

15.0 mm TL

59a

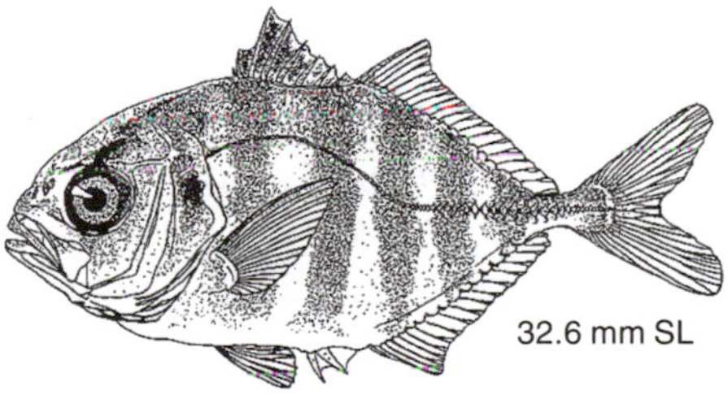

32.6 mm SL

59b

60a. A barbel (or barbels) or pores may be visible at tip of lower jaw 61
60b. No barbel(s) or pores visible at tip of lower jaw 64

61a. A single barbel at tip of lower jaw 62
61b. Several barbels at tip of lower jaw 63

62a. Body uniformally dusky .... *Menticirrhus saxatilis*
62b. Body uniformally dusky with paler venter and abdomen .... *Menticirrhus americanus*

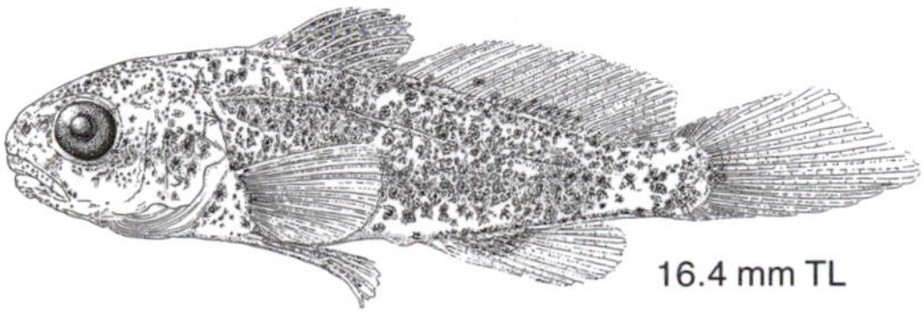

16.4 mm TL

62a

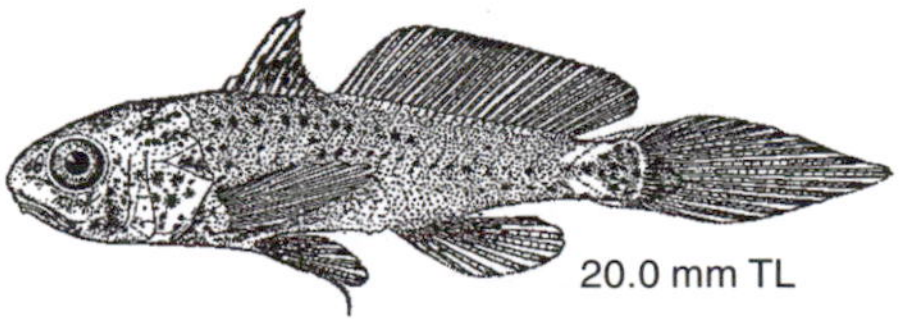

20.0 mm TL

62b

63a. Anal fin with two spines, seven to eight rays; caudal fin asymmetrical (lower lobe longer); body with series of spots or blotches along back and sides; multiple barbels tiny (or indistinct) .... *Micropogonias undulatus*

63b. Anal fin with two spines, six rays; caudal fin slightly rounded; series of dark bars from nape to base of caudal; no ventral pigment; multiple barbels larger .... *Pogonias cromis*

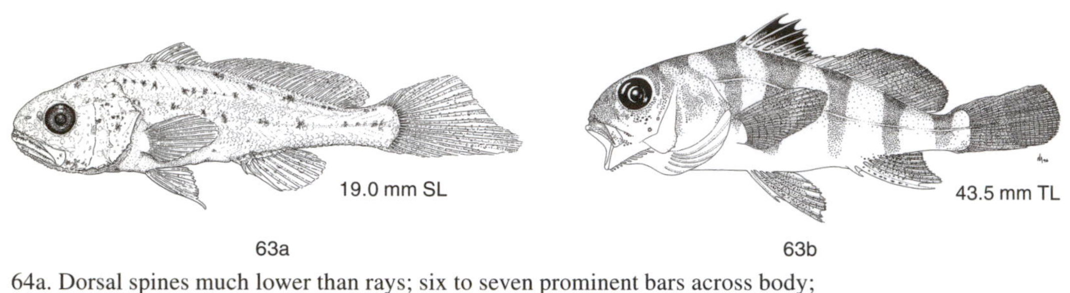

63a 63b

64a. Dorsal spines much lower than rays; six to seven prominent bars across body; caudal fin deeply forked .... *Seriola zonata*

64b. Dorsal spines as high as, or higher than, rays; bars present or absent on body; caudal fin not deeply forked 65

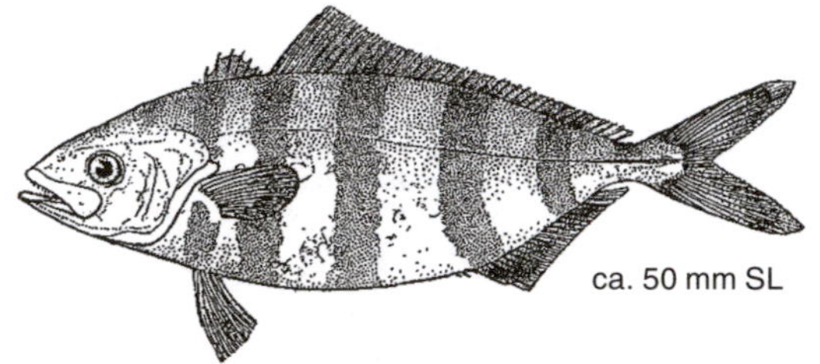

64a

65a. Caudal fin squared off or very slightly forked; a dark spot on body behind gill cover; anal fin with two spines, 12–13 rays .... *Leiostomus xanthurus*

65b. Caudal fin slightly rounded; no prominent dark spot behind gill cover; anal fin with two spines, 6 or 8–10 rays 66

65c. Caudal fin with elongate central rays; no prominent spot behind gill cover; anal fin with two spines, 7–8 or 10–12 rays 67

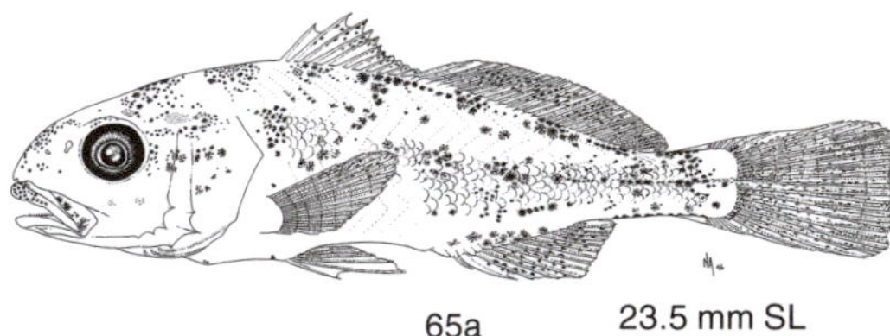

65a

66a. Anal fin with two spines, six rays; series of dark bars from nape to base of caudal; multiple barbels under tip of lower jaw may be visible .... *Pogonias cromis*

66b. Anal fin with two spines, 8–10 rays; body silvery gray over-all; barbels never present .... *Bairdiella chrysoura*

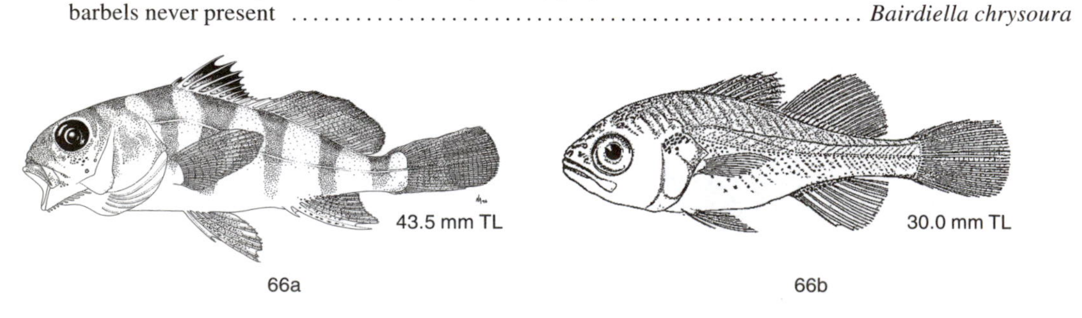

66a 66b

67a. Anal fin with two spines, 10–12 rays; caudal fin symmetrical; body with dusky vertical bars; tip of lower jaw protrudes beyond snout tip .... *Cynoscion regalis*

67b. Anal fin with two spines, seven to eight rays; caudal fin symmetrical; band or spot at base of caudal fin and several blotches along sides; snout longer than lower jaw .... *Sciaenops ocellatus*

67c. Anal fin with two spines, seven to eight rays; caudal fin asymmetrical (lower lobe longer); color dusky overall or with series of spots or wavy lines; snout longer than lower jaw 68

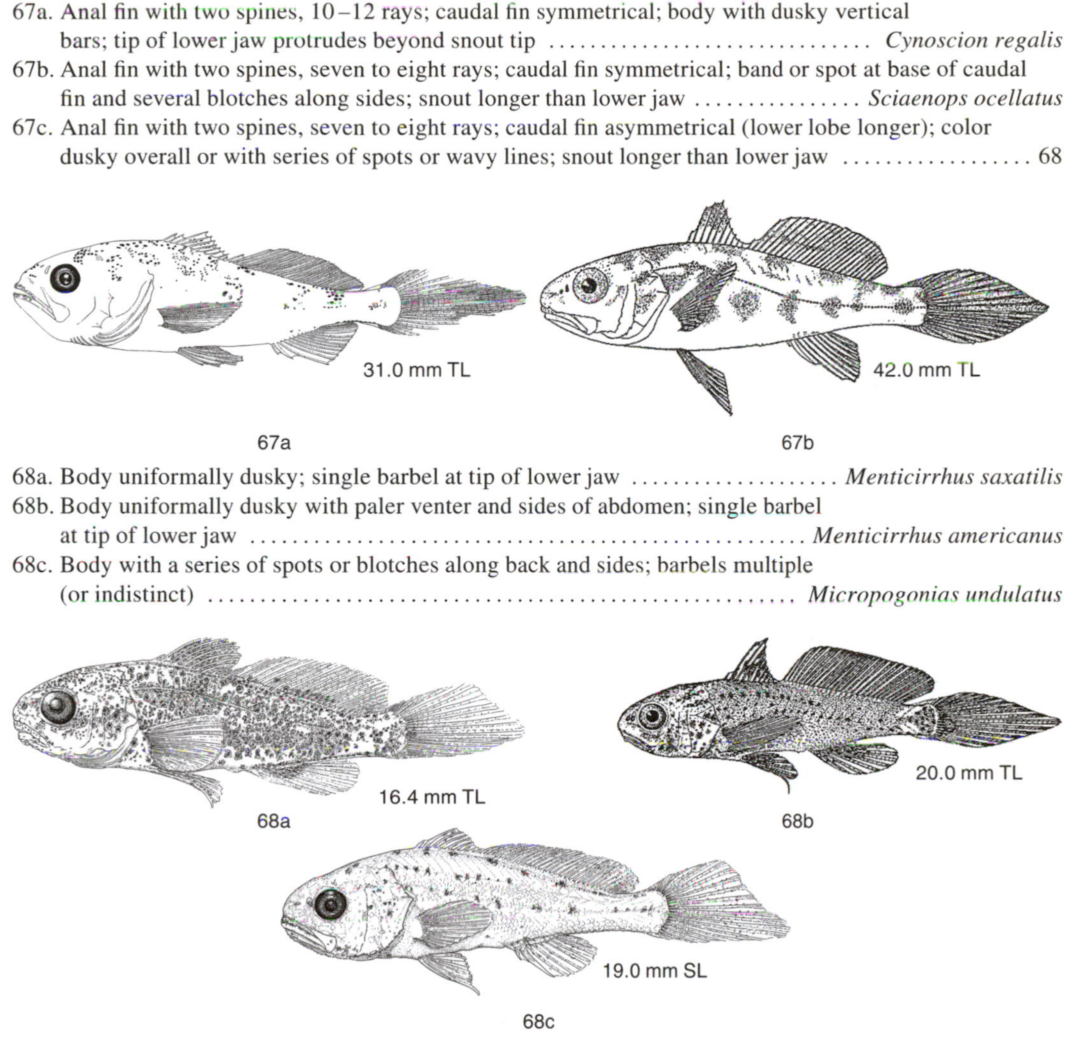

67a 67b

68a. Body uniformally dusky; single barbel at tip of lower jaw .... *Menticirrhus saxatilis*

68b. Body uniformally dusky with paler venter and sides of abdomen; single barbel at tip of lower jaw .... *Menticirrhus americanus*

68c. Body with a series of spots or blotches along back and sides; barbels multiple (or indistinct) .... *Micropogonias undulatus*

68a 68b

68c

69a. Body slim; second anal fin spine same thickness as, or only slightly thicker than, first and third .... *Morone saxatilis*

69b. Body deeper; second anal fin spine stouter than first or third .... *Morone americana*

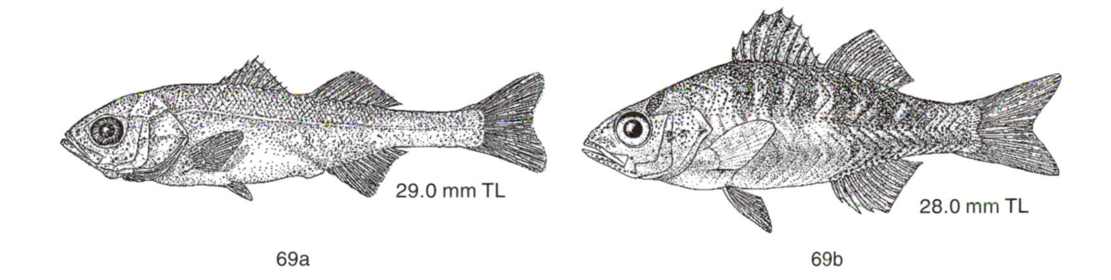

69a 69b

Appendix B. Conversion Factors for Total (TL), Standard (SL), and Fork (FL) Lengths for Young-of-the-Year of Estuarine Species

Species	N	TL mean	TL min	TL max	SL mean	SL min	SL max	FL* mean	FL* min	FL* max	Conversion factors
Alosa aestivalis	55[†]	74.6	41	143	60.7	34	111	66.4	38	127	TL = 1.251 (SL) − 1.301, R^2 = 0.992
											TL = 1.160 (FL) − 2.371, R^2 = 0.996
A. mediocris	17[†]	42.5	27	110	35.5	23	89	39.2	25	95	TL = 1.249 (SL) − 1.756, R^2 = 0.999
											TL = 1.162 (FL) − 2.041, R^2 = 0.998
A. pseudoharengus	53[†]	72.9	28	187	59.5	23	154	64.5	26	165	TL = 1.222 (SL) + 0.261, R^2 = 0.999
											TL = 1.142 (FL) − 0.804, R^2 = 0.999
A. sapidissima	50[†]	89.8	35	154	75.4	29	130	82.2	33	144	TL = 1.163 (SL) + 2.133, R^2 = 0.996
											TL = 1.100 (FL) − 0.631, R^2 = 0.994
Brevoortia tyrannus	176[‡]	42.9	18.2	238	36.0	16.0	191	—	—	—	TL = 1.254 (SL) − 2.266, R^2 = 0.995
	53[†]	62.5	24	146	—	—	—	55.3	21	127	TL = 1.141 (FL) − 0.615, R^2 = 0.998
Clupea harengus	229[§]	65.9	31	90.0	54.9	27.0	74.0	—	—	—	TL = 1.144 (SL) + 3.045, R^2 = 0.972
	223[§]	65.9	48	90.0	—	—	—	60.5	41	81	TL = 1.094 (FL) − 0.204, R^2 = 0.983
Anchoa hepsetus	46[†]	101.7	71	131	85.6	60	108	91.8	65	118	TL = 1.188 (SL) − 0.023, R^2 = 0.986
											TL = 1.141 (FL) − 3.069, R^2 = 0.986
A. mitchilli	40[†]	45.6	23	83	37.9	20	71	41.8	22	76	TL = 1.167 (SL) + 1.363, R^2 = 0.931
											TL = 1.069 (FL) + 0.900, R^2 = 0.928
Osmerus mordax	26[†]	61.7	31	82	49.8	24	68	113.8	66	144	TL = 1.164 (SL) + 1.479, R^2 = 0.984
											TL = 1.102 (FL) − 2.875, R^2 = 0.979
Synodus foetens	14[†]	103.5	50	160	90.7	44	140	96.4	47	147	TL = 1.156 (SL) − 1.388, R^2 = 0.989
											TL = 1.074 (FL) + 0.031, R^2 = 0.997
Pollachius virens	113[‡]	82.4	44.0	142	75.5	39.0	129	—	—	—	TL = 1.084 (SL) + 0.587, R^2 = 0.992
	45[†]	81.2	51.0	142	—	—	—	78.4	48	136	TL = 1.042 (FL) − 0.447, R^2 = 0.998
Urophycis chuss	19[§]	46.6	22.4	150	40.6	19.2	132	—	—	—	TL = 1.159 (SL) − 0.511, R^2 = 0.999
U. regia	31[§]	72.9	38.7	285	62.8	33.7	215	—	—	—	TL = 1.309 (SL) − 9.328, R^2 = 0.987
Opsanus tau	36[§]	82.0	18.8	191	70.1	15.0	165	—	—	—	TL = 1.150 (SL) + 1.287, R^2 = 0.996
Strongylura marina	16[†]	150.9	69	284	142.0	65	265	141.2	52	226	TL = 1.072 (SL) − 1.315, R^2 = 0.990
											TL = 1.269 (BL) + 5.921, R^2 = 0.994
Cyprinodon variegatus	80[§]	29.1	15.0	49	24.1	12.0	41	—	—	—	TL = 1.190 (SL) + 0.486, R^2 = 0.984
Fundulus heteroclitus	84[§]	42.1	6.2	93	35.2	5.1	79	—	—	—	TL = 1.188 (SL) + 0.356, R^2 = 0.997

F. majalis	608[§]	31.1	8.0	134	26.4	7.0	117	—	—	—	TL = 1.140 (SL) + 1.037, R^2 = 0.997
Menidia beryllina	46[†]	52.6	20	80	44.0	16	67	48.7	18	73	TL = 1.162 (SL) + 1.468, R^2 = 0.989 TL = 1.060 (FL) + 1.021, R^2 = 0.997
M. menidia	40[§]	64.3	6.7	102	55.9	5.5	90	57.6	41	111	TL = 1.107 (SL) + 2.407, R^2 = 0.986 TL = 1.047 (FL) + 1.131, R^2 = 0.993
Membras martinica	36[†]	72.3	38	112	60.1	33	93	68.4	36	106	TL = 1.124 (SL) + 4.722, R^2 = 0.989 TL = 1.035 (FL) + 1.498, R^2 = 0.995
Apeltes quadracus	50[†]	34.9	23	46	31.0	20	41	—	—	—	TL = 1.122 (SL) + 0.061, R^2 = 0.993
Gasterosteus aculeatus	68[§]	53.0	13.4	80.0	46.5	11.9	70.0	—	—	—	TL = 1.130 (SL) + 0.445, R^2 = 0.996
Myoxocephalus aenaeus	26[†]	61.7	31	82	49.8	24	68	—	—	—	TL = 1.177 (SL) − 3.085, R^2 = 0.971
Morone americana	60[†]	81.8	26	275	65.7	21	230	77.1	24	266	TL = 1.206 (SL) + 2.520, R^2 = 0.997 TL = 1.040 (FL) + 1.651, R^2 = 0.998
Centropristis striata	2929[§]	51.3	13.8	251	42.4	12.3	226	—	—	—	TL = 1.194 (SL) + 0.701, R^2 = 0.993
Caranx hippos	44[§]	83.5	60	123	67.2	47	98	74.4	55	107	TL = 1.240 (SL) + 0.192, R^2 = 0.993 TL = 1.162 (FL) − 2.961, R^2 = 0.988
Pomatomus saltatrix	101[‡]	117.5	30.0	225	95.7	24	183	—	—	—	TL = 1.233 (SL) − 0.503, R^2 = 0.995
	40[†]	128.3	48	225	—	—	—	117.1	46	202	TL = 1.118 (FL) − 2.698, R^2 = 0.997
Lutjanus griseus	7[†]	48.7	15	79	39.6	11	64	—	—	—	TL = 1.184 (SL) + 1.850, R^2 = 0.996
Stenotomus chrysops	30[†]	68.5	55	90	55.3	45	73	63.9	51	82	TL = 1.109 (SL) + 7.151, R^2 = 0.950 TL = 1.089 (FL) − 1.070, R^2 = 0.976
Bairdiella chrysoura	52[§]	62.4	20.0	149	51.0	17.0	126	—	—	—	TL = 1.187 (SL) + 1.946, R^2 = 0.989
Cynoscion regalis	210[§]	90.8	2.3	652	75.8	2.2	543	—	—	—	TL = 1.120 (SL) − 0.058, R^2 = 0.993
Leiostomus xanthurus	369[‡]	76.9	9.7	147	60.7	5.7	118	—	—	—	TL = 1.275 (SL) − 0.542, R^2 = 0.995
	28[†]	83.9	45	145	—	—	—	81.6	44	140	TL = 1.050 (FL) − 1.794, R^2 = 0.999
Menticirrhus saxatilis	98[§]	51.6	11.8	348	41.9	9.5	290	—	—	—	TL = 1.204 (SL) + 1.149, R^2 = 0.999
Micropogonias undulatus	58[§]	14.2	7.8	55	11.0	6.4	42	—	—	—	TL = 1.322 (SL) − 0.348, R^2 = 0.993
Pogonias cromis	19[†]	91.5	42	150	73.2	34	119	—	—	—	TL = 1.216 (SL) + 2.479, R^2 = 0.988
Chaetodon ocellatus	203[§]	35.4	11.0	77.0	28.9	7.0	63.0	—	—	—	TL = 1.140 (SL) + 2.414, R^2 = 0.994
Tautoga onitis	2471[§]	66.3	6.8	200	56.1	5.4	174	—	—	—	TL = 1.146 (SL) + 2.045, R^2 = 0.993
Tautogolabrus adspersus	1252[§]	43.8	11.8	113	37.1	9.8	98	—	—	—	TL = 1.152 (SL) + 0.999, R^2 = 0.992

Continued

Appendix B. *Continued*

Species	N	TL mean	TL min	TL max	SL mean	SL min	SL max	FL* mean	FL* min	FL* max	Conversion factors
Mugil cephalus	79[†]	69.6	35	133	56.2	26	109	66.0	32	124	TL = 1.238 (SL) + 0.110, $R^2 = 0.997$ TL = 1.078 (FL) − 1.558, $R^2 = 0.998$
M. curema	74[‡]	79.4	25.0	140	62.9	20	115	—	—	—	TL = 1.282 (SL) − 1.315, $R^2 = 0.996$
	34[†]	78.6	32	140	—	—	—	73.9	30	131	TL = 1.071 (FL) − 0.315, $R^2 = 0.998$
Sphyraena borealis	22[†]	79.0	52	137	70.0	37	123	74.0	42	128	TL = 1.102 (SL) + 1.853, $R^2 = 0.982$ TL = 1.057 (FL) + 0.808, $R^2 = 0.987$
Astroscopus guttatus	32[‡]	49.9	11.1	147	37.7	8.2	113	—	—	—	TL = 1.326 (SL) − 0.109, $R^2 = 0.997$
Hypsoblennius hentz	11[†]	70.5	21	115	59.6	16	97	—	—	—	TL = 1.173 (SL) + 0.521, $R^2 = 0.990$
Gobionellus boleosoma	23[§]	22.8	14	36	17.9	11	28	—	—	—	TL = 1.264 (SL) + 0.146, $R^2 = 0.984$
Gobiosoma bosc	157[§]	22.8	6.3	60	19.0	5.4	50.0	—	—	—	TL = 1.182 (SL) + 0.365, $R^2 = 0.996$
G. ginsburgi	232[§]	25.8	5.2	47	21.2	4.3	40	—	—	—	TL = 1.214 (SL) + 0.144, $R^2 = 0.991$
Ammodytes americanus	50[§]	15.8	10.5	21.7	14.4	9.6	19.3	—	—	—	TL = 1.059 (SL) + 1.385, $R^2 = 0.999$
Peprilus alepidotus	9[†]	68.3	52	90	51.4	41	70	59.0	46	78	TL = 1.271 (SL) + 2.963, $R^2 = 0.955$ TL = 1.184 (FL) − 1.496, $R^2 = 0.995$
P. triacanthus	30[§]	38.4	14	138	29.2	10	107	33.0	11	113	TL = 1.273 (SL) + 1.180, $R^2 = 0.998$ TL = 1.216 (FL) + 1.754, $R^2 = 0.995$
Prionotus carolinus	178[§]	28.4	8.9	162	22.5	7.5	128	—	—	—	TL = 1.263 (SL) − 0.044, $R^2 = 0.997$
P. evolans	89[§]	21.1	8.7	75	16.8	7.9	59	—	—	—	TL = 1.268 (SL) − 0.125, $R^2 = 0.987$
Etropus microstomus	503[§]	29.1	9.8	97.7	23.9	8.6	81.5	—	—	—	TL = 1.183 (SL) + 0.843, $R^2 = 0.991$
Paralichthys dentatus	639[§]	171.8	11.7	312	143.0	7.0	267	—	—	—	TL = 1.196 (SL) + 0.747, $R^2 = 0.998$
Scophthalmus aquosus	127[§]	30.9	8.4	291	24.9	7.1	232	—	—	—	TL = 1.248 (SL) − 0.226, $R^2 = 0.999$
Pseudopleuronectes americanus	421[§]	64.9	8.0	225	53.9	6.4	186	—	—	—	TL = 1.213 (SL) − 0.447, $R^2 = 0.994$
Sphoeroides maculatus	43[§]	27.9	5.7	89	22.4	4.7	72	—	—	—	TL = 1.210 (SL) + 0.762, $R^2 = 0.999$

[*] For *Strongylura marina* this column represents BL (body length).
[†] From collections of Academy of Natural Sciences of Philadelphia.
[‡] Combined Rutgers University Marine Field Station and Academy of Natural Sciences of Philadelphia collections.
[§] From Rutgers University Marine Field Station collections.

Literature Cited

Abbe, G. R. 1967. An evaluation of the distribution of fish populations of the Delaware River estuary. Master's thesis (unpubl.) University of Delaware, Newark.

Able, K. W. 1978. Ichthyoplankton of the St. Lawrence Estuary: Composition, distribution and abundance. J. Fish. Res. Board Can. 35: 1518–1531.

———. 1984a. Variation in spawning site selection of the mummichog, *Fundulus heteroclitus.* Copeia 1984(2): 522–525.

———. 1984b. Cyprinodontiformes: Development. Pp. 362–368 *in* H. G. Moser, W. J. Richards, D. M. Cohen, M. P. Fahay, A. W. Kendall, Jr., and S. L. Richardson, eds. *Ontogeny and systematics of fishes.* Am. Soc. Ichthyol. Herpetol. Spec. Publ. No. 1. Allen Press, Lawrence, Kans.

———. 1990. Life history patterns of New Jersey salt marsh killifishes. Bull. N. J. Acad. Sci. 35(2): 23–30.

———. 1992. Checklist of New Jersey saltwater fishes. Bull. N. J. Acad. Sci. 37(1): 1–11.

Able, K. W., and M. Castagna. 1975. Aspects of an undescribed reproductive behavior in *Fundulus heteroclitus* (Pisces: Cyprinodontidae) from Virginia. Chesapeake Sci. 16(4): 282–284.

Able, K. W., and J. D. Felley. 1986. Geographical variation in *Fundulus heteroclitus:* tests for concordance between egg and adult morphologies. Am. Zool. 26: 145–157.

Able, K. W., and S. M. Hagan. 1995. Fishes in the vicinity of Beach Haven Ridge: annual and seasonal patterns of abundance during the early 1970s. Rutgers Univ. Inst. Mar. Coastal Sci. Tech. Rep. 95–24.

Able, K. W., and L. S. Hales, Jr. 1997. Movements of juvenile black sea bass, *Centropristis striata* (Linnaeus), in a southern New Jersey estuary. J. Exp. Mar. Biol. Ecol. 213: 153–167.

Able, K. W., and S. C. Kaiser. 1994. Synthesis of summer flounder habitat parameters. NOAA Coastal Ocean Program Decision Analysis Series No. 1. NOAA Coastal Ocean Office, Silver Spring, Md.

Able, K. W., and A. L. Studholme. 1994. Habitat quality in the New York/New Jersey Harbor estuary: An evaluation of pier effects on fishes. Progress Report to the Hudson River Foundation, New York.

Able, K. W., C. B. Grimes, R. A. Cooper, and J. R. Uzmann. 1982. Burrow construction and behavior of tilefish, *Lopholatilus chamaeleonticeps,* in the Hudson Submarine Canyon. Environ. Biol. Fishes 7(3): 199–205.

Able, K. W., C. Talbot, and J. K. Shisler. 1983. The spotfin killifish, *Fundulus luciae,* is common in New Jersey salt marshes. Bull. N.J. Acad. Sci. 28(1): 7–10.

Able, K. W., K. L. Heck, Jr., M. P. Fahay, and C. T. Roman. 1988. Use of salt-marsh peat reefs by small juvenile lobsters on Cape Cod, Massachusetts. Estuaries 11(2): 83–86.

Able, K. W., K. A. Wilson, and K. L. Heck, Jr. 1989. Fishes of vegetated habitats in New Jersey estuaries: composition, distribution, and abundance based on quantitative sampling. Rutgers Univ. Cent. Coastal Environ. Stud. Publ. No. 1041.

Able, K. W., R. E. Matheson, W. W. Morse, M. P. Fahay, and G. Shepherd. 1990. Patterns of summer flounder (*Paralichthys dentatus*) early life history in the Mid-Atlantic Bight and New Jersey estuaries. U.S. Fish. Full. 88(1): 1–12.

Able, K. W., R. Hoden, D. A. Witting, and J. B. Durand. 1992. Physical parameters of the Great Bay–Mullica River Estuary (with a list of research publications). Rutgers Univ. Inst. Mar. Coastal Sci. Tech. Rep. 92–06.

Able, K. W., M. P. Fahay, and G. R. Shepherd. 1995a. Early life history of black sea bass *Centropristis striata* in the Mid-Atlantic Bight and a New Jersey estuary. U.S. Fish. Bull. 93: 429–445.

Able, K. W., A. L. Studholme, and J. P. Manderson. 1995b. Habitat quality in the New York/New Jersey Harbor estuary: An evaluation of pier effects in fishes. Final Report to the Hudson River Foundation. Hudson River Foundation, New York.

Able, K. W., R. Lathrop, and M. P. De Luca. 1996a. Background for research and monitoring in the Mullica River–Great Bay Estuary. Rutgers Univ. Inst. Mar. Coastal Sci. Tech. Rep. 96–07.

Able, K. W., D. A. Witting, R. S. McBride, R. A.

Rountree, and K. J. Smith. 1996b. Fishes of polyhaline estuarine shores in Great Bay–Little Egg Harbor, New Jersey: a case study of seasonal and habitat influences. Pp. 335–353 *in* K. F. Nordstrom and C. T. Roman, eds. *Estuarine shores: Evolution, environments and human alterations.* Wiley, Chichester, England.

Able, K. W., A. Kustka, D. Witting, K. Smith, R. Rountree, and R. McBride. 1997. Fishes of Great Bay, New Jersey: larvae and juveniles collected by nightlighting. Rutgers Univ. Inst. Mar. Coastal Sci. Tech. Rep. 97–05.

Ahlstrom, E. H., K. Amaoka, D. A. Hensley, H. G. Moser, and B. Y. Sumida. 1984. Pleuronectiformes: Development. Pp. 640–670 *in* H. G. Moser, W. J. Richards, D. M. Cohen, M. P. Fahay, A. W. Kendall, Jr., and S. L. Richardson, eds. *Ontogeny and systematics of fishes.* Am. Soc. Ichthyol. Herpetol. Spec. Publ. No. 1. Allen Press, Lawrence, Kans.

Ahrenholz, D. W., J. F. Guthrie, and C. W. Krouse. 1989. Results of abundance surveys of juvenile Atlantic and gulf menhaden, *Brevoortia tyrannus* and *B. patronus.* Natl. Ocean. Atmos. Adm. Tech. Rep. Natl. Mar. Fish. Serv. Southeast Cent. 84.

Allen, D. M., J. P. Clymer, and S. S. Herman. 1978. *Fishes of the Hereford Inlet Estuary, southern New Jersey.* Wetlands Institute, Lehigh University, Bethlehem, Pa.

Alvarez-Lajonchere, L. 1976. Contribucion al estudio del ciclo de vida de *Mugil curema* Valenciennes in Cuvier and Valenciennes, 1836 (Pisces, Mugilidae). Cienc. Ser. 8 Invest. Mar. (Havana) No. 28.

Anderson, E. D. 1982. Red hake, *Urophycis chuss.* MESA (Mar. Ecosyst. Anal.) N. Y. Bight Atlas Monogr. 15 : 74–76.

Anderson, W. W. 1957. Early development, spawning, growth and occurrence of the silver mullet (*Mugil curema*) along the south Atlantic coast of the United States. U.S. Fish. Bull. 57 : 397–414.

———. 1958. Larval development, growth and spawning of striped mullet (*Mugil cephalus*) along the south Atlantic coast of the United States. U.S. Fish. Bull. 58 : 501–519.

———. 1968. Fishes taken during shrimp trawling along the south Atlantic coast of the United States, 1931–1935. U.S. Fish Wildl. Serv. Spec. Sci. Rep. Fish. 570.

Anderson, W. W., J. W. Gehringer, and F. H. Berry. 1966. Family Synodontidae: Lizard fishes. Pp. 30–102 *in Fishes of the western North Atlantic.* Sears Found. Mar. Res. Mem. 1 (Part 5). Yale University, New Haven, Conn.

Anthony, V. C. 1982. Atlantic Herring, *Clupea harengus harengus.* MESA (Mar. Ecosyst. Anal.) N. Y. Bight Atlas Monogr. 15 : 63–65.

Arias, A. M., and P. Drake. 1990. *Estados juveniles de la ictiofauna en los Canos de las Salinas de la Bahia de Cadiz.* Instituto de Ciencias Marinas de Andalucia, Cadiz, Spain.

Armstrong, P. B., and J. S. Child. 1965. Stages in the normal development of *Fundulus heteroclitus.* Biol. Bull. 128 : 143–168.

Arnold, E. L., and J. R. Thompson. 1958. Offshore spawning of the striped mullet *Mugil cephalus* in the Gulf of Mexico. Copeia 1958 : 705–707.

Arve, J. 1960. Preliminary report on attracting fish by oyster-shell plantings in Chincoteaque Bay, Maryland. Chesapeake Sci. 1 : 58–65.

Ashton, D. E., M. T. Barbour, J. R. Fletcher, V. R. Kranz, B. D. Lorenz, and G. C. Slawson. 1975. Overview environmental study of electric power development in the Delaware River Basin, 1975–1989: aquatic ecology baseline. Prepared for Delaware River Basin Electric Utility Group. NUS Corp., Pittsburgh, Pa.

Bailey, K. M., and E. D. Houde. 1989. Predation on eggs and larvae of marine fishes and the recruitment problem. Adv. Mar. Biol. 25 : 1–83.

Balon, E. K. 1984. Reflections on some decisive events in the early life of fishes. Trans. Am. Fish. Soc. 133 : 178–185.

———. 1990. Epigenesis of an epigeneticist: the development of some alternative concepts on the early ontogeny and evolution of fishes. Guelph Ichthyol. Rev. 1 : 1–48.

Barans, C. A. 1969. Distribution, growth and behavior of the spotted hake in the Chesapeake Bight. Master's thesis. College of William and Mary, Gloucester Point, Va.

———. 1972. Spotted hake, *Urophycis regia* of the York River and Lower Chesapeake Bay. Chesapeake Sci. 13(1) : 59–62.

Bardach, J. E., and J. Case. 1965. Sensory capabilities of the modified fins of squirrel hake (*Urophycis chuss*) and searobins (*Prionotus carolinus* and *P. evolans*). Copeia 1965 : 194–206.

Barkman, R. C., D. A. Bengtson, and A. D. Beck. 1981. Daily growth of juvenile fish (*Menidia menidia*) in the natural habitat compared with juveniles reared in the laboratory. Rapp. P.-V. Reun. Cons. Int. Explor. Mer 178 : 324–326.

Barnes, R. S. K. 1980. *Coastal lagoons: The natural history of a neglected habitat.* Cambridge University Press, Cambridge.

Barsukov, V. V. 1959. The wolffish (Anarhichadidae). Fauna SSSR Moscow N.S. 73.

Bath, D. W., and J. M. O'Connor. 1982. The biology of white perch, *Morone americana,* in the Hudson River estuary. U.S. Fish. Bull. 80 : 599–610.

Beacham, T. D. 1983. Variability in size or age at sexual maturity of white hake, pollock, longfin hake, and silver hake in the Canadian Maritimes area of the Northwest Atlantic Ocean. Can. Tech. Rep. Fish. Aquat. Sci. 1157.

Beacham, T. D., and S. J. Nepszy. 1980. Some aspects of the biology of white hake, *Urophycis tenuis,* in the southern Gulf of St. Lawrence. J. Northwest. Atl. Fish. Sci. 1 : 49–54.

Bean, T. H. 1887. Report on the fishes observed in Great Egg Harbor Bay, New Jersey, during the summer of 1887. U.S. Comm. Fish. Bull. 7(1887) : 129–154.

Beardsley, R. C., and C. N. Flagg. 1976. The water structure, mean currents and shelf water/slop water front on the New England continental shelf. Mem. Soc. R. Sci. Liège 6(10) : 209–225.

Beck, S. 1995. White perch. Pp. 235–243 *in* L. E. Dove and R. M. Nyman, eds. *Living resources of the Delaware Estuary.* Delaware Estuary Program.

Beck, W. R., and W. H. Massmann. 1951. Migratory behavior of the rainwater fish, *Lucania parva,* in the York River, Virginia. Copeia 1951(2) : 176.

Beckley, L. E. 1984. The ichthyofauna of the Sundays Estuary, South Africa, with particular reference to the juvenile marine component. Estuaries 7(3) : 248–258.

———. 1985. Coastal zone utilization by juvenile fish in the eastern cape, South Africa. Ph.D. diss. University of Cape Town, Cape Town.

———. 1986. The ichthyoplankton assemblage of Algoa Bay nearshore region in relation to coastal zone utilization by juvenile fish. S. Afr. J. Zool. 21(3) : 245–252.

Bejda, A. J., B. A. Phelan, and A. L. Studholme. 1992. The effect of dissolved oxygen on the growth of young-of-the-year winter flounder, *Pseudopleuronectes americanus.* Environ. Biol. Fishes 34 : 321–327.

Bejda, A. J., A. L. Studholme, and B. L. Olla. 1987. Behavioral responses of red hake, *Urophycis chuss,* to decreasing concentrations of dissolved oxygen. Environ. Biol. Fishes 19(4) : 261–268.

Bengston, D. A. 1984. Resource partitioning by *Menidia menidia* and *Menidia beryllina* (Osteichthyes: Atherinidae). Mar. Ecol. Prog. Ser. 18 : 21–30.

Bengtson, D. A., and R. C. Barkman. 1981. Growth of postlarval Atlantic silversides in four temperature regimes. Prog. Fish-Cult. 43 : 146–148.

Berg, D. L., and J. S. Levinton. 1985. The biology of the Hudson-Raritan estuary, with emphasis on fishes. U.S. Dep. Commer. Natl. Ocean. Atmos. Adm./Natl. Ocean Serv. Tech. Memo. NOS OMA 16 : 1–170.

Berrien, P., and J. Sibunka. In press. Distribution patterns of fish eggs in the United States Northeast Continental Shelf Ecosystem. 1977–1987. Natl. Ocean. Atmos. Adm. Tech. Rep.

Berry, F. H. 1959. Young jack crevalles (*Caranx* species) off the southeastern Atlantic coast of the United States. U.S. Fish. Bull. 59 : 417–535.

———. 1964. Review and emendation of family Clupeidae. Copeia 1964 : 720–730.

Bertin, L. 1956. *Eels: a biological study.* Cleaver-Hume Press, London.

Beverton, R. J. H., and T. C. Iles. 1992. Mortality rates of 0-group plaice (*Pleuronectes platessa* L.), dab (*Limanda limanda* L.) and turbot (*Scophthalmus maximus* L.) in European waters. III. Density-dependence of mortality rates of 0-group plaice and some demographic implications. Neth. J. Sea Res. 29(1–3) : 61–79.

Bigelow, H. B., and W. C. Schroeder. 1953. Fishes of the Gulf of Maine. U.S. Fish Wildl. Serv. Fish. Bull. 74 : 1–576.

Bigelow, H. B., and M. Sears. 1935. Studies of the waters on the continental shelf, Cape Cod to Chesapeake Bay. II. Salinity. Pap. Phys. Oceanogr. Meteorol. 4 : 1–94.

Bigelow, H. B., and W. W. Welsh. 1925. Fishes of the Gulf of Maine. Bull. U.S. Bur. Fish. 40.

Bisbal, G. A., and D. A. Bengtson. 1995. Effects of delayed feeding on survival and growth of summer flounder *Paralichthys dentatus* larvae. Mar. Ecol. Prog. Ser. 121 : 301–306.

Blaber, S. J. M. 1980. Fish of the Trinity Inlet system of north Queensland with notes on the ecology of fish faunas of tropical Indo-Pacific estuaries. Aust. J. Mar. Freshw. Res. 31 : 137–146.

Blaber, S. J. M., and T. G. Blaber. 1980. Factors affecting the distribution of juvenile estuarine and inshore fish. J. Fish Biol. 17 : 143–162.

Blaxter, J. H. S., ed. 1974. The early life history of fish. Proceedings of an international symposium at the Dunstaffnage Marine Research Laboratory of the Scottish Marine Biological Association, Oban, Scotland, 17–23 May, 1973. Springer, New York.

———. 1988. Pattern and variety in development. Pp. 1–58 *in* W. S. Hoard and D. J. Randall, eds. *Fish physiology. Vol. XI, Part A. Eggs and larvae.* Academic Press, New York.

Bleakney, J. S. 1963. First record of the fish *Po-*

gonias cromis from Canadian waters. Copeia 1963(1):173.

Boesch, D. F., and R. E. Turner. 1984. Dependence of fishery species on salt marshes: the role of food and refuge. Estuaries 7:460–468.

Böhlke, J. E., and C. C. G. Chaplin. 1968. *Fishes of the Bahamas and adjacent tropical waters.* University of Texas Press, Austin.

Boicourt, W. C. 1982. Estuarine larval retention mechanisms on two scales. Pp. 445–457 *in* V. S. Kennedy, ed. *Estuarine comparisons.* Academic Press, New York.

Booth, D. J. 1995. Juvenile groups in a coral-reef damselfish: density-dependent effects on individual fitness and population demography. Ecology 76(1):91–106.

Booth, R. A. 1967. A description of the larval stages of the tomcod, *Microgadus tomcod,* with comments on its spawning ecology. Unpubl. Ph.D. diss. University of Connecticut, Storrs.

Boreman, J., S. J. Correia, and D. B. Witherell. 1993. Effects of changes in age-0 survival and fishing mortality on egg production of winter flounder in Cape Cod Bay. Am. Fish. Soc. Symp. 14:39–45.

Boughton, D. A., B. B. Collette, and A. R. McCune. 1991. Heterochrony in jaw morphology of needlefishes (Teleostei: Belonidae). Syst. Zool. 40(3):329–354.

Bourne, D. W., and J. J. Govoni. 1988. Distribution of fish eggs and larvae and patterns of water circulation in Narragansett Bay, 1972–1973. Am. Fish. Soc. Symp. 3:132–148.

Bowen, B. W., and J. C. Avise. 1990. Genetic structure of Atlantic and Gulf of Mexico populations of sea bass, menhaden, and sturgeon: influence of zoogeographic factors and life-history patterns. Mar. Biol. 107:371–381.

Bowers-Altman, J. 1993. Wildlife in New Jersey: The striped cusk eel. N.J. Outdoors Fall: 64–65.

Boyd, W. A. 1976. Hydrography. Pp. 25–34 *in* D. Merriman and L. M. Thorpe, eds. *The Connecticut River ecology study.* Monogr. No. 1. American Fisheries Society, Washington, D.C.

Boynton, W. R., E. M. Setzler, K. V. Wood, H. H. Zion, and M. Homer. 1987. Final report of Potomac River fisheries study: Ichthyoplankton and juvenile investigations. Univ. Md. Center for Environmental and Estuarine Studies, Chesapeake Biol. Lab. Ref. No. 77–169.

Bozeman, E. L., Jr., and J. M. Dean. 1980. The abundance of estuarine larval and juvenile fish in a South Carolina intertidal creek. Estuaries 3: 89–97.

Breder, C. M., Jr. 1936. All modern conveniences: a note on the nest architecture of the four-spined stickleback. Bull. N.Y. Zool. Soc. 39:72–76.

———. 1940. The spawning of *Mugil cephalus* on the Florida west coast. Copeia 1940:138–139.

———. 1948. *Field book of marine fishes of the Atlantic Coast from Labrador to Texas.* Putnam, New York.

———. 1962. Effects of a hurricane on the small fishes of a shallow bay. Copeia 1962(2):459.

Breder, C. M., Jr., and E. Clark. 1947. A contribution to the visceral anatomy, development, and relationship of the Plectognathi. Bull. Am. Mus. Nat Hist. 88(5):291–319.

Breder, C. M., Jr., and D. E. Rosen. 1966. *Modes of reproduction in fishes.* T.F.H. Publications, Jersey City, N.J.

Breitburg, D. L. 1989. Demersal schooling prior to settlement by larvae of the naked goby. Environ. Biol. Fishes 26:97–103.

———. 1991. Settlement patterns and presettlement behavior of the naked goby, *Gobiosoma bosci,* a temperate oyster reef fish. Mar. Biol. 109: 213–221.

———. 1992. Episodic hypoxia in Chesapeake Bay: interacting effects of recruitment, behavior, and physical disturbance. Ecol. Monogr. 62: 525–546.

Breitburg, D. L., N. Steinberg, S. DuBeau, C. Cooksey, and E. D. Houde. 1994. Effects of low dissolved oxygen on predation on estuarine fish larvae. Mar. Ecol. Prog. Ser. 104:235–246.

Breitburg, D. L., M. A. Palmer, and T. Loher. 1995. Larval distributions and the spatial patterns of settlement of an oyster reef fish: responses to flow and structure. Mar. Ecol. Prog. Ser. 125: 45–60.

Bridges, D. W. 1971. The pattern of scale development in juvenile Atlantic croaker (*Micropogonias undulatus*). Copeia 1971(2):331–332.

Briggs, J. C. 1958. A list of Florida fishes and their distribution. Bull. Fla. State Mus. 2:225–318.

———. 1960. Fishes of worldwide (circumtropical) distribution. Copeia 1960:171–180.

Briggs, P. T., and J. S. O'Connor. 1971. Comparison of shore zone fishes over naturally vegetated and sand-filled bottoms in Great South Bay. N.Y. Fish Game J. 18:15–41.

Broadhead, G. C. 1953. Investigations of the black mullet, *Mugil cephalus,* L., in northwest Florida. Fla. Board Conserv. Tech. Ser. No. 7.

———. 1958. Growth of the black mullet (*Mugil cephalus* L.) in west and northwest Florida. Fla. Board Conserv. Tech. Ser. No. 25.

Bumpus, D. F. 1973. A description of the circulation

on the continental shelf of the East Coast of the U.S. Prog. Oceanogr. 6 : 111–157.

Burger, J., ed. 1994. *Before and after an oil spill: the Arthur Kill.* Rutgers University Press, New Brunswick, N.J.

Burgess, W. E. 1978. *Butterflyfishes of the world.* T.F.H. Publications, Jersey City, N.J.

Burke, J. S. 1995. Role of feeding and prey distribution of summer and southern flounder in selection of estuarine nursery habitats. J. Fish. Biol. 47 : 355–366.

Burnett, J., S. H. Clark, and L. O'Brien. 1984. A preliminary assessment of white hake in the Gulf of Maine-Georges Bank area. Natl. Mar. Fish Serv., Northeast Fish. Sci. Cent. Woods Hole Lab. Ref. Doc. No. 84–31 : 1–33.

Burr, B. M., and F. J. Schwartz. 1986. Occurrence, growth, and food habits of the spotted hake, *Urophycis regia,* in the Cape Fear estuary and adjacent Atlantic Ocean, North Carolina. Northeast Gulf Sci. 8(2) : 115–127.

Byrne, D. M. 1978. Life history of the spotfin killifish, *Fundulus luciae* (Pisces: Cyprinodontidae), in Fox Creek Marsh, Virginia. Estuaries 4 : 211–227.

Campana, S. E. 1996. Year-class strength and growth rate in young Atlantic cod *Gadus morhua.* Mar. Ecol. Prog. Ser. 135 : 21–26.

Campana, S. E., K. T. Frank, P. C. F. Hurley, P. A. Koeller, F. H. Page, and P. C. Smith. 1989. Survival and abundance of young Atlantic cod (*Gadus morhua*) and haddock (*Melanogrammus aeglefinus*) as indicatory of year-class strength. Can. J. Fish. Aquat. Sci. 46 (Suppl. 1) : 171–182.

Campbell, B. C., and K. W. Able. In press. Life history characteristics of the northern pipefish, *Syngnathus fuscus* in southern New Jersey. Estuaries (in press).

Cantelmo, F. R., and C. H. Wahtola. 1992. Aquatic habitat impacts of pile-supported and other structures in the lower Hudson River. Pp. 59–75 *in* W. Wise, D. J. Suszkowski, and J. R. Waldman, eds. *Proceedings: Conference of impacts of New York Harbor development on aquatic resources.* Hudson River Foundation, New York.

Carr, M. H. 1991. Habitat selection and recruitment of an assemblage of temperate zone reef fishes. J. Exp. Mar. Biol. Ecol. 146 : 113–137.

Casterlin, M. E., and W. W. Reynolds. 1979. Diel activity patterns of the smooth dogfish shark, *Mustelus canis.* Bull. Mar. Sci. 29(3) : 440–442.

Castonguay, M., and J. D. McCleave. 1987. Vertical distributions, diel and ontogenetic vertical migrations and net avoidance of leptocephali of *Anguilla* and other common species in the Sargasso Sea. J. Plankton Res. 9(1) : 195–214.

Castro, J. I. 1983. *The sharks of North American waters.* Texas A&M University Press, College Station.

Castro, L. R., and R. K. Cowen. 1991. Environmental factors affecting the early life history of bay anchovy *Anchoa mitchilli* in Great South Bay, New York. Mar. Ecol. Prog. Ser. 76 : 235–247.

Chacko, P. I. 1949. Bionomics of the Indian paradise fish, *Macropodus cupanus,* in the waters of Madras. Proc. 36th Indian Sci. Congr. Pt. 3 : 164.

Chambers, J. R. 1992. Coastal degradation and fish population losses. Pp. 45–51 *in* R. H. Stroud, ed. *Stemming the tide of coastal fish habitat loss.* Proceedings of a symposium on conservation of coastal fish habitat, Baltimore, Md., March 7–9, 1991. National Coalition for Marine Conservation, Savannah, Ga.

Chambers, R. C. 1997. Environmental influences on egg and propagule sizes in marine fishes. Pp. 63–102 *in* R. C. Chambers and E. A. Trippel, eds. *Early life history and recruitment in fish populations.* Chapman and Hall, New York.

Chambers, R. C., and W. C. Leggett. 1987. Size and age at metamorphosis in marine fishes: an analysis of laboratory-reared winter flounder (*Pseudopleuronectes americanus*) with a review of variation in other species. Can. J. Fish. Aquat. Sci. 44 : 1936–1947.

———. 1992. Possible causes and consequences of variation in age and size at metamorphosis in flatfishes (Pleuronectiformes): an analysis at the individual, population, and species level. Neth. J. Sea Res. 29 : 7–24.

Chambers, R. C., W. C. Leggett, and J. A. Brown. 1988. Variation in and among life history traits of laboratory-reared winter flounder *Pseudopleuronectes americanus.* Mar. Ecol. Prog. Ser. 47 : 1–15.

Chambers, R. C., and E. A. Trippel, eds. 1997. *Early life history and recruitment of fish populations.* Chapman and Hall, London.

Chao, L. N., and J. A. Musick. 1977. Life history, feeding habits, and functional morphology of juvenile sciaenid fishes in the York River Estuary, Virginia. U.S. Fish. Bull. 75 : 657–702.

Chapleau, F., and A. Keast. 1988. A phylogenetic reassessment of the monophyletic status of of the family Soleidae, with comments on the suborder Soleoidei (Pisces: Pleuronectiformes). Can. J. Zool. 66 : 2797–2810.

Charnell, R. L., and D. V. Hansen. 1974. Summary and analysis of physical oceanography data col-

lected in the New York Bight Apex. Mar. Ecosyst. Anal. Rep. No. 74–3.
Cheney, R. E. 1978. Oceanographic observations in the western North Atlantic during FOX 1, May 1978. Tech. Note 3700–79–78, V. 5 U.S. Naval Oceanographic Office, Washington, D.C.
Chiarella, L. A., and D. O. Conover. 1990. Spawning season and first-year growth of adult bluefish from the New York Bight. Trans. Am. Fish. Soc. 119:455–462.
Chittenden, M. E., Jr. 1969. Life history and ecology of the American shad, *Alosa sapidissima,* in the Delaware River. Ph.D. diss. Rutgers University, New Brunswick, N.J.
———. 1971. Status of the striped bass, *Morone saxatilis,* in the Delaware River. Chesapeake Sci. 12:131–136.
———. 1972a. Responses of young American shad, *Alosa sapidissima,* to low temperatures. Trans. Am. Fish. Soc. 101(4):680–685.
———. 1972b. Salinity tolerance of young blueback herring, *Alosa aestivalis.* Trans. Am. Fish. Soc. 101(1):123–125.
———. 1973a. Effects of handling on oxygen requirements of American shad (*Alosa sapidissima*). J. Fish. Res. Board Can. 30:105–110.
———. 1973b. Salinity tolerance of young American shad, *Alosa sapidissima.* Chesapeake Sci. 14(3):207–210.
———. 1975. Dynamics of American shad, *Alosa sapidissima,* runs in the Delaware River. U.S. Fish. Bull. 73(3):487–494.
———. 1976. Weight loss, mortality, feeding and duration of residence of adult American shad, *Alosa sapidissima,* in fresh water. U.S. Fish. Bull. 74(1):151–157.
Chizmadia, P. A., M. J. Kennish, and V. L. Ohori. 1984. Physical description of Barnegat Bay. Pp. 1–28 *in* M. J. Kennish and R. A. Lutz, eds. *Ecology of Barnegat Bay, New Jersey.* Lecture Notes on Coastal and Estuarine Studies. Springer, New York.
Cianci, J. M. 1969. Larval development of the alewife and the glut herring. M.S. thesis. University of Connecticut, Storrs.
Clark, J. 1967. Fish and man: Conflict in the Atlantic estuaries. Am. Littoral Soc. Spec. Publ. 5.
Clark, J., W. G. Smith, A. W. Kendall, Jr., and M. P. Fahay. 1969. Studies of estuarine dependence of Atlantic coastal fishes. Data Report I: Northern Section, Cape Cod to Cape Lookout. RV *Dolphin* cruises 1965–1966: Zooplankton volumes, midwater trawl collections, temperatures and salinities. U.S. Bur. Sport Fish. Wildl. Tech. Pap. 28.
Clark, J. C. 1994. Impoundment estuarine interactions. Federal aid in Fisheries Restoration Project F-44-R Annual Report. Delaware Division of Fish and Wildlife, Dover, Del.
Clay, D., W. T. Stobo, B. Beck, and P. C. F. Hurley. 1989. Growth of juvenile pollock (*Pollachius virens* L.) along the Atlantic coast of Canada with inferences of inshore-offshore movements. J. Northwest Atl. Fish. Sci. 9:37–43.
Clemmer, G. H., and F. J. Schwartz. 1964. Age, growth and weight relationships of the striped killifish, *Fundulus majalis,* near Solomons, Maryland. Trans. Am. Fish. Soc. 93(2):197–198.
Clymer, J. P. 1978. The distribution, trophic dynamics and competitive interactions of three salt marsh killifishes (Pisces: Cyprinodontidae). Ph.D. diss. Lehigh University, Bethlehem, Pa.
Cohen, D. M., T. Inada, T. Iwamoto, and N. Scialabba. 1990. Gadiform fishes of the world (Order Gadiformes). An annotated and illustrated catalogue of cods, hakes, grenadiers and other gadiform fishes known to date. FAO Fish. Synopsis. 125, Vol. 10.
Cole, C. F., R. J. Essig, and O. R. Sainelle. 1980. Biological investigation of the alewife population, Parker River, Massachusetts. U.S. Natl. Mar. Fish. Serv. Final Rep. No. 11, 884-32-50-80-CR.
Collette, B. B. 1966. *Belonion,* a new genus of freshwater needlefishes from South America. Am. Mus. Novit. 2274.
———. 1968. *Strongylura timucu* (Walbaum): a valid species of western Atlantic needlefish. Copeia 1:189–192.
———. 1986. Resilience of the fish assemblage in New England tidepools. U.S. Fish. Bull. 84(1):200–204.
Collette, B. B., and K. E. Hartel. 1988. An annotated list of the fishes of Massachusetts Bay. Natl. Ocean. Atmos. Adm. Tech. Memo. Natl. Mar. Fish. Serv., Northeast Fish. Sci. Cent. 51.
Collette, B. B., and G. K. McPhee, eds. In prep. Revision of Bigelow and Schroeder's "Fishes of the Gulf of Maine." Smithsonian Institution Press, Washington, D.C.
Collette, B. B., G. E. McGowen, N. V. Parin, and S. Mito. 1984a. Beloniformes: development and relationships. Pp. 335–354 *in* H. G. Moser, W. J. Richards, D. M. Cohen, M. P. Fahay, A. W. Kendall, Jr., and S. L. Richardson, eds. *Ontogeny and systematics of fishes.* Am. Soc. Ichthyol. Herpetol. Spec. Publ. No. 1. Allen Press, Lawrence, Kans.
Collette, B. B., T. Potthoff, W. J. Richards, S. Ueyanagi, J. L. Russo, and Y. Nishikawa. 1984b. Scombroidei: development and relationships.

Pp. 591–619 *in* H. G. Moser, W. J. Richards, D. M. Cohen, M. P. Fahay, A. W. Kendall, Jr., and S. L. Richardson, eds. *Ontogeny and systematics of fishes.* Am. Soc. Ichthyol. Herpetol. Spec. Publ. No. 1. Allen Press, Lawrence, Kans.

Collins, M. R. 1985a. Species profiles: life histories and environmental requirements of coastal fishes and invertebrates (south Florida)—white mullet. U.S. Fish Wildl. Ser. Biol. Rep. 82(11.39). U.S. Army Corps of Engineers, TR EL-82-4.

———. 1985b. Species profiles: life histories and environmental requirements of coastal fishes and invertebrates (south Florida)—striped mullet. U.S. Fish Wildl. Ser. Biol. Rep. 82(11.34). U.S. Army Corps of Engineers, TR EL-82-4.

Collins, M. R., and B. W. Stender. 1987. Larval king mackerel (*Scomberomorus cavalla*), Spanish mackerel (*S. maculatus*) and bluefish (*Pomatomus saltatrix*) off the southeast coast of the United States, 1973–1980. Bull. Mar. Sci. 41: 822–834.

———. 1989. Larval striped mullet (*Mugil cephalus*) and white mullet (*Mugil curema*) off the southeastern United States. Bull. Mar. Sci. 45(3): 580–589.

Colton, J. B., Jr. 1972. Temperature trends and the distribution in continental shelf waters, Nova Scotia to Long Island. U.S. Fish. Bull. 70(3): 637–658.

Colton, J. B., and R. R. Marak. 1969. Guide for identifying the common planktonic fish eggs and larvae of continental shelf waters, Cape Sable to Block Island. Natl. Mar. Fish. Serv. Woods Hole Biol. Lab. Ref. Doc. 69–9.

Colton, J. B., W. G. Smith, A. W. Kendall, Jr., P. L. Berrien, and M. P. Fahay. 1979. Principal spawning areas and times of marine fishes, Cape Sable to Cape Hatteras. U.S. Fish. Bull. 76: 911–915.

Comyns, B. H., and G. C. Grant. 1993. Identification and distribution of *Urophycis* and *Phycis* (Pisces: Gadidae) larvae and pelagic juveniles in the U.S. Middle Atlantic Bight. U.S. Fish. Bull. 91: 210–223.

Conover, D. O. 1992. Seasonality and the scheduling of life history at different latitudes. J. Fish Biol. 41(B): 161–178.

Conover, D. O., and B. E. Kynard. 1984. Field and laboratory observations of spawning periodicity and behavior of a northern population of the Atlantic silverside, *Menidia menidia* (Pisces: Atherinidae). Environ. Biol. Fishes 11: 161–171.

Conover, D. O., and S. A. Murawski. 1982. Offshore winter migration of the Atlantic silverside, *Menidia menidia.* U.S. Fish. Bull. 80: 145–150.

Conover, D. O., and T. M. C. Present. 1990. Countergradient variation in growth rate: compensation for length of the growing season among Atlantic silversides from different latitudes. Oecologia 83: 316–324.

Conover, N. R. 1958. Investigations of white perch *Morone americana* (Gmelin) in Albemarle Sound and the Lower Roanoke River, North Carolina. Master's thesis. North Carolina State College, Raleigh.

Cook, S. K. 1988. Physical oceanography of the Middle Atlantic Bight. Pp. 1–49 *in* A. L. Pacheco, ed. *Characterization of the Middle Atlantic Water Management Unit of the Northeast Regional Action Plan.* Natl. Ocean. Atmos. Adm. Tech. Memo.

Cooper, J. A. 1996. Monophyly and intrarelationships of the family Pleuronetidae (Pleuronectiformes), with a revised classification. Ph.D. diss. University of Ottawa, Ottawa, Canada.

Cooper, J. E. 1978. Identification of eggs, larvae and juveniles of the rainbow smelt, *Osmerus mordax,* with comparisons to larval alewife, *Alosa pseudoharengus,* and gizzard shad, *Dorosoma cepedianum.* Trans. Am. Fish. Soc. 107: 56–62.

Cooper, R. A. 1961. Early life history and spawning migration of the alewife. Master's thesis. University of Rhode Island, Kingston.

———. 1966. Migration and population estimation of the tautog, *Tautoga onitis* (Linnaeus), from Rhode Island. Trans. Am. Fish. Soc. 95: 239–247.

———. 1967. Age and growth of tautog, *Tautoga onitis* (Linnaeus), from Rhode Island. Trans. Am. Fish. Soc. 96: 134–142.

Cooper, R. A., P. Valentine, J. R. Uzmann, and R. A. Slater. 1987. Submarine canyons. Pp. 52–63 *in* R. H. Backus and D. W. Bourne, eds. *Georges Bank.* MIT Press, Cambridge, Mass.

Coorey, D. N., K. W. Able, and J. K. Shisler. 1985. Life history and food habits of *Menidia beryllina* in a New Jersey salt marsh. Bull. N.J. Acad. Sci. 30(1): 29–37.

Copp, G. H., and V. Kovac. 1996. When do fish with indirect development become juveniles? Can. J. Fish. Aquat. Sci. 53: 746–752.

Cowan, J. H., Jr., and R. S. Birdsong. 1985. Seasonal occurrence of larval and juvenile fishes in a Virginia Atlantic coast estuary with emphasis on drums (Family Sciaenidae.) Estuaries 8(1): 48–59.

Cowen, R. K., and S. Sponaugle. 1997. Relationships between early life history traits and recruitment among coral reef fishes. Pp. 423–450 *in* R. C. Chambers and E. A. Trippel, eds. *Early life history and recruitment in fish populations.* Chapman and Hall, London.

Cowen, R. K., L. Chiarella, C. Gomez, and M. Bell. 1991. Distribution, age and lateral plate variation of larval sticklebacks (*Gasterosteus*) off the Atlantic coast of New Jersey, New York and southern New England. Can J. Fish. Aquat. Sci. 48: 1679–1684.

Cowen, R. K., J. A. Hare, and M. P. Fahay. 1993. Beyond hydrography: Can physical processes explain larval fish assemblages within the Middle Atlantic Bight? Bull. Mar. Sci. 53(2): 567–587.

Cox, P. 1916. Are migrating eels deterred by a range of lights? Report on experimental tests. Contrib. Can. Biol. Sessional Pap. 38a: 115–118.

Crawford, R. E., and C. G. Carey. 1985. Retention of winter flounder larvae within a Rhode Island salt pond. Estuaries 8(2B): 217–227.

Croker, R. A. 1965. Planktonic fish eggs and larvae of Sandy Hook estuary. Chesapeake Sci. 6(2): 92–95.

Csanady, G. T. 1978. Wind effects on surface to bottom fronts. J. Geophys. Res. 83(C9): 4633–4640.

Csanady, G. T., and P. Hamilton. 1988. Circulation of slopewater. Continental Shelf Res. 8(5–7): 565–624.

Curran, M. C. 1992. The behavioral physiology of labroid fishes. Ph.D. diss. Massachusetts Institute of Technology/Woods Hole Oceanographic Institution, WHOI-92–41.

Cushing, D. H. 1974. The possible density-dependence of larval mortality and adult mortality in fishes. Pp. 103–111 *in* J. H. S. Blaxter, ed. *The early life history of fish.* Springer, New York.

———. 1996. *Towards a science of recruitment in fish populations.* Ecology Institute, Oldendorf/Luhe, Germany.

Dahlberg, M. D. 1970. Atlantic and Gulf of Mexico menhadens, genus *Brevoortia* (Pisces: Clupeidae). Bull. Fla. State Mus. Biol. Sci. 15(3).

———. 1975. Guide to coastal fishes of Georgia and nearby states. University of Georgia Press, Athens.

Dahlberg, M. D., and J. C. Conyers. 1973. An ecological study of *Gobiosoma bosci* and *G. ginsburgi* (Pisces, Gobiidae) on the Georgia coast. U.S. Fish. Bull. 71(1): 279–287.

Dahlgren, U. 1927. The life history of the fish *Astroscopus* (the "stargazer"). Sci. Monthly 24: 348–365.

Daiber, F. C. 1957. Sea trout research. Estuarine Bull. 2(5): 1–6.

Daiber, F. C., and R. W. Smith. 1970. An analysis of fish populations in the Delaware Bay area. Annu. Dingell-Johnson Rep., 1969–1970, Proj. F-13-R-12 (Unpubl. ms.)

Dando, P. R. 1984. Reproduction in estuarine fish. Pp. 155–170 *in* W. Potts and R. J. Wootton, eds. *Fish reproduction: strategies and tactics.* Academic Press, New York.

Daniel, L. B., III, and J. E. Graves. 1994. Morphometric and genetic identification of eggs and spring-spawning sciaenids in lower Chesapeake Bay. U.S. Fish. Bull. 92: 254–261.

Dawson, C. E. 1966. Studies of the gobies (Pisces: Gobiidae) of Mississippi Sound and adjacent waters. I. *Gobiosoma.* Am. Midl. Nat. 76(2): 379–409.

———. 1982. Family Syngnathidae. Pp. 1–172 *in Fishes of the western North Atlantic.* Sears Found. Mar. Res. Mem. 1(Part 8). Yale University, New Haven, Conn.

Day, J. H. 1981. *Estuarine ecology with particular reference to southern Africa.* A. A. Balkema, Rotterdam.

Day, J. W., Jr., C. A. S. Hall, W. M. Kemp, and A. Yanez-Arancibia. 1989. *Estuarine ecology.* Wiley, New York.

Deegan, L., and J. W. Day. 1986. Coastal fishery habitat requirements. Pp. 44–52 *in* A. Yanez-Arancibia and D. Pauly, eds. *Recruitment processes in tropical coastal demersal communities.* Ocean Science in Relation to Living Resources (OSLR), International Recruitment Project (IREP). IOC-FAO-UNESCO Workshop OSLR/IREP Project. Vol. 44. UNESCO, Paris.

de Lafontaine, Y., T. Lambert, G. R. Lilly, W. D. McKone, and R. J. Miller, eds. 1992. Juvenile stages: The missing link in fisheries research. Report of a workshop. Can. Tech. Rep. Fish. Aquatic Sci. 1890.

Denoncourt, R. F., J. C. Fisher, and K. M. Rapp. 1978. A freshwater population of the mummichog, *Fundulus heteroclitus,* from the Susquehanna River drainage in Pennsylvania. Estuaries 1: 269–272.

Derickson, W. K. 1970. The shore zone fishes of Rehoboth and Indian River bays of Delaware. Master's thesis. University of Delaware, Newark.

Derickson, W. K., and K. S. Price. 1973. The fishes of the shore zone of Rehoboth and Indian River bays, Delaware. Trans. Am. Fish. Soc. 102: 552–562.

de Sylva, D. P., F. A. Kalber, Jr., and C. N. Shuster. 1962. Fishes and ecological conditions in the shore zone of the Delaware River Estuary, with notes on other species collected in deeper water. Univ. Del. Mar. Lab. Inf. Ser. Publ. 5: 1–164.

Deuel, D. G., J. R. Clark, and A. J. Mansueti. 1966. Description of embryonic and early larval stages of the bluefish, *Pomatomus saltatrix.* Trans. Am. Fish. Soc. 95:264–271.

Dew, C. B. 1976. A contribution to the life history of the cunner, *Tautogolabrus adspersus,* in Fishers Island Sound, Connecticut. Chesapeake Sci. 17:101–113.

———. 1991. Early life history and population dynamics of Atlantic tomcod (*Microgadus tomcod*) in the Hudson River estuary, New York. Ph.D. diss. City University of New York, New York.

———. 1995. The nonrandom size distribution and size-selective transport of age-0 Atlantic tomcod (*Microgadus tomcod*) in the lower Hudson River estuary. Can. J. Fish. Aquat. Sci. 52:2353–2366.

Dew, C. B., and J. H. Hecht. 1976. Observations on the population dynamics of Atlantic tomcod (*Microgadus tomcod*) in the Hudson River Estuary. Proceedings of the 4th symposium on Hudson River ecology. Paper 25. Hudson River Environmental Society, Bronx, N.Y.

———. 1994a. Hatching, estuarine transport, and distribution of larval and early juvenile Atlantic tomcod, *Microgadus tomcod,* in the Hudson River. Estuaries 17(2):472–488.

———. 1994b. Recruitment, growth, mortality, and biomass production of larval and early juvenile Atlantic tomcod in the Hudson River Estuary. Trans. Am. Fish. Soc. 123:681–702.

DiMichele, C., and D. A. Powers. 1982. LDH-B genotype-specific hatching times of *Fundulus heteroclitus* embryos. Nature 296:563–564.

DiMichele, L., and M. H. Taylor. 1980. The environmental control of hatching in *Fundulus heteroclitus.* J. Exp. Zool. 214:181–187.

Ditty, J. G. 1989. Separating early larvae of sciaenids from the western North Atlantic: a review and comparison of larvae off Louisiana and Atlantic coast of the U.S. Bull. Mar. Sci. 44(3):1083–1105.

Ditty, J. G., and F. M. Truesdale. 1983. Comparative larval development of *Peprilus burti, P. triacanthus* and *P. paru* (Pisces: Stromateidae) from the western North Atlantic. Copeia 1983:397–406.

Doherty, P. J. 1991. Spatial and temporal patterns in recruitment. Pp. 271–287 *in* P. F. Sale, ed. *The ecology of fishes on coral reefs.* Academic Press, San Diego, Calif.

Dorf, B. A., and J. C. Powell. 1997. Distribution, abundance, and habitat characteristics of juvenile tautog (*Tautoga onitis,* family Labridae) in Narragansett Bay, Rhode Island, 1988–1992. Estuaries 20(3):589–600.

Dorsey, S. E., E. D. Houde, and J. C. Gamble. 1996. Cohort abundances and daily variability in mortality of eggs and yolk-sac larvae of bay anchovy, *Anchoa mitchilli,* in Chesapeake Bay. U.S. Fish. Bull. 94:257–267.

Dove, L. E., and R. M. Nyman. 1995. *Living resources of the Delaware estuary.* The Delaware Estuary Program.

Dovel, W. L. 1960. Larval development of the oyster toadfish, *Opsanus tau.* Chesapeake Sci. 1(3–4): 187–195.

———. 1968. Fish eggs and larvae. In: Biological and geological research on the effects of dredging spoil disposal in the upper Chesapeake Bay. Eighth Prog. Rep. No. 68-2-B. To: U.S. Army Corps Eng., Contr. No. 14-16-005-2096. Univ. Md. Nat. Resour. Inst. Contrib. Ref. No. 68-2-B.

———. 1971. Fish eggs and larvae of the upper Chesapeake Bay. Univ. Md. Nat. Resour. Inst. Contrib. No. 460, Spec. Sci. Rep. 4.

———. 1981. Ichthyoplankton of the lower Hudson Estuary, New York. N.Y. Fish Game J. 28(1): 21–39.

Dovel, W. L., T. A. Mihursky, and A. J. McErlean. 1969. Life history aspects of the hogchoker, *Trinectes maculatus,* in the Patuxent River estuary, Maryland. Chesapeake Sci. 10(2):104–119.

Doyle, M. J. 1977. A morphological staging system for the larval development of the herring, *Clupea harengus* L. J. Mar. Biol. Assoc. U.K. 57: 859–867.

Duggins, C. F., Jr., A. A. Karlin, and K. G. Relyea. 1983. Electrophoretic comparison of *Cyprinodon variegatus* Lacépède and *Cyprinodon hubbsi* Carr, with comments on the genus *Cyprinodon* (Atheriniformes: Cyprinodontidae). Northeast Gulf Sci. 6(2):99–107.

Dunn, J. R., and A. C. Matarese. 1984. Gadidae: development and relationships. Pp. 283–299 *in* H. G. Moser, W. J. Richards, D. M. Cohen, M. P. Fahay, A. W. Kendall, Jr., and S. L. Richardson, eds. *Ontogeny and systematics of fishes.* Am. Soc. Ichthyol. Herpetol. Spec. Publ. No. 1. Allen Press, Lawrence, Kans.

DuPaul, W. D., and J. D. McEachran. 1973. Age and growth of the butterfish, *Peprilus triacanthus,* in the lower York River. Chesapeake Sci. 14(3): 205–207.

Durand, J. B. 1984. Nitrogen distribution in New Jersey coastal bays. Pp. 29–51 *in* M. J. Kennish and R. A. Lutz, eds. *Ecology of Barnegat Bay, New Jersey.* Lecture Notes on Coastal and Estuarine Studies. Springer, New York.

Durand, J. B., and R. J. Nadeau. 1972. Water resources development in the Mullica River Basin. Part 1. Biological evaluation of the Mullica

River–Great Bay Estuary. Water Resource Institute, Rutgers University, New Brunswick, N.J.

Durski, S. 1996. Detection of the Hudson-Raritan Estuary plume by satellite remote sensing. Bull. N.J. Acad. Sci. 41(1): 15–20.

Eales, J. G. 1968. The eel fisheries of eastern Canada. Fish. Res. Board Can. Bull. 166.

Edel, R. K. 1975. The induction of maturation of female American eels through hormone injections. Helgol. Wiss. Meeresunters. 27(2): 131–138.

Edwards, R. L., and K. O. Emory. 1968. The view from a storied sub. The *Alvin* off Norfolk, Va. Commer. Fish. Rev. 30(8–9): 48–55.

Edwards, R. L., R. Livingstone, Jr., and P. E. Hamer. 1962. Winter water temperatures and an annotated list of fishes—Nantucket Shoals to Cape Hatteras. *Albatross III* Cruise No. 126. Spec. Sci. Rep. Fisheries. 397: 1–31.

Eklund, A. M., and T. E. Targett. 1990. Reproductive seasonality of fishes inhabiting hard bottom areas in the Middle Atlantic Bight. Copeia 1990: 1180–1194.

———. 1991. Seasonality of fish catch rates and species composition from the hard bottom trap fishery in the Middle Atlantic Bight (U.S. east coast). Fish. Res. 12(1991): 1–22.

Elliott, J. M. 1989. The critical-period concept for juvenile survival and its relevance for population regulation in young sea trout, *Salmo trutta.* J. Fish. Biol. 35A: 91–98.

Elliott, E. M., D. Jimenez, C. O. Anderson, Jr., and D. J. Brown. 1979. Ichthyoplankton abundance and distribution in Beverly-Salem Harbor, March 1975 through February 1977. Massachusetts Division of Marine Fisheries, Boston.

Ennis, G. P. 1969. Occurrences of the little sculpin, *Myoxocephalus aenaeus,* in Newfoundland waters. J. Fish. Res. Board Can. 27: 1689–1694.

Epifanio, C. E. 1995. Transport of blue crab (*Callinectes sapidus*) larvae in the waters off Mid-Atlantic states. Bull. Mar. Sci. 57(3): 713–725.

Epifanio, C. E., A. K. Masse, and R. W. Garvine. 1989. Transport of blue crab larvae by surface currents off Delaware Bay, USA. Mar. Ecol. Progr. Ser. 54: 35–41.

Epperly, S. P. 1984. Fishes of the Pamlico-Albemarle Peninsula, N.C. Area utilization and potential impacts. Spec. Sci. Rep. No. 42, CEIP Report No. 23, North Carolina Department of Natural Resources and Community Development, Division of Marine Fishes, Morehead City, N.C.

Epperly, S. P., and S. W. Ross. 1986. Characterization of the North Carolina Pamlico-Albemarle estuarine complex. Natl. Ocean. Atmos. Adm. Tech. Memo. NMFS-SEFC-175: 1–55.

Eschmeyer, W. N. 1990. Catalog of the genera of recent fishes. California Academy of Sciences, San Francisco.

Fahay, M. P. 1975. An annotated list of larval and juvenile fishes captured with surface-towed meter net in the South Atlantic Bight during four RV *Dolphin* cruises between May 1967 and February 1968. Natl. Ocean. Atmos. Adm. Tech. Rep. Natl. Mar. Fish. Serv. SSRF-685: 1–39.

———. 1983. Guide to the early stages of marine fishes occurring in the western North Atlantic Ocean, Cape Hatteras to the southern Scotian Shelf. J. Northwest. Atl. Fish. Sci. 4: 1–423.

———. 1987. Larval and juvenile hakes (*Phycis–Urophycis* sp.) examined during a study of the white hake, *Urophycis tenuis* (Mitchill), in the Georges Bank–Gulf of Maine area. Dep. Comm. Natl. Ocean. Atmos. Adm. NMFS Northeast Fish. Cent. Sandy Hook Lab. Rep. No. 87–03.

———. 1992. Development and distribution of cusk eel eggs and larvae in the Middle Atlantic Bight with a description of *Ophidion robinsi* n. sp. (Teleostei: Ophidiidae). Copeia 1992(3): 799–819.

———. 1993. The early life history stages of New Jersey's saltwater fishes: sources of information. Bull. N.J. Acad. Sci. 38(2): 1–16.

Fahay, M. P., and K. W. Able. 1989. The white hake, *Urophycis tenuis* in the Gulf of Maine: spawning seasonality, habitat use and growth in young-of-the-year, and relationships to the Scotian Shelf population. Can. J. Zool. 67: 1715–1724.

Fahay, M. P., and D. F. Markle. 1984. Gadiformes: development and relationships. Pp. 265–283 *in* H. G. Moser, W. J. Richards, D. M. Cohen, M. P. Fahay, A. W. Kendall, Jr., and S. L. Richardson, eds. *Ontogeny and systematics of fishes.* Am. Soc. Ichthyol. Herpetol. Spec. Publ. No. 1. Allen Press, Lawrence, Kans.

Fahy, W. E. 1978. The influence of crowding upon the total number of vertebrae developing in *Fundulus majalis* (Walbaum). J. Cons. Int. Explor. Mer 38(2): 252–256.

———. 1979. The influence of temperature change on number of anal fin rays developing in *Fundulus majalis.* J. Cons. Int. Explor. Mer 38(3): 280–285.

———. 1980. The influence of temperature change on number of dorsal fin rays developing in *Fundulus majalis* (Walbaum). J. Cons. Int. Explor. Mer 39(1): 104–109.

———. 1982. The influence of temperature change

on number of pectoral fin rays developing in *Fundulus majalis* (Walbaum). J. Cons. Int. Explor. Mer 40(1): 21–26.

Fay, C. W., R. J. Neves, and G. B. Pardue. 1983. Striped bass. Species profiles: life histories and environmental requirements of coastal fishes and invertebrates (Mid-Atlantic). National Coastal Ecosystem Team, U.S. Fish and Wildlife Service, Washington, D.C.

Ferraro, S. P. 1980. Daily time of spawning of 12 fishes in the Peconic Bays, New York. U.S. Fish. Bull. 78(2): 455–464.

Fine, M. L. 1978. Seasonal and geographical variation of the mating call of the oyster toadfish *Opsanus tau* L. Oecologia 36: 45–57.

Finkelstein, S. L. 1969a. Age and growth of scup in the waters of eastern Long Island Sound. N.Y. Fish Game J. 16(1): 84–110.

———. 1969b. Age at maturity of scup from New York waters. N.Y. Fish Game J. 16(2): 224–237.

Fish, M. P. 1954. The character and significance of sound production among fishes of the western North Atlantic. Bull. Bingham Oceanogr. Collect. Yale Univ. 4(4): 1–109.

Fitzgerald, J., and J. L. Chamberlin. 1981. Anticyclonic warm core Gulf Stream eddies off the northeastern United States during 1979. Ann. Biol. Copenh. 36: 44–51.

Flescher, D. D. 1980. Guide to some trawl-caught marine fishes from Maine to Cape Hatteras, North Carolina. NOAA (Natl. Ocean. Atmos. Adm.) Tech. Rep. NMFS Circ. 431: 1–34.

Fletcher, G. L. 1977. Circannual cycles of blood plasma freezing point and Na+ and Cl− concentrations in Newfoundland winter flounder (*Pseudopleuronectes americanus*): correlation with water temperature and photoperiod. Can. J. Zool. 55: 789–795.

Flores-Coto, C., and S. M. Warlen. 1993. Spawning time, growth, and recruitment of larval spot *Leiostomus xanthurus* into a North Carolina estuary. U.S. Fish. Bull. 91(1): 8–22.

Foerster, J. W., and S. L. Goodbred. 1978. Evidence for a resident alewife population in the northern Chesapeake Bay. Estuarine Coastal Mar. Sci. 7: 437–444.

Forrester, G. E. 1990. Factors influencing the juvenile demography of a coral reef fish. Ecology 71(5): 1666–1681.

Foster, N. R. 1967. Comparative studies on the biology of killifishes (Pisces: Cyprinodontidae). Unpubl. Ph.D. diss. Cornell University, Ithaca, N.Y.

———. 1974. Belonidae—needlefishes. Pp. 125–126 *in* A. J. Lippson and R. L. Moran, eds. Manual for identification of early developmental stages of fishes of the Potomac River estuary. Md. Power Plant Siting Prog. Misc. Rep. No. 13.

Fowler, H. W. 1906. The fishes of New Jersey. Annu. Rep. N.J. State Mus. 1905(II): 35–477.

———. 1945. A study of the fishes of the Southern Piedmont and coastal plain. Acad. Nat. Sci. Phila. Monogr. 7.

Friedland, K. D., D. W. Ahrenholz, and J. F. Guthrie. 1996. Formation and seasonal evolution of Atlantic menhaden juvenile nurseries in coastal estuaries. Estuaries 19(1): 105–114.

Friedland, K. D., and L. W. Haas. 1988. Emigration of juvenile Atlantic menhaden, *Brevoortia tyrannus* (Pisces: Clupeidae), from the York River Estuary. Estuaries 11: 45–50.

Frisbie, C. M. 1961. Young black drum, *Pogonias cromis,* in tidal fresh and brackish waters, especially in the Chesapeake and Delaware Bay areas. Chesapeake Sci. 2(1–2): 94–100.

Fritz, E. S., W. H. Meredith, and V. A. Lotrich. 1975. Fall and winter movements and activity level of the mummichog, *Fundulus heteroclitus,* in a tidal creek. Chesapeake Sci. 16: 211–215.

Fritz, R. L. 1965. Autumn distribution of groundfish species in the Gulf of Maine and adjacent waters, 1955–1961. Am. Geol. Soc. Ser. Atlas Mar. Environ. Folio 10: 1–48.

Fritzsche, R. A. 1978. *Development of fishes of the Mid-Atlantic Bight: An atlas of egg, larval and juvenile Stages. Vol. V: Chaetodontidae through Ophidiidae.* Prepared for U.S. Fish and Wildlife Service. Chesapeake Biological Laboratory, Center for Environmental and Estuarine Studies, University of Maryland.

Fuiman, L. A., and D. M. Higgs. 1997. Ontogeny, growth and the recruitment process. Pp. 225–250 *in* R. C. Chambers and E. A. Trippel, eds. *Early life history and recruitment in fish populations.* Chapman and Hall, London.

Funderburk, S. L., S. J. Jordan, J. A. Mihursky, and D. Riley. 1991. Habitat requirements for Chesapeake Bay living resources. Living Research Subcommittee, Chesapeake Bay Program, Solomons, Md.

Garlo, E. V., C. B. Milstein, and A. E. Jahn. 1979. Impact of hypoxic conditions in the vicinity of Little Egg Inlet, New Jersey in summer, 1976. Estuarine Coastal Mar. Sci. 8: 421–432.

Gauldie, R. W. 1996. Ages estimated from average microincrement width in otoliths of *Macruronus novaezelandiae* are verified by length mode progressions. Bull. Mar. Sci. 59(3): 498–507.

Gibbs, R. H., Jr. 1959. A synopsis of the post larva

of western Atlantic lizard-fishes (Synodontidae). Copeia 1959:232–236.

Gilbert, C. R. 1986. Species profiles: life histories and environmental requirements of coastal fishes and invertebrates (South Florida) southern, Gulf, and summer flounders. U.S. Fish Wildl. Serv. Biol. Rep. 82(11,54).

Gilhen, J. 1972. The white mullet, *Mugil curema,* added to and the striped mullet, *M. cephalus,* deleted from the Canadian Atlantic fish fauna. Can. Field-Nat. 86:74–77.

Gill, T. 1904. State ichthyology of Massachusetts. Science N.S. 20(506):321–338.

Gilmore, R. G., L. H. Bullock, and F. H. Berry. 1978. Hypothermal mortality in marine fishes, south-central Florida, January, 1977. Northeast Gulf Sci. 2(2):77–97.

Ginsburg, I. 1933. A revision of the genus *Gobiosoma* (Family Gobiidae). Bull. Bingham Oceanogr. Collect. Yale Univ. 4(5):1–59.

Gleason, T. R., and D. A. Bengston. 1996. Growth, survival and size-selective predation mortality of larval and juvenile island silversides, *Menidia beryllina* (Pisces: Atherinidae). J. Exp. Mar. Biol. Ecol. 199:165–177.

Glenn, S. M., M. F. Crowley, D. B. Haidvogel, and Y. T. Song. 1996. Underwater observatory captures coastal upwelling events off New Jersey. Eos Trans. Am. Geophys. Union 77(25): 233–236.

Gonzalez-Villasenor, L. I., and D. A. Powers. 1990. Mitochondrial DNA restriction site polymorphisms in the teleost *Fundulus heteroclitus* support secondary intergradation. Evolution 44(1): 27–37.

Good, R. E., and N. F. Good. 1984. The Pinelands National Reserve: an ecosystem approach to management. Bioscience 34(3):169–173.

Goode, G. B., and T. H. Bean. 1896. Oceanic ichthyology: Deep-sea and pelagic fishes of the world. Spec. Bull. U.S. Natl. Mus. 2:1–553.

Gooley, B. R., and F. H. Lesser. 1977. The history of the use of *Gambusia affinis* (Baird and Girard) in New Jersey. Proc. Annu. Meet. N.J. Mosq. Control Assoc. 64:154–159.

Goshorn, D. M., and C. E. Epifanio. 1991. Diet of larval weakfish and prey abundance in Delaware Bay. Trans. Am. Fish. Soc. 120:684–692.

Gosline, W. A. 1948. Speciation in the fishes of the genus *Menidia.* Evolution 2(4):306–313.

Govoni, J. J. 1987. The ontogeny of dentition in *Leiostomus xanthurus.* Copeia 1987:1041–1046.

———. 1993. Flux of larval fishes across frontal boundaries: examples from the Mississippi River plume front and the western Gulf Stream front in winter. Bull. Mar. Sci. 53(2):538–566.

Grabe, S. A. 1978. Food and feeding habits of juvenile Atlantic tomcod, *Microgadus tomcod,* from Haverstraw Bay, Hudson River. U.S. Fish. Bull. 76(1):89–94.

Graham, J. J. 1956. Observations on the alewife in freshwater. Univ. Toronto Biol. Ser. No. 62.

Graham, J. J., B. J. Joule, and C. L. Crosby. 1984. Characteristics of the Atlantic herring (*Clupea harengus* L.) spawning population along the Maine coast, inferred from larval studies. J. Northwest. Atl. Fish. Sci. 5(2):131–142.

Gray, G. A., and H. E. Winn. 1961. Reproductive ecology and sound production of the toadfish, *Opsanus tau.* Ecology 42(2):274–282.

Greeley, J. R. 1938. Section II. Fishes and habitat conditions of the shore zone based upon July and August seining investigations. Pp. 72–91 *in* A Biological Survey of the Salt Waters of Long Island, 1938. Part II. N.Y. State Conserv. Dep. Suppl. 28th Annu. Rep.

Greenfield, D. W. 1968. Observations on the behavior of the basketweave cusk-eel *Otophidium scrippsi* Hubbs. Cal. Fish Game 54:108–114.

Griffith, R. W. 1974. Environment and salinity tolerance in the genus *Fundulus.* Copeia 1974(2): 319–329.

Grimes, C. B., K. W. Able, and R. S. Jones. 1986. Tilefish, *Lopholatilus chamaeleonticeps,* habitat, behavior and community structure in Mid-Atlantic and southern New England waters. Environ. Biol. Fishes 15(4):273–292.

Griswold, C. A., and T. W. McKenney. 1984. Larval development of the scup, *Stenotomus chrysops* (Pisces: Sparidae). U.S. Fish. Bull. 82(1):77–84.

Grosslein, M. D., and T. R. Azarovitz, eds. 1982. *Fish distribution.* (Mar. Ecosyst. Anal.) N.Y. Bight Atlas Monogr. 15.

Gudger, E. W. 1910. Habits and life history of the toadfish (*Opsanus tau*). Bull. U.S. Bur. Fish. 28: 1071–1109.

———. 1927. The nest and nesting habits of the butterfish or gunnel, *Pholis gunnellus.* Am. Mus. Nat. Hist. 27:65–72.

Gulland, J. A. 1965. Survival of the youngest stages of fish and its relation to year-class strength. Spec. Publ. Int. Comm. Northwest Atl. Fish. 6: 365–371.

Gunter, G. 1941. Relative numbers of shallow water fishes of the northern Gulf of Mexico, with some records of rare fishes from the Texas coast. Am. Midl. Nat. 26(1):194–200.

———. 1945. Studies on marine fishes of Texas. Publ. Inst. Mar. Sci. Univ. Tex. 1:1–190.

———. 1950. Distribution and abundance of fishes on the Aransas National Wildlife Refuge, with life history notes. Publ. Inst. Mar. Sci. Univ. Tex. 1(2): 89–101.

———. 1961. Some relations of estuarine organisms to salinity. Limnol. Oceanogr. 6: 182–190.

———. 1967. Some relationships of estuaries to the fisheries of the Gulf of Mexico. Am. Assoc. Adv. Sci., Publ. 83: 621–638.

Gutherz, E. J. 1967. Field guide to the flatfishes of the family Bothidae in the Western North Atlantic. U.S. Fish Wildl. Serv. Bur. Commer. Fish. Circ. 263.

Haas, R. E., and C. W. Recksiek. 1995. Age verification of winter flounder in Narragansett Bay. Trans. Am. Fish. Soc. 124: 103–111.

Haedrich, R. L. 1983. Estuarine fishes. Pp. 183–207 *in* B. H. Ketchum, ed. *Ecosystems of the world. Vol. 26. Estuaries and enclosed seas.* Elsevier, New York.

Haedrich, R. L., and C. A. S. Hall. 1976. Fishes and estuaries. Oceanus 19(5): 55–63.

Haedrich, R. L., G. T. Rowe, and P. T. Polloni. 1980. The megabenthic fauna in the deep sea south of New England, USA. Mar. Biol. 57: 165–179.

Hales, L. S., Jr., and K. W. Able. 1995. Effects of oxygen concentration on somatic and otolith growth rates of juvenile black sea bass, *Centropristis striata.* Pp. 135–153 *in* D. H. Secor et al., eds. *Recent developments in fish otolith research.* Belle W. Baruch Library in Marine Science No. 19. University of South Carolina Press, Columbia.

Hales, L. S., Jr., and K. W. Able. In press. Overwinter mortality, growth and behavior of young-of-the-year of four coastal marine fishes in New Jersey (USA) waters. Mar. Biol.

Hamer, P. E. 1959. Age and growth of the bluefish. Master's thesis. Rutgers University, New Brunswick, N.J.

———. 1970. Studies of the scup, *Stenotomus chrysops,* in the Middle Atlantic Bight. N.J. Div. Fish Game Shellfish Misc. Rep. No. 5M.

———. 1972. Phase III. Use studies. Pp. 115–141 *in* Studies of the Mullica River-Great Bay Estuary. N.J. Dept. Environ. Protect. Div. Fish Game Shellfish Bur. Fish. Misc. Rep. No. 6M-1.

Hardy, C. C., E. R. Baylor, and P. Moskowitz. 1976. Sea surface circulation in the northwest apex of the New York Bight—with appendix: Bottom drift over the continental shelf. (Natl. Ocean. Atmos. Adm.) Tech. Memo. (Mar. Ecosyst. Anal.)-13.

Hardy, J. D., Jr. 1978a. Development of fishes of the Mid-Atlantic Bight: An atlas of egg, larval and juvenile stages. Vol. II. Anquillidae through Syngnathidae. Biological Services Program, U.S. Fish and Wildlife Service FWS/OBS-789/12.455. U.S. Government Printing Office, Washington.

———. 1978b. Development of fishes of the Mid-Atlantic Bight: An atlas of egg, larval and juvenile stages. Vol. III. Aphredoderidae through Rachycentridae. Biological Services Program, U.S. Fish and Wildlife Service FWS/OBS-78/12. U.S. Government Printing Office, Washington.

Hardy, J. D., Jr., and L. L. Hudson. 1975. Descriptions of the eggs and juveniles of the Atlantic tomcod, *Microgadus tomcod.* Univ. Md. Nat. Resour. Inst. Chesapeake Biol. Lab. Ref. 75–11.

Hare, J. A., and R. K. Cowen. 1993. Ecological and evolutionary implications of the larval transport and reproductive strategy of bluefish (*Pomatomus saltatrix*). Mar. Ecol. Progr. Ser. 98: 1–16.

———. 1994. Ontogeny and otolith microstructure of bluefish (*Pomatomus saltatrix*) (Pisces: Pomatomidae). Mar. Biol. 118: 541–550.

———. 1996. Transport mechanisms of larval and pelagic juvenile bluefish (*Pomatomus saltatrix*) from South Atlantic Bight spawning grounds to Middle Atlantic Bight nursery habitats. Limnol. Oceanogr. 41: 1264–1280.

Hare, J. A., M. P. Fahay, and R. K. Cowen. In prep. Springtime ichthyoplankton of the Slope Sea: larval assemblages, relationship to hydrography and implications for larval transport.

Harmic, J. L. 1958. Some aspects of the development and ecology of the pelagic phase of the gray squeteague, *Cynoscion regalis* (Bloch and Schneider), in the Delaware estuary. Ph.D. diss. University of Delaware, Newark.

Haro, A. J., and W. H. Krueger. 1988. Pigmentation, size, and migration of elvers (*Anguilla rostrata* [Lesueur]) in a coastal Rhode Island stream. Can. J. Zool. 66: 2528–2533.

Hastings, R. W., and R. E. Good. 1977. Population analysis of the fishes of a freshwater tidal tributary of the lower Delaware River. Bull. N.J. Acad. Sci. 22(2): 13–20.

Hauser, W. J. 1975. Occurrence of two leptocephali in an estuary. U.S. Fish. Bull. 77: 444–445.

Havey, K. A. 1973. Production of juvenile alewives at Love Lake, Washington County, Maine. Trans. Am. Fish. Soc. 102(2): 434–437.

Heck, K. L., Jr., K. W. Able, M. P. Fahay, and C. T. Roman. 1989. Fishes and decapod crustaceans of Cape Cod eelgrass meadows: Species composition and seasonal abundance patterns. Estuaries 12: 59–65.

Heintzelman, D. S., ed. 1971. Rare or endangered fish and wildlife of New Jersey. N.J. State Mus. Sci. Notes 4.

Hempel, G. 1965. On the importance of larval survival for the population dynamics of marine fish food. Calif. Coop. Ocean. Fish. Invest. Rep. 10: 13–23.

Herke, W. H. 1977. Life history concepts of motile estuarine-dependent species should be reevaluated. Self-published, Baton Rouge, La. Library of Congress No. 77-90015.

Herman, S. S. 1963. Planktonic fish eggs and larvae of Narragansett Bay. Limnol. Oceanogr. 8: 103–109.

Hettler, W. F. 1984. Description of eggs, larvae and early juveniles of gulf menhaden, *Brevoortia patronus,* and comparisons with Atlantic menhaden, *B. tyrannus,* and yellowfin menhaden, *B. smithi.* U.S. Fish. Bull. 82: 85–95.

———. 1989. Nekton use of regularly flooded saltmarsh cordgrass habitat in North Carolina, USA. Mar. Ecol. Prog. Ser. 56: 111–118.

Hettler, W. F., Jr., and D. L. Barker. 1993. Distribution and abundance of larval fishes at two North Carolina inlets. Estuarine Coastal Shelf Sci. 37: 161–179.

Hickey, C. R., Jr., A. D. Sosnow, and J. W. Lester. 1975. Pound net catches of warm-water fishes at Montauk, New York. N.Y. Fish Game J. 22(1): 38–50.

Hicks, D. C., and J. R. Miller. 1980. Meteorological forcing and bottom water movement off the northern New Jersey coast. Estuarine Coastal Mar. Sci. II: 563–571.

Higgins, E. 1928. Progress report in biological inquiries, 1926. U.S. Comm. Fish. Rep. 1927 8: 515–559.

Higham, J. R., and W. R. Nicholson. 1964. Sexual maturation and spawning of Atlantic menhaden. U.S. Fish. Bull. 63(2): 255–271.

Hildebrand, S. F. 1919. Notes on the life history of the minnows, *Gambusia affinis* and *Cyprinodon variegatus.* U.S. Bur. Fish. Rep. 34(1914): 409–429.

———. 1943. A review of the American anchovies (Family Engraulidae). Bull. Bingham Oceanogr. Collect. Yale Univ. 8(2): 1–165.

———. 1963. Part 3. Family Clupeidae. Pp. 257–454 *in* Y. H. Olsen, ed. *Fishes of the western North Atlantic.* Sears Foundation for Marine Research, Yale University, New Haven, Conn.

Hildebrand, S. F., and L. E. Cable. 1930. Development and life history of fourteen teleostean fishes at Beaufort, N.C. U.S. Bur. Fish. Bull. 46(1093): 383–488.

———. 1934. Reproduction and development of whitings or kingfishes, drums, spot, croaker, and weakfishes of sea trouts family Sciaenidae of the Atlantic coast of the United States. U.S. Bur. Fish. Bull. 48: 41–117.

———. 1938. Further notes on the development and life history of some teleosts at Beaufort, N.C. U.S. Bur. Fish. Bull. 48: 505–642.

Hildebrand, S. F., and W. C. Schroeder. 1928. Fishes of Chesapeake Bay. U.S. Fish Wildl. Serv. Fish. Bull. 53 Pt. 1, IV, Bur. Fish. Doc. 1024.

Himchak, P. J. 1981. Monitoring of the striped bass population in New Jersey: Spawning and recruitment of the striped bass, *Morone saxatilis,* in the Delaware River. N.J. Dep. Environ. Protect. Div. Fish Game Wildl. Bur. Mar. Fish. Final Rep. April 1, 1980 to March 31, 1981.

———. 1982a. Monitoring of the spring ichthyoplankton in the Delaware River. N.J. Dep. Environ. Protect. Div. Fish Game Wildl. Bur. Mar. Fish. Nacote Creek Res. Stn. Misc. Rep. No. 55M: 1–13.

———. 1982b. Distribution and abundance of larval and young finfishes in the Maurice River and in waterways near Atlantic City, New Jersey. Master's thesis. Rutgers University, New Brunswick, N.J.

———. 1983. Spring ichthyoplankton in the Raritan/South and the Great Egg Harbor Rivers, New Jersey. N.J. Dep. Environ. Protect. Div. Fish Game Wildl. Bur. Mar. Fish. Tech. Ser. 83–2: 1–17.

———. 1984. Spring ichthyoplankton in the Passaic River, New Jersey. N.J. Dep. Environ. Protect. Div. Fish Game Wildl. Bur. Mar. Fish. Tech. Ser. 84–1.

Himchak, P. J., and R. Allen. 1985. Spring ichthyoplankton in the Navesink/swimming River, New Jersey. N.J. Dep. Environ. Protect. Div. Fish Game Wildl. Bur. Mar. Fish. Tech. Ser. 85–1: 1–11.

Hines, A. H., K. E. Osgood, and J. J. Miklas. 1985. Semilunar reproductive cycles in *Fundulus heteroclitus* (Pisces: Cyprinodontidae) in an area without lunar tidal cycles. U.S. Fish. Bull. 83(3): 467–472.

Hiyama, Y. 1953. Thermotaxis of eel fry in stage of ascending river mouth. Jpn. J. Ichthyol. 2: 23–30.

Hjort, J. 1914. Fluctuations in the great fisheries of northern Europe. Cons. Perm. Int. Explor. Mer Rapp. P.-V. Vol. 20.

———. 1926. Fluctuations in the year classes of important food fishes. J. Cons. Perm. Int. Explor. Mer 1: 5–38.

Hoese, H. D. 1965. Spawning of marine fishes in the Port Aransas, Texas area as determined by the distribution of young and larvae. Ph.D. diss. University of Texas, Austin.

Hoese, H. D., and R. H. Moore. 1977. Fishes of the Gulf of Mexico–Texas, Louisiana, and adjacent waters. Texas A&M University Press, College Station.

Hoff, H. K. 1976. The life history of the striped bass, *Morone saxatilis* (Walbaum), in the Great Bay–Mullica River estuary and the vicinity of Little Egg Inlet. Pp. 43–53 *in* D. L. Thomas et al., eds. *Ecological studies in the bays and waterways near Little Egg Inlet and in the ocean in the vicinity of the proposed site for the Atlantic Generating Station, New Jersey.* Progress Report for the period January–December 1975. Ichthyological Associates, Ithaca, N.Y.

Hoff, J. G. 1971. Mass mortality of the crevalle jack, *Caranx hippos* (Linnaeus) on the Atlantic coast of Massachusetts. Chesapeake Sci. 12:49.

Hoff, J. G., and R. M. Ibara. 1977. Factors affecting the seasonal abundance, composition and diversity of fishes in a southeastern New England estuary. Estuarine Coastal Mar. Sci. 5:665–678.

Holliday, F. G. T. 1971. Salinity: Animas-Fishes. Pp. 997–1033 *in* O. Kinne, ed. *Marine ecology.* Vol. 1, Part 2. Wiley-Interscience, London.

Hollis, E. H. 1967. An investigation of striped bass in Maryland. Final Rep. Fed. Aid Fish. Rest. Proj. F-3-R.

Holt, G. J., A. G. Johnson, C. R. Arnold, W. A. Fable, Jr., and T. D. Williams. 1981. Description of eggs and larvae of laboratory reared red drum, *Sciaenops ocellata.* Copeia 1981:751–756.

Hood, P. B., K. W. Able, and C. B. Grimes. 1988. Biology of the conger eel *Conger oceanicus* in the Mid-Atlantic Bight. I. Distribution, age, growth and reproduction. Mar. Biol. 98: 587–596.

Horn, M. H. 1970a. Systematics and biology of the stromateoid fishes of the genus [*Peprilus*]. Bull. Mus. Comp. Zool. Harv. Univ. 149(5):165–261.

———. 1970b. The swimbladder as a juvenile organ in stromateoid fishes. Breviora 359.

Horn, M. H., and L. G. Allen. 1976. Numbers of species and faunal resemblance of marine fishes in California bays and estuaries. Bull. South. Calif. Acad. Sci. 75(2):159–170.

Horseman, L. O., and C. A. Shirey. 1974. Age, growth and distribution of white perch in the Chesapeake and Delaware Canal. Vol. I, Part B *in* Ecological studies in the vicinity of the proposed Summit Power Station, January through December 1973.

Hoss, D. E., and G. W. Thayer. 1993. The importance of habitat to the early life history of estuarine dependent fishes. Am. Fish. Soc. Symp. 14: 147–158.

Hostetter, E. B., and T. A. Munroe. 1993. Age, growth and reproduction of tautog *Tautoga onitis* (Labridae: Perciformes) from coastal waters of Virginia. U.S. Fish. Bull. 91:45–64.

Houde, E. D. 1972. Development and early life history of the northern sennet, *Sphyraena borealis* Dekay (Pisces: Sphyraenidae) reared in the laboratory. U.S. Fish. Bull. 70:185–196.

———. 1987. Fish early life dynamics and recruitment variability. Am. Fish. Soc. Symp. 2:17–29.

Houde, E. D., S. A. Berkeley, J. J. Klinovsky, and R. C. Schekter. 1976. Culture of larvae of the white mullet (*Mugil curema*). Aquaculture 8: 365–370.

Houde, E. D., J. C. Gamble, S. E. Dorsey, and J. H. Cowan, Jr. 1994. Drifting mesocosms: the influence of gelatinous zooplankton on mortality of bay anchovy, *Anchoa mitchilli,* eggs and yolk-sac larvae. Int. Counc. Explor. Sea J. Mar. Sci. 51: 383–394.

Houghton, R. W., F. Aikman III, and H. W. Ou. 1988. Shelf-slope frontal structure and cross-shelf exchange at the New England shelf-break. Continental Shelf Res. 8(5–7):687–710.

Howe, A. B. 1971. Biological investigation of the Atlantic tomcod, *Microgadus tomcod* (Walbaum), in the Weweantic River estuary, Massachusetts, 1967. Master's thesis. University of Massachusetts, Amherst.

Howe, A. B., P. G. Coates, and D. E. Pierce. 1976. Winter flounder estuarine year-class abundance, mortality and recruitment. Trans. Am. Fish. Soc. 105(6):647–657.

Huber, M. E. 1978. Adult spawning success and emigration of juvenile alewives (*Alosa pseudoharengus*) from the Parker River, Massachusetts. Master's thesis. University of Massachusetts, Amherst.

Hubbs, C. L. 1963. *Chaetodon aya* and related deep-living butterflyfishes: their variation, distribution and synonymy. Bull. Mar. Sci. 13(1):133–192.

Hubbs, C. L., and R. R. Miller. 1965. Studies on cyprinodont fishes. XXII. Variation in *Lucania parva,* its establishment in western United States, and description of a new species from an interior basin in Coahuila, Mexico. Misc. Publ. Mus. Zool. Univ. Mich. 127:1–104.

Hunt von Herbing, I., and W. Hunte. 1991. Spawning and recruitment of the bluehead wrasse *Thalassoma bifasciatum* in Barbados. Mar. Ecol. Prog. Ser. 72:49–58.

Huntsman, A. G. 1922. The fishes of the Bay of Fundy. Contrib. Can. Biol. 1921(3):49–72.

Iles, T. D., and M. Sinclair. 1982. Atlantic herring: stock discreteness and abundance. Science 215: 627–633.

Ingham, M., ed. 1982. Summary of the physical oceanographic processes and features pertinent to pollution distribution in the coastal and offshore waters of the northeastern United States, Virginia to Maine. (Natl. Ocean. Atmos. Adm.) Tech. Memo Natl. Mar. Fish. Serv. F/Northeast Cent. 17:1–165.

Jackson, C. F. 1953. Northward occurrence of southern fishes (*Fundulus, Mugil, Pomatomus*) in coastal waters of New Hampshire. Copeia 1953:192.

Jacot, A. P. 1920. Age, growth and scale characters of the mullets, *Mugil cephalus* and *Mugil curema.* Trans. Am. Microsc. Soc. 34(3):199–229.

Jeffries, H. P. 1960. Winter occurrences of *Anguilla rostrata* elvers in New England and Middle Atlantic estuaries. Limnol. Oceanogr. 5(3): 338–340.

Jeffries, H. P., and W. C. Johnson. 1974. Seasonal distributions of bottom fishes in the Narragansett Bay area: Seven-year variations in the abundance of winter flounder (*Pseudopleuronectes americanus*). J. Fish. Res. Board Can. 31:1057–1066.

Johns, D. M., W. H. Howell, and G. Klein-MacPhee. 1981. Yolk utilization and growth to yolk-sac absorption in summer flounder (*Paralichthys dentatus*) larvae at constant and cyclic temperatures. Mar. Biol. 63:301–308.

Johnson, G. D. 1978. *Development of fishes of the Mid-Atlantic Bight: An atlas of egg, larval and juvenile stages. Vol. IV: Carangidae through Ephippidae.* Chesapeake Biological Laboratory, Center for Environmental and Estuarine Studies, University of Maryland. Prepared for U.S. Fish and Wildlife Service. U.S. Government Printing Office, Washington.

Johnson, M. S. 1974. Comparative geographic variation in *Menidia.* Evolution 28:607–618.

Jones, G. P. 1987a. Competitive interactions among adults and juveniles in a coral reef fish. Ecology 68:1534–1547.

———. 1987b. Some interactions between residents and recruits in two coral reef fishes. J. Exp. Mar. Biol. Ecol. 114:169–182.

———. 1990. The importance of recruitment to the dynamics of a coral reef fish population. Ecology 71:1691–1698.

———. 1991. Postrecruitment processes in the ecology of coral reef fish populations: a multifactorial perspective. Pp. 294–328 *in* P. F. Sale, ed. *The ecology of fishes on coral reefs.* Academic Press, San Diego, Calif.

Jones, J. W., and H. B. N. Hynes. 1950. The age and growth of *Gasterosteus aculeatus, Pygosteus pungitius* and *Spinachia vulgaris,* as shown by their otoliths. J. Anim. Ecol. 19:59–73.

Jones, P. W., F. D. Martin, and J. D. Hardy, Jr. 1978. *Development of fishes of the Mid-Atlantic Bight: An atlas of egg, larval and juvenile stages. Vol. I: Acipenseridae through Ictaluridae.* Chesapeake Biological Laboratory, Center for Environmental and Estuarine Studies, University of Maryland. Prepared for U.S. Fish and Wildlife Service. U.S. Government Printing Office, Washington.

Joseph, E. B. 1972. Status of sciaenid stocks of the middle Atlantic coast. Chesapeake Sci. 13(2): 87–100.

Joseph, E. B., and J. Davis. 1965. A preliminary assessment of the river herring stocks of lower Chesapeake Bay. Va. Inst. Mar. Sci. Spec. Sci. Rep. No. 51.

Joseph, E. B., W. H. Massmann, and J. J. Norcross. 1964. The pelagic eggs and larval stages of the black drum from Chesapeake Bay. Copeia 1964: 425–434.

Juanes, F., R. E. Marks, K. A. McKown, and D. O. Conover. 1993. Predation by age-0 bluefish on age-0 anadromous fishes in the Hudson River estuary. Trans. Am. Fish. Soc. 122:348–356.

Juanes, F., J. A. Buckel, and D. O. Conover. 1994. Accelerating the onset of piscivory: intersection of predator and prey phenologies. J. Fish. Biol. 45(Suppl. A):41–54.

Juanes, F., J. A. Hare, and A. G. Miskiewicz. 1996. Comparing early life history strategies of *Pomatomus saltatrix:* a global approach. Mar. Freshw. Res. 47:365–379.

June, F. C., and J. L. Chamberlain. 1959. The role of the estuary in the life history and biology of the Atlantic menhaden. Proc. Gulf Caribb. Fish. Inst. 11(1958):41–45, 55–57.

Kahnle, A. 1987. Bottom trawl survey of juvenile fishes in the Hudson River estuary. Pp. 85–123 *in* Hudson River Fisheries Research Unit, 1986 Annual Report. Bureau of Fisheries, Division of Fish and Wildlife, N.Y. State Department of Environmental Conservation, New Paltz, N.Y.

Kahnle, A., and K. Hattala. 1988. Bottom trawl survey of juvenile fishes in the Hudson River Estuary. Summary Report for 1981–1986. Hudson River Fisheries Research Unit, Bureau of Fisheries, Region 3, N.Y. State Department of Environmental Conservation, New Paltz, N.Y.

Kaufman, L., J. Ebersole, J. Beets, and C. C. McIvor. 1992. A key phase in the recruitment dynamics of coral reef fishes: post-settlement transition. Environ. Biol. Fishes 34 : 109–118.

Kawahara, S. 1978. Age and growth of butterfish, *Peprilus triacanthus* (Peck), in ICNAF Subarea 5 and Statistical Area 6. Int. Comm. Northwest Atl. Fish. Select. Pap. 3 : 73–78.

Keefe, M., and K. W. Able. 1992. Habitat quality in New Jersey estuaries: habitat-specific growth rates of juvenile summer flounder in vegetated habitats. Final Rep. N.J. Dep. Environ. Protect. Energy. New Brunswick, N.J.

———. 1993. Patterns of metamorphosis in summer flounder, *Paralichthys dentatus.* J. Fish. Biol. 42 : 713–728.

———. 1994. Contributions of abiotic and biotic factors to settlement in summer flounder, *Paralichthys dentatus*. Copeia 1994(2) : 458–465.

Keirans, W. J., Jr. 1977. An immunochemically assisted ichthyoplankton survey with elaboration of species specific antigens of fish egg vitellins: southern New Jersey barrier island-lagoon complex. Ph.D. diss. Lehigh University, Bethlehem, Pa.

Keirans, W. J., Jr., S. S. Herman, and R. G. Marlsberger. 1986. Differentiation of *Prionotus carolinus* and *Prionotus evolans* eggs in Hereford Inlet Estuary, southern New Jersey, using immunodiffusion. U.S. Fish. Bull. 84(1) : 63–68.

Keller, G. H., D. Lamberg, G. Rowe, and N. Staresinic. 1973. Bottom currents in the Hudson Canyon. Science 180 : 191–193.

Kellogg, R. L. 1982. Temperature requirements for the survival and early development of the anadromous alewife. Prog. Fish-Cult. 44 : 63–73.

Kendall, A. W., Jr. 1972. Descriptions of black sea bass, *Centropristis striata* (Linnaeus), larvae and their occurrence north of Cape Lookout, North Carolina, in 1966. U.S. Fish. Bull. 70 : 1243–1260.

Kendall, A. W., Jr. 1979. Morphological comparisons of North American sea bass larvae (Pisces: Serranidae). (Natl. Ocean. Atmos. Adm.) Tech. Rep. NMFS Circ. 428.

Kendall, A. W., Jr., and L. P. Mercer. 1982. Black sea bass, *Centropristis striata.* MESA (Mar. Ecosyst. Anal.) New York Bight Atlas Monogr. 15 : 82–83.

Kendall, A. W., Jr., and N. A. Naplin. 1981. Diel-depth distribution of summer ichthyoplankton in the middle Atlantic bight. U.S. Fish. Bull. 79(4) : 705–726.

Kendall, A. W., Jr., and J. W. Reintjes. 1975. Geographic and hydrographic distribution of Atlantic menhaden eggs and larvae along the middle Atlantic coast from R/V *Dolphin* cruises, 1965–1966. U.S. Fish. Bull. 73 : 317–335.

Kendall, A. W., Jr., and L. A. Walford. 1979. Sources and distribution of bluefish, *Pomatomus saltatrix,* larvae and juveniles off the east coast of the United States. U.S. Fish. Bull. 77 : 213–227.

Kendall, W. C. 1914. The fishes of Maine. Pp. 1–198 *in* Proceedings of the Portland Society of Natural History, Vol. III. Portland, Maine.

Kennish, M. J. 1992. *Ecology of estuaries: Anthropogenic effects.* CRC Press, Boca Raton, Fla.

Kennish, M. J., and R. A. Lutz, eds. 1984. *Ecology of Barnegat Bay, New Jersey.* Springer, New York.

Kennish, M. J., and R. K. Olsson. 1975. Effects of thermal discharges on the microstructural growth of *Mercenaria mercenaria.* Environ. Geol. 1 : 41–63.

Kernehan, R. J., R. E. Smith, S. L. Tyler, and M. L. Brewster. 1977. Ichthyoplankton. Volume II in Ecological studies in the vicinity of the proposed Summit Power Station. January 1975 through December 1975. Ichthyological Associates, Summit, Del.

Ketchum, B. H., and N. Corwin. 1964. The persistence of "winter" water on the continental shelf south of Long Island, New York. Limnol. Oceanogr. 9(4) : 467–475.

Kilby, J. D. 1949. A preliminary report on the young striped mullet *Mugil cephalus* L. in two gulf coastal areas of Florida. Q. J. Fla. Acad. Sci. 11 : 7–23.

Kissil, G. W. 1969. Contribution to the life history of the alewife, (*Alosa pseudoharengus*) (Wilson), in Connecticut. Ph.D. diss. University of Connecticut, Storrs.

———. 1974. Spawning of the anadromous alewife in Bride Lake, Connecticut. Trans. Am. Fish. Soc. 103 : 312–317.

Klauda, R. J., J. B. McLaren, R. E. Schmidt, and W. P. Dey. 1988. Life history of white perch in the Hudson River Estuary. Am. Fish. Soc. Monogr. 4 : 69–88.

Kneib, R. T. 1984. Patterns in the utilization of the intertidal salt marsh by larvae and juveniles of *Fundulus heteroclitus* (Linnaeus) and *Fundulus luciae* (Baird). J. Exp. Mar. Biol. Ecol. 83 : 41–51.

———. 1986. Size-specific patterns in the reproductive cycle of the killifish, *Fundulus heteroclitus* (Pisces: Fundulidae) from Sapelo Island, Georgia. Copeia 1986 : 342–351.

Kneib, R. T., and A. E. Stiven. 1978. Growth, reproduction and feeding of *Fundulus heteroclitus* (L.)

in a North Carolina salt marsh. J. Exp. Mar. Biol. Ecol. 31: 121–140.

Kramer, S. H. 1991. Growth, mortality and movements of juvenile California halibut *Paralichthys californicus* in shallow coastal and bay habitats of San Diego County, California. U.S. Fish. Bull. 89: 195–207.

Krumholz, L. A. 1948. Reproduction in the western mosquito fish *Gambusia affinis affinis* (Baird and Girard) and its use in mosquito control. Ecol. Monogr. 18: 1–43.

Kuntz, A. 1914a. The embryology and larval development of *Bairdiella chrysura* and *Anchovia mitchilli.* Bur. Fish. Doc. No. 795. Bull. U.S. Bur. Fish. 33(1913).

———. 1914b. Notes on the habits, morphology of the reproductive organ and embryology of the viviparous fish *Gambusia affinis.* Bull. U.S. Bur. Fish. 33(1913): 181–190.

———. 1916. Notes on the embryology and larval development of five species of teleostean fishes. Bull. U.S. Bur. Fish. 34(1914): 407–429.

Kuntz, A., and L. Radcliffe. 1917. Notes on the embryology and larval development of twelve teleostean fishes. Bull. U.S. Bur. Fish. 35: 87–134.

Lake, T. 1983. Poor man's marlin—Atlantic needlefish. Conservationist 38(1): 32–36.

Lang, K. L., F. P. Almeida, G. R. Bolz, and M. P. Fahay. 1996. The use of otolith microstructure in resolving issues of first year growth and spawning seasonality of white hake, *Urophycis tenuis,* in the Gulf of Maine–Georges Bank region. U.S. Fish. Bull. 94(1): 170–175.

Lankford, T. E., Jr., and T. E. Targett. 1994. Suitability of estuarine nursery zones for juvenile weakfish (*Cynoscion regalis*): effects of temperature and salinity on feeding, growth and survival. Mar. Biol. 119: 611–620.

Laprise, R., and J. J. Dodson. 1989. Ontogenetic changes in the longitudinal distribution of two species of larval fish in a turbid well-mixed estuary. J. Fish. Biol. (1989) 35(Suppl. A): 39–47.

Laroche, J. L., and J. Davis. 1973. Age, growth, and reproduction of the northern puffer, *Sphoeroides maculatus.* U.S. Fish. Bull. 71(4): 955–963.

Laroche, W. A. 1981. Development of larval smooth flounder, *Liopsetta putnami,* with a redescription of development of winter flounder, *Pseudopleuronectes americanus* (family Pleuronectidae). U.S. Fish. Bull. 78(4): 897–909.

Laroche, W. A., W. F. Smith-Vaniz, and S. L. Richardson. 1984. Carangidae: development. Pp. 510–522 *in* H. G. Moser, W. J. Richards, D. M. Cohen, M. P. Fahay, A. W. Kendall, Jr., and S. L. Richardson, eds. *Ontogeny and systematics of fishes.* Am. Soc. Ichthyol. Herpetol. Spec. Publ. No. 1. Allen Press, Lawrence, Kans.

Lasker, R., and K. Sherman. 1981. The early life history of fish: recent studies. Second ICES Symposium, Woods Hole, Mass., 2–5 April 1979. Rapp. P.-V. Reun. 178: 1–607.

Lassiter, R. R. 1962. Life history aspects of the bluefish, *Pomatomus saltatrix* (Linnaeus), from the coast of North Carolina. Master's thesis, North Carolina State College, Raleigh.

Laurence, G. C. 1975. Laboratory growth and metabolism of the winter flounder *Pseudopleuronectes americanus* from hatching through metamorphosis at three temperatures. Mar. Biol. 32: 223–229.

Lavenda, N. 1949. Sexual differences and normal protogynous hermaphroditism in the Atlantic sea bass *Centropristis striata.* Copeia 1949: 185–194.

Lawler, Matusky and Skelly Engineers. 1975. 1974 Hudson River aquatic ecology studies. Bowline Point and Lovett Generating Stations. Prepared for Orange and Rockland Utilities, Inc.

———. 1982. Age, growth and population dynamics of white perch (*Morone americana,* Gmelin) in Haverstraw Bay, Hudson River, New York. Prepared for Orange and Rockland Utilities, Inc.

Lazzari, M. A., and K. W. Able. 1990. Northern pipefish, *Syngnathus fuscus,* occurrences over the Mid-Atlantic Bight continental shelf: evidence of seasonal migration. Environ. Biol. Fishes 27: 177–185.

Lazzari, M. A., K. W. Able, and M. P. Fahay. 1989. Life history and food habits of the grubby, *Myoxocephalus aenaeus* (Cottidae), in a Cape Cod estuary. Copeia 1989(1): 7–12.

Lee, D. S., C. R. Gilbert, C. H. Hocutt, R. E. Jenkins, D. E. McAllister, and J. R. Stauffer, Jr. 1980. *Atlas of North American freshwater fishes.* North Carolina State Museum of Natural History, Raleigh.

Leggett, W. C. 1976. The American shad (*Alosa sapidissima*), with special reference to its migration and population dynamics in the Connecticut River. Am. Fish. Soc. Monogr. 1: 169–225.

———. 1977. Ocean migration rates of American shad (*Alosa sapidissima*). J. Fish. Res. Board Can. 34: 1422–1426.

———. 1986. The dependence of fish larval survival on food and predator densities. Pp. 117–137 *in* S. Skreslet, ed. *The role of freshwater outflow in coastal marine ecosystems.* NATO ASI Series, Vol. G7. Springer, Berlin.

Leggett, W. C., and J. E. Carscadden. 1978. Latitudinal variation in reproductive characteristics of

American shad (*Alosa sapidissima*): evidence for population specific life history strategies in fish. J. Fish. Res. Board Can. 35 : 1469–1478.

Leim, A. H. 1924. The life history of the shad (*Alosa sapidissima*) (Wilson) with special reference to the factors limiting its abundance. Contrib. Can. Biol. 2 : 163–284.

Leis, J. M. 1984. Tetradontiformes: relationships. Pp. 459–463 *in* H. G. Moser, W. J. Richards, D. M. Cohen, M. P. Fahay, A. W. Kendall, Jr., and S. L. Richardson, eds. *Ontogeny and systematics of fishes.* Am. Soc. Ichthyol. Herpetol. Spec. Publ. No. 1. Allen Press, Lawrence, Kans.

Lenanton, R. C. J. 1982. Alternative non-estuarine nursery habitats for some commercially and recreationally important fish species of southwestern Australia. Aust. J. Mar. Freshw. Res. 33 : 881–900.

Lenanton, R. C. J., and E. P. Hodgkin. 1985. Life history strategies of fish in some temperate Australian estuaries. Pp. 267–284 *in* A. Yanez-Arancibia, ed. *Fish community ecology in estuaries and coastal lagoons: Towards an ecosystem integration.* UNAM Press, Mexico City.

Lenanton, R. C. J., and I. C. Potter. 1987. Contributions of estuaries to commercial fisheries in temperate western Australia and the concept of estuarine dependence. Estuaries 10(1) : 28–35.

Leslie, A. J., Jr., and D. J. Stewart. 1986. Systematics and distribution ecology of *Etropus* (Pisces, Bothidae) on the Atlantic coast of the United States with a description of a new species. Copeia 1986(1) : 140–156.

Levin, P. S. 1991. Effects of microhabitat on recruitment in a Gulf of Maine reef fish. Mar. Ecol. Prog. Ser. 75 : 183–189.

———. 1993. Habitat structure, conspecific presence and spatial variation in the recruitment of a temperate reef fish. Oecologia 94 : 176–185.

Levy, A., K. W. Able, C. B. Grimes, and P. Hood. 1988. Biology of the conger eel *Conger oceanicus* in the Mid-Atlantic Bight. II. Foods and feeding ecology. Mar. Biol. 98 : 597–600.

Lewis, R. M. 1965. The effect of minimum temperature on the survival of larval Atlantic menhaden, *Brevoortia tyrannus.* Trans. Am. Fish. Soc. 94(4) : 409–412.

Lewis, R. M., E. P. Wilkins, and H. R. Gordy. 1972. A description of young Atlantic menhaden, *Brevoortia tyrannus,* in the White Oak River estuary, North Carolina. U.S. Fish. Bull. 70(1) : 115–118.

Li, S., and J. A. Mathias. 1987. The critical period of high mortality of larvae fish—a discussion based on current research. Chin. J. Oceanol. Limnol. 5(1) : 80–96.

Limburg, K. E. 1995. Otolith strontium traces environmental history of subyearling American shad *Alosa sapidissima.* Mar. Ecol. Progr. Ser. 119 : 25–35.

Linton, J. R., and B. L. Soloff. 1964. The physiology of the brood pouch of the male seahorse, *Hippocampus erectus.* Bull. Mar. Sci. Gulf Carib. 14 : 45–61.

Lippson, A. J., and R. L. Lippson. 1984. Life in the Chesapeake Bay. Johns Hopkins University Press, Baltimore.

Lippson, A. J., and R. L. Moran. 1974. Manual for identification of early developmental stages of fishes of the Potomac River Estuary. Md. Dep. Nat. Resour. Power Plant Siting Program PPSP-MP-13.

Livingston, R. J. 1975. Long-term fluctuations of epibenthic fish and invertebrate populations in Apalachicola Bay, Florida. U.S. Fish. Bull. 74 : 311–321.

Loesch, J. G. 1969. A study of the blueback herring in Connecticut waters. Ph.D. diss. University of Connecticut, Storrs.

Loesch, J. G., and W. A. Lund, Jr. 1977. A contribution to the life history of the blueback herring, *Alosa aestivalis.* Trans. Am. Fish. Soc. 106 : 583–589.

Longley, W. H., and S. F. Hildebrand. 1941. Systematic catalogue of the fishes of Tortugas, Florida. Papers from Tortugas Laboratory. Carnegie Inst. Wash. Publ. 535.34 : 1–317.

Loos, J. J., and E. S. Perry. 1991. Larval migration and mortality rates of bay anchovy in the Patuxent River. Pp. 65–76 *in* R. D. Hoyt, ed. Larval fish recruitment and research in the Americas. Natl. Ocean. Atmos. Adm. Tech. Rep. NMFS 95.

Lotrich, V. A. 1975. Summer home range and movements of *Fundulus heteroclitus* in a tidal creek. Ecology 56 : 191–198.

Lowerre-Barbieri, S. K., M. E. Chittenden, Jr., and L. R. Barbieri. 1996a. The multiple spawning pattern of weakfish in the Chesapeake Bay and Middle Atlantic Bight. J. Fish Biol. 48 : 1139–1163.

———. 1996b. Variable spawning activity and annual fecundity of weakfish in Chesapeake Bay. Trans. Am. Fish. Soc. 125 : 532–545.

Luczkovich, J. L., and B. L. Olla. 1983. Feeding behavior, prey consumption, and growth of juvenile red hake. Trans. Am. Fish. Soc. 112 : 629–637.

Lund, W. A., Jr., and B. C. Marcy, Jr. 1975. Early development of the grubby, *Myoxocephalus aenaeus* (Mitchill). Biol. Bull. (Woods Hole) 149 : 373–383.

Luo, J., and J. A. Musick. 1991. Reproductive biology of the bay anchovy, *Anchoa mitchilli,* in Chesapeake Bay. Trans. Am. Fish. Soc. 120: 701–710.

Lux, F. E., and F. E. Nichy. 1971. Number and lengths, by season, of fishes caught with an otter trawl near Woods Hole, Massachusetts, September 1961 to December 1962. U.S. Dep. Comm. Spec. Sci. Rep. 622: 1–14.

Lux, F. E., and C. L. Wheeler. 1992. Larval and juvenile fishes caught in a neuston survey of Buzzards Bay, Massachusetts in 1979. Natl. Ocean. Atmos. Adm. Natl. Mar. Fish. Serv. Northeast Fish. Sci. Cent. Ref. Doc. 92–09: 1–12.

McAllister, D. E. 1960. Sand-hiding behavior in young white hake. Can. Field Nat. 74(4): 177.

McBride, R. A., and T. F. Moslow. 1991. Origin, evolution and distribution of shoreface sand ridges, Atlantic inner shelf, U.S.A. Mar. Geol. 97: 57–85.

McBride, R. S. 1994. Comparative ecology and life history of two temperate, northwestern Atlantic searobins, *Prionotus carolinus* and *P. evolans* (Pisces: Triglidae). Ph.D. diss. Rutgers, the State University of New Jersey, New Brunswick, N.J.

———. 1995. Perennial occurrence and fast growth rates by crevalle jacks (Carangidae: *Caranx hippos*) in the Hudson River estuary. Pp. VI–1–34 *in* E. A. Blair and J. R. Waldman, eds. *Final Reports of the Tibor T. Polgar Fellowship Program, 1994.* Hudson River Foundation, N.Y.

McBride, R. S., and K. W. Able. 1994. Reproductive seasonality, distribution, and abundance of *Prionotus carolinus* and *P. evolans* (Pisces: Triglidae) in the New York Bight. Estuarine Coastal Shelf Sci. 38: 173–188.

———. 1998. Ecology and fate of butterflyfishes, *Chaetodon* spp., in the temperate northwestern Atlantic. Bull. Mar. Sci. In press.

McBride, R. S., and D. O. Conover. 1991. Recruitment of young-of-the-year bluefish *Pomatomus saltatrix* to the New York Bight: Variation in abundance and growth of spring- and summer-spawned cohorts. Mar. Ecol. Progr. Ser. 78: 205–216.

McBride, R. S., J. L. Ross, and D. O. Conover. 1993. Recruitment of bluefish *Pomatomus saltatrix* to estuaries of the U.S. South Atlantic Bight. Fish. Bull. U.S. 91: 389–395.

McBride, R. S., M. D. Scherer, and J. D. Powell. 1995. Correlated variations in abundance, size, growth and loss rates of age-0 bluefish in a southern New England estuary. Trans. Am. Fish. Soc. 124: 898–910.

McBride, R. S., J. B. O'Gorman, and K. W. Able. In press. Seasonal movements, size-structure, and interannual abundance of searobins (Triglidae: *Prionotus*) in the temperate, northwestern Atlantic. U.S. Fish. Bull.

McBride, R. S., M. P. Fahay, J. A. Hare, and K. W. Able. In review. Larval biology and settlement of searobins (*Prionotus carolinus* and *P. evolans*) in the Mid-Atlantic Bight. Copeia

McCambridge, J. T., Jr., and R. W. Alden, III. 1984. Growth of juvenile spot, *Leiostomus xanthurus* Lacepède, in the nursery region of the James River, Virginia. Estuaries 7(4B): 478–486.

McCleave, J. D., and M. J. Miller. 1994. Spawning of *Conger oceanicus* and *Conger triporiceps* (Congridae) in the Sargasso Sea and subsequent distribution of leptocephali. Environ. Biol. Fishes 39: 339–355.

McDermott, V. J. 1971. Study of the ichthyoplankton associated with two of New Jersey's coastal inlets. N.J. Dep. Environ. Protect. Div. Fish Game Shellfish Bur. Fish. Misc. Rep. No. 7M.

McDowall, I. 1985. River estuaries in the life cycles of New Zealand fish species. Pp. 557 *in* A. Yanez-Arancibia, ed. *Fish community ecology in estuaries and coastal Lagoons: Towards an ecosystem integration.* UNAM Press, Mexico City.

McEachran, J. D., and J. Davis. 1970. Age and growth of the striped searobin. Trans. Am. Fish. Soc. 99(2): 343–352.

McGovern, J. C., and J. E. Olney. 1996. Factors affecting survival of early life stages and subsequent recruitment of striped bass on the Pamunkey River, Virginia. Can. J. Fish. Aquat. Sci. 53: 1713–1726.

McGurk, M. S. 1984. Effects of delayed feeding and temperature on the age of irreversible starvation and on the rates of growth and mortality of Pacific herring larvae. Mar. Biol. 84: 13–26.

McHugh, J. L. 1966. Management of estuarine fishes. Am. Fish. Soc. Spec. Publ. 3: 133–154.

———. 1967. Estuarine nekton. Am. Assoc. Adv. Sci. Publ. 83: 581–620.

McHugh, J. L., R. T. Oglesby, and A. L. Pacheco. 1959. Length, weight and age composition of the menhaden catch in Virginia waters. Limnol. Oceanogr. 4(2): 145–162.

McIvor, C. C., and W. E. Odum. 1988. Food, predation risk and microhabitat selection in a marsh fish assemblage. Ecology 69: 1341–1351.

MacKenzie, C. L., Jr. 1992. *The fisheries of Raritan Bay.* Rutgers University Press, New Brunswick, N.J.

MacKenzie, C. L., Jr., and L. L. Stehlik. 1988. Past and present distributions of soft clams and eelgrass in Raritan Bay. Bull. N.J. Acad. Sci. 33(2): 61–62.

McKenzie, R. A. 1964. Smelt life history and fishery in the Miramichi River, New Brunswick. Fish. Res. Board Can. Bull. 144.

McKown, K. A. 1991. An investigation of the movements and growth of the 1989 Hudson River year class. A Study of the striped bass in the marine District of New York VI. U.S. Dep. Commer. Natl. Mar. Fish. Serv. Natl. Ocean. Atmos. Adm. Anadromous Fish Conservation Act P.L. 89–304. Completion Report. New York, Project AFC-14.

McLaren, J. B., T. H. Peck, W. P. Dey, and M. Gardinier. 1988. Biology of Atlantic tomcod in the Hudson River estuary. Trans. Am. Fish. Soc. Monogr. 4 : 102–112.

Major, P. F. 1978. Aspects of estuarine intertidal ecology of juvenile striped mullet, *Mugil cephalus,* in Hawaii. U.S. Fish. Bull. 76 : 299–313.

Malchoff, M. H. 1993. Age, growth and distribution of cunner (*Tautogolabrus adspersus*) and tautog (*Tautoga onitis*) larvae in the New York Bight: a single season analysis. Master's thesis. Bard College, Annandale-on-Hudson, New York.

Malloy, K. D., and T. E. Targett. 1991. Feeding, growth and survival of juvenile summer flounder *Paralichthys dentatus:* experimental analysis of the effects of temperature and salinity. Mar. Ecol. Prog. Ser. 72(3) : 213–223.

Mann, D. A., J. Bowers-Altman, and R. A. Rountree. 1997. Sounds produced by the striped cuskeel *Ophidion marginatum* (Ophidiidae) during courtship and spawning. Copeia 1997(3) : 610–612.

Mansueti, A. J., and J. D. Hardy. 1967. *Development of fishes of the Chesapeake Bay region: An atlas of egg, larval, and juvenile stages. Part I.* Natural Resources Institute, University of Maryland, Port City Press, Baltimore.

Mansueti, R. J. 1955. Natural history of the American shad in Maryland waters. Md. Tidewater News 11 (Suppl. 4) : 1–2.

———. 1958. Eggs, larvae, and young of the striped bass, *Roccus saltatrix.* Md. Dep. Res. Educ. Contrib. 112 Chesapeake Biological Laboratory, Solomons, Md.

———. 1961. Movements, reproduction and mortality of the white perch, *Roccus americanus,* in the Patuxent estuary, Maryland. Chesapeake Sci. 2 : 142–205.

———. 1962. Eggs, larvae and young of the hickory shad, *Alosa mediocris,* with comments on its ecology in the estuary. Chesapeake Sci. 3(30) : 173–205.

———. 1963. Symbiotic behavior between small fishes and jellyfishes, with new data on that between the stromateid, *Peprilus alepidotus,* and the scyphomedusa, *Chrysaora quinquecirrha.* Copeia 1963(1) : 40–80.

———. 1964. Eggs, larvae, and young of the white perch, *Roccus americanus,* with comments on its ecology in the estuary. Chesapeake Sci. 5 (1–2) : 3–45.

Mansueti, R. J., and R. Pauly. 1956. Age and growth of the northern hogchoker, *Trinectes maculatus,* in the Patuxent River, Maryland. Copeia 1956(1) : 60–62.

Marcellus, K. L. 1972. Fishes of Barnegat Bay, New Jersey with particular reference to seasonal influences and possible effects of thermal discharges. Ph.D. diss. Rutgers University, New Brunswick, N.J.

Marcy, B. C., Jr. 1969. Age determination from scales of *Alosa pseudoharengus* and *Alosa aestivalis* in Connecticut waters. Trans. Am. Fish. Soc. 98 : 621–630.

———. 1972. Spawning of the American shad, *Alosa sapidissima,* in the lower Connecticut River. Chesapeake Sci. 13 : 116–119.

———. 1973. Vulnerability and survival of young Connecticut River fish entrained at a nuclear power plant. J. Fish. Res. Board Can. 30(8) : 1195–1203.

———. 1976a. Fishes of the lower Connecticut River and the effects of the Connecticut Yankee plant. Pp. 61–113 *in* D. Merriman and L. M. Thorpe, eds. *The Connecticut River ecological study.* American Fisheries Society, Monogr. No. 1. Washington, D.C.

———. 1976b. Planktonic fish eggs and larvae of the lower Connecticut River and the effects of the Connecticut Yankee plant. Pp. 115–139 *in* D. Merriman and L. M. Thorpe, eds. *The Connecticut River ecological study.* American Fisheries Society, Monogr. No. 1. Washington, D.C.

Marcy, B. C., Jr., and P. M. Jacobson. 1976. Early life history studies of American shad in the lower Connecticut River and the effects of the Connecticut Yankee Plant. Pp. 141–168 *in* D. Merriman and L. M. Thorpe, eds. *The Connecticut River ecological study.* American Fisheries Society, Monogr. No. 1. Washington, D.C.

Marcy, B. C., Jr., and F. P. Richards. 1974. Age and growth of the white perch, *Morone americana,* in the lower Connecticut River. Trans. Am. Fish. Soc. 103 : 117–120.

Marguiles, D. 1988. Effects of food concentrations and temperature on development, growth and survival of white perch, *Morone americana,* eggs and larvae. U.S. Fish. Bull. 87(1) : 63–72.

Markle, D. F. 1982. Identification of larval and juvenile Canadian Atlantic gadoids with comments on the systematics of gadid subfamilies. Can. J. Zool. 60(12): 3420–3438.

Markle, D. F., and L. A. Frost. 1985. Comparative morphology, seasonality, and a key to planktonic fish eggs from the Nova Scotian shelf. Can. J. Zool. 63: 246–257.

Markle, D. F., and J. A. Musick. 1974. Benthic-slope fishes found at 900 m depth along a transect in the western N. Atlantic Ocean. Mar. Biol. 26: 225–233.

Markle, D. F., D. A. Methven, and L. J. Coates-Markle. 1982. Aspects of spatial and temporal co-occurrence in the life history stages of the sibling hakes, *Urophycis chuss* (Walbaum 1792) and *Urophycis tenuis* (Mitchill 1815) (Pisces: Gadidae). Can. J. Zool. 60(9): 2057–2078.

Marks, R. E., and D. O. Conover. 1993. Ontogenetic shift in the diet of young-of-the-year bluefish, *Pomatomus saltatrix,* during the oceanic phase of the early life history. U.S. Fish. Bull. 91: 97–106.

Marteinsdottir, G. 1991. Early life history of the mummichog (*Fundulus heteroclitus*): Egg size variation and its significance in reproduction and survival of eggs and larvae. Ph.D. diss. Rutgers University, New Brunswick, N.J.

Marteinsdottir, G., and K. W. Able. 1988. Geographic variation in egg size among populations of the mummichog, *Fundulus heteroclitus* (Pisces: Fundulidae). Copeia 1988(2): 471–478.

———. 1992. Influence of egg size on embryos and larvae of *Fundulus heteroclitus* (L.). J. Fish. Biol. 41: 883–896.

Martin, F. D., and G. E. Drewry. 1978. *Development of fishes of the Mid-Atlantic Bight: An atlas of egg, larval and juvenile stages. Vol. VI: Stromateidae through Ogcocephalidae.* Chesapeake Biological Laboratory, Center for Environmental and Estuarine Studies, University of Maryland. Prepared for U.S. Fish and Wildlife Service. FWS/OBS-78/12. U.S. Government Printing Office, Washington.

Massmann, W. H. 1954. Marine fishes in fresh and brackish waters of Virginia rivers. Ecology 35(1): 75–78.

———. 1957. New and recent records for fishes in Chesapeake Bay. Copeia 1957(2): 156–157.

Massmann, W. H., J. P. Whitcomb, and A. L. Pacheco. 1958. Distribution and abundance of gray weakfish in the York River system, Virginia. Trans. N. Am. Wildl. Natl. Resour. Conf. 23: 361–369.

Massmann, W. H., J. J. Norcross, and E. B. Joseph. 1961. Fishes and fish larvae collected from Atlantic plankton cruises of R/V *Pathfinder,* Dec. 1959–Dec. 1960. Va. Fish. Lab. Spec. Rep. 26.

Massmann, W. H., E. B. Joseph, and J. J. Norcross. 1962. Fishes and fish larvae collected from Atlantic plankton cruises of R/V *Pathfinder,* March 1961–March 1962. Va. Inst. Mar. Sci. Spec. Sci. Rep. 33.

Massmann, W. H., J. J. Norcross, and E. B. Joseph. 1963. Distribution of larvae of the naked goby, *Gobiosoma bosci,* in the York River. 4(3): 120–125.

Matallanas, J., and G. Riba. 1980. Aspects biologicos de *Ophidion barbatum* Linnaeus, 1758 y *O. rochei* Muller, 1845 (Pisces, Ophidiidae) de la costa catalana. Invest. Pesq. 44: 399–406.

Matarese, A. C., A. W. Kendall, Jr., D. M. Blood, and B. M. Vinter. 1989. Laboratory guide to early life history stages of northeast Pacific fishes. NOAA Tech. Rep. NMFS 80.

May, R. C. 1974. Larval mortality in marine fishes and the critical period concept. Pp. 3–19 *in* J. S. Blaxter, ed. *The early life history of fish.* Springer, New York.

Mayo, R. K. 1974. Population structure, movement, and fecundity of the anadromous alewife (Wilson), in the Parker River, Massachusetts, 1971–1972. Master's thesis. University of Massachusetts, Amherst.

———. 1982a. Blueback herring, *Alosa aestivalis.* (Mar. Ecosyst. Anal.) New York Bight Atlas Monogr. 15: 54–57.

———. 1982b. Alewife, *Alosa pseudoharengus.* (Mar. Ecosyst. Anal.) New York Bight Atlas Monogr. 15: 57–59.

Mayo, R. K., J. M. McGlade, and S. H. Clark. 1989. Patterns of exploitation and biological status of pollock (*Pollachius virens* L.) in the Scotian Shelf, Georges Bank, and Gulf of Maine area. J. Northwest. Atl. Fish. Sci. 9: 13–36.

Meldrim, J. W., and J. J. Gift. 1971. Temperature preference, avoidance and shock experiments with estuarine fishes. Ichthyol. Assoc. Bull. No. 7. Ichthyological Associates, Ithaca, N.Y.

Mercer, L. P. 1973. The comparative ecology of two species of pipefish (Syngnathidae) in the York River, Virginia. Master's thesis. College of William and Mary, Gloucester Point, Va.

———. 1978. The reproductive biology and population dynamics of black sea bass, (*Centropristis striata*). Ph.D. diss. College of William and Mary, Gloucester Point, Va.

———. 1983. A biological and fisheries profile of weakfish, *Cynoscion regalis.* Div. Mar. Fish. Spec. Sci. Rep. No. 39. North Carolina Department of Natural Resources and Community Development, Morehead City, N.C.

———. 1987. Fishery management plan for spot. Fisheries Management Report of the Atlantic States Marine Fisheries Commission, No. 11. North Carolina Department of Natural Resources and Community Development, Morehead City, N.C.

Meredith, W. H., and V. A. Lotrich. 1979. Production dynamics of a tidal creek population of *Fundulus heteroclitus* (Linnaeus). Estuarine Coastal Mar. Sci. 8 : 99–118.

Merriman, D. 1947. Notes on the mid-summer ichthyofauna of a Connecticut beach at different tide levels. Copeia 1947(4) : 281–286.

Merriman, D., and R. C. Sclar. 1952. The pelagic fish eggs and larvae of Block Island Sound. Bull. Bingham Oceanogr. Collect. Yale Univ. 13 : 165–219.

Merriman, D., and L. M. Thorpe, eds. 1976. *The Connecticut River ecological study.* Am. Fish. Soc. Monogr. 1.

Merriman, D., and H. E. Warfel. 1948. Studies on the marine resources of southern New England. VII. Analysis of a fish population. Bull. Bingham Oceanogr. Collect. Yale Univ. 11(4): 131–164.

Merriner, J. V. 1976. Aspects of the reproductive biology of the weakfish, *Cynoscion regalis* (Sciaenidae), in North Carolina. U.S. Fish. Bull. 74 : 18–26.

Methven, D. A. 1985. Identification and development of larval and juvenile *Urophycis chuss, U. tenuis* and *Phycis chesteri* (Pisces, Gadidae) from the Northwest Atlantic. J. Northwest. Atl. Fish. Sci. 6 : 9–20.

Middaugh, D. P., and T. Takita. 1983. Tidal and diurnal spawning cues in the Atlantic silverside, *Menidia menidia.* Environ. Biol. Fishes 8 : 97–104.

Miller, J. M. 1984. Habitat choices in estuarine fish: do they have any? Pp. 337–352 *in* B. J. Copeland et al., eds. Research for managing the nation's estuaries: proceedings of a conference in Raleigh, North Carolina, March 13–15, 1984. University of North Carolina Sea Grant College Publication UNC-SG-84-08.

Miller, J. M. 1994. An overview of the second flatfish symposium: recruitment in flatfish. Neth. J. Sea Res. 32(2) : 103–106.

Miller, J. P. 1995. American shad. Pp. 251–257 *in* L. E. Dove and R. M. Nyman, eds. *Living resources of the Delaware estuary.* The Delaware Estuary Program.

Miller, M. J. 1995. Species assemblages of leptocephali in the Sargasso Sea and Florida Current. Mar. Ecol. Prog. Ser. 121 : 11–26.

Miller, M. J., and J. D. McCleave. 1994. Species assemblages of leptocephali in the Subtropical Convergence Zone of the Sargasso Sea. J. Mar. Res. 52 : 743–772.

Miller, R. J. 1959. A review of the seabasses of the genus *Centropristes* (Serranidae). Tulane Stud. Zool. 7 : 33–68.

Miller, R. R. 1972. Threatened freshwater fishes of the United States. Trans. Am. Fish. Soc. 101 : 239–252.

Miller, R. V. 1969. Continental migrations of fishes. Underwater Nat. 6(1) : 15–24.

Milstein, C. B. 1981. Abundance and distribution of juvenile *Alosa* species off southern New Jersey. Trans. Am. Fish. Soc. 110 : 306–309.

Milstein, C. B., and D. L. Thomas. 1976a. Fishes new or uncommon to the New Jersey coast. Chesapeake Sci. 17(3) : 198–204.

———. 1976b. First record of the cubera snapper (*Lutjanus cyanopterus*) for New Jersey. Bull. N.J. Acad. Sci. 21(1) : 13.

Milstein, C. B., D. L. Thomas, E. V. Garlo, et al. 1977. Summary of ecological studies for 1972–1975 in the bays and other waterways near Little Egg Inlet and in the ocean in the vicinity of the proposed site for the Atlantic Generation Station, New Jersey. Bull. No. 18. Ichthyological Associates, Ithaca, N.Y.

Minello, T. J., R. J. Zimmerman, and R. J. Martinez. 1987. Fish predation on juvenile brown shrimp, *Penaeus aztecus* Ives: effects of turbidity and substratum on predation rates. U.S. Fish. Bull. 85 : 59–70.

Moe, M. A., Jr. 1976. Rearing Atlantic angelfish. Mar. Aquar. 7 : 17–26.

Monaco, M. E., T. A. Lowery, and R. L. Emmett. 1992. Assemblages of U.S. west coast estuaries based on the distribution of fishes. J. Biogeogr. 19 : 251–267.

Monteleone, D. M. 1992. Seasonality and abundance of ichthyoplankton in Great South Bay, New York. Estuaries 15(2) : 230–238.

Moore, E. 1947. Studies on the marine resources of southern New England. VI: The sand flounder, *Lophosetta aquosa* (Mitchill); a general study of the species with special emphasis on age determination by mean of scales and otoliths. Bull. Bingham Oceanogr. Collect. Yale Univ. 11(3): 1–79.

Morgan, R. P., II, and V. J. Rasin, Jr. 1982. Influence of temperature and salinity on development of white perch eggs. Trans. Am. Fish. Soc. 111(3): 396–398.

Morin, R. P., and K. W. Able. 1983. Pattern of geographic variation in the egg morphology of the fundulid, *Fundulus heteroclitus.* Copeia 1983(3): 726–740.

Moring, J. R., and M. E. Moring. 1986. A late leptocephalus stage of a conger eel, *Conger oceanicus,* found in a tidepool. Copeia 1986(1): 222–223.

Morrill, A. D. 1895. The pectoral appendages of *Prionotus* and their innervation. J. Morphol. 11: 17–192.

Morrow, J. E., Jr. 1951. Studies on the marine resources of southern New England. VIII. The biology of the longhorn sculpin, *Myoxocephalus octodecimspinosus* Mitchill, with a discussion of the southern New England "trash" fishery. Bull. Bingham Oceanogr. Coll. Yale Univ. 13(2): 1–89.

Morse, W. W. 1978. Biological and fisheries data on scup, *Stenotomus chrysops* (Linnaeus). Sandy Hook Laboratory, N.E. Fish. Cent. Nat. Mar. Fish. Serv. Natl. Ocean. Atmos. Adm. Tech. Ser. Rep. No. 12: 1–41.

———. 1980. Maturity, spawning, and fecundity of Atlantic croaker, *Micropogonias undulatus,* occurring north of Cape Hatteras, North Carolina. U.S. Fish. Bull. 78(1): 190–195.

———. 1981. Reproduction of the summer flounder, *Paralichthys dentatus* (L.). J. Fish. Biol. 19: 189–203.

———. 1982. Scup, *Stenotomus chrysops.* (Mar. Ecosyst. Anal.) New York Bight Atlas Monogr. 15: 89–91.

———. 1989. Catchability, growth and mortality of larval fishes. U.S. Fish. Bull. 87(3): 417–446.

Morse, W. W., and K. W. Able. 1995. Distribution and life history of windowpane, *Scophthalmus aquosus,* off the northeastern United States. U.S. Fish. Bull. 93: 675–693.

Morse, W. W., M. P. Fahay, and W. G. Smith. 1987. MARMAP surveys of the continental shelf from Cape Hatteras, North Carolina to Cape Sable, Nova Scotia (1977–1984). Atlas No. 2. Annual distribution patterns of fish larvae. Natl. Ocean. Atmos. Adm. Tech. Memo. NMFS-F Natl. Mar. Fish. Serv.-F/Northeast. Fish. Cent. 47: 1–215.

Morton, T. 1989. Species profiles: life histories and environmental requirements of coastal fishes and invertebrates (Mid-Atlantic)—bay anchovy. U.S. Fish Wildl. Serv. Biol. Rep. 82(11.97): 1–13.

Moser, H. G. 1981. Morphological and functional aspects of marine fish larvae. Pp. 89–131 *in* R. Lasker, ed. *Marine fish larvae, morphology, ecology and relation to fisheries.* University of Washington Press, Seattle.

Moser, H. G., ed. 1996. The early stages of fishes in the California Current region. California Cooperative Oceanic Fisheries Investigation, Atlas No. 33.

Moser, H. G., W. J. Richards, D. M. Cohen, M. P. Fahay, A. W. Kendall, and S. L. Richardson, eds. 1984. Ontogeny and systematics of fishes. Am. Soc. Ichthyol. Herpetol. Spec. Publ. No. 1. Allen Press, Lawrence, Kans.

Moss, S. A. 1972. Tooth replacement and body growth rates in the smooth dogfish, *Mustelus canis* (Mitchill). Copeia 1972(4): 808–811.

———. 1973. The responses of planehead filefish, *Monacanthus hispidus* (Linnaeus), to low temperature. Chesapeake Sci. 14: 300–303.

Mulkana, M. S. 1966. The growth and feeding habits of juvenile fishes in two Rhode Island estuaries. Gulf Res. Rep. 2: 97–167.

Mullen, D. M., C. W. Fay, and J. R. Moring. 1986. Species Profiles: life histories and environmental requirements of coastal fishes and invertebrates (North Atlantic). Alewife/blueback herring. U.S. Fish Wildl. Serv. Coastal Ecology Group, Waterways Exp. Stn. Biol. Rep. 82 (11.56).

Mulligan, T. J., and R. W. Chapman. 1989. Mitochondrial DNA analysis of Chesapeake white perch, *Morone americana.* Copeia 1989: 679–688.

Munroe, T. A., and R. A. Lotspeich. 1979. Some lifehistory aspects of the seaboard goby (*Gobiosoma ginsburgi*) in Rhode Island. Estuaries 2(1): 22–27.

Murawski, W. S. 1969. The distribution of striped bass, *Roccus saxatilis,* eggs and larvae in the lower Delaware River. N.J. Dep. Conserv. Econ. Develop. Bur. Fish. Misc. Rep. No. 1M, Nacote Creek, N.J.

Murawski, W. S., and P. J. Festa. 1979. Creel census of the summer flounder, *Paralichthys dentatus* sportfishery in Great Bay, New Jersey. N.J. Dep. Environ. Protect. Div. Fish Game Shellfish. Nacote Creek Res. Stn. N.J. Tech. Rep. No. 19M. Nacote Creek, N.J.

Murdy, E. O., R. S. Birdsong, and J. A. Musick. 1997. *Fishes of Chesapeake Bay.* Smithsonian Institution Press, Washington, D.C.

Musick, J. A. 1969. The comparative biology of two American Atlantic hakes, *Urophycis chuss* and *U. tenuis* (Pisces: Gadidae). Ph.D. diss. Harvard University, Cambridge, Mass.

———. 1972. Fishes of Chesapeake Bay and the adjacent coastal plain. Pp. 175–212 *in* M. L. Wass et al., comp., A check list of the biota of lower Chesapeake Bay, with inclusions from the upper bay and the Virginian Sea. Va. Inst. Mar. Sci. Spec. Sci. Rep. No. 65. Gloucester Point, Va.

———. 1974. Seasonal distribution of sibling hakes, *Urophycis chuss* and *U. tenuis* (Pisces: Gadidae) in New England. U.S. Fish. Bull. 72(2): 481–495.

Nelson, T. A., P. E. Gadd, and T. L. Clarke. 1978. Wind-induced current flow in the upper Hudson Valley. J. Geophys. Res. 83(C12): 6073–6082.

Nelson, T. C. 1928. On the association of the common goby (*Gobiosoma bosci*) with the oyster, including a case of parasitism. Nat. Hist. 28 (1): 78–84.

Nepszy, S. J. 1968. On the biology of the hake (*Urophycis tenuis* Mitchill) in the southern Gulf of St. Lawrence. Master's thesis. McGill University, Montreal.

Nero, L. 1976. The natural history of the naked goby (*Gobiosoma bosci*) (Perciformes: Gobiidae). Master's thesis. Old Dominion University, Norfolk, Va.

Netzel, J., and E. Stanek. 1966. Some biological characteristics of blueback and alewife from Georges Bank, July and October, 1964. Int. Comm. Northwest Atl. Fish. Res. Bull. No. 3.

Neuman, M. 1996. Evidence of upwelling along the New Jersey coastline and the south shore of Long Island, New York. Bull. N.J. Acad. Sci. 41(1): 7–13.

Neuman, M. J., and K. W. Able. 1998. Experimental evidence of sediment preference by early life history stages of windowpane, (*Scophthalmus aquosus*). J. Sea Res.

Neves, R. J. 1981. Offshore distribution of alewife, *Alosa pseudoharengus,* and blueback herring, *Alosa aestivalis,* along the Atlantic coast. U.S. Fish. Bull. 79: 473–485.

Neville, W. C., and G. B. Talbot. 1964. The fishery for scup with special reference to fluctuations in yield and their causes. U.S. Fish Wildl. Serv. Spec. Sci. Rep. Fish. 459.

Newberger, T. A., and E. D. Houde. 1995. Population biology of bay anchovy *Anchoa mitchilli* in the mid-Chesapeake. Mar. Ecol. Prog. Ser. 116: 25–37.

Newman, H. H. 1908. The process of heredity as exhibited by the development of *Fundulus* hybrids. J. Exp. Zool. 5: 503–561.

———. 1909. The question of viviparity in *Fundulus majalis.* Science 30: 769–771.

———. 1914. Modes on inheritance in teleost hybrids. J. Exp. Zool. 16: 447–499.

Nichols, J. T., and C. M. Breder. 1927. The marine fishes of New York and southern New England. Zoologica (N.Y.) 9(1): 1–192.

Nicholson, W. R. 1971. Coastal movements of Atlantic menhaden as inferred from changes in age and length distributions. Trans. Am. Fish. Soc. 100(4): 708–716.

Nixon, S. W. 1980. Between coastal marshes and coastal waters—a review of twenty years of speculation and research on the role of salt marshes in estuarine productivity and water chemistry. Pp. 437–525 *in* P. Hamilton and K. B. MacDonald, eds. *Estuarine and wetland processes.* Plenum, New York.

———. 1985. Overview. *In* Anon. Narragansett Bay: Issues, Resources, Status and Management Natl. Ocean. Atmos. Adm. Estuary-of-the-Month Seminar Ser. No. 1.

Nizinski, M. S., B. B. Collette, and B. B. Washington. 1990. Separation of two species of sand lances, *Ammodytes americanus* and *A. dubius,* in the western North Atlantic. U.S. Fish. Bull. 88: 241–255.

NOAA. 1985. National estuarine inventory: data atlas. Vol. 1. Physical and Hydrologic Characteristics. NOAA/NOS Strategic Assessment Branch, Rockville, Md.

Norcross, B. L., and D. A. Bodolus. 1991. Hypothetical northern spawning limit and larval transport of spot. Natl. Ocean. Atmos. Adm. Tech. Rep. Natl. Mar. Fish. Serv. 95: 77–88.

Norcross, J. J., S. L. Richardson, W. H. Massmann, and E. B. Joseph. 1974. Development of young bluefish (*Pomatomus saltatrix*) and the distribution of eggs and young in Virginia coastal waters. Trans. Am. Fish. Soc. 103: 477–497.

Norden, C. R. 1967. Development and identification of the larval alewife in Lake Michigan. Proc. Conf. Great Lakes Res. 10: 70–78.

Nordlie, F. G., W. A. Szelistowski, and W. C. Nordlie. 1982. Ontogenesis of osmotic regulation in the striped mullet, *Mugil cephalus* L. J. Fish. Biol. 20: 79–86.

Nyman, R. M., and D. O. Conover. 1988. The relation between spawning season and the recruitment of young-of-the-year bluefish, *Pomatomus saltatrix,* to New York. U.S. Fish. Bull. 86: 237–250.

O'Brien, L., J. Burnett, and R. K. Mayo. 1993. Maturation of nineteen species of finfish off the northeast coast of the United States, 1985–1990. Natl. Ocean. Atmos. Adm. Tech. Rep. Natl. Mar. Fish. Serv. 113: 1–66.

Odum, W. E. 1970. Utilization of the direct grazing and plant detritus food chains by the striped mullet *Mugil cephalus.* Pp. 222–240 *in* J. J. Steele, ed. *Marine food chains.* Oliver and Boyd, Edinburgh.

Officer, C. B., R. B. Biggs, J. L. Taft, L. E. Cronin, M. A. Tyler, and W. R. Boynton. 1984. Chesapeake Bay anoxia: origin, development and significance. Science 223: 22–27.

O'Herron, J. C., II, T. Lloyd, and K. Laidig. 1994. A survey of fish in the Delaware Estuary from the area of the Chesapeake and Delaware Canal to Trenton. Prepared for Delaware Estuary Program, EPA-Region III, Philadelphia, Pa.

Okiyama, M., ed. 1988. *An atlas of early stage of fishes in Japan.* Tokai University Press, Tokyo.

Olla, B. L., A. J. Bejda, and A. D. Martin. 1974. Daily activity, movements, feeding, and seasonal occurrence in the tautog, *Tautoga onitis.* U.S. Fish. Bull. 72: 27–35.

———. 1979. Seasonal dispersal and habitat selection of cunner, *Tautogolabrus adspersus,* and young tautog, *Tautoga onitis,* in Fire Island inlet, Long Island, New York. U.S. Fish. Bull. 77: 255–261.

Olney, J. E. 1983. Eggs and early larvae of the bay anchovy, *Anchoa mitchilli,* and the weakfish, *Cynoscion regalis,* in lower Chesapeake Bay with notes on associated ichthyoplankton. Estuaries 6: 20–35.

Olney, J. E., and G. W. Boehlert. 1988. Nearshore ichthyoplankton associated with seagrass beds in the lower Chesapeake Bay. Mar. Ecol. Progr. Ser. 45: 33–43.

Olney, J. E., G. C. Grant, F. E. Schultz, C. L. Cooper, and J. Hageman. 1983. Pterygiophore-interdigitation patterns in larvae of four *Morone* species. Trans. Am. Fish. Soc. 112: 525–531.

Orth, R. J., and K. L. Heck, Jr. 1980. Structural components of eelgrass (*Zostera marina*) meadows in the lower Chesapeake Bay—fishes. Estuaries 3(4): 278–288.

Oviatt, C. A., and S. W. Nixon. 1973. The demersal fish of Narragansett Bay: an analysis of community structure, distribution and abundance. Estuarine Coastal Mar. Sci. I: 361–378.

Pacheco, A. L. 1962. Age and growth of spot in lower Chesapeake Bay, with notes on distribution and abundance of juveniles in the York River system. Chesapeake Sci. 3(1): 18–28.

———. 1973. Proceedings of a workshop on egg, larval, and juvenile stages of fish in Atlantic coast estuaries. Held at Bears Bluff Laboratories, Wadmalaw Island, South Carolina, June 1968. Natl. Mar. Fish. Serv. Mid. Atl. Coast. Fish. Cent. Tech. Publ. No. 1. Highlands, N.J.

———. 1983. Seasonal occurrence of finfish and larger invertebrates at three sites in lower New York Harbor, 1981–1982. Nat. Ocean. Atmos. Adm. Natl. Mar. Fish. Serv. Northeast Fish. Cent., Sandy Hook Laboratory. Final Report to N.Y. District Corps of Engineers. Support Agreement NYD82-88(C).

———. 1984. Seasonal occurrence of finfish and larger invertebrates at eight locations in Lower and Sandy Hook Bays, 1982–1983. Natl. Ocean. Atmos. Adm. Natl. Mar. Fish. Serv. Northeast Fish. Cent., Sandy Hook Laboratory. Final Report to N.Y. District Corps of Engineers. Support Agreement NYD 83-46(C).

Pacheco, A. L., and G. C. Grant. 1965. Studies of the early life histories of Atlantic menhaden in estuarine nurseries. Part 1. Seasonal occurrence of juvenile menhaden and other small fishes in a tributary creek of Indian River, Delaware, 1957–1958. U.S. Fish Wild. Serv. Spec. Sci. Rep. 504.

———. 1973. Immature fishes associated with larval Atlantic menhaden at Indian River inlet, Delaware, 1958–61. Pp. 78–87 *in* A. L. Pacheco, ed. Proceedings of a workshop on egg, larval and juvenile stages of fish in Atlantic coast estuaries. Natl. Mar. Fish. Serv., Mid. Atl. Coast. Cent. Tech. Publ. No. 1. Highlands, N.J.

Pape, E., III. 1981. A drifter study of the Lagrainian mean circulation of Delaware Bay and adjacent shelf waters. Master's thesis, University of Delaware, Newark. SG-18-81. Delaware Sea Grant College Program, Newark, Del.

Paperno, R. 1991. Spatial and temporal patterns of growth and mortality of juvenile weakfish (*Cynoscion regalis*) in Delaware Bay: Assessment using otolith microincrement analysis. Ph.D. diss. University of Delaware, Newark.

Pardue, G. B. 1983. Habitat suitability index models: alewife and blueback herring. U.S. Dep. Int. Fish Wildl. Serv. FWS/OBS-82/10.58.

Parenti, L. R. 1981. A phylogenetic and biogeographic analysis of cyprinodontiform fishes (Teleostei, Atherinomorpha). Bull. Am. Mus. Nat. Hist. 168: 335–557.

Parr, A. E. 1931. A practical revision of the western Atlantic species of the genus *Citharichthys* (including *Etropus*). With observations on the Pacific *Citharichthys crossotus* and *C. spilopterus.* Bull. Bingham Oceanogr. Collect. Yale Univ. 41: 1–24.

———. 1933. A geographical ecological analysis of the seasonal changes in water along the Atlantic coast of the U.S. Bull. Bingham Oceanogr. Collect. Yale Univ. 4: 1–90.

Patrick, R. 1994. Rivers of the United States. Vol. I. Estuaries. Wiley, New York.

Pearcy, W. G. 1962a. Ecology of an estuarine population of winter flounder, *Pseudopleuronectes americanus* (Walbaum). II. Distribution and Dynamics of larvae. Bull. Bingham Oceanogr. Collect. Yale Univ. 18(1): 16–37.

———. 1962b. Distribution and origin of demersal eggs within the order Pleuronectiformes. J. Cons. Int. Explor. Mer 27: 232–235.

Pearcy, W. G., and S. W. Richards. 1962. Distribution and ecology of fishes of the Mystic River estuary, Connecticut. Ecology 43: 248–259.

Pearson, J. C. 1929. Natural history and conservation of the redfish and other commercial sciaenids on the Texas coast. U.S. Bur. Fish. Bull. 44: 129–214.

———. 1932. Winter trawl fishing off the Virginia and North Carolina coast. U.S. Bur. Fish. Invest. Rep. 10.

———. 1941. The young of some marine fishes taken in lower Chesapeake Bay, Virginia, with special reference to the gray sea trout, *Cynoscion regalis* (Bloch). U.S. Fish. Bull. 36: 77–102.

Pentilla, J., and L. M. Dery. 1988. Age determination methods for Northwest Atlantic species. Natl. Ocean. Atmos. Adm. Tech. Rep. Natl. Mar. Fish. Serv. 72.

Pepin, P. 1991. Effect of temperature and size on development, mortality, and survival rates of the pelagic early life history stages of marine fish. Can. J. Fish. Aquat. Sci. 48: 503–518.

Perlmutter, A. 1939. A biological survey of the salt waters of Long Island, 1938. Part II, Sect. I. An ecological survey of young fishes and eggs identified from tow-net collections. N.Y. Conserv. Dep. Saltwater Survey, 1938(15): 11–71.

———. 1963. Observations on fishes of the genus *Gasterosteus* in the waters of Long Island, New York. Copeia 1963: 168–174.

Perlmutter, A., E. E. Schmidt, and E. Leff. 1967. Distribution and abundance of fish along the shores of the lower Hudson River during the summer of 1965. N.Y. Fish Game J. 14(1): 47–75.

Peterson, R. H., P. H. Johansen, and J. L. Metcalfe. 1980. Observations on early life stages of Atlantic tomcod, *Microgadus tomcod.* U.S. Fish. Bull. 78(1): 147–158.

Pfeiler, E. 1986. Towards an explanation of the developmental strategy in leptocephalus larvae of marine teleost fishes. Environ. Biol. Fishes 15: 3–13.

Pfeiler, E., and J. J. Govoni. 1993. Metabolic rates in early life history stages of elopomorph fishes. Biol. Bull. 185: 277–283.

Phelan, B. A. 1992. Winter flounder movements in the inner New York Bight. Trans. Am. Fish. Soc. 121: 777–784.

Phillips, R. J. 1914. Kingfish at Corson's Inlet, New Jersey. Copeia 1914(9): 3–4.

Picard, P., J. J. Dodson, and G. L. FitzGerald. 1990. The comparative ecology of the threespine (*Gasterosteus aculeatus*) and blackspotted sticklebacks (*G. wheatlandi*) (Pisces: Gasterosteidae) in three sub-habitats of the middle Saint Lawrence estuary, Canada. Can. J. Zool. 68: 1202–1208.

Pierce, D. E., and A. B. Howe. 1977. A further study on winter flounder group identification off Massachusetts. Trans. Am. Fish. Soc. 106: 131–139.

Post, J. R., and D. O. Evans. 1989. Size-dependent overwinter mortality of young-of-the-year yellow perch (*Perca flavescens*): laboratory, in situ enclosure, and field experiments. Can. J. Fish. Aquat. Sci. 46: 1958–1968.

Powell, A. B., and H. R. Gordy. 1980. Egg and larval development of the spot, *Leiostomus xanthurus* (Sciaenidae). U.S. Fish. Bull. 78: 701–714.

Powell, A. B., and T. Henley. 1995. Egg and larval development of laboratory-reared gulf flounder, *Paralichthys albigutta,* and southern flounder, *P. lethostigma* (Pisces, Paralichthyidae). U.S. Fish. Bull. 93: 504–515.

Powell, A. B., and F. J. Schwartz. 1977. Distribution of paralichthid flounders (Bothidae: *Paralichthys*) in North Carolina estuaries. Chesapeake Sci. 18(11): 334–339.

Power, J. H., and J. D. McCleave. 1983. Simulation of the North Atlantic Ocean drift of *Anguilla* leptocephali. U.S. Fish. Bull. 81: 483–500.

Powers, D. A., M. Smith, I. Gonzalez-Villasenor, L. DiMichele, D. Crawford, G. Bernardi, and T. Lauerman. 1993. A multidisciplinary approach to the selection/neutralist controversy using the model teleost, *Fundulus heteroclitus.* Pp. 43–108 *in* D. Futuyma and J. Antonovics, eds. Oxford Surveys in Evolutionary Biology, Vol. 9. Oxford University Press, New York.

Powles, H. 1980. Descriptions of larval silver perch, *Bairdiella chrysoura,* banded drum, *Larimus fasciatus,* and star drum, *Stellifer lanceolatus* (Sciaenidae). U.S. Fish. Bull. 78: 119–136.

———. 1981. Distribution and movements of neustonic young estuarine dependant (*Mugil* spp.) and estuarine independent (*Coryphaena* spp.) fishes off the southeastern United States. Rapp. P.-V. Reun. Cons. Int. Explor. Mer 178: 207–210.

Powles, H., and B. W. Stender. 1978. Taxonomic data on the early life history stages of Sciaenidae of the South Atlantic Bight of the United States. S.C. Mar. Resour. Cent. Tech. Rep. No. 31.

Psuty, N. P., M. P. De Luca, R. Lathrop, K. W. Able, S. Whitney, and J. F. Grassle. 1993. The Mullica River–Great Bay National Estuarine Research Reserve: A unique opportunity for research, preservation and management. Proceedings of

Coastal Zone '93, July 1993. New Orleans, La.

Public Service Electric and Gas Co. 1984. White perch (*Morone americana*): a synthesis of information on natural history, with reference to occurrence in the Delaware River and Estuary and involvement with the Salem Generating Station. Salem Nuclear Generating Station 316(b) Demonstration Appendix X. Newark, N.J.

Pyle, A. B. 1964. Some aspects of the life history of *Cyprinodon variegatus* Lacépède 1803, in New Jersey and its reaction to environmental changes. Master's thesis. Rutgers University, New Brunswick, N.J.

Quinn, T. P., and J. T. Light. 1989. Occurrence of threespine sticklebacks (*Gasterosteus aculeatus*) in the open North Pacific Ocean: migration or drift? Can. J. Zool. 67:2850–2852.

Radtke, R. L., and J. M. Dean. 1982. Increment formation in the otoliths of embryos, larvae and juveniles of the mummichog, *Fundulus heteroclitus.* U.S. Fish. Bull. 80(2):41–55.

Radtke, R. L., M. L. Fine, and J. Bell. 1985. Somatic and otolith growth in the oyster toadfish (*Opsanus tau* L.) J. Exp. Mar. Biol. Ecol. 90: 259–275.

Rangely, R. W., and D. L. Kramer. 1995a. Use of rocky intertidal habitats by juvenile pollock *Pollachius virens.* Mar. Ecol. Prog. Ser. 126:9–17.

———. 1995b. Tidal effects on habitat selection and aggregation by juvenile pollock *Pollachius virens* in the rocky intertidal zone. Mar. Ecol. Prog. Ser. 126:19–29.

Rathjen, W. F., and L. C. Miller. 1957. Aspects of the early life history of the striped bass (*Roccus saxatilis*) in the Hudson River. N.Y. Fish Game J. 4(1):43–60.

Ray, G. C. 1991. Coastal-zone biodiversity patterns: principles of landscape ecology may help explain the processes underlying coastal diversity. BioScience 41(7):490–498.

———. 1997. Do the metapopulations dynamics of estuarine fishes influence the stability of shelf ecosystems? Bull. Mar. Sci. 60:1040–1049.

Reid, G. K., Jr. 1955. A summer study of the biology and ecology of East Bay, Texas. Part II. The fish fauna of East Bay, the Gulf Beach, and summary. Tex. J. Sci. 7(4):430–453.

Reintjes, J. W. 1969. Synopsis of biological data on the Atlantic menhaden, *Brevoortia tyrannus.* U.S. Fish. Wildl. Serv. Circ. 320. FAO Spec. Synop. 42.

Reintjes, J. W., and A. L. Pacheco. 1966. The relation of menhaden to estuaries. Am. Fish. Soc. Spec. Publ. 3:50–58.

Reisman, H. 1963. Reproductive behavior of *Apeltes quadracus,* including some comparisons with other gasterosteid fishes. Copeia 1963:191–192.

Relyea, K. G. 1983. A systematic study of two species complexes of the genus *Fundulus* (Pisces: Cyprinodontidae). Bull. Fla. State Mus. Biol. Sci. 29(1):1–64.

Richards, C. E. 1973. Age, growth and distribution of the black drum (*Pogonias cromis*) in Virginia. Trans. Am. Fish. Soc. 102(3):584–590.

Richards, C. E., and M. Castagna. 1970. Marine fishes of Virginia's eastern shore (inlet and marsh, seaside waters). Chesapeake Sci. 11(4): 235–248.

Richards, S. W. 1959. Pelagic fish eggs and larvae. VI. In: Oceanography of Long Island Sound. Bull. Bingham Oceanogr. Collect. Yale Univ. 17: 95–124.

———. 1963. The demersal fish population of Long Island Sound. III. Food of juveniles from a mud locality (Station 3A). Bull. Bingham. Oceanogr. Collect. Yale Univ. 18:73–93.

Richards, S. W., J. M. Mann, and J. A. Walker. 1979. Comparison of spawning seasons, age, growth rates and food of two sympatric species of searobins, *Prionotus carolinus* and *Prionotus evolans,* from Long Island Sound. Estuaries 2(4): 255–268.

Richards, S. W., and A. M. McBean. 1966. Comparison of postlarval juveniles of *Fundulus heteroclitus* and *Fundulus majalis* (Pisces: Cyprinodontidae). Trans. Am. Fish. Soc. 95(2):218–226.

Richards, W. J., and V. P. Saksena. 1980. Description of larvae and early juveniles of laboratory-reared gray snapper, *Lutjanus griseus* (Linnaeus) (Pisces, Lutjanidae). Bull. Mar. Sci. 30:516–521.

Richardson, S. L., and E. B. Joseph. 1973. Larvae and young of western North Atlantic bothid flatfishes *Etropus microstomus* and *Citharichthys arctifrons* in the Chesapeake Bight. U.S. Fish. Bull. 71:735–767.

Richkus, W. A. 1974. Factors influencing the seasonal and daily patterns of alewife migration in a Rhode Island River. J. Fish. Res. Board Can. 31: 1485–1497.

———. 1975. Migratory behavior and growth of juvenile anadromous alewives, *Alosa pseudoharengus,* in a Rhode Island drainage. Trans. Am. Fish. Soc. 104:483–493.

Riley, G. A. 1956. Physical Oceanography. *In* Oceanography of Long Island Sound. Bull. Bingham Oceanogr. Collect. Yale Univ. 15:15–46.

Rivas, R. 1963. Subgenera and species groups in

the poeciliid fish genus *Gambusia poey.* Copeia 1963:331–347.

Robbins, T. W. 1969. A systematic study of the silversides, *Membras* (Bonaparte) and *Menidia* (Linnaeus), (Atherinidae, Teleostei). Ph.D. diss., Cornell University, Ithaca, N.Y.

Roberts, S. C. 1978. Biological and fisheries data on northern searobin, *Prionotus carolinus* (Linnaeus). U.S. Dep. Commer. Natl. Ocean. Atmos. Adm. Natl. Mar. Fish. Serv. Northeast Fish. Cent. Sandy Hook Lab. Tech. Ser. Rep. No. 13.

Roberts-Goodwin, S. C. 1981. Biological and fisheries data on striped searobin, *Prionotus evolans* (Linnaeus). U.S. Dep. Commer. Natl. Ocean. Atmos. Adm. Natl. Mar. Fish. Serv. Northeast Fish. Cent. Sandy Hook Lab. Tech. Ser. Rep. No. 25.

Robertson, D. R. 1988a. Abundances of surgeonfishes on patch-reef in Caribbean Panama: Due to settlement, or post-settlement events? Mar. Biol. 97:495–501.

———. 1988b. Extreme variation in settlement of the Caribbean triggerfish *Balistes vetula* in Panama. Copeia 1988:698–703.

Robins, C. R., R. M. Bailey, C. E. Bond, J. R. Brooker, E. A. Lachner, R. N. Lea, and W. B. Scott. 1991. Common and scientific names of fishes from the United States and Canada. Am. Fish. Soc. Spec. Publ. No. 20.

Robins, C. R., D. M. Cohen, and C. H. Robins. 1979. The eels, *Anguilla* and *Histiobranchus,* photographed on the floor of the deep Atlantic in the Bahamas. Bull. Mar. Sci. 29(3):401–405.

Robins, C. R., and G. C. Ray. 1986. *A field guide to Atlantic Coast fishes of North America.* Houghton Mifflin, Boston.

Roelke, D. L., and S. M. Sogard. 1993. Gender-based differences in habitat selection and activity level in the northern pipefish (*Syngnathus fuscus*). Copeia 2:528–532.

Roff, D. A. 1988. The evolution of migration and some life history parameters in marine fishes. Environ. Biol. Fishes 22(2):133–146.

Rohde, F. C., and V. J. Schuler. 1974. Abundance and distribution of fishes in Appoquinimink and Alloway creeks. Pp. 395–453 *in* V. J. Schuler, ed. An ecological study of the Delaware River in the vicinity of Artificial Island. Progress report for the period January through December 1972. Ichthyological Associates, Ithaca, N.Y.

Roman, C. T., and K. W. Able. 1989. An ecological analysis of Nauset Marsh, Cape Cod National Seashore. Department of Interior, National Park Service, Cooperative Research Unit of Rutgers University, New Brunswick, N.J.

Rosen, D. E., and R. M. Bailey. 1963. The poeciliid fishes (Cyprinodontiformes), their structure, zoogeography and systematics. Bull. Am. Mus. Nat. Hist. 126:1–176.

Ross, S. W., and S. P. Epperly. 1985. Utilization of shallow estuarine nursery areas by fishes in Pamlico Sound and adjacent tributaries, North Carolina. Pp. 207–232 *in* A. Yanez-Arancibia, ed. *Fish community ecology in estuaries and coastal lagoons: towards an ecosystem integration.* UNAM Press, Mexico City.

Rotunno, T. K. 1992. Species identification and temporal spawning patterns of butterfish, *Peprilus* spp., in the south and mid-Atlantic bights. Master's thesis. Marine Science Research Center, State University of New York, Stony Brook.

Rountree, R. A. 1992. Fish and macroinvertebrate community structure and habitat use patterns in salt marsh creeks of southern New Jersey, with a discussion of marsh carbon export. Ph.D. diss. Rutgers University, New Brunswick, N.J.

Rountree, R. A., and K. W. Able. 1992a. Fauna of polyhaline subtidal marsh creeks in southern New Jersey: composition, abundance and biomass. Estuaries 15(2):171–185.

———. 1992b. Foraging habits, growth, and temporal patterns of salt-marsh creek habitat use by young-of-year summer flounder in New Jersey. Trans. Am. Fish. Soc. 121:765–776.

———. 1993. Diel variation in decapod crustacean and fish assemblages in New Jersey polyhaline marsh creeks. Estuarine Coastal Shelf Sci. 37: 181–201.

———. 1996. Seasonal abundance, growth and foraging habits of juvenile smooth dogfish, *Mustelus canis,* in a New Jersey estuary. U.S. Fish. Bull. 94(3):522–534.

———. 1997. Noctural fish use of New Jersey marsh creek and adjacent bay shoal habitats. Estuarine Coastal Shelf Sci. 44:703–711.

Rountree, R. A., K. J. Smith, and K. W. Able. 1992. Length frequency data for fishes and turtles from polyhaline subtidal and intertidal creeks in southern New Jersey. Institute of Marine and Coastal Science, Rutgers University, New Brunswick, N.J. Tech. Rep. Contrib. No. 92–34.

Rowe, P. M., and C. E. Epifanio. 1994a. Tidal stream transport of weakfish larvae in Delaware Bay, USA. Mar. Ecol. Progr. Ser. 110:105–114.

———. 1994b. Flux and transport of larval weakfish in Delaware Bay, USA. Mar. Ecol. Progr. Ser. 110:115–120.

Ruiz, G. M., A. H. Hines, and M. H. Posey. 1993. Shallow water as a refuge habitat for fish and crustaceans in non-vegetated estuaries: an example from Chesapeake Bay. Mar. Ecol. Prog. Ser. 99:1–16.

Ryder, J. A. 1882. Development of the silver gar (*Belone longirostris*) with observations on the genesis of the blood in embryo fishes, and a comparison of fish ova with those of other vertebrates. U.S. Fish. Comm. Bull. 1:283–302.

———. 1887. On the development of osseus fishes, including marine and freshwater forms. U.S. Dep. Comm. Fish. Rep. 13(1885):489–603.

Saila, S. B., and R. G. Lough. 1981. Mortality and growth estimation from size data—an application to some Atlantic herring larvae. Rapp. P.-V. Reun. Cons. Int. Explor. Mer 178:7–14.

Saksena, V. P., and E. B. Joseph. 1972. Dissolved oxygen requirements of newly hatched larvae of the striped blenny (*Chasmodes bosquianus*), the naked goby (*Gobiosoma bosci*) and the skilletfish (*Gobiesox strumosus*). Chesapeake Sci. 12: 23–28.

Samaritan, J. M., and R. E. Schmidt. 1982. Aspects of the life history of a freshwater population of the mummichog, *Fundulus heteroclitus* (Pisces: Cyprinodontidae), in the Bronx River, New York, U.S.A. Hydrobiologia 94:149–154.

Saucerman, S. E., and L. A. Deegan. 1991. Lateral and cross-channel movement of young-of-the-year winter flounder (*Pseudopleuronectes americanus*) in Waquot Bay, Massachusetts. Estuaries 14(4):440–446.

Sawyer, P. J. 1967. Intertidal life-history of the rock gunnel, *Pholis gunnellus,* in the western Atlantic. Copeia 1967(1):55–61.

Scarlett, P. G. 1991. Relative abundance of winter flounder (*Pseudopleuronectes americanus*) in nine inshore areas of New Jersey. Bull. N.J. Acad. Sci. 36(2):1–5.

Scarlett, P. G., and R. L. Allen. 1992. Temporal and spatial distribution of winter flounder (*Pleuronectes americanus*) spawning in Manasquan River, New Jersey. Bull. N.J. Acad. Sci. 37 (1):13–17.

Schaefer, R. H. 1965. Age and growth of the northern kingfish in New York waters. N.Y. Fish Game J. 12(2):191–216.

———. 1967. Species composition, size and seasonal abundance of fish in the surf waters of Long Island. N.Y. Fish Game J. 14(1):1–46.

Scherer, M. D. 1972. The biology of the blueback herring, *Alosa aestivalis* (Mitchill) in the Connecticut River above the Holyoke Dam, Holyoke, Massachusetts. Master's thesis. University of Massachusetts, Amherst.

Schmelz, G. W. 1970. Some effects of temperature and salinity on the life processes of the striped killifish, *Fundulus majalis* (Walbaum). Ph.D. diss. University of Delaware, Newark.

Schmidt, E. J. 1905. The pelagic post-larval stages of the Atlantic species of *Gadus,* Part I. Medd. Dan. Fisk. Havunders. Ser. Fisk. 1(4): 1–77.

———. 1906. The pelagic post-larval stages of the Atlantic species of *Gadus,* Part II. Medd. Dan. Fisk. Havunders. Ser. Fisk. 2(2):1–20.

Schmidt, J. 1931. Eels and conger eels of the North Atlantic. Nature 128:602–604.

Schoedinger, S. E., and C. E. Epifanio. 1997. Growth, development and survival of larval *Tautoga onitis* (Linnaeus) in large laboratory containers. J. Exp. Mar. Biol. Ecol. 210:143–155.

Schroeder, W. C. 1933. Unique records of the brier skate and rock eel from New England. Bull. Boston Soc. Nat. Hist. 66:5–6.

Schubel, J. R. 1986. Long Island Sound in time and space: an overview. Pp. 1–21 *in* V. R. Gibson and M. S. Connor, eds. *Long Island Sound: Issues, resources, status and management seminar proceedings.* Natl. Ocean. Atmos. Adm. Estuary-of-the-Month Seminar Series No. 3. U.S. Dep. Commerce, Washington, D.C.

Schubel, J. R., and D. J. Hirschberg. 1978. Estuarine graveyards, climatic change, and the importance of the estuarine environment. Pp. 285–303 *in* M. L. Wiley, ed. *Estuarine interactions.* Academic Press, New York.

Schubel, J. R., C. F. Smith, and T. S. Y. Koo. 1977. Thermal effects of powerplant entrainment on survival of larval fishes: a laboratory assessment. Chesapeake Sci. 18(3):290–298.

Schubel, J. R., and J. C. S. Wang. 1973. The effects of suspended sediment on the hatching success of *Perca flavescens* (yellow perch), *Morone americana* (white perch), *Morone saxatilis* (striped bass), and *Alosa pseudoharengus* (alewife) eggs. Chesapeake Bay Institute, Johns Hopkins University Ref. No. 73-3.

Schuler, V. J. 1971. An ecological study of the Delaware River in the vicinity of Artificial Island. Progress report for the period January–December 1970. For Public Service Electric and Gas Company, Salem Nuclear Generating Station. Ichthyological Associates, Ithaca, N.Y.

———. 1974. An ecological study of the Delaware River in the vicinity of Artificial Island. Progress Report for the period January–December 1972. Ichthyological Associates, Ithaca, N.Y.

Schultz, E. T., K. E. Reynolds, and D. O. Conover. 1996. Countergradient variation in growth among newly hatched *Fundulus heteroclitus:*

geographic differences revealed by common-environment experiments. Funct. Ecol. 10: 366–374.

Schwartz, F. J. 1961. Fishes of Chincoteague and Sinepuxent bays. Am. Midl. Nat. 65(2): 384–408.

———. 1964a. Fishes of Isle of Wight and Assawoman Bays near Ocean City, Maryland. Chesapeake Sci. 5(4): 172–193.

———. 1964b. Effects of winter water conditions on fifteen species of captive marine fishes. Am. Midl. Nat. 71(2): 434–444.

———. 1965a. Age, growth and egg complement of the stickleback *Apeltes quadracus* at Solomons, Maryland. Chesapeake Sci. 6(2): 116–118.

———. 1965b. Movements of the oyster toadfish (Pisces: Batrachoididae) about Solomons, Maryland. Chesapeake Sci. 5(1): 155–159.

Schwartz, F. J., and B. W. Dutcher. 1963. Age, growth and food of the oyster toadfish near Solomons, Maryland. Trans. Am. Fish. Soc. 92: 170–173.

Scott, J. S. 1982. Depth, temperature and salinity preferences of common fishes of the Scotian Shelf. J. Northwest. Atl. Fish. Sci. 3: 29–39.

Scott, J., and G. Csanady. 1976. Nearshore currents off Long Island. J. Geophys. Res. 18(30): 5401–5409.

Scott, W. B., and E. J. Crossman. 1973. *Freshwater fishes of Canada.* Fish. Res. Board Can. Bull. 184.

Scott, W. B., and M. C. Scott. 1988. *Atlantic fishes of Canada.* Can. Bull. Fish. Aquat. Sci. 219.

Secor, D. H., J. M. Dean, and S. E. Campana, eds. 1995. *Recent developments in fish otolith research.* Belle W. Baruch Library in Marine Science, No. 19. University of South Carolina Press, Columbia.

Seligman, E. G., Jr. 1951. *Cyprinodon variegatus riverendi* (poey) and other aquatic notes. Aquarium 20: 234–236.

Setzler-Hamilton, E. A. 1991. White perch, *Morone americana. In* S. L. Funderburk, S. J. Jordan, J. A. Mihursky, and D. Riley, eds. *Habitat requirements for Chesapeake Bay living resources.* Prepared for Chesapeake Bay Program. Chesapeake Research Consortium, Solomons, Md.

Shenker, J. M., D. J. Hepner, P. E. Frere, L. E. Currence, and W. W. Wakefield. 1983. Upriver migration and abundance of naked goby (*Gobiosoma bosci*) larvae in the Patuxent River Estuary, Maryland. Estuaries 6(1): 35–42.

Shepherd, G. R. 1991. Meristic and morphometric variation in black sea bass north of Cape Hatteras, North Carolina. N. Am. J. Fish. Manage. 11: 139–148.

Shepherd, G. R., and C. B. Grimes. 1983. Geographic and historic variations in growth of weakfish, *Cynoscion regalis,* in the Middle Atlantic Bight. U.S. Fish. Bull. 81(4): 803–813.

———. 1984. Reproduction of weakfish, *Cynoscion regalis,* in the New York Bight and evidence for geographically specific life history characteristics. U.S. Fish. Bull. 82(3): 501–511.

Shepherd, G. R., and M. Terceiro. 1994. The summer flounder, scup, and black sea bass fishery of the Middle Atlantic Bight and southern New England waters. Natl. Ocean. Atmos. Adm. Tech. Rep. Natl. Mar. Fish. Serv. 122: 1–13.

Sherman, K. 1980. MARMAP, a fisheries ecosystem study in the Northwest Atlantic: fluctuations in ichthyoplankton-zooplankton components and their potential for impact on the system. Pp. 9–37 *in* F. P. Diemer, F. J. Vernberg, and D. R. Mirkes, eds. *Advanced concepts in ocean measurements for marine biology.* Belle W. Baruch Institute for Marine Biology and Coastal Research. University of South Carolina Press, Columbia.

Sherman, K. 1988. Ichthtyoplankton surveys: a strategy for monitoring fisheries change in a large marine ecosystem. Pp. 3–9 *in* W. G. Smith, ed. *An analysis and evaluation of ichthyoplankton survey data from the northeast continental shelf ecosystem.* Natl. Ocean. Atmos. Adm. Tech. Memo. Natl. Mar. Fish. Serv.-F/Northeast Cent.-57.

Sherman, K., W. Smith, W. Morse, M. Berman, J. Green, and L. Esjymont. 1984. Spawning strategies of fishes in relation to circulation, phytoplankton production, and pulses in zooplankton off the northeastern United States. Mar. Ecol. Progr. Ser. 18: 1–19.

Shima, M. 1989. Oceanic transport of the early life history stages of bluefish (*Pomatomus saltatrix*) from Cape Hatteras to the Mid-Atlantic Bight. Master's thesis. State University of New York, Stony Brook.

Shipp, R. L. 1974. The puffer fishes (Tetraodontidae) of the Atlantic Ocean. Publ. Gulf Coast Res. Lab. Mus. No. 4, Gulf Coast Research Laboratory, Ocean Springs, Miss.

Shipp, R. L., and R. W. Yerger. 1969. Status, characters, and distribution of the northern and southern puffers of the genus *Sphoeroides.* Copeia 1969: 425–433.

Shulman, M. J., and J. C. Ogden. 1987. What controls tropical reef fish populations: recruitment or benthic mortality: An example in the Caribbean reef fish *Haemulon flavolineatum.* Mar. Ecol. Prog. Ser. 39: 233–242.

Shuster, C. N., Jr. 1959. A biological evaluation of the Delaware River estuary. Univ. Del. Mar. Lab. Inf. Ser. Publ. (3).

Sibunka, J. D., and A. L. Pacheco. 1981. Biological and fisheries data on northern puffer, *Sphoeroides maculatus* (Bloch and Schneider). U.S. Dep. Commer. Natl. Ocean. Atmos. Adm. Natl. Mar. Fish. Serv. Northeast Fish. Cent. Sandy Hook Lab. Tech. Ser. Rep. No. 26.

Sibunka, J. D., and M. J. Silverman. 1984. MARMAP surveys of the continental shelf from Cape Hatteras, North Carolina to Cape Sable, Nova Scotia (1977–1983). Atlas No. 1. Summary of operations. Natl. Ocean. Atmos. Adm. Tech. Memo. Natl. Mar. Fish. Serv.-F/Northeast Fish Cent.-33.

———. 1989. MARMAP surveys of the continental shelf from Cape Hatteras, North Carolina to Cape Sable, Nova Scotia (1984–87). Atlas No. 3. Summary of operations. Natl. Ocean. Atmos. Adm. Tech. Memo. Natl. Mar. Fish. Serv.-F/NEC-68.

Silverman, M. J. 1975. Scale development in the bluefish, *Pomatomus saltatrix*. Trans. Am. Fish. Soc. 104(4): 773–774.

Simmons, E. G. 1957. An ecological survey of the upper Laguna Madre of Texas. Publ. Inst. Mar. Sci. Univ. Tex. 4(2): 156–202.

Simmons, E. G., and J. P. Breuer. 1962. A study of redfish, *Sciaenops ocellatus* Linnaeus, and black drum, *Pogonias cromis* Linnaeus. Publ. Inst. Mar. Sci. Univ. Tex. 8: 184–211.

Sissenwine, M. P. 1984. Why do fish populations vary? Pp. 59–94 *in* R. M. Day, ed. *Exploitation of marine communities*. Dahlem Konferenzen. Springer, New York.

Sissenwine, M. P., E. B. Cohen, and M. D. Grosslein. 1984. Structure of the Georges Bank ecosystem. Rapp. P.-V. Reun. Cons. Int. Explor. Mer 183: 243–254.

Sisson, R. T. 1974. The growth and movement of scup (*Stenotomus chrysops*) in Narragansett Bay, Rhode Island and along the Atlantic coast. Completion Rep. PL88-309, Project 3-K38-R-3. Rhode Island Department of Natural Resources, Providence.

Smigielski, A. S. 1975. Hormone induced spawnings of the summer flounder and rearing of the larvae in the laboratory. Prog. Fish-Cult. 37: 3–8.

Smigielski, A. S., T. A. Halavik, L. J. Buckley, S. M. Drew, and G. C. Laurence. 1984. Spawning, embryo development and growth of the American sand lance *Ammodytes americanus* in the laboratory. Mar. Ecol. Prog. Ser. 14: 287–292.

Smith, B. A. 1971. The fishes of four low-salinity tidal tributaries of the Delaware River estuary. Bull. No. 5. Ichthyological Associates, Ithaca, N.Y.

Smith, C. L. 1985. The inland fishes of New York State. N.Y. State Dep. Environment and Conservation, Albany, N.Y.

Smith, D. G. 1968. The occurrence of larvae of the American eel, *Anguilla rostrata,* in the Straits of Florida and nearby areas. Bull. Mar. Sci. 18: 280–293.

———. 1984. Elopiformes, Notacanthiformes and Anguilliformes: relationships. Pp. 94–102 *in* H. G. Moser, W. J. Richards, D. M. Cohen, M. P. Fahay, A. W. Kendall, Jr., and S. L. Richardson, eds. *Ontogeny and systematics of fishes.* Am. Soc. Ichthyol. Herpetol. Spec. Publ. No. 1. Allen Press, Lawrence, Kans.

Smith, D. G. 1989a. Family Congridae: leptocephali. Pp. 723–763 *in Fishes of the Western North Atlantic.* Sears Foundation for Marine Research Part 9, Vol. 2. Yale University, New Haven, Conn.

———. 1989b. Family Anguillidae: leptocephali. Pp. 898–899 *in Fishes of the Western North Atlantic.* Sears Foundation for Marine Research Part 9, Vol. 2. Yale University, New Haven, Conn.

———. 1989c. Family Muraenidae: leptocephali. Pp. 900–916 *in Fishes of the Western North Atlantic.* Sears Foundation for Marine Research Part 9, Vol. 2. Yale University, New Haven, Conn.

Smith, K. J. 1995. Processes regulating habitat use by salt marsh nekton in a southern New Jersey estuary. Ph.D. diss. Rutgers University, New Brunswick, N.J.

Smith, K. J., and K. W. Able. 1994. Salt-marsh tide pools as winter refuges for the mummichog, *Fundulus heteroclitus,* in New Jersey. Estuaries 17(1B): 226–234.

———. In prep. Role of vegetation in salt marsh surface pools: patterns of faunal composition, abundance and growth. Estuarine Coastal Shelf Sci.

Smith, M. W., and J. W. Saunders. 1955. The American eel in certain freshwaters of the maritime provinces of Canada. J. Fish. Res. Board Can. 12: 238–269.

Smith, M. W., M. C. Glimcher, and D. A. Powers. 1992. Differential introgression of nuclear alleles between subspecies of the teleost *Fundulus heteroclitus.* Mol. Mar. Biol. Biotechnol. 1: 226–238.

Smith, R. E. 1966. Foreward. Am. Fish. Soc. Spec. Publ. 3: vii–viii.

Smith, W. A. 1887. Domestic habits of butterfish

(*Blennius gunnellus* Linn.). Proc. Trans. Nat. Hist. Soc. Glasgow N.S. 1:137–140.

Smith, W. G., P. Berrien, and T. Potthoff. 1994. Spawning patterns of bluefish, *Pomatomus saltatrix,* in the northeast continental shelf ecosystem. Bull. Mar. Sci. 54(1):8–16.

Smith, W. G., and M. P. Fahay. 1970. Description of eggs and larvae of the summer flounder, *Paralichthys dentatus.* U.S. Fish. Wildl. Serv. Res. Rep. 75:1–21.

Smith, W. G., and W. W. Morse. 1988. Seasonal distribution, abundance and diversity patterns of fish eggs and larvae in the Middle Atlantic Bight. Pp. 177–189 *in* A. Pacheco, ed. Characterization of the Middle Atlantic Water management Unit of the Northeast Regional Action Plan. Natl. Ocean. Atmos. Adm. Tech. Memo. Natl. Mar. Fish. Serv.-F/Northeast Fish. Cent.-56.

———. 1990. Larval distribution patterns: evidence for the collapse/recolonization of Atlantic herring on Georges Bank. Int. Counc. Explor. Sea C.M. 1990/H:17:1–16.

Smith, W. G., and A. Wells. 1977. Biological and fisheries data on striped bass, *Morone saxatilis* (Walbaum). U.S. Dep. Commer. NOAA NMFS Northeast Fish. Cent. Sandy Hook Lab. Tech. Ser. Rep. No. 4.

Smith, W. G., J. D. Sibunka, and A. Wells. 1975. Seasonal distribution of larval flatfishes (Pleuronectiformes) on the continental shelf between Cape Cod, Massachusetts, and Cape Lookout, North Carolina, 1965–1966. Natl. Ocean. Atmos. Adm. Tech. Rep. Natl. Mar. Fish. Serv. Spec. Sci. Rep. Fish.-691.

Smith, W. G., A. Wells, and D. G. McMillan. 1979. The distribution and abundance of ichthyoplankton in the Middle Atlantic Bight as determined from coastal surveys and site-specific studies, 1965–1976. U.S. Dep. Commer. NOAA NMFS Northeast Fish. Cent. Sandy Hook Lab. Rep. No. SHL 79-02.

Sogard, S. M. 1989. Colonization of artificial seagrass by fishes and decapod crustaceans: importance of proximity to natural eelgrass. J. Exp. Mar. Biol. Ecol. 133:15–37.

———. 1991. Interpretation of otolith microstructure in juvenile winter flounder (*Pseudopleuronectes americanus*): ontogentic development, daily increment validation, and somatic growth relationships. 48(10):1862–1871.

———. 1992. Variability in growth rates of juvenile fishes in different estuarine habitats. Mar. Ecol. Prog. Ser. 85:35–53.

Sogard, S. M., and K. W. Able. 1991. A comparison of eelgrass, sea lettuce macroalgae, and marsh creeks as habitat for epibenthic fishes and decapods. Estuarine Coast. Shelf Sci. 33:501–519.

———. 1994. Diel variation in immigration of fishes and decapod crustaceans to artificial seagrass habitat. Estuaries 17(3):622–630.

Sogard, S. M., K. W. Able, and M. P. Fahay. 1992. Early life history of the tautog *Tautoga onitis* in the Mid-Atlantic Bight. U.S. Fish. Bull. 90: 529–539.

Springer, S., and H. R. Bullis, Jr. 1956. Collections by the Oregon in the Gulf of Mexico. List of crustaceans, mollusks, and fishes identified from collections made by the exploratory fishing vessel Oregon in the Gulf of Mexico and adjacent seas 1950 through 1955. U.S. Fish. Wildl. Serv. Spec. Sci. Rep. Fish. 196.

Springer, V. G., and K. D. Woodburn. 1960. An ecological study of the fishes of the Tampa Bay area. Profess. Pap. Ser. No. 1. Florida State Board of Conservation Marine Lab.

Stahl, L., J. Koczan, and D. Swift. 1974. Anatomy of a shoreface-connected sand ridge on the New Jersey shelf: implications for the genesis of the shelf surficial sand sheet. Geology 2:117–120.

Stanley, J. C., and D. S. Danie. 1983. Species profile: Life histories and environmental requirements of coastal fishes and invertebrates (North America): white perch. U.S. Fish Wildl. Serv. Div. Biol. Serv. FWS/OBS-82/11.7.

Stanne, S. P., R. G. Panetta, and B. E. Forish. 1996. *The Hudson: An illustrated guide to the living river.* Rutgers University Press, New Brunswick, N.J.

Starck, W. A. 1964. A contribution to the biology of the gray snapper, *Lutjanus griseus* (Linnaeus), in the vicinity of Lower Matecumbe Key, Florida. Ph.D. diss. University of Miami.

———. 1971. Biology of the gray snapper, *Lutjanus griseus* (Linnaeus), in the Florida Keys. Pp. 11–150 *in* W. A. Starck and R. E. Schroeder, eds. Investigations on the Gray Snapper, *Lutjanus griseus.* Stud. Trop. Oceangr. (Miami) 10:1–224.

Steimle, F. W., and C. J. Sindermann. 1978. Review of oxygen depletion and associated mass mortalities of shellfish in the Middle Atlantic Bight in 1976. Mar. Fish. Rev. 40(12):17–26.

Steiner, W. W., and B. L. Olla. 1985. Behavioral responses of prejuvenile red hake, *Urophycis chuss,* to experimental thermoclines. Environ. Biol. Fishes 14(2/3):167–173.

Steiner, W. W., J. J. Luczkovich, and B. L. Olla. 1982. Activity, shelter usage, growth and recruitment of juvenile red hake, *Urophycis chuss.* Mar. Ecol. Prog. Ser. 7:125–135.

Stevenson, R. A., Jr. 1958. The biology of the anchovies *Anchoa mitchilli* Cuvier and Valinciennes 1848 and *Anchoa hepsetus hepsetus* Linnaeus 1758 in Delaware Bay. Master's thesis. University of Delaware, Newark.

Stoecker, R. R., J. Collura, and P. J. Fallon, Jr. 1992. Aquatic studies at the Hudson River Center site. Pp. 407–427 *in* C. L. Smith, ed. *Estuarine research in the 1980s.* The Hudson River Environmental Society, Seventh Symposium on Hudson River Ecology. State University of New York Press, Albany.

Stone, S. L., T. A. Lowery, J. D. Field, C. D. Williams, D. M. Nelson, S. H. Jury, M. E. Monaco, and L. Andreasen. 1994. Distribution and abundance of fishes and invertebrates in Mid-Atlantic estuaries. Estuarine Living Marine Resources Rep. No. 12. Natl. Ocean. Atmos. Adm./NOS Strategic Environmental Assessments Division, Silver Spring, Md.

St. Pierre, R. A., and J. Davis. 1972. Age, growth and mortality of the white perch, *Morone americana,* in the James and York Rivers. Chesapeake Sci. 13:272–281.

Stroud, R. H. 1971. Introduction to symposium. Pp. 3–8 *in* P. A. Douglas and R. H. Stroud, eds. A symposium on the biological significance of estuaries. Sport Fishing Institute, Washington, D.C.

Subrahmanyam, C. B. 1964. Eggs and early development of a carngid from Madras. J. Mar. Biol. Assoc. India 6(10):142–146.

Sullivan, W. E. 1915. A description of the young stages of the winter flounder, *Pseudopleuronectes americanus* (Walbaum). Trans. Am. Fish. Soc. 44(2):125–136.

Sumner, F. B., R. C. Osburn, and L. J. Cole. 1913. A catalogue of the marine fauna. *In*: A biological survey of the waters of Woods Hole and vicinity. Bull. Bur. Fish. XXXI, 1911, Part II, Sect. III: 549–794.

Swanson, R., and C. Parker. 1988. Physical environmental factors contributing to recurring hypoxia in the New York Bight. Trans. Am. Fish. Soc. 117:37–47.

Swanson, R. L., and C. J. Sindermann. 1979. Oxygen depletion and associated benthic mortalities in the New York Bight, 1976. Natl. Ocean. Atmos. Adm. Prof. Pap. 11.

Swiecicki, D. P., and T. R. Tatham. 1977. Ichthyoplankton. Pp. 115–147 *in* C. B. Milstein, D. L. Thomas, et al., eds., Summary of Ecological Studies for 1972–1975 in the Bays and other Waterways near Little Egg Inlet and in the Ocean in the Vicinity of the Proposed Site for the Atlantic Generation Station, New Jersey. Bull. No. 18. Ichthyological Associates, Ithaca, N.Y.

Sylvester, J. R., C. E. Nash, and C. E. Emberson. 1974. Preliminary study of temperature tolerance in juvenile Hawaiian mullet (*Mugil cephalus*). Prog. Fish-Cult. 36:99–100.

Szedlmayer, S. T., and K. W. Able. 1992. Validation studies of daily increment formation for larval and juvenile summer flounder, *Paralichthys dentatus.* Can. J. Fish. Aquat. Sci. 49(9):1856–1862.

———. 1993. Ultrasonic telemetry of age-0 summer flounder, *Paralichthys dentatus,* movements in a southern New Jersey estuary. Copeia 1993(3):728–736.

———. 1996. Patterns of seasonal availability and habitat use by fishes and decapod crustaceans in a southern New Jersey estuary. Estuaries 19(3).

Szedlmayer, S. T., M. P. Weinstein, and J. A. Musick. 1990. Differential growth among cohorts of age-0 weakfish, *Cynoscion regalis,* in Chesapeake Bay. U.S. Fish. Bull. 88:745–752.

Szedlmayer, S. T., K. W. Able, J. A. Musick, and M. P. Weinstein. 1991. Are scale circuli deposited daily in juvenile weakfish, *Cynoscion regalis*? Environ. Biol. Fishes 31:87–94.

Szedlmayer, S. T., K. W. Able, and R. A. Rountree. 1992. Growth and temperature-induced mortality of young-of-the-year summer flounder (*Paralichthys dentatus*) in southern New Jersey. Copeia 1992(1):120–128.

Tagatz, M. E., and D. L. Dudley. 1961. Seasonal occurrence of marine fishes in four shore habitats near Beaufort, N.C., 1957–1960. U.S. Fish. Wildl. Serv., Spec. Sci. Rep. Fish. No. 390.

Talbot, C. W., and K. W. Able. 1984. Composition and distribution of larval fishes in New Jersey high marshes. Estuaries 7(4A):434–443.

Talbot, G. B., and J. E. Sykes. 1958. Atlantic coastal migrations of American shad. U.S. Fish. Bull. 58:473–490.

Talbot, C. W., K. W. Able, J. K. Shisler, and D. Coorey. 1980. Seasonal variation in the composition of fresh and brackish water fishes of New Jersey mosquito control impoundments. Proc. Annu. Meet. N.J. Mosq. Control Assoc. 67:50–63.

Talbot, C. W., K. W. Able, and J. K. Shisler. 1986. Fish species composition in New Jersey salt marshes: effects of alterations for mosquito control. Trans. Am. Fish. Soc. 115:269–278.

Tatham, T. R., D. P. Swiecicki, and F. C. Stoop. 1974. Ichthyoplankton. Pp. 88–106 *in* D. L. Thomas, et al., eds. Ecological studies in the bays and other waterways near Little Egg Inlet and in the ocean in the vicinity of the proposed site for the Atlantic Generating Station, N.J. Vol. 1. Fishes. Ichthyological Associates, Ithaca, N.Y.

Tatham, T. R., D. J. Danila, D. L. Thomas, et al. 1977. *Ecological studies for the Oyster Creek Generating Station.* Progress Report for the period September 1975–August 1976. Vol. 1. Fin- and Shellfish. Ichthyological Associates, Ithaca, N.Y.

Tatham, T. R., D. L. Thomas, and D. J. Danila. 1984. Fishes of Barnegat Bay, New Jersey. Pp. 241–280 *in* M. J. Kennish and R. A. Lutz, eds. *Ecology of Barnegat Bay*, New Jersey. Springer, New York.

Taylor, M. H., and L. DiMichele. 1983. Spawning site utilization in a Delaware population of *Fundulus heteroclitus* (Pisces: Cypriniodontidae). Copeia 1983 : 719–725.

Taylor, M. H., L. DiMichele, and G. J. Leach. 1977. Egg stranding in the life cycle of the mummichog, *Fundulus heteroclitus.* Copeia 1977 : 397–399.

Taylor, M. H., G. J. Leach, L. DiMichele, W. M. Levitan, and W. F. Jacob. 1979. Lunar spawning cycle in the mummichog, *Fundulus heteroclitus* (Pisces: Cyprinodontidae). Copeia 1979 : 291–297.

Taylor, M. H., L. DiMichele, R. T. Kneib, and S. Bradford. 1982. Comparison of reproductive strategies in Georgia and Massachusetts populations of *Fundulus heteroclitus.* Am. Zool. 191 : 821.

Teal, J. M. 1986. The ecology of regularly flooded salt marshes of New England: A community profile. U.S. Fish. Wild. Serv. Biol. Rep. 85(7.4).

Teixeira, R. 1995. Reproductive and feeding biology of selected syngnathids (Pisces: Teleostei) of the Western Atlantic. Ph.D. diss. College of William and Mary, Williamsburg, Va.

Tesch, F. W. 1977. The eel: Biology and management of anguillid eels. Chapman and Hall, London.

Texas Instruments, Inc. 1973. Hudson River ecology study in the area of Indian Point. First annual report to Consolidated Edison Co. of New York, Inc.

Thomas, D. L. 1971. The early life history and ecology of six species of drum (Sciaenidae) in the lower Delaware River, a brackish tidal estuary. An ecological study of the Delaware River in the vicinity of Artificial Island, pt. III. Bull. No. 3. Ichthyological Associates, Ithaca, N.Y.

Thomas, D. L., and C. B. Milstein. 1973. Ecological studies in the bays and other waterways near Little Egg Inlet and in the ocean in the vicinity of the proposed site for the Atlantic Generating Station, New Jersey. Progress Report for Period January–December 1972, Part 1. Ichthyological Associates, Ithaca, N.Y.

———. 1974. A comparison of fishes collected by trawl along the coast of southern NJ from 1929 to 1933 with those taken in the vicinity of Little Egg Inlet in 1972 and 1973. Pp. 70–75 *in* D. L. Thomas et al., eds. Ecological studies in the bays and other waterways near Little Egg Inlet and in the ocean in the vicinity of the proposed site for the Atlantic Generating Station, NJ. Vol. 1. Fishes. Ichthyological Associates, Ithaca, N.Y.

Thomas, D. L., and B. A. Smith. 1973. Studies of young of the black drum, *Pogonias cromis,* in low salinity waters of the Delaware Estuary. Chesapeake Sci. 14(2) : 124–130.

Thomas, D. L., C. B. Milstein, T. R. Tatham, E. Van Eps, D. K. Stauble, and H. K. Hoff. 1972. Ecological considerations for ocean sites off New Jersey for proposed nuclear generating stations. Vol. 1, Part 1. Ecological studies in the vicinity of ocean sites 7 and 8. Ichthyological Associates, Ithaca, N.Y.

Thomas, D. L., C. B. Milstein, T. R. Tatham, R. C. Bieder, F. J. Margraf, D. J. Danila, H. K. Hoff, E. A. Illjes, M. M. McCullough, and F. A. Swiecicki. 1974. Fishes. Pp. 115–131 *in* Ecological studies in the bays and other waterways near Little Egg Inlet and in the ocean in the vicinity of the proposed site for the Atlantic Generating Station, NJ. Vol. 1. Ichthyological Associates, Ithaca, N.Y.

Thomas, D. L., C. B. Milstein, T. R. Tatham, R. C. Bieder, D. J. Danila, H. K. Hoff, D. P. Swiecicki, R. P. Smith, G. J. Miller, J. J. Gift, and M. C. Wyllie. 1975. Fishes. Pp. 25–142 *in* Ecological studies in the bays and other waterways near Little Egg Inlet and in the ocean in the vicinity of the proposed site for the Atlantic Generating Station, N.J. Progress Rep. for Jan–Dec 1974. Vol. 1. Ichthyological Associates, Ithaca, N.Y.

Thomas, M. L. H. 1968. Overwintering of American lobsters, *Homarus americanus,* in burrows in Bideford River, Prince Edward Island. J. Fish. Res. Board Can. 25 : 2725–2727.

Thorpe, L. A. 1991. Aspects of the biology of windowpane flounder, *Scophthalmus aquosus,* in the northwest Atlantic Ocean. Master's thesis. University of Massachusetts, Amherst.

Thunberg, B. E. 1971. Olfaction in parent stream selection by the alewife. Anim. Behav. 19 : 217–225.

Tracy, H. C. 1910. Annotated list of fishes known to inhabit the waters of Rhode Island. Pp. 35–176 *in* 40th Annu. Rep. Comm. Inland Fisheries, Rhode Island. Providence, R.I.

———. 1959. Stages in the development of the anatomy of motility of the toadfish (*Opsanus tau*). J. Comp. Neurol. 11 : 27–82.

Tucker, J. W., Jr. 1982. Larval development of *Citharichthys cornutus, C. gymnorhinus, C. spilopterus,* and *Etropus crossotus* (Bothidae) with notes on larval occurrence. U.S. Fish. Bull. 80 : 35–73.

———. 1989. Energy utilization in bay anchovy, *Anchoa mitchilli,* and black sea bass, *Centropristis striata striata,* eggs and larvae. U.S. Fish. Bull. 87(2) : 279–293.

Tupper, M. 1994. Settlement and post-settlement processes in the population regulation of a temperate reef fish: the role of energy. Ph.D. diss. Dalhousie University, Halifax, Canada.

Tupper, M., and R. G. Boutilier. 1995a. Effects of conspecific density on settlement, growth and post-settlement survival of a temperate reef fish. J. Exp. Mar. Biol. Ecol. 191 : 209–222.

———. 1995b. Size and priority at settlement determine growth and competitive success of newly settled Atlantic cod. Mar. Ecol. Prog. Ser. 118 : 295–300.

———. 1995c. Effects of habitat on settlement, growth and postsettlement survival of Atlantic cod (*Gadus morhua*). Can. J. Fish. Aquat. Sci. 52 : 1834–1841.

———. In review. Effects of habitat on settlement, growth and post-settlement survival of a temperate reef fish. Mar. Ecol. Prog. Ser.

Tupper, M., and W. Hunte. 1994. Recruitment dynamics of coral reef fishes in Barbados. Mar. Ecol. Prog. Ser. 108 : 225–235.

Vari, R. P. 1982. Subfamily Hippocampinae. Pp. 173–189 *in Fishes of the Western North Atlantic.* Sears Found. Mar. Res. Mem. 1(8). Yale University, New Haven, Conn.

Veer, H. W. van der, and M. J. N. Bergman. 1987. Predation by crustaceans on a newly settled 0-group plaice (*Pleuronectes platessa*) population in the western Wadden Sea. Mar. Ecol. Prog. Ser. 35 : 203–215.

Victor, B. C. 1986. Larval settlement and juvenile mortality in a recruitment-limited coral reef fish population. Ecol. Monogr. 56 : 145–160.

Vincent, A., I. Ahnesjo, A. Berglund, and G. Rosenqvist. 1992. Pipefish and seahorses: Are they all sex role reversed? Trends Ecol. Evol. 7 : 237–241.

Virginia Institute of Marine Sciences. 1962. Fishes and fish larvae collected from Atlantic plankton cruises of R/V *Pathfinder,* March 1961–March 1962. Spec. Sci. Rep. No. 33. College of William and Mary, Gloucester Point, Va.

Vladykov, V. D. 1966. Remarks on the American eel (*Anguilla rostrata* LeSueur). Sizes of elvers entering streams; the relative abundance of adult males and females and present economic importance of eels in North America. Int. Ver. Theor. Angew. Limnol. Verh. 16 : 1007–1017.

———. 1971. Homing of the American eel, *Anguilla rostrata,* as evidenced by returns of transplanted tagged eels in New Brunswick. Can. Field Nat. 85(3) : 241–248.

von Alt, C. J., and J. F. Grassle. 1992. LEO-15: an unmanned long term environmental observatory. OCEANS 92 Newport, R.I.

Vouglitois, J. J. 1983. The ichthyofauna of Barnegat Bay, New Jersey: relationships between long term temperature fluctuations and the population dynamics and life history of temperate estuarine fishes during a five-year period, 1976–1980. Master's thesis. Rutgers University, New Brunswick, N.J.

Vouglitois, J. J., K. W. Able, R. J. Kurtz, and K. A. Tighe. 1987. Life history and population dynamics of the bay anchovy in New Jersey. Trans. Am. Fish. Soc. 116(2) : 141–153.

Walburg, C. H. 1960. Abundance and life history of the shad, St. Johns River, Florida. U.S. Fish. Bull. 60(177) : 487–501.

Waldman, J. R., D. J. Dunning, Q. E. Ross, and M. T. Mattson. 1990. Range dynamics of Hudson River striped bass along the Atlantic coast. Trans. Am. Fish. Soc. 119 : 910–919.

Walford, L. A. 1966. The estuary as a habitat for fishery organisms, Introduction. Am. Fish. Soc. Spec. Publ. No. 3 : 15.

Wallace, J. H., H. M. Kok, L. E. Beckley, B. Bennett, S. J. M. Blaber, and A. K. Whitfield. 1984. South African estuaries and their importance to fishes. S.-Afr. Tydskr. Wet. 80 : 203–207.

Wallace, R. A., and K. Selman. 1981. The reproductive activity of *Fundulus heteroclitus* females from Woods Hole, Massachusetts, as compared with more southern locations. Copeia 1 : 212–215.

Wallace, R. C. 1971. Age, growth, year class strength and survival rates of white perch, *Morone americana* (Gmelin), in the Delaware River in the vicinity of Artificial Island. Chesapeake Sci. 12 : 205–218.

Walline, P. D. 1985. Growth of larval walleye pollock related to domains within the southeast Bering Sea. Mar. Ecol. Prog. Ser. 21 : 197–203.

Wang, S. B., and E. D. Houde. 1995. Distribution,

relative abundance, biomass and production of bay anchovy *Anchoa mitchilli* in Chesapeake Bay. Mar. Ecol. Prog. Ser. 121: 27–38.

Wang, J. C., and R. J. Kernehan. 1979. *Fishes of the Delaware estuaries: A guide to the early life histories.* EA Communications, Towson, Md.

Warfel, H. E., and D. Merriman. 1944. Studies on the marine resources of southern New England. I. An analysis of the fish population of the shore zone. Bull. Bingham Oceanogr. Collect. Yale Univ. IX(2): 1–91.

Waring, G. T. 1975. Preliminary analysis of the status of butterfish in ICNAF Subarea 5 and Statistical Area 6. Int. Comm. North Am. Fish. Res. Doc. 75/14.

Waring, G., and S. Murawski. 1982. Butterfish. MESA (Mar. Ecosyst. Anal.) N.Y. Bight Atlas Monogr. 15: 105–107.

Warinner, J. E., J. P. Miller, and J. Davis. 1969. Distribution of juvenile river herring in the Potomac River. Proc. Annu. Conf. Southeast Assoc. Game Fish Comm. 23: 384–388.

Warlen, S. M. 1963. Some aspects of the life history of *Cyprinodon variegatus,* Lacépède (1803), in Southern Delaware. Master's thesis. University of Delaware, Newark.

Warlen, S. M., K. W. Able, and E. Laban. (In press). Recruitment of larval Atlantic menhaden to North Carolina and New Jersey estuaries: evidence for larval transport northward along the east coast of the United States. U.S. Fish. Bull.

Watson, J. F. 1968. The early life history of the American shad, *Alosa sapidissima* (Wilson), in the Connecticut River above Holyoke, Massachusetts. Master's thesis. University of Massachusetts, Amherst.

Weinstein, M. P. 1979. Shallow marsh habitats as primary nurseries for fishes and shellfish, Cape Fear River, North Carolina. U.S. Fish. Bull. 77: 339–357.

———. 1983. Population dynamics of an estuarine dependent fish, the spot (*Leiostomus xanthurus*), along a tidal creek-seagrass coenocline. Can. J. Fish. Aquat. Sci. 40: 1633–1638.

———. 1985. Distributional ecology of fishes inhabiting warm-temperate and tropical estuaries: community relationships and implications. Pp. 285–310 *in* A. Yanez-Arancibia, ed. Fish community ecology in estuaries and coastal lagoons: towards an ecosystem integration. UNAM Press, Mexico City.

Weinstein, M. P., and M. P. Walters. 1981. Growth, survival and production in young-of-the-year populations of *Leiostomus xanthurus* Lacépède residing in tidal creeks. Estuaries 4(3): 185–197.

Weinstein, M. P., L. Scott, S. P. O'Neil, R. C. Siegfriend, and S. T. Szedlmayer. 1984. Population dynamics of spot, *Leiostomus xanthurus,* in polyhaline tidal creeks of the York River estuary, Virginia. Estuaries 7(4A): 444–450.

Weisberg, S. B. 1986. Competition and coexistence among four estuarine species of *Fundulus.* Am. Zool. 26: 249–257.

Weisberg, S. B., and W. H. Burton. 1993. Spring distribution and abundance of ichthyoplankton in the tidal Delaware River. U.S. Fish. Bull. 91(4): 788–797.

Weisberg, S. B., P. Himchak, T. Baum, H. T. Wilson, and R. Allen. 1996. Temporal trends in abundance of fish in the tidal Delaware River. Estuaries 19(3): 723–729.

Welsh, W. W. 1915. Notes on the habits of the young of the squirrel hake and sea snail. Copeia 1915(18): 2–3.

Welsh, W. W., and C. M. Breder. 1922. A contribution to the life history of the puffer, *Spheroides maculatus* (Schneider). Zoologica (N.Y.) 2: 261–276.

———. 1923. Contributions to the life histories of Sciaenidae of the eastern U.S. coast. U.S. Bur. Fish. Bull. 39: 141–201.

Wenner, C. A. 1973. Occurrence of American eels, *Anguilla rostrata,* in water overlying the eastern North American continental shelf. J. Fish. Res. Board Can. 30(11): 1752–1755.

Wenner, C. A., and J. A. Musick. 1974. Fecundity and gonad observations of the American eel, *Anguilla rostrata,* migrating from Chesapeake Bay, Virginia. J. Fish. Res. Board Can. 31(8): 1387–1391.

Wenner, C. A., and G. R. Sedberry. 1989. Species composition, distribution and relative abundance of fishes in the coastal habitat off the southeastern United States. U.S. Dep. Commer. Natl. Ocean. Atmos. Adm. Tech. Rep. Natl. Mar. Fish. Serv. 79.

Wenner, C. A., W. A. Roumillat, and C. W. Waltz. 1986. Contributions to the life history of black sea bass, *Centropristis striata,* off the southeastern United States. U.S. Fish. Bull. 84(3): 723–741.

Werme, C. E. 1981. Resource partitioning in a salt marsh fish community. Ph.D. diss. Boston University, Boston.

West, R. J., and R. J. King. 1996. Marine, brackish, and freshwater fish communities in the vegetated and bare shallows of an Australian coastal river. Estuaries 19(1): 31–41.

Wheatland, S. B. 1956. Pelagic fish eggs and larvae. Chap. VII *in* Oceanography of Long Island

Sound, 1952–1954. Bull. Bingham Oceanogr. Collect. Yale Univ. 15: 234–314.

Wheeler, A. 1969. The fishes of the British Isles and north-west Europe. Michigan State University Press, East Lansing.

White, E. G. 1918. The origin of the electric organs in *Astroscopus guttatus*. Carnegie Inst. Wash. Pap. Dept. Mar. Biol. 12(6): 139–172.

Whitfield, A. K. 1994. An estuary-association classification for the fishes of southern Africa. S.-Afr. Tydskr. Wet. 90: 411–417.

Whitfield, A. K., and S. J. M. Blaber. 1978. Food and feeding ecology of piscivorous fishes at Lake St. Lucia, Zuzuland. J. Fish. Biol. 13: 675–691.

Whitworth, W. R., D. R. Gibbons, J. H. Heuer, W. E. Johns, and R. E. Schmidt. 1975. A general survey of the fisheries resources of the Thames River watershed, Connecticut. Conn. Storrs Agric. Exp. Stn. Bull. 435: 1–37.

Wicklund, R. 1970. A puffer kill related to nocturnal behavior and adverse environmental changes. Underw. Nat. 6(3): 28–29.

Wicklund, R. I., S. J. Wilk, and L. Ogren. 1968. Observations on wintering locations of the northern pipefish and spotted seahorse. Underwater Nat. 5(2): 26–28.

Wilk, S. J. 1976. The weakfish—a wide ranging species. Atl. States Mar. Fish. Comm. Mar. Atl. Coast Fish. Leaf. No. 18.

———. 1977. Biological and fisheries data on bluefish, *Pomatomus saltatrix* (Linnaeus). Sandy Hook, NMFS Tech. Rep. No. 11.

———. 1982. Bluefish, *Pomatomus saltatrix*. MESA (Mar. Ecosyst. Anal.) N.Y. Bight Atlas, Monogr. 15: 86–89.

Wilk, S. J., and M. J. Silverman. 1976a. Fish and hydrographic collections made by the research vessels Dolphin and Delaware II during 1968–72 from New York to Florida. Natl. Ocean. Atmos. Adm. Tech. Rep. Natl. Mar. Fish. Serv. Spec. Sci. Rep. Fish. 697: 1–159.

———. 1976b. Summer benthic fish fauna of Sandy Hook Bay, New Jersey 1976. Natl. Ocean. Atmos. Adm. Tech. Rep. Natl. Mar. Fish. Serv. Spec. Sci. Rep. Fish.-698.

Wilk, S. J., W. W. Morse, D. E. Ralph, and E. J. Steady. 1975. Life history aspects of New York Bight finfishes (June 1974–June 1975). Annu. Rep. NOAA NMFS Middle Atlantic Coastal Fish. Cent. Sandy Hook Lab. Highlands, N.J.

Wilk, S. J., W. J. Clifford, and D. J. Christensen. 1979. The recreational fishery for pollock (*Pollachius virens*) in southern New England and the Middle Atlantic. Natl. Mar. Fish. Serv. Northeast Fish. Cent. Sandy Hook Lab. Ref. Doc. No. 79–31.

Wilk, S. J., W. W. Morse, and L. L. Stehlik. 1990. Annual cycles of gonad-somatic indices as indicators of spawning activity for selected species of finfish collected from the New York Bight. U.S. Fish. Bull. 88: 775–786.

Wilk, S. J., R. A. Pikanowski, A. J. Pacheco, D. G. McMillan, B. A. Phelan, and L. L. Stehlik. 1992. Fish and megainvertebrates collected in the New York Bight apex during the 12-mile dumpsite recovery study, July 1986–September 1989. Natl. Ocean. Atmos. Adm. Tech. Mem. NMFS-F/NEC-90: 1–78.

Williams, D. D., and J. C. Delbeck. 1989. Biology of the threespine stickleback, *Gasterosteus aculeatus*, and the blackspotted stickleback, *G. wheatlandi*, during their marine pelagic phase in the Bay of Fundy, Canada. Environ. Bio. Fish. 24(1): 33–41.

Williams, D. M., S. English, and M. J. Milicich. 1994. Annual recruitment surveys of coral reef fishes are good indicators of patterns of settlement. Bull. Mar. Sci. 54(1): 314–331.

Williams, G. C. 1975. Viable embryogenesis of the winter flounder, *Pseudopleuronectes americanus* from 1.8° to 15° C. Mar. Biol. 33: 71–74.

Wilson, C. A., J. M. Dean, and R. Radtke. 1982. Age, growth rate and feeding habits of the oyster toadfish, *Opsanus tau* (Linnaeus) in South Carolina. J. Exp. Mar. Biol. Ecol. 62: 251–259.

Wilson, H. V. 1891. The embryology of the sea bass (*Serranus atrarius*). U.S. Fish. Comm. Bull. 9: 209–277.

Winters, G. H., J. A. Moores, and R. Chaulk. 1973. Northern range extension and probable spawning of gaspereau in the Newfoundland area. J. Fish. Res. Board Can. 30: 860–861.

Witting, D. A. 1995. Influence of invertebrate predators on survival and growth of juvenile winter flounder. Ph.D. diss. Rutgers University, New Brunswick, N.J.

Witting, D. A., and K. W. Able. 1993. Effects of body size on probability of predation for juvenile summer and winter flounder based on laboratory experiments. U.S. Fish. Bull. 91: 577–581.

———. 1995. Predation by sevenspine bay shrimp, *Crangon septemspinosa*, on winter flounder, *Pleuronectes americanus*, during settlement: laboratory observations. Mar. Ecol. Prog. Ser. 123: 23–31.

———. In review. Settlement, growth, and dispersal of juvenile winter flounder (*Pseudopleuronectes americanus*) in a southern New Jersey estuary. Can. J. Fish. Aquat. Sci.

Witting, D. A., K. W. Able, and M. P. Fahay. In review. Larval fishes of a Middle Atlantic Bight estuary: assemblage structure and temporal stability. Can. J. Fish. Aquat. Sci.

Woodhead, P. M. J., S. S. McCafferty, and M. A. O'Hare. 1987. Assessments of the fish community of the Lower Hudson-Raritan Estuary complex. Prepared for N.Y. District, U.S. Army Corps of Engineers under a support agreement with the National Parks Service and Rutgers University. Spec. Rep. 81. Ref. 87–5. State University of New York, Stony Brook.

Wooten, M. C., K. T. Scribner, and M. H. Smith. 1988. Genetic variability and systematics of *Gambusia* in the southeastern United States. Copeia 1988 : 293–289.

Wooten, R. J. 1976. *The biology of sticklebacks.* Academic Press, New York.

Wyanski, D. M. 1988. Depth and substrate characteristics of age 0 summer flounder (*Paralichthys dentatus*) habitat in Virginia estuaries. Master's thesis. College of William and Mary, Williamsburg, Va.

Yañez-Arancibia, A., ed. 1985. *Fish community ecology in estuaries and coastal lagoons: towards an ecosystem integration.* Contribution 592 from the Institute de Ciencias del Mar y Limnologia. UNAM Press, Mexico City.

Young, B. H., K. A. McKown, V. J. Vecchio, K. Hattala, and J. D. Sicluna. 1991. A study of the striped bass in the Marine District of New York VI. Anadromous Fish Conservation Act P.L. 89-304. Completion Report. New York. Project AFC-14. N.Y. State Department of Environmental Conservation.

Youson, J. H. 1988. First metamorphosis. Pp. 135–196 *in* W. S. Hoar and D. J. Randall, eds. *Fish physiology,* Vol. 11B. Academic Press, San Diego, Calif.

Yuschak, P. 1985. Fecundity, eggs, larvae and osteological development of the striped searobin, (*Prionotus evolans*) (Pisces, Triglidae). J. Northwest Atl. Fish. Sci. 6 : 65–85.

Yuschak, P., and W. A. Lund. 1984. Eggs, larvae and osteological development of the northern searobin, *Prionotus carolinus* (Pisces, Triglidae). J. Northwest Atl. Fish. Sci. 5 : 1–15.

Zastrow, C. E., E. D. Houde, and L. G. Morin. 1991. Spawning, fecundity, hatch-date frequency and young-of-the-year growth of bay anchovy, *Anchoa mitchilli,* in mid-Chesapeake Bay. Mar. Ecol. Prog. Ser. 73 : 161–171.

Zich, H. E. 1977. The collection of existing information and field investigation of anadromous clupeid spawning in New Jersey. N.J. Div. Fish Game Shellfish. Misc. Rep. 41. Trenton, N.J.

Index

ABOUT THE AUTHORS

Michael P. Fahay was originally hired by the U.S. Fish and Wildlife Service to study estuarine dependence of Atlantic coastal fishes. Interests over the past 30 years have focused on egg and larval stages occurring over the continental shelf between Nova Scotia and Miami, Florida, while employed at the National Marine Fisheries Service's Sandy Hook Laboratory. Although he initially concentrated on the taxonomy and systematics of these early stages, he has recently begun to join data sets derived from oceanic and estuarine sampling in order to better describe early life-history patterns.

Kenneth W. Able is a Rutgers University professor in the Institute of Marine and Coastal Sciences. He is also director of the Rutgers University Marine Field Station located within the salt marshes of Great Bay in southern New Jersey. He has worked on fishes in estuaries in Canada, Massachusetts, New Jersey, Virginia, and Florida. Much of his research and teaching has focused on the life history, ecology, and behavior of estuarine fishes.